Harmonic Maass Forms and Mock Modular Forms: Theory and Applications

American Mathematical Society

Colloquium Publications
Volume 64

Harmonic Maass Forms and Mock Modular Forms: Theory and Applications

Kathrin Bringmann
Amanda Folsom
Ken Ono
Larry Rolen

AMERICAN MATHEMATICAL SOCIETY

2010 *Mathematics Subject Classification*. Primary 11F03, 11F11, 11F27, 11F30, 11F37, 11F50.

For additional information and updates on this book, visit
www.ams.org/bookpages/coll-64

Library of Congress Cataloging-in-Publication Data

Names: Bringmann, Kathrin, author. | Folsom, Amanda, 1979- author. | Ono, Ken, 1968- author. | Rolen, Larry, author.
Title: Harmonic Maass forms and mock modular forms : theory and applications / Kathrin Bringmann, Amanda Folsom, Ken Ono, Larry Rolen.
Description: Providence, Rhode Island : American Mathematical Society, 2017. | Series: American Mathematical Society colloquium publications ; volume 64 | Includes bibliographical references and index.
Identifiers: LCCN 2017026415 | ISBN 9781470419448 (alk. paper)
Subjects: LCSH: Forms, Modular. | Forms (Mathematics) | Number theory. | AMS: Number theory – Discontinuous groups and automorphic forms – Modular and automorphic functions. msc | Number theory – Discontinuous groups and automorphic forms – Holomorphic modular forms of integral weight. msc | Number theory – Discontinuous groups and automorphic forms – Theta series; Weil representation; theta correspondences. msc | Number theory – Discontinuous groups and automorphic forms – Fourier coefficients of automorphic forms. msc | Number theory – Discontinuous groups and automorphic forms – Forms of half-integer weight; nonholomorphic modular forms. msc | Number theory – Discontinuous groups and automorphic forms – Jacobi forms. msc
Classification: LCC QA567.2.M63 H37 2017 | DDC 512.7–dc23 LC record available at https://lccn.loc.gov/2017026415

Softcover ISBN 978-1-4704-7984-8

∞ The paper used in this book is acid-free and falls within the guidelines
established to ensure permanence and durability.
Visit the AMS home page at `http://www.ams.org/`

11 10 9 8 7 6 5 4 3 2   30 29 28 27 26 25

Contents

Preface

Modular forms are central objects in contemporary mathematics. They are meromorphic functions $f\colon \mathbb{H} \mapsto \mathbb{C}$ which satisfy

$$f\left(\frac{a\tau + b}{c\tau + d}\right) = (c\tau + d)^k f(\tau)$$

for every matrix $\left(\begin{smallmatrix} a & b \\ c & d \end{smallmatrix}\right) \in \Gamma$ and $\tau \in \mathbb{H}$, where Γ is a subgroup of $\mathrm{SL}_2(\mathbb{Z})$ and the weight k is generally in $\frac{1}{2}\mathbb{Z}$. There are various types of modular forms which arise naturally in mathematics. *Modular functions* have weight $k = 0$. *Cusp forms* are those holomorphic modular forms which vanish at the cusps of Γ. *Weakly holomorphic forms* are permitted to have poles provided that they are supported at cusps.

There are many facets of these functions which are of importance in mathematics. The study of their Fourier expansions has driven research in the "Langlands Program" via the development of the theory of Galois representations and progress on the Ramanujan-Petersson Conjecture. The values of these functions appear in explicit class field theory. Their L-functions are devices which bridge analysis and arithmetic geometry.

The "web of modularity" is breathtaking. Indeed, modular forms play central roles in algebraic number theory, algebraic topology, arithmetic geometry, combinatorics, number theory, representation theory, and mathematical physics. In the last few decades, modular forms have been featured in fantastic achievements such as progress on the Birch and Swinnerton-Dyer Conjecture, mirror symmetry, Monstrous Moonshine, and the proof of Fermat's Last Theorem. These works are dramatic examples which illustrate the evolution of mathematics. It would have been nearly impossible to prophesize them fifty years ago.

This book is about a generalization of the theory of modular forms and the corresponding extension of their web of applications. This is the theory of harmonic Maass forms and mock modular forms. Instead of traveling back in time to the 1960s, the first glimpses of harmonic Maass forms and mock modular forms can be found in much older work, namely the enigmatic "deathbed" letter that Ramanujan wrote to G. H. Hardy in 1920 (cf. pages 220-224 of [**54**]):

"I am extremely sorry for not writing you a single letter up to now...I discovered very interesting functions recently which I call "Mock" ϑ-functions. Unlike the "False" ϑ-functions (studied partially by Prof. Rogers in his interesting paper) they enter into mathematics as beautifully as the ordinary theta functions. I am sending you with this letter some examples."

The letter contained 17 examples including:

$$f(q) := 1 + \sum_{n=1}^{\infty} \frac{q^{n^2}}{(1+q)^2(1+q^2)^2\cdots(1+q^n)^2},$$

$$\omega(q) := \sum_{n=0}^{\infty} \frac{q^{2n^2+2n}}{(1-q)^2(1-q^3)^2\cdots(1-q^{2n+1})^2},$$

$$\lambda(q) := \sum_{n=1}^{\infty} \frac{(-1)^n(1-q)(1-q^3)\cdots(1-q^{2n-1})q^n}{(1+q)(1+q^2)\cdots(1+q^{n-1})}.$$

For eight decades, very little was understood about Ramanujan's mock theta functions. Despite dozens of papers on them, a comprehensive theory which explained them and their role in mathematics remained elusive. Finally, Zwegers [**528, 529**] recognized that Ramanujan had discovered glimpses of special families of nonholomorphic modular forms. More precisely, Ramanujan's mock theta functions turned out to be holomorphic parts of these modular forms. For this reason, mathematicians now refer to the holomorphic parts of such modular forms as *mock modular forms.*

Zwegers' work fit Ramanujan's mock theta functions beautifully into a theory which involves basic hypergeometric series, indefinite theta functions, and an extension of the theory of Jacobi forms as developed by Eichler and Zagier in their seminal monograph [**191**].

At almost the same time, Bruinier and Funke [**121**] wrote an important paper on the theory of geometric theta lifts. In their work they defined the notion of a harmonic Maass form. The nonholomorphic modular forms constructed by Zwegers turned out to be weight $1/2$ harmonic Maass forms. This coincidental development ignited research on harmonic Maass forms and mock modular forms. This book represents a survey of this research. This work includes the development of general theory about harmonic Maass forms and mock modular forms, as well as the applications of this theory within the context of the web of modularity.

There have been a number of expository survey articles on mock modular forms by two of the authors, Duke, and Zagier [**166, 195, 198, 407, 408, 520**]. Furthermore, the books by Bruinier [**119**] and M. R. Murty and V. K. Murty [**392**] include nice treatments of some aspects of the theory of harmonic Maass forms and mock modular forms. This book is intended to serve as a uniform and somewhat comprehensive introduction to the subject for graduate students and research mathematicians. We assume that readers are familiar with the classical theory of modular forms which is contained in books such as [**162, 282, 316, 388, 405, 451, 455**]. There have also been a number of conferences, schools, and workshops devoted to the subject. The reader is encouraged to view the exercises [**197**] assembled for the 2013 Arizona Winter School, and notes which accompanied the 2016 "School on mock modular forms and related topics" at Kyushu University [**438**].

We conclude with a brief description of the contents of this book. For the convenience of the reader, we begin in Part 1 by recalling much of the standard theory of elliptic functions, theta functions, Jacobi forms, and classical Maass forms. The idea is to provide a comprehensive and self-contained treatment of these subjects in order to provide a suitable foundation for learning the theory of harmonic Maass forms. Part 2 contains the framework of the theory of harmonic Maass forms, including a treatment of Zwegers' celebrated Ph.D. thesis which has not been published elsewhere. Part 3 includes a sampling of some of the most interesting and exciting applications of the theory of harmonic Maass forms. These applications include a discussion of Ramanujan's original mock theta functions, the theory of partitions, the theory of singular moduli, Borcherds products, the arithmetic of elliptic curves, the representation theory of infinite dimensional affine Kac-Moody Lie algebras, and generalized Moonshine.

Kathrin Bringmann, Amanda Folsom, Ken Ono, and Larry Rolen
May 30, 2017

Acknowledgments

The authors are grateful for numerous helpful discussions and comments from Claudia Alfes-Neumann, Nickolas Andersen, Victor Aricheta, Olivia Beckwith, Lea Beneish, Manjul Bhargava, Jan Bruinier, Nikolaos Diamantis, John F. R. Duncan, Stephan Ehlen, Solomon Friedberg, Jens Funke, Michael Griffin, Pavel Guerzhoy, Kazuhiro Hikami, Özlem Imamoğlu, Paul Jenkins, Seokho Jin, Ben Kane, Jonas Kaszian, Byungchan Kim, Matthew Krauel, Stephen Kudla, Yingkun Li, Steffen Löbrich, Madeline Locus Dawsey, Jeremy Lovejoy, Jan Manschot, Michael Mertens, Stephen D. Miller, Steven J. Miller, Jackson Morrow, Boris Pioline, Martin Raum, Olav Richter, Peter Sarnak, Markus Schwagenscheidt, J.-P. Serre, Arul Shankar, Jesse Thorner, Sarah Trebat-Leder, Ian Wagner, Michael Woodbury, Don Zagier, and Sander Zwegers.

The authors thank the Asa Griggs Candler Fund, DFG, DMV, NSF, and the University of Cologne for their generous support. The research of the first author was supported by the Alfried Krupp Prize for Young University Teachers of the Krupp Foundation, and the research leading to these results received funding from the European Research Council under the European Union's Seventh Framework Programme (FP/2007-2013)/ERC Grant agreement n. 335220 - AQSER. The second author is grateful for the support of NSF grant DMS-1449679. The third author is grateful for the support of NSF grants DMS-1157289 and DMS-1601306. The fourth author thanks the University of Cologne and the DFG for their generous support via the University of Cologne postdoc grant DFG Grant D-72133-G-403-151001011, funded under the Institutional Strategy of the University of Cologne within the German Excellence Initiative

Part 1

Background

CHAPTER 1

Elliptic Functions

This book is about *harmonic Maass forms and functions*, real-analytic functions which generalize elliptic modular forms. The classical approach to the theory of modular forms typically begins with the work of Eisenstein, Jacobi, and Weierstrass on elliptic functions, doubly periodic meromorphic functions on $\mathbb{C}$. The theory of elliptic functions is generally developed by making use of the Weierstrass $\wp$-function, a fundamental example of an elliptic function. The $\wp$-function leads to the Eisenstein series, which in turn naturally lead to the analytic theory of elliptic modular forms.

It has recently been observed **[244]** that examples of harmonic Maass functions also arise naturally from the theory of elliptic functions. The fundamental function in this context is the negative of the antiderivative of the $\wp$-function, the Weierstrass ζ-function. This function is not doubly periodic. However, Eisenstein found that it could be "corrected" to obtain a doubly periodic function, which is no longer meromorphic. This observation anticipated the theory of harmonic Maass forms in the special case of lattices that arise from the theory of periods for modular elliptic curves. For modular elliptic curves $E/\mathbb{Q}$, Eisenstein's corrected Weierstrass ζ-function, combined with the work of Eichler and Shimura, gives rise to canonical harmonic Maass functions, functions which we shall show encode much deep arithmetic information (cf. Chapter 19 below).

In view of the important role of elliptic functions, we begin by briefly recalling the main features of the theory. We refer the reader to standard texts such as **[318, 342, 445, 463, 492]** for a more detailed account of the theory.

DEFINITION 1.1. A **lattice** Λ in $\mathbb{C}$ is any free subgroup of rank 2 over $\mathbb{Z}$ which generates $\mathbb{C}$ over $\mathbb{R}$.

Let Λ be a lattice in $\mathbb{C}$, and suppose that $\{\omega_1, \omega_2\}$ is a basis (that is, $\Lambda = \mathbb{Z}\omega_1 \oplus \mathbb{Z}\omega_2$). We denote this by $\Lambda = [\omega_1, \omega_2]$.

DEFINITION 1.2. A function f is an **elliptic function** with respect to the lattice Λ if f is meromorphic on $\mathbb{C}$ and periodic with respect to Λ. That is,

$$f(z+\omega) = f(z)$$

for all $\omega \in \Lambda$ and $z \in \mathbb{C}$. A function which satisfies such a periodicity relation for some lattice Λ is also called **doubly periodic**.

REMARK. The Riemann surface $\mathbb{C}/\Lambda$ has genus one, and hence is a torus. The theory of Eichler and Shimura on periods of modular forms makes it possible to directly compute lattices Λ if they are associated to modular elliptic curves. We make use of this fact in Section 6.2 below when we use the Weierstrass ζ-function to produce harmonic Maass forms from modular elliptic curves.

It is often useful to consider representatives for $\mathbb{C}/\Lambda$ in $\mathbb{C}$.

DEFINITION 1.3. For a lattice $\Lambda = [\omega_1, \omega_2]$ and any $z \in \mathbb{C}$, the **fundamental parallelogram** $P_{\Lambda,z}$ for Λ and z is the set of all points

$$P_{\Lambda,z} := \{z + r\omega_1 + s\omega_2 \; : \; 0 \le r, s < 1\}.$$

Elliptic functions are constrained by standard facts from complex analysis. We let $\mathrm{ord}_w(f)$ denote the order of a function f at the point $w \in \mathbb{C}$, and we let the *order* of f be its number of poles (counting multiplicity) in any $P_{\Lambda,z}$. We have the following result which is often referred to as Liouville's Theorem.

THEOREM 1.4 (Liouville). *Suppose that f is an elliptic function with respect to a lattice Λ. Then the following are true:*

i) *If f is holomorphic, then it is constant.*
ii) *If f is non-constant, then for every $z \in \mathbb{C}$, we have that*

$$\sum_{w \in P_{\Lambda,z}} \mathrm{ord}_w(f) = 0.$$

iii) *If the function f is non-constant, then it must have at least order 2 in $\mathbb{C}/\Lambda$.*

REMARK. By the Residue Theorem, the sum of the residues of an elliptic function on the torus $\mathbb{C}/\Lambda$ is zero.

1.1. Eisenstein series

For the remainder of this chapter, we assume that $\Lambda = [\omega_1, \omega_2]$, where we have ordered the basis of the lattice to satisfy $\mathrm{Im}(\frac{\omega_1}{\omega_2}) > 0$. The Eisenstein series, sums of reciprocals of even powers of the non-zero elements of a lattice Λ, are the fundamental building blocks in the classical treatment of the theory of modular forms.

DEFINITION 1.5. For any integer $k \ge 2$, the **Eisenstein series of weight $2k$ for Λ**, denoted $G_{2k}(\Lambda)$, is defined by

$$G_{2k}(\Lambda) := \sum_{\omega \in \Lambda \setminus \{0\}} \frac{1}{\omega^{2k}}.$$

REMARK. It is not difficult to see that an odd weight Eisenstein series $G_{2k+1}(\Lambda)$ (defined analogously) must be identically equal to zero by letting $\omega \mapsto -\omega$. The case of weight $2k = 2$ Eisenstein series is particularly interesting. Although the series $G_2(\Lambda)$ is not defined due to problems with convergence, we see in this chapter that weight 2 Eisenstein series can be suitably defined, and they play an important role in the theory of modular forms.

It is customary to let $\tau = \tau_\Lambda := \frac{\omega_1}{\omega_2} \in \mathbb{H}$, where $\mathbb{H}$ is the *upper-half of the complex plane*

$$\mathbb{H} := \{u + iv \; : \; u, v \in \mathbb{R}, v > 0\}.$$

The *inhomogeneous Eisenstein series* of weight $2k$, also denoted $G_{2k}(\tau)$ for convenience, is defined by

$$G_{2k}(\tau) := \omega_2^{2k} G_{2k}([\omega_1, \omega_2]).$$

As a function of τ, it follows that

$$G_{2k}(\tau) = \sum_{(m,n)\in\mathbb{Z}^2\setminus\{(0,0)\}} \frac{1}{(m\tau+n)^{2k}}.$$

These Eisenstein series admit Fourier expansions with number-theoretically interesting coefficients. To describe this more precisely, for $k \in \mathbb{N}_0$, the *k-th divisor function* $\sigma_k(n)$ is defined for $n \in \mathbb{N}$ by

$$\sigma_k(n) := \sum_{\substack{d|n \\ d>0}} d^k.$$

For non-negative integers n, the *Bernoulli numbers* B_n are given by the generating function

$$\frac{w}{e^w - 1} =: \sum_{n=0}^{\infty} \frac{B_n}{n!} w^n.$$

We have the following standard theorem about the inhomogeneous Eisenstein series G_{2k}, which includes their Fourier expansions in $q := e(\tau) = e^{2\pi i\tau}$ (note that throughout, we let $e(\alpha) := e^{2\pi i\alpha}$).

THEOREM 1.6. *For any integer $k \geq 2$ and $\tau \in \mathbb{H}$, we have that*

$$G_{2k}(\tau) = \frac{2(2\pi i)^{2k}}{(2k-1)!}\left(-\frac{B_{2k}}{4k} + \sum_{n=1}^{\infty} \sigma_{2k-1}(n)q^n\right).$$

Moreover, we have that $G_{2k} \in M_{2k}$, the space of weight $2k$ holomorphic modular forms on $\mathrm{SL}_2(\mathbb{Z})$.

From Theorem 1.6 it becomes clear that it is natural to define the *normalized Eisenstein series* E_{2k} for integers $k \geq 2$ as

$$E_{2k}(\tau) := -\frac{(2k)!}{(2\pi i)^{2k} B_{2k}} G_{2k}(\tau). \tag{1.1}$$

We then immediately obtain the Fourier expansion of the normalized Eisenstein series E_{2k}.

COROLLARY 1.7. *With notation and hypotheses as above, we have that*

$$E_{2k}(\tau) = 1 - \frac{4k}{B_{2k}} \sum_{n=1}^{\infty} \sigma_{2k-1}(n)q^n.$$

It is well known that these Eisenstein series play central roles in the theory of modular forms and provide some of the first examples of modular forms. Next we explain the role of the Eisenstein series $G_{2k}(\Lambda)$ in the theory of elliptic functions. In particular, they are the coefficients of the Weierstrass ℘-function.

1.2. Weierstrass ℘-function

As we have explained, elliptic functions must have order at least 2. Therefore, it is natural to try to construct an order 2 elliptic function, one with exactly two poles at 0. The Weierstrass ℘-function is such a function. It is defined as follows.

DEFINITION 1.8. The **Weierstrass $\wp$-function** $\wp_\Lambda(z)$ for Λ is defined for all $z \in \mathbb{C}$ by

$$\wp_\Lambda(z) := \frac{1}{z^2} + \sum_{\omega \in \Lambda \setminus \{0\}} \left(\frac{1}{(z-\omega)^2} - \frac{1}{\omega^2} \right).$$

REMARK. Strictly speaking, there is a single $\wp$-function depending on both z and the lattice, despite the fact that at times we will be viewing it as a function in a single variable in z only with the lattice fixed.

The following well-known theorem gives the main properties of $\wp_\Lambda$ as an elliptic function.

THEOREM 1.9. *The Weierstrass $\wp$-function $\wp_\Lambda$ is an even elliptic function with respect to Λ. In particular, the following are true:*

i) *For all $\omega \in \Lambda$ and $z \in \mathbb{C}$, we have that*

$$\wp_\Lambda(z+\omega) = \wp_\Lambda(z).$$

ii) *Apart from double poles at $z = \omega \in \Lambda$, $\wp_\Lambda$ is holomorphic.*

The Eisenstein series defined in the previous subsection are useful for describing the Laurent expansions of $\wp_\Lambda$. To make this precise, let $N_\Lambda := \min_{\omega \in \Lambda \setminus \{0\}} \{|\omega|\}$. Then we have the following Eisenstein series expansion for $\wp_\Lambda$.

LEMMA 1.10. *For $z \in \mathbb{C}$ satisfying $0 < |z| < N_\Lambda$, we have that*

$$\wp_\Lambda(z) = \frac{1}{z^2} + \sum_{n=1}^{\infty} (2n+1) G_{2n+2}(\Lambda) z^{2n}.$$

SKETCH OF PROOF. The proof follows easily by noticing that for $\omega \in \Lambda \setminus \{0\}$ and $0 < |z| < N_\Lambda$, we have

$$\frac{1}{(z-\omega)^2} = \frac{1}{\omega^2 \left(1 - \frac{z}{\omega}\right)^2} = \frac{1}{\omega^2} \left(1 + \sum_{n=1}^{\infty} (n+1) \left(\frac{z}{\omega}\right)^n \right),$$

which in turn implies that

$$\frac{1}{(z-\omega)^2} - \frac{1}{\omega^2} = \sum_{n=1}^{\infty} \frac{(n+1) z^n}{\omega^{n+2}}.$$

The claimed Laurent expansion follows by summing over ω and using Definition 1.5 (and the fact that odd weight Eisenstein series are identically equal to 0). □

REMARK. It is often convenient to consider the q-series expansion of $\wp_\Lambda$ if $\Lambda = \Lambda_\tau := \mathbb{Z} + \mathbb{Z}\tau$. Let $q = e^{2\pi i \tau}$ with $\tau \in \mathbb{H}$. Then, for $z \in \mathbb{C}$, we have (for example, cf. Theorem I.6.2 of **[463]**)

$$\frac{1}{(2\pi i)^2} \wp_\Lambda(z) = \sum_{m \in \mathbb{Z}} \frac{e^{2\pi i z} q^m}{(1 - e^{2\pi i z} q^m)^2} + \frac{1}{12} - 2 \sum_{n=1}^{\infty} \frac{q^n}{(1-q^n)^2}. \tag{1.2}$$

We make important use of this q-series identity in connection with the Weierstrass ζ-function.

One of the most important features of $\wp_\Lambda$ is that it satisfies a simple differential equation, one whose arithmetic properties encode the 2-division values of $\wp_\Lambda$. Namely, we have the following theorem.

THEOREM 1.11. *The following are true:*

i) *The function $\wp_\Lambda$ satisfies the differential equation*

$$(\wp'_\Lambda(z))^2 = 4\wp_\Lambda(z)^3 - g_2(\Lambda)\wp_\Lambda(z) - g_3(\Lambda),$$

where $g_2 := 60G_4$ and $g_3 := 140G_6$.

ii) *The 2-division values $e_1 := \wp_\Lambda(\frac{\omega_1}{2})$, $e_2 := \wp_\Lambda(\frac{\omega_2}{2})$, and $e_3 := \wp_\Lambda(\frac{\omega_1+\omega_2}{2})$ are distinct, and they satisfy*

$$\begin{aligned} 4\wp_\Lambda(z)^3 &- 60G_4(\Lambda)\wp_\Lambda(z) - 140G_6(\Lambda) \\ &= 4\left(\wp_\Lambda(z) - e_1\right)\left(\wp_\Lambda(z) - e_2\right)\left(\wp_\Lambda(z) - e_3\right). \end{aligned}$$

Here and throughout, we let $\wp'_\Lambda(z) := \frac{\partial}{\partial z}\wp_\Lambda(z)$.

REMARK. In fact, $\mathbb{C}(\wp, \wp')/\mathbb{C}(\wp)$ is algebraic of degree 2 and is the prototype of what is called an *elliptic function field*, which is an important concept in algebraic number theory and algebraic geometry (cf. Section 6.1 and specifically Example 6.1.4 of [**473**]).

SKETCH OF PROOF. i) Using Lemma 1.10, we have that

$$\wp'_\Lambda(z) = -2\sum_{\omega\in\Lambda} \frac{1}{(z-\omega)^3} = -\frac{2}{z^3} + 6G_4(\Lambda)z + 20G_6(\Lambda)z^3 + \cdots,$$

and it is an order 3 elliptic function. Its square

$$(\wp'_\Lambda(z))^2 = \frac{4}{z^6} - \frac{24G_4(\Lambda)}{z^2} - 80G_6(\Lambda) + \cdots$$

is an order 6 elliptic function, which one finds equals

$$4\wp_\Lambda(z)^3 - 60G_4(\Lambda)\wp_\Lambda(z) - 140G_6(\Lambda).$$

Indeed, the difference between these two elliptic functions has no poles, and so must be a constant. By considering constant terms of the z-series expansions, we see that this constant must be zero.

ii) By periodicity, we have for each $\omega \in \Lambda$ that

$$\wp'_\Lambda\left(\frac{\omega}{2}\right) = \wp'_\Lambda\left(-\frac{\omega}{2}\right).$$

Since $\wp'_\Lambda$ is odd, it follows that each half-period is a zero of $\wp'_\Lambda$. Since $\wp'_\Lambda$ has order 3, the half-periods $\omega_1/2, \omega_2/2$, and $(\omega_1+\omega_2)/2$ are its only zeros in the fundamental parallelogram $P_{\Lambda,0}$. Moreover, they are simple zeros.

To see that the e_j are distinct, observe that each half-period is a double zero of $\wp_\Lambda - e_j$. This follows since the half-periods are simple zeros of $\wp'_\Lambda$. To see that $e_1 \neq e_2$, suppose on the contrary that $e_1 = e_2$. Then $\wp_\Lambda - e_1$ would have double zeros at both $\omega_1/2$ and $\omega_2/2$, which implies that $\wp_\Lambda - e_1$ is an elliptic function with order at least 4. □

The following theorem explains the fundamental role of Weierstrass $\wp$-function in the theory of elliptic functions. All elliptic functions with respect to Λ are naturally generated from this function.

THEOREM 1.12. *The following are true:*

i) *The even elliptic functions with respect to Λ are the rational functions in $\wp_\Lambda$.*

ii) *The field of all elliptic functions with respect to Λ is given by $\mathbb{C}(\wp_\Lambda, \wp'_\Lambda)$.*

SKETCH OF PROOF. Any elliptic function f can be written as the sum

$$f(z) = \frac{f(z) + f(-z)}{2} + \frac{f(z) - f(-z)}{2}$$

of an even elliptic and an odd elliptic function. If f is odd, then $f\wp'_\Lambda$ is an even elliptic function, so it suffices to prove i). It is not difficult to construct an elliptic function

$$g(z) := \prod_{j=1}^{m} \left(\wp_\Lambda(z) - \wp_\Lambda(u_j)\right)^{n_j}$$

which has the same divisor as f, which in turn means that g/f is a constant by Theorem 1.4. □

We conclude this section with the following theorem which gives addition laws, which foreshadow deeper connections with the theory of elliptic curves.

THEOREM 1.13. *For all $z, w \in \mathbb{C}$ such that $z, w, z \pm w \notin \Lambda$, we have that*

$$\wp_\Lambda(z+w) + \wp_\Lambda(z) + \wp_\Lambda(w) = \frac{1}{4}\left(\frac{\wp'_\Lambda(z) - \wp'_\Lambda(w)}{\wp_\Lambda(z) - \wp_\Lambda(w)}\right)^2.$$

In particular, we have for $z \notin \frac{1}{2}\Lambda$

$$\wp_\Lambda(2z) = -2\wp_\Lambda(z) + \frac{1}{4}\left(\frac{\wp''_\Lambda(z)}{\wp'_\Lambda(z)}\right)^2.$$

REMARK. The map $z \mapsto (\wp_\Lambda(z), \wp'_\Lambda(z))$ can be viewed in terms of the arithmetic of the elliptic curve that corresponds to the differential equation in Theorem 1.11. Theorem 1.13 gives this realization of $\mathbb{C}/\Lambda$ in terms of the arithmetic of the elliptic curve

$$\left\{(X,Y) \in \mathbb{C} \times \mathbb{C} \,:\, Y^2 = 4X^3 - 60G_4(\Lambda)X - 140G_6(\Lambda)\right\}.$$

1.3. Weierstrass ζ-function

Here we discuss the Weierstrass ζ-function, a function that we use in Section 6.2 below to produce canonical harmonic Maass functions associated to modular elliptic curves.

DEFINITION 1.14. The **Weierstrass ζ-function**, $\zeta_\Lambda(z)$, with respect to a lattice Λ, is defined for $z \notin \Lambda$ by

$$\zeta_\Lambda(z) := \frac{1}{z} + \sum_{\omega \in \Lambda \setminus \{0\}} \left(\frac{1}{z-\omega} + \frac{1}{\omega} + \frac{z}{\omega^2}\right).$$

One can show that ζ_Λ can be expressed in terms of the Eisenstein series by arguing as in the proof of Lemma 1.10.

LEMMA 1.15. *We have that*

$$\zeta_\Lambda(z) = \frac{1}{z} - \sum_{n=1}^{\infty} G_{2n+2}(\Lambda) z^{2n+1}.$$

Moreover, we have that $\zeta'_\Lambda = -\wp_\Lambda$.

The Weierstrass ζ-function is not an elliptic function. Indeed, there are no single order elliptic functions. Eisenstein showed how to "correct" these functions to obtain doubly periodic functions at the expense of holomorphicity. This correction anticipates the theory of harmonic Maass forms (cf. Section 6.2).

To make this precise, we let

$$S_\Lambda := \lim_{s\to 0^+} \sum_{\omega\in\Lambda\setminus\{0\}} \frac{1}{\omega^2|\omega|^{2s}}.$$

This is a value of a real-analytic weight 2 form. That is, the complex number S_Λ arises from the *nonholomorphic weight 2 Eisenstein series* (cf. Section 6.1), defined by ($\tau = u + iv$ with $u, v \in \mathbb{R}$)

$$E_2^*(\tau) := E_2(\tau) - \frac{3}{\pi v}, \tag{1.3}$$

where

$$E_2(\tau) := 1 - 24\sum_{n=1}^{\infty} \sigma_1(n)q^n. \tag{1.4}$$

Let $\mathrm{vol}(\Lambda) := \mathrm{vol}(P_{\Lambda,z})$ be the volume of a fundamental parallelogram of the lattice Λ. Note that if $\Lambda = \Lambda_\tau$, then $\mathrm{vol}(\Lambda) = v$. Using these quantities, we have Eisenstein's corrected ζ-function (cf. Chapter VI of **[492]**)

$$\zeta_\Lambda^*(z) := \zeta_\Lambda(z) - S_\Lambda z - \frac{\pi}{\mathrm{vol}(\Lambda)}\overline{z}, \tag{1.5}$$

for which Eisenstein showed the following result.

THEOREM 1.16. *Assume the notation above. Then ζ_Λ^* is a nonholomorphic function for which*

$$\zeta_\Lambda^*(z+\omega) = \zeta_\Lambda^*(z)$$

for all $\omega \in \Lambda$.

Here we offer a proof due to Zagier **[506]**. For an alternate proof, see Section 6.2.

SKETCH OF PROOF. We assume that $\Lambda = \Lambda_\tau$ with $\tau \in \mathbb{H}$. Using the definition of ζ_Λ, one can show (using, e.g., Theorem I.6.3 of **[463]**) that

$$\zeta_\Lambda^\times(z) := \zeta_\Lambda(z) - \frac{\pi^2}{3}E_2(\tau)z = -\pi i \sum_{m\in\mathbb{Z}} \left(\frac{1+e^{2\pi i z}q^m}{1-e^{2\pi i z}q^m} - \mathrm{sgn}\left(m+\tfrac{1}{2}\right)\right), \tag{1.6}$$

where $q = e^{2\pi i\tau}$ and where for $x \in \mathbb{R}$, $\mathrm{sgn}(x)$ denotes the *sign function*

$$\mathrm{sgn}(x) := \begin{cases} +1 & \text{if } x > 0, \\ 0 & \text{if } x = 0, \\ -1 & \text{if } x < 0. \end{cases} \tag{1.7}$$

This then makes it apparent that $\zeta_\Lambda^\times(z+1) = \zeta_\Lambda^\times(z)$ and

$$\zeta_\Lambda^\times(z+\tau) = \zeta_\Lambda^\times(z) - 2\pi i. \tag{1.8}$$

Therefore, it follows that

$$\zeta_\Lambda(z) - \frac{\pi^2}{3}\left(-\frac{3}{\pi v} + E_2(\tau)\right)z - \frac{\pi\overline{z}}{v} = \zeta_\Lambda^*$$

is Λ-invariant, as claimed. $\square$

1.4. Eichler integrals of weight 2 newforms

In order to construct the Weierstrass mock modular forms in Section 6.2 below, we also require important properties of the so-called *Eichler integrals* attached to any modular elliptic curve over $\mathbb{Q}$. Although these functions are not modular, their properties play a central role. Namely, these Eichler integrals determine the lattices which give the analytic parameterizations of modular elliptic curves. As we shall see in Chapter 12 below, the general theory of Eichler integrals of integer weight cusp forms gives even more information with deep connections to the theory of critical L-values, as well as to the mock modular world.

In this section, we sketch some of the most important properties of Eichler integrals attached to weight 2 newforms. This gives rise to the classical Eichler-Shimura theory in weight 2, which provides a converse to the Modularity Theorem for rational elliptic curves (which is further discussed as Theorem 19.6 below). That is, we describe how to algorithmically construct elliptic curves (at least, over $\mathbb{C}$) from weight 2 newforms, with a larger focus on examples than on rigorous proofs.

To give this picture, we first define for any $F \in S_2(\Gamma_0(N)))$ (where $S_k(\Gamma)$ denotes the space of weight k holomorphic cusp forms on Γ and $\Gamma_0(N)$ is the usual congruence subgroup of $\mathrm{SL}_2(\mathbb{Z})$) and any fixed $\tau_0 \in \mathbb{H}$, the *(holomorphic) Eichler integral*

$$\mathcal{E}_F(\tau) := \int_{\tau_0}^{\tau} F(w)dw.$$

Since F is analytic, this is independent of the choice of path for the integral. The key property of $\mathcal{E}_F$ is that it is "nearly" a modular form of weight 0. More specifically, if $\gamma \in \Gamma_0(N)$, then the one-form $F(w)dw$ is $\Gamma_0(N)$-invariant, and so

$$\begin{aligned}\mathcal{E}_F(\gamma\tau) &= \int_{\gamma\tau_0}^{\gamma\tau} F(w)dw + \int_{\tau_0}^{\gamma\tau_0} F(w)dw \\ &= \int_{\tau_0}^{\tau} F(w)dw + \int_{\tau_0}^{\gamma\tau_0} F(w)dw = \mathcal{E}_F(\tau) + \int_{\tau_0}^{\gamma\tau_0} F(w)dw.\end{aligned}$$

Thus, we directly find that the function $\Psi_F \colon \Gamma_0(N) \to \mathbb{C}$ defined by

$$\Psi_F(\gamma) := \mathcal{E}_F(\gamma\tau) - \mathcal{E}_F(\tau) = \int_{\tau_0}^{\gamma\tau_0} F(w)dw$$

is independent of τ and τ_0. Moreover, a simple exercise shows that Ψ_F is an (additive) homomorphism of groups, and it is not difficult to prove that $\Psi_F(\gamma) = 0$ if the absolute value of the trace $|\mathrm{tr}(\gamma)|$ is less than or equal to two (for example, cf. Proposition 11.1 of [**311**]).

The central fact that we have been aiming towards is that the image of Ψ_F, that is, the set of "errors to modularity" of $\mathcal{E}_F$, is a lattice in $\mathbb{C}$. This plays a crucial role in the proof of Theorem 6.4 below. To see why this might be true, we consider the smallest level where weight 2 cusp forms exist, namely we suppose $N = 11$. The example which follows was computed and elegantly described by Knapp in Chapter XI of [**311**], and here we roughly follow his exposition. It is well known that $S_2(\Gamma_0(11))$ is one-dimensional. To see this, one can for example note that if we consider *Dedekind's eta-function*, the weight 1/2 cusp form

$$\eta(\tau) := q^{\frac{1}{24}} \prod_{n=1}^{\infty} (1 - q^n), \tag{1.9}$$

then the function $f(\tau) := \eta(\tau)^2\eta(11\tau)^2$ lies in $S_2(\Gamma_0(11))$. However, f has no zeros on $\mathbb{H}$ and only simple zeros at the cusps, so if $g \in S_2(\Gamma_0(11))$, then g/f is a weight zero holomorphic modular form of level 11 i.e., it lies in the space $M_0(\Gamma_0(11))$, which is well known to consist of only constant functions. This is in accordance with the Modularity Theorem as there is only one isogeny class of elliptic curves over $\mathbb{Q}$ of conductor 11 (as was shown in an important early paper providing evidence for the Taniyama-Shimura-Weil Conjecture [**3**]).

Now we want to show how one can obtain the lattice associated to the corresponding elliptic curve. In the case of our level 11 function f, this is straightforward and instructional. Namely, the group $\Gamma_0(11)$ is generated by

$$J := \begin{pmatrix} -1 & 0 \\ 0 & -1 \end{pmatrix}, \quad T := \begin{pmatrix} 1 & 1 \\ 0 & 1 \end{pmatrix}, \quad M := \begin{pmatrix} 8 & 1 \\ -33 & -4 \end{pmatrix}, \quad R := \begin{pmatrix} 9 & 1 \\ -55 & -6 \end{pmatrix}.$$

As mentioned above, it is clear that Ψ_f maps both J and T to zero (since the absolute value of both of their traces is two). Thus, the image of Ψ_f in $\mathbb{C}$ is generated by $\Psi_f(M)$ and $\Psi_f(R)$. These numbers can be effectively computed, and as noted by Knapp, making a careful choice of τ_0 to speed up computations easily shows that

$$\omega_1 := \Psi_f(M) \approx -0.232177875650357 - 0.101000467297158i,$$
$$\omega_2 := \Psi_f(R) \approx -0.202000934594317i.$$

Clearly, the complex numbers ω_1 and ω_2 are linearly independent over $\mathbb{R}$ and thus define a lattice $\Lambda_f := \mathbb{Z}\omega_1 + \mathbb{Z}\omega_2$. Hence, it follows that the Eichler integral $\mathcal{E}_f$ actually descends to a map from $\Gamma_0(11)\backslash\mathbb{H}$ to $\mathbb{C}/\Lambda_f$. If we carefully keep track of the behavior at cusps, we see that we have constructed an analytic map from

$$X_0(11) := \Gamma_0(11)\backslash\mathbb{H}^*$$

to $\mathbb{C}/\Lambda_f$, where $\mathbb{H}^* := \mathbb{H} \cup \mathbb{Q} \cup \{i\infty\}$ is the compactified upper half-plane. That is, we have constructed a complex map from $X_0(11)$ onto an elliptic curve, which we denote by E_f.

As the Modularity Theorem predicts, this elliptic curve may be defined over $\mathbb{Q}$, as can the surjection from the modular curve onto this elliptic curve (the existence of a rational map from a modular curve onto a given elliptic curve is in fact an equivalent condition for modularity of the elliptic curve). This may be observed as follows. We first substitute the numerically computed values for ω_1 and ω_2 into the definition of the Weierstrass invariants of Λ_f to find:

$$g_2(\Lambda_f) \approx 64419.8788704687 - 0.0000000003i,$$
$$g_3(\Lambda_f) \approx -5699399.99557174 + 0.00000002i,$$

where we recall that $g_2 = 60G_4$ and $g_3 = 140G_6$. Hence, the j-invariant of the curve is approximately

$$j(\Lambda_f) \approx -757.672637860052 + 0.000000000008i.$$

After a little effort, one finds that this j-invariant is approximately

$$j(\Lambda_f) \approx -\frac{2^{12}31^3}{11^5},$$

and in fact this is the correct value of $j(\Lambda_f)$. Assuming this to be true, this implies that E_f may be defined over $\mathbb{Q}$. In fact, there are infinitely many $\mathbb{Q}$-rational structures on E_f which are isomorphic over $\mathbb{C}$, and they are all quadratic twists of each other (meaning that they become isomorphic over quadratic extensions of $\mathbb{Q}$). As worked out in Chapter XI, Section I of [**311**], it is elementary to explicitly compute which rational elliptic curves can have this j-invariant, the simplest of which is given by

$$E\colon Y^2 + Y = X^3 - X^2 - 10X - 20.$$

Another deep connection exists between E and f. It turns out that they have identical L-functions, as can be shown using an important interpretation of the Hecke operators and Hecke eigenvalues of f in terms of actions on the homology of the modular curve.

The above properties in the case of level 11 are not coincidental. In fact, as shown by Eichler and Shimura, similar results hold for all weight 2 newforms. We may state a simplified version of their theorem as follows (cf. Theorem 11.74 of [**311**]).

THEOREM 1.17 (Eichler and Shimura). *Let $F \in S_2(\Gamma_0(N))$ be a (normalized) newform. Suppose further that the Fourier coefficients of F lie in $\mathbb{Z}$. Then the following are true:*

i) *The set*
$$\Lambda_F := \{\Psi_F(\gamma) : \gamma \in \Gamma_0(N)\}$$
is a lattice in $\mathbb{C}$, and $\mathbb{C}/\Lambda_F$ may be defined over $\mathbb{Q}$.

ii) *Denote the rational elliptic curve whose existence is implied by part* i) *by E_F. Then the Hasse-Weil L-function of E (cf.* (19.3)*) coincides with the Hecke L-function of F (cf.* (12.5)*) as an Euler product except possibly at finitely many primes.*

CHAPTER 2

Theta Functions and Holomorphic Jacobi Forms

The theory of elliptic functions and modular forms is closely related to the theory of *Jacobi forms*, two variable functions which simultaneously capture both of their essences. The theory of Jacobi forms, which has its origins in the work of Jacobi on theta functions, was developed in its present form by Eichler and Zagier [**191**]. Jacobi forms were originally defined by Eichler in order to prove the Saito-Kurokawa conjecture on $\Gamma_0(N)$, which led to the study of the general framework and a refinement of the levels of the modular forms arising in the Saito-Kurokawa conjecture in [**191**] (cf. Section 2.8 below for more on this and related topics).

In this book we discuss similar functions which serve as some of the fundamental building blocks of half-integral weight harmonic Maass forms. In particular, the theory of Jacobi forms provides a nice way of packaging together families of modular forms into single modular objects via their Taylor expansions, and also their specializations to torsion points, as we discuss in detail in this chapter. The first of these ideas is similar to how the Weierstrass $\wp$-function, which we shall see is in fact a meromorphic Jacobi form, encodes the Eisenstein series of level one, as shown in Chapter 1. Before discussing harmonic Maass forms, it is important to first recall here the main aspects of the Eichler-Zagier theory of holomorphic Jacobi forms, which is described in greater detail in [**158, 191**]. We note that throughout this chapter, q denotes $e^{2\pi i\tau}$ and we set $\zeta := e^{2\pi iz}$, where $\tau \in \mathbb{H}$ and $z \in \mathbb{C}$.

2.1. Jacobi theta functions

In order to motivate the construction of Jacobi forms, we begin by discussing the theory of classical theta functions, which also arise in examples throughout this book. For more facts on these theta functions, we refer the reader to [**390**].

DEFINITION 2.1. Define for $z \in \mathbb{C}$ and $\tau \in \mathbb{H}$ the **Jacobi theta function** $\vartheta(z)$

$$\vartheta(z) = \vartheta(z;\tau) := \sum_{n\in\frac{1}{2}+\mathbb{Z}} e^{\pi i n^2\tau + 2\pi i n\left(z+\frac{1}{2}\right)}.$$

This function is an odd holomorphic Jacobi form of weight $1/2$ and index $1/2$ ("odd" meaning odd as a function of z); cf. Definition 2.11 and the remarks that follow. To be more precise, we have the following transformation laws.

LEMMA 2.2. *We have*

$$\begin{aligned}\vartheta(z+1) &= -\vartheta(z),\\ \vartheta(z+\tau) &= -e^{-\pi i\tau-2\pi iz}\vartheta(z),\\ \vartheta(z;\tau+1) &= e^{\frac{\pi i}{4}}\vartheta(z;\tau),\\ \vartheta\left(\frac{z}{\tau};-\frac{1}{\tau}\right) &= -i\sqrt{-i\tau}e^{\frac{\pi iz^2}{\tau}}\vartheta(z;\tau).\end{aligned}$$

In light of the first two of these transformations, the Jacobi theta function may be viewed as an elliptic analogue of the function $e(z) := e^{2\pi iz}$, which satisfies $e(z+n) = e(z)$ for all $n \in \mathbb{Z}$. One of the most important properties of the Jacobi theta function is that it factors into the following infinite product which is typically referred to as the *Jacobi Triple Product Identity.*

LEMMA 2.3. *We have that*

$$\vartheta(z;\tau) = -iq^{\frac{1}{8}}\zeta^{-\frac{1}{2}}\prod_{n=1}^{\infty}(1-q^n)\left(1-\zeta q^{n-1}\right)\left(1-\zeta^{-1}q^n\right).$$

REMARK. Lemma 2.3 has various nice applications. Firstly one directly sees that $z \mapsto \vartheta(z;\tau)$ has simple zeros exactly in $\mathbb{Z}+\mathbb{Z}\tau$. Moreover, we have that

$$\vartheta'(0;\tau) := \left[\frac{\partial}{\partial z}\vartheta(z;\tau)\right]_{z=0} = -2\pi\eta(\tau)^3, \tag{2.1}$$

where

$$\eta(\tau) = q^{\frac{1}{24}}\prod_{n=1}^{\infty}(1-q^n) \tag{2.2}$$

is Dedekind's eta-function.

We also make frequent use of the classical weight $1/2$ and $3/2$ theta functions, which in the case of weight $1/2$ provide a basis of all modular forms (thanks to the Serre-Stark basis theorem given below as Theorem 2.8) and which in weight $3/2$ are especially important for us in Chapter 8 below as they describe the shadows of mock theta functions built out of Zwegers' μ-function. The connection to the Jacobi theta function above and to the Jacobi forms below, which is made evident by an inspection of the defining formulas, is that these functions are specializations of Jacobi forms where z is fixed. We see below that this is a general procedure allowing one to produce modular forms from Jacobi forms.

DEFINITION 2.4. Define for a Dirichlet character $\chi \pmod{N}$ ($N \in \mathbb{N}$) the **twisted theta function θ_χ** by

$$\theta_\chi(\tau) := \sum_{n\in\mathbb{Z}}\chi(n)n^\nu q^{n^2},$$

where $\nu = 0$ if χ is even and $\nu = 1$ if χ is odd, respectively.

EXAMPLE 2.5. Choosing $N = 1$ and $\chi = 1$ gives us the classical theta function

$$\theta(\tau) := \sum_{n\in\mathbb{Z}} q^{n^2}.$$

Gauss proved that if we define $r(n)$ by

$$\sum_{n=0}^{\infty} r(n)q^n := \theta(\tau)^3 = 1 + 6q + 12q^2 + 8q^3 + 6q^4 + \cdots,$$

then we have

$$r(n) = \begin{cases} 12H(4n) & \text{if } n \equiv 1, 2 \pmod 4, \\ 24H(n) & \text{if } n \equiv 3 \pmod 8, \\ r\left(\frac{n}{4}\right) & \text{if } n \equiv 0 \pmod 4, \\ 0 & \text{if } n \equiv 7 \pmod 8, \end{cases} \tag{2.3}$$

where $H(n)$ denotes the Hurwitz class numbers of discriminant $-n$. We will see in Theorem 6.3 below that the generating function for $H(n)$ is in fact a mock modular form (cf. Definition 5.16).

Let $M_k(\Gamma_0(N), \chi)$ be the space of weight k modular forms with respect to $\Gamma_0(N)$ with Dirichlet character χ.

THEOREM 2.6. *If χ is a primitive Dirichlet character with conductor c_χ and ν is as above, then*

$$\theta_\chi \in M_{\nu+\frac{1}{2}}\left(\Gamma_0\left(4c_\chi^2\right), \chi\chi_4^\nu\right)$$

with χ_4 the unique nontrivial Dirichlet character modulo 4. If χ is odd, then θ_χ is a cusp form.

Serre and Stark **[450]** showed that for weight 1/2 these theta functions are the building blocks of all modular forms. To describe this, we make the following definition.

DEFINITION 2.7. If χ is a Dirichlet character, then consider its canonical factorization

$$\chi = \prod_{\substack{p \mid c_\chi \\ p \text{ prime}}} \chi_p,$$

where χ_p is a Dirichlet character whose conductor is the highest power of p dividing c_χ. The character χ is called **totally even** if $\chi_p(-1) = 1$ for every prime $p|c_\chi$. Moreover, for $N \in \mathbb{N}$ we define the set $\Omega(4N, \chi)$, which consists of all pairs (ψ, t), where $t \in \mathbb{N}$, and $\psi = \psi_t$ is any even primitive Dirichlet character with conductor $c = c_\chi$ satisfying $c^2 t | N$, as well as the identity $\chi(n) = \psi(n)(\frac{t}{n})$, for every $n \in \mathbb{Z}$ with $\gcd(n, 4N) = 1$.

THEOREM 2.8 (Serre and Stark Basis Theorem). *Let $N \in \mathbb{N}$ and χ be an even Dirichlet character* $\pmod{4N}$.

i) *A basis for $M_{\frac{1}{2}}(\Gamma_0(4N), \chi)$ is given by*

$$\{\theta_\psi(t\tau) : (\psi, t) \in \Omega(4N, \chi)\}.$$

ii) *A basis for $S_{\frac{1}{2}}(\Gamma_0(4N), \chi)$ is given by*

$$\{\theta_\psi(t\tau) : (\psi, t) \in \Omega(4N, \chi), \psi \text{ not totally even}\}.$$

The modularity of θ_χ can be proved using certain theta functions with congruence conditions due to Shimura **[456]**.

DEFINITION 2.9. Define for $\nu \in \{0,1\}$, $h \in \mathbb{Z}$, $N \in \mathbb{N}$, $A \in \mathbb{N}$ with $A|N$, $N|hA$

$$\theta_\nu(A,h,N;\tau) := \sum_{\substack{m\in\mathbb{Z} \\ m\equiv h \pmod{N}}} m^\nu q^{\frac{Am^2}{2N^2}}.$$

Using Poisson summation, one can show the following.

THEOREM 2.10. *Assuming the notation above, the following are true:*

i) *We have the translation property*

$$\theta_\nu(A,h,N;\tau+2) = e\left(\frac{h^2A}{N^2}\right)\theta_\nu(A,h,N;\tau).$$

ii) *We have the modular inversion property*

$$\theta_\nu\left(A,h,N;-\frac{1}{\tau}\right) = (-i)^\nu(-i\tau)^{\frac{1}{2}+\nu}A^{-\frac{1}{2}} \sum_{\substack{k \pmod{N} \\ Ak\equiv 0 \pmod{N}}} e\left(\frac{Akh}{N^2}\right)\theta_\nu(A,k,N;\tau).$$

iii) *For $c \in \mathbb{N}$, we have*

$$\theta_\nu(A,h,N;\tau) = \sum_{\substack{g \pmod{cN} \\ g\equiv h \pmod{N}}} \theta_\nu(cA,g,cN;c\tau).$$

2.2. Basic facts on Jacobi forms

We now turn to the definition of Jacobi forms, which place examples such as Jacobi's theta function ϑ in a general framework. As we see in the definition below, it is natural to think of the variable z as the elliptic variable and the variable τ as the modular variable. Using these conventions, we have the following definition.

DEFINITION 2.11. Let k and m be positive integers. A **holomorphic Jacobi form** of weight k and index m on $\mathrm{SL}_2(\mathbb{Z})$ is a holomorphic function $\phi\colon \mathbb{C}\times\mathbb{H}\to\mathbb{C}$ satisfying:

i) For all $\left(\begin{smallmatrix} a & b \\ c & d\end{smallmatrix}\right) \in \mathrm{SL}_2(\mathbb{Z})$, we have that

$$\phi\left(\frac{z}{c\tau+d};\frac{a\tau+b}{c\tau+d}\right) = (c\tau+d)^k e\left(\frac{cmz^2}{c\tau+d}\right)\phi(z;\tau).$$

ii) For all $(\lambda,\mu)\in\mathbb{Z}^2$, we have that

$$\phi(z+\lambda\tau+\mu;\tau) = e\left(-m\left(\lambda^2\tau+2\lambda z\right)\right)\phi(z;\tau).$$

iii) In terms of $q=e(\tau)$ and $\zeta=e(z)$, the function ϕ has a Fourier expansion of the form

$$\phi(z;\tau) = \sum_{\substack{n,r\in\mathbb{Z} \\ 4mn\geq r^2}} c_\phi(n,r)q^n\zeta^r.$$

If additionally in iii) ϕ satisfies the condition $c_\phi(n,r)=0$ if $4mn=r^2$, then ϕ is called a **Jacobi cusp form**.

REMARK. One can also consider Jacobi forms of non-positive index. This will be studied in greater detail in Section 11.3, Subsection 20.2.3, and Subsection 20.2.4.

REMARK. To ease notation, throughout we simply write $\Sigma_{n,r}$ to denote the sum in iii), where we require that $4mn \geq r^2$.

REMARK. Although we only define holomorphic Jacobi forms of full level (i.e., on $\mathrm{SL}_2(\mathbb{Z})$), the theory of forms on finite index subgroups of $\mathrm{SL}_2(\mathbb{Z})$ can be developed analogously in the obvious way.

REMARK. Strictly speaking, the Jacobi theta function from Definition 2.1 does not transform as in Definition 2.11 due to the fact that it has weight and index equal to $1/2$. Definition 2.11 may be suitably modified to allow half-integral weight and index, however, as well as multiplier systems, in a similar manner as one ordinarily considers in the context of modular forms (cf. (4.3)).

The holomorphic (resp. cuspidal) Jacobi forms of weight k and index m naturally form $\mathbb{C}$-vector spaces which we denote by $J_{k,m}$ (resp. $J_{k,m}^{\mathrm{cusp}}$). The next lemma, which follows from Definition 2.11 ii), exhibits dependence among the Fourier coefficients $c_\phi(n,r)$, a fact we make use of later in this section.

LEMMA 2.12. *Let $\phi \in J_{k,m}$ have Fourier coefficients $c_\phi(n,r)$. Then the coefficients $c_\phi(n,r)$ depend only on $4nm - r^2$ and $r \pmod{2m}$. Moreover, if k is even and m is either equal to 1 or is prime, then the coefficients $c_\phi(n,r)$ depend only on $4nm - r^2$.*

REMARK. If m is either equal to 1 or is prime, we write $c_\phi\left(4nm - r^2\right) := c_\phi(n,r)$. Note also that for ease of notation, in the remainder of this chapter we sometimes suppress dependence on ϕ and write c for c_ϕ when the context is clear.

In direct analogy with the theory of classical modular forms, we have the following fundamental fact about these spaces.

THEOREM 2.13. *For positive integers k and m, the $\mathbb{C}$-vector spaces $J_{k,m}$ and $J_{k,m}^{\mathrm{cusp}}$ are finite dimensional.*

In fact, Theorem 2.13 can be made precise; dimensions of spaces of Jacobi forms can be given in terms of the dimensions of spaces of modular forms. For example, for positive integers m and integers $k \geq 3$, we have that

$$\dim(J_{k,m}) = \begin{cases} \displaystyle\sum_{j=0}^{m} \left(\dim(M_{k+2j}) - \left\lceil \frac{j^2}{4m} \right\rceil\right) & \text{if } k \text{ is even,} \\ \displaystyle\sum_{j=1}^{m-1} \left(\dim(M_{k+2j-1}) - \left\lceil \frac{j^2}{4m} \right\rceil\right) & \text{if } k \text{ is odd.} \end{cases} \tag{2.4}$$

In particular, note that (2.4) implies that there are no non-zero Jacobi forms if $m = 1$ and k is odd. We refer the interested reader to **[191]** for a proof of (2.4). Using (2.4), one can give $\dim(J_{k,m})$ explicitly in terms of class numbers. To describe this, for integers $d \notin \{-3, -4\}$, we let $h(d)$ be the class number of positive definite binary quadratic forms of discriminant d and set $h(-3) = 1/3$ and $h(-4) = 1/2$. Further, we let b_m denote the largest integer whose square divides m, and for $x \in \mathbb{R}$ set

$$((x)) := x - \frac{1}{2}\left(\lceil x \rceil + \lfloor x \rfloor\right).$$

Eichler and Zagier gave the following expression for dimensions of spaces of Jacobi forms in Section 10 of **[191]**, where it is noted that a similar formula can be established for odd k.

COROLLARY 2.14. *Assume the notation as above. For even integers $k \geq 4$ and positive integers m, we have that*

$$\dim(J_{k,m}) = -\frac{1}{4}\sum_{d|4m} h(-d) + \frac{2k-3}{24}(m+1) + \frac{1}{2}\left(\left(\frac{m}{4}\right)\right) + \frac{b_m}{4} + \kappa_1(k,m) + \kappa_2(k,m) + \kappa(m),$$

where

$$\kappa_1(k,m) := \begin{cases} \frac{1}{3} & \text{if } m \equiv k \not\equiv 2 \pmod 3, \\ -\frac{1}{3} & \text{if } m \not\equiv k \equiv 2 \pmod 3, \\ 0 & \text{otherwise,} \end{cases}$$

$$\kappa_2(k,m) := \begin{cases} \frac{1}{4} & \text{if } m \text{ even, } k \equiv 0 \pmod 4, \\ -\frac{1}{4} & \text{if } m \text{ even, } k \equiv 2 \pmod 4, \\ 0 & \text{if } m \text{ odd,} \end{cases}$$

$$\kappa(m) := \begin{cases} \frac{1}{4} & \text{if } 4 \nmid m, \\ \frac{1}{2} & \text{if } 4|m. \end{cases}$$

We postpone the proof of Theorem 2.13 until the end of this section, and first establish some essential facts. We begin with the following observation: in view of the transformation laws in Definition 2.11, it is natural to suspect that specializations of Jacobi forms can be ordinary modular forms. This is indeed the case. By letting z be a torsion point with respect to the lattice $\mathbb{Z} + \mathbb{Z}\tau$, we obtain a modular form. More precisely, we have the following theorem.

THEOREM 2.15. *Suppose that $\phi \in J_{k,m}$. If $\lambda, \mu \in \mathbb{Q}$, then the function*

$$g(\tau) := e\left(m\lambda^2\tau\right)\phi(\lambda\tau + \mu; \tau)$$

is a weight k holomorphic modular form on some finite index subgroup of $\mathrm{SL}_2(\mathbb{Z})$ which depends on λ, μ, and m. Moreover, if $\phi \in J_{k,m}^{\mathrm{cusp}}$, then g is a cusp form.

REMARK. The subgroup referred to in Theorem 2.15 is explicitly given in (2.8) below.

SKETCH OF PROOF OF THEOREM 2.15. Consider the set of triples

$$G^J := \{(\gamma, \boldsymbol{\alpha}, \zeta) \in \mathrm{SL}_2(\mathbb{R}) \times \mathbb{R}^2 \times \mathbb{C} \;:\; |\zeta| = 1\}.$$

One can show that G^J is a group with action defined by

$$(\gamma, \boldsymbol{\alpha}, \zeta) \cdot (\gamma', \boldsymbol{\alpha}', \zeta') := (\gamma\gamma', \boldsymbol{\alpha}\gamma' + \boldsymbol{\alpha}', \zeta\zeta' e(\det \boldsymbol{\beta})), \tag{2.5}$$

where

$$\boldsymbol{\beta} = \boldsymbol{\beta}_{\boldsymbol{\alpha},\gamma',\boldsymbol{\alpha}'} := \begin{pmatrix} \boldsymbol{\alpha}\gamma' \\ \boldsymbol{\alpha}' \end{pmatrix}.$$

If we let $\gamma = \left(\begin{smallmatrix} a & b \\ c & d \end{smallmatrix}\right) \in \mathrm{SL}_2(\mathbb{R})$ and $\boldsymbol{\alpha} = (\lambda, \mu) \in \mathbb{R}^2$, for positive integers k and m, this group acts on the set $\{\phi : \mathbb{C} \times \mathbb{H} \to \mathbb{C}\}$ by

$$\begin{aligned}(2.6)\qquad &\Big(\phi \,\big|_{k,m} (\gamma, \boldsymbol{\alpha}, \zeta)\Big)(z;\tau) \\ &:= \zeta^m (c\tau + d)^{-k} e\left(\frac{-cm(z + \lambda\tau + \mu)^2}{c\tau + d} + m\left(\lambda^2\tau + 2\lambda z + \lambda\mu\right)\right) \\ &\quad \times \phi\left(\frac{z + \lambda\tau + \mu}{c\tau + d}; \frac{a\tau + b}{c\tau + d}\right).\end{aligned}$$

We often write $|$ for $|_{k,m}$ and omit dependence on k and m whenever the context is clear. Let $\phi \in J_{k,m}$ and define for $\boldsymbol{\alpha} = (\lambda, \mu) \in \mathbb{Q}^2$ the function

$$\phi_{\boldsymbol{\alpha}}(\tau) := \left(\phi \mid (I_2, \boldsymbol{\alpha}, 1)\right)(0;\tau) = e\left(m\lambda^2\tau + m\lambda\mu\right)\phi(\lambda\tau + \mu; \tau),$$

where $I_2 := \left(\begin{smallmatrix} 1 & 0 \\ 0 & 1 \end{smallmatrix}\right)$. The function $\phi_{\boldsymbol{\alpha}}$ is, up to a constant, the function g defined in Theorem 2.15. If $\gamma \in \mathrm{SL}_2(\mathbb{Z})$, we have that

$$\begin{aligned}(c\tau + d)^{-k}\phi_{\boldsymbol{\alpha}}(\gamma\tau) &= \left(\phi \mid (I_2, \boldsymbol{\alpha}, 1) \mid (\gamma, \mathbf{0}, 1)\right)(0;\tau) \\ (2.7)\qquad &= \left(\phi \mid (I_2, \boldsymbol{\alpha}\gamma, 1)\right)(0;\tau) = \phi_{\boldsymbol{\alpha}\gamma}(\tau),\end{aligned}$$

where we have used (2.6) and the fact that $\phi \in J_{k,m}$. It is not difficult to show that

$$\phi_{\boldsymbol{\alpha}+(n_1,n_2)}(\tau) = e(m(\lambda n_2 - \mu n_1))\phi_{\boldsymbol{\alpha}}(\tau)$$

if $(n_1, n_2) \in \mathbb{Z}^2$. From this fact, and (2.7), we have that $\phi_{\boldsymbol{\alpha}}$ is modular with respect to the (congruence) subgroup

$$\begin{aligned}(2.8)\quad &\left\{\gamma \in \mathrm{SL}_2(\mathbb{Z}) \,:\, \boldsymbol{\alpha}\gamma \equiv \boldsymbol{\alpha} \pmod{\mathbb{Z}^2},\ m\det\begin{pmatrix} \boldsymbol{\alpha} \\ \boldsymbol{\alpha}\gamma \end{pmatrix} \in \mathbb{Z}\right\} \\ &= \left\{\begin{pmatrix} a & b \\ c & d \end{pmatrix} \in \mathrm{SL}_2(\mathbb{Z}) \,:\, (a-1)\lambda + c\mu, b\lambda + (d-1)\mu,\right. \\ &\qquad\qquad\qquad \left. m\left(c\mu^2 + (a-d)\lambda\mu - b\lambda^2\right) \in \mathbb{Z}\right\},\end{aligned}$$

which depends only on $\boldsymbol{\alpha}$ and m. (We have corrected a minor typographical error from **[191]** in the second expression given in (2.8).) Finally, a direct calculation using Definition 2.11 iii) shows that g is holomorphic and is a cusp form if $\phi \in J_{k,m}^{\mathrm{cusp}}$. □

REMARK. The group G^J appearing in the proof of Theorem 2.15 is the *real Jacobi group*. It is also common to define Jacobi forms using (2.6).

We now establish an "elliptic-like" property of holomorphic Jacobi forms. Namely, we show that the number of zeros (including multiplicities) of a non-trivial form in a fundamental parallelogram (recall Definition 1.3)

$$P_{[\tau,1],w} = \{w + r\tau + s \ :\ 0 \le r, s < 1\}$$

is bounded linearly in the index of the form.

THEOREM 2.16. *Let $\phi \in J_{k,m}$, and fix $\tau \in \mathbb{H}$ and $w \in \mathbb{C}$. Then the function $h(z) = h_\tau(z) := \phi(z;\tau)$ either has $2m$ zeros (including multiplicities) in the fundamental parallelogram $P_{[\tau,1],w}$ or it is identically zero.*

PROOF. Suppose that h is not identically equal to zero. Using Def. 2.11 ii), it is not difficult to show that $g(z) := h'(z)/h(z)$ is invariant under $z \mapsto z+1$. Thus, we have that

$$\frac{1}{2\pi i}\int_{\partial P_{[\tau,1],w}} g(z)dz = \frac{1}{2\pi i}\left(\int_w^{w+1} - \int_{w+\tau}^{w+1+\tau}\right) g(z)dz$$
$$= \frac{1}{2\pi i}\int_w^{w+1} g(z)dz - \frac{1}{2\pi i}\int_w^{w+1} g(z+\tau)dz.$$

Now using Definition 2.11 ii) again, we find that $g(z+\tau) = g(z) - 4\pi i m$. Hence, we have that

$$\frac{1}{2\pi i}\int_{\partial P_{[\tau,1],w}} g(z)dz = 2m.$$

On the other hand, this integral counts the zeros (including multiplicities) of h in $P_{[\tau,1],w}$ by the Residue Theorem. □

We use Theorem 2.16 in the proof of the finite dimensionality of the spaces $J_{k,m}$ and $J_{k,m}^{\text{cusp}}$.

PROOF OF THEOREM 2.13. Let $\phi \in J_{k,m}$ and choose $2m+1$ pairs of rational numbers $(\lambda_j, \mu_j) \in \mathbb{Q}^2$, $1 \leq j \leq 2m+1$, which satisfy $(\lambda_j, \mu_j) \not\equiv (\lambda_k, \mu_k) \pmod{\mathbb{Z}^2}$ for $j \neq k$. Then by Theorem 2.15, for each j, the function

$$g_j(\tau) := e\left(m\lambda_j^2\tau\right)\phi\left(\lambda_j\tau + \mu_j; \tau\right)$$

is in $M_k(\Gamma_j)$ for some subgroup Γ_j. Consider the map

$$\phi\colon J_{k,m} \to \bigoplus_{j=1}^{2m+1} M_k(\Gamma_j), \qquad \phi \mapsto (g_1, g_2, \ldots, g_{2m+1}), \tag{2.9}$$

and suppose some $\phi \neq 0$ maps to $(g_1, g_2, \ldots, g_{2m+1}) = (0, 0, \ldots, 0)$. Then, because $(\lambda_j, \mu_j) \not\equiv (\lambda_k, \mu_k) \pmod{\mathbb{Z}^2}$ for $j \neq k$, we have that $\lambda_j\tau + \mu_j \not\equiv \lambda_k\tau + \mu_k \pmod{\Lambda_\tau}$ in the fundamental parallelogram $P_{[\tau,1],0}$; hence the $2m+1$ pairs (λ_j, μ_j) give rise to $2m+1$ zeros of the function $h(z) = h_\tau(z) := \phi(z;\tau)$ in $P_{[\tau,1],0}$. This contradicts Theorem 2.16 and implies that the map (2.9) is injective. Therefore, we find that

$$\dim(J_{k,m}^{\text{cusp}}) \leq \dim(J_{k,m}) \leq \sum_{j=1}^{2m+1} \dim(M_k(\Gamma_j)) < \infty.$$

The finiteness follows from the finite dimensionality of spaces of holomorphic modular forms. □

REMARK. The proof of Theorem 2.13 shows that $J_{k,m} = 0$ for $k \leq 0$, except if $k = m = 0$, in which case $J_{k,m}$ consists of the constant functions.

2.3. Examples of Jacobi forms

Here we discuss natural examples of Jacobi forms. We revisit Jacobi's theta functions, discuss Jacobi-Eisenstein series, and explain the connection with the Weierstrass $\wp$-function.

2.3.1. The Jacobi theta function. Although we focus on integer weight Jacobi forms in this chapter, Jacobi forms of half-integral weight and index can be defined in an analogous way. Moreover, the prototype for the theory is the Jacobi theta function (cf. Section 2.1)

$$\begin{aligned}\vartheta(z;\tau) &= \sum_{n\in\frac{1}{2}+\mathbb{Z}} e^{\pi i n^2\tau + 2\pi i n\left(z+\frac{1}{2}\right)} \\ &= -iq^{\frac{1}{8}}\zeta^{-\frac{1}{2}}\prod_{n=1}^{\infty}\left(1-q^n\right)\left(1-\zeta q^{n-1}\right)\left(1-\zeta^{-1}q^n\right),\end{aligned}$$

which is a Jacobi form of weight $1/2$ and index $1/2$. Indeed, the transformation laws in Lemma 2.2 correspond nicely with the transformation laws in Definition 2.11.

2.3.2. Jacobi-Eisenstein series. In analogy with the holomorphic (modular) Eisenstein series, we have the holomorphic Jacobi-Eisenstein series. They are Jacobi forms which are constructed by averaging (as a convergent sum) over the action of the Jacobi group.

We begin by defining the *Jacobi group* Γ^J. It is the semidirect product

$$\Gamma^J := \mathrm{SL}_2(\mathbb{Z}) \ltimes \mathbb{Z}^2 \subseteq G^J, \tag{2.10}$$

endowed with the group law from (2.5),

$$(\gamma,(\lambda,\mu))(\gamma',(\lambda',\mu')) := (\gamma\gamma',(\lambda,\mu)\gamma'+(\lambda',\mu')),$$

for pairs $(\gamma,(\lambda,\mu)),(\gamma',(\lambda',\mu')) \in \mathrm{SL}_2(\mathbb{Z})\times\mathbb{Z}^2$. For positive integers k and m, the group Γ^J acts on functions $\phi:\mathbb{C}\times\mathbb{H}\to\mathbb{C}$ as described in (2.6) by the *slash operator*

$$\begin{aligned}&\left(\phi\big|_{k,m}(\gamma,\mathbf{n})\right)(z;\tau) \\ &:= (c\tau+d)^{-k}e\left(\frac{-cm(z+\lambda\tau+\mu)^2}{c\tau+d}+m\left(\lambda^2\tau+2\lambda z\right)\right)\phi\left(\frac{z+\lambda\tau+\mu}{c\tau+d};\frac{a\tau+b}{c\tau+d}\right),\end{aligned}$$

with $\gamma=\left(\begin{smallmatrix}a&b\\c&d\end{smallmatrix}\right)\in\mathrm{SL}_2(\mathbb{Z})$ and $\boldsymbol{n}=(\lambda,\mu)\in\mathbb{Z}^2$. If $\phi\in J_{k,m}$, then Definition 2.11 i) and ii) imply that

$$\left(\phi\big|_{k,m}(\gamma,\mathbf{n})\right)(z;\tau)=\phi(z;\tau).$$

REMARK. We refer to a function ϕ that satisfies Definition 2.11 as a weight k index m holomorphic Jacobi form on Γ^J.

The Jacobi-Eisenstein series are constructed by a suitable summation over this action. To this end, we have an analog of the "group of translations" in Γ^J, the group

$$\Gamma_\infty^J := \left\{(\gamma,\mathbf{n})\in\Gamma^J\ :\ 1|_{k,m}(\gamma,\mathbf{n})=1\right\}=\left\{\left(\pm\begin{pmatrix}1&n\\0&1\end{pmatrix},(0,\mu)\right):n,\mu\in\mathbb{Z}\right\}, \tag{2.11}$$

where $1\colon\mathbb{C}\times\mathbb{H}\to\mathbb{C}$ denotes the constant function $1(z;\tau)=1$. Armed with these definitions, we can now define the (holomorphic) Jacobi-Eisenstein series.

DEFINITION 2.17. For integers $k\geq 2$ and positive integers m, the **Jacobi-Eisenstein series** $E_{2k,m}(z;\tau)$ is defined by

$$E_{2k,m}(z;\tau) := \sum_{(\gamma,\mathbf{n})\in\Gamma_\infty^J\backslash\Gamma^J}\left(1\big|_{2k,m}(\gamma,\mathbf{n})\right)(z;\tau).$$

REMARK. Jacobi-Eisenstein series with multipliers, on subgroups of Γ^J, and of odd weights $2k+1$ can be defined analogously. It is not difficult to show that the holomorphic Eisenstein series of odd weights $E_{2k+1,m}$ are identically equal to zero, as is the case with the holomorphic Eisenstein series E_{2k+1} defined in Chapter 1. However, as discussed in Section 2 of **[191]**, similarly as in the modular form case "twisted" Jacobi-Eisenstein series of odd weight may be defined which do not vanish and which span appropriate spaces of Jacobi forms along with Jacobi cusp forms.

REMARK. One can show, using Definition 2.17, that the Jacobi-Eisenstein series may be expressed for integers $k \geq 2$ and $m \geq 1$ as the sum

(2.12)
$$E_{2k,m}(z;\tau) = \frac{1}{2} \sum_{\substack{c,d\in\mathbb{Z} \\ \gcd(c,d)=1}} \sum_{\lambda\in\mathbb{Z}} (c\tau+d)^{-2k} e\left(m\lambda^2 \frac{a\tau+b}{c\tau+d} + \frac{2m\lambda z}{c\tau+d} - \frac{cmz^2}{c\tau+d}\right),$$

where $a, b \in \mathbb{Z}$ are chosen so that $ad - bc = 1$.

REMARK. Similar to the Eisenstein series, one can define *(Jacobi) Poincaré series of exponential type* by averaging suitable exponential functions. To be more precise, we let

$$P_{k,m;(n,r)}(z;\tau) := \sum_{\gamma\in\Gamma^J_\infty\backslash\Gamma^J} e^{2\pi i(n\tau+rz)}\big|_{k,m}\gamma.$$

In **[325]** Kohnen investigated these Poincaré series and computed their Fourier expansion.

The Jacobi-Eisenstein series are examples of holomorphic Jacobi forms. Namely, we have the following theorem.

THEOREM 2.18. *If $k \geq 2$ and $m \geq 1$ are integers, then the Jacobi-Eisenstein series $E_{2k,m}$ are (non-cuspidal) holomorphic Jacobi forms of weight $2k$ and index m on Γ^J.*

SKETCH OF PROOF. For $k \geq 2$ and $m \in \mathbb{N}$, we have that the series defining $E_{2k,m}$ is absolutely convergent and is invariant with respect to the slash operator $|_{2k,m}$ on Γ^J. Using the Fourier expansions of Theorem 2.19 below, it follows that the Jacobi-Eisenstein series are holomorphic Jacobi forms, as well as that they are non-cuspidal. □

Here we describe the Fourier expansions of Jacobi-Eisenstein series. For this, we define certain L-series. If D_0 is either a fundamental discriminant or equal to 1, then we let

$$L_{D_0}(s) := L(\chi_{D_0}, s),$$

where $\chi_{D_0} := \left(\frac{D_0}{\cdot}\right)$. More generally, for $D = D_0 f^2 \in \mathbb{Z}$, where $D_0 := \operatorname{disc}(\mathbb{Q}(\sqrt{D}))$ and $f \in \mathbb{N}$, define

(2.13)
$$L_D(s) := \begin{cases} 0 & \text{if } D \not\equiv 0,1 \pmod 4, \\ \zeta(2s-1) & \text{if } D = 0, \\ L_{D_0}(s) \sum_{d|f} \mu(d) \left(\frac{D_0}{d}\right) d^{-s} \sigma_{1-2s}\left(\frac{f}{d}\right) & \text{if } D \equiv 0,1 \pmod 4, D \neq 0. \end{cases}$$

Here, μ is the Möbius function, σ_r is the r-th divisor function, and $\left(\frac{\cdot}{\cdot}\right)$ the Kronecker symbol. Furthermore, $\zeta(s)$ is the *Riemann zeta-function*

$$\zeta(s) := \sum_{n=1}^{\infty} \frac{1}{n^s}, \tag{2.14}$$

defined for $\mathrm{Re}(s) > 1$. One critical feature of $\zeta(s)$ is that it has an analytic continuation and satisfies a functional equation; see Section 3.5 below for more about this function.

Cohen **[150]** used these L-series to define (generalized) *class numbers*

$$H(k-1, |D|) = L_D(2-k) \tag{2.15}$$

for $D < 0$ and even k. In particular, the values $H(1, |D|)$ are Hurwitz class numbers, which count the class number of quadratic forms of discriminant $-|D|$, where each class $\mathcal{C}$ is weighted with multiplicity $1/\mathrm{Aut}(\mathcal{C})$. The values $H(k-1, |D|)$ for $D < 0, k$ even are known to be non-zero rational numbers and emerge as (factors of) the Fourier coefficients of the Jacobi-Eisenstein series. To describe this, we fix an index m and define for $n, r \in \mathbb{Z}$ with $4nm \geq r^2$ the numbers

$$e_{2k,m}(n,r) := \begin{cases} 1 & \text{if } 4nm = r^2, r \equiv 0 \pmod{2m}, \\ 0 & \text{if } 4nm = r^2, r \not\equiv 0 \pmod{2m}, \\ \frac{H(2k-1, 4nm-r^2)}{\zeta(3-4k)} \prod_{p \mid m} \rho_p & \text{if } 4nm > r^2. \end{cases}$$

Here the ρ_p are certain rational Euler factors, and so the $e_{2k,m}(n,r)$ are $\mathbb{Q}$-rational.

THEOREM 2.19. *Let m be a positive integer, and let $k \geq 2$ be an integer. Then the Fourier expansion of $E_{2k,m}$ is given by*

$$E_{2k,m}(z;\tau) = \sum_{\substack{n,r \in \mathbb{Z} \\ 4nm \geq r^2}} e_{2k,m}(n,r) q^n \zeta^r.$$

REMARK. In Theorem 2.19, for primes p dividing the index m, ρ_p is a certain Euler factor; while not explicitly given in **[191]**, the authors indicate that it may be deduced from their work. For square-free m, for example,

$$e_{2k,m}(n,r) = \frac{1}{\sigma_{2k-1}(m)\zeta(3-4k)} \sum_{d \mid \gcd(m,n,r)} d^{2k-1} H\left(2k-1, \frac{4nm-r^2}{d^2}\right).$$

REMARK. We revisit Theorem 2.19 in Section 2.4 below, where we discuss the relationship between Jacobi forms and half-integral weight modular forms. We show that the Fourier expansion in Theorem 2.19 can be used to give a new proof of the Fourier expansion of Cohen's half-integral weight Eisenstein series (see Cohen's original paper **[150]** for more on these important series).

REMARK. The Jacobi-Eisenstein series $E_{2,m}$ as defined fail to converge; however, analogous to the nonholomorphic ("almost holomorphic") modular form E_2^* (cf. Section 6.1 below), one can define a nonholomorphic Jacobi-Eisenstein series of weight 2. In particular, the authors in **[147, 148]** studied the nonholomorphic weight 2 form

$$E_{2,1}^*(z;\tau;s) := \frac{1}{2} \sum_{\substack{c,d\in\mathbb{Z} \\ \gcd(c,d)=1}} \sum_{\lambda\in\mathbb{Z}} \frac{e\left(\lambda^2 \frac{a\tau+b}{c\tau+d} + 2\lambda \frac{z}{c\tau+d} - \frac{cz^2}{c\tau+d}\right)}{(c\tau+d)^2|c\tau+d|^{2s}},$$

which can be analytically continued to the entire s-plane. The authors showed that the holomorphic part of this function at $s = 0$ is a 2-variable generating function for Hurwitz class numbers:

$$E_{2,1}(z;\tau) = -12 \sum_{\substack{n,r\in\mathbb{Z} \\ 4n\geq r^2}} H\left(4n - r^2\right) q^n \zeta^r.$$

Proof of Theorem 2.19. We split the sum (2.12) into two parts, one involving terms with $c \neq 0$ and one involving terms with $c = 0$. When $c = 0$, we find the sum $\sum_{\lambda\in\mathbb{Z}} q^{m\lambda^2}\zeta^{2m\lambda}$. Using the identity ($c \neq 0$)

$$\lambda^2 \frac{a\tau+b}{c\tau+d} + \frac{2\lambda z}{c\tau+d} - \frac{cz^2}{c\tau+d} = -\frac{c\left(z - \frac{\lambda}{c}\right)^2}{c\tau+d} + \frac{a\lambda^2}{c},$$

we find that the contribution to (2.12) from the terms with $c \neq 0$ can be written as

$$\sum_{c=1}^{\infty} c^{-2k} \sum_{\substack{d \pmod{c} \\ \gcd(c,d)=1}} \sum_{\lambda \pmod{c}} e\left(\frac{m\overline{d}\lambda^2}{c}\right) F_{2k,m}\left(z - \frac{\lambda}{c}; \tau + \frac{d}{c}\right),$$

where $\overline{d}$ is an integer satisfying $d\overline{d} \equiv 1 \pmod{c}$, and

$$F_{2k,m}(z;\tau) := \sum_{a,b\in\mathbb{Z}} (\tau+a)^{-2k} e\left(-m\frac{(z+b)^2}{\tau+a}\right)$$

is $\mathbb{Z}$-periodic in z and τ. Poisson summation gives that

$$F_{2k,m}(z;\tau) = \sum_{n,r\in\mathbb{Z}} \alpha(n,r) q^n \zeta^r,$$

where

$$\begin{aligned} \alpha(n,r) &= \int_{\mathrm{Im}(\tau)=r_1} \tau^{-2k} e(-n\tau) \int_{\mathrm{Im}(z)=r_2} e\left(-m\frac{z^2}{\tau} - rz\right) dz d\tau \\ &= \begin{cases} 0 & \text{if } r^2 \geq 4nm, \\ \dfrac{(-1)^k \pi^{2k-\frac{1}{2}} m^{1-2k}}{4^{k-1}\Gamma\left(2k - \frac{1}{2}\right)} \left(4nm - r^2\right)^{2k-\frac{3}{2}} & \text{if } r^2 < 4nm, \end{cases} \end{aligned}$$

where here the lines of integration are in terms of any real numbers r_1 and r_2 with $r_1 > 0$ and where Γ denotes the usual Gamma-function. Thus, we have that

$$E_{2k,m}(z;\tau) = \sum_{\substack{n,r\in\mathbb{Z} \\ 4nm>r^2}} e_{2k,m}(n,r) q^n \zeta^r,$$

where

$$e_{2k,m}(n,r) = \frac{(-1)^k \pi^{2k-\frac{1}{2}} m^{1-2k}}{4^{k-1}\Gamma\left(2k-\frac{1}{2}\right)} \left(4nm - r^2\right)^{2k-\frac{3}{2}} \sum_{c=1}^{\infty} c^{-2k} \times \sum_{\substack{\lambda,d \pmod{c} \\ \gcd(c,d)=1}} e\left(\frac{m\bar{d}\lambda^2 - r\lambda + nd}{c}\right). \tag{2.16}$$

The inner double sum in (2.16) is equal to

$$\sum_{a|c} \mu\left(\frac{c}{a}\right) a \sum_{\substack{\lambda \pmod{c} \\ Q(\lambda)\equiv 0 \pmod{a}}} 1, \tag{2.17}$$

where $Q(\lambda) = Q_{m,n,r}(\lambda) := m\lambda^2 + r\lambda + n$ and μ is the Möbius function. Now the condition $Q(\lambda) \equiv 0 \pmod{a}$ depends only on $\lambda \pmod{a}$, allowing further simplification of the sum in (2.17). The triple sum in (2.16) may be rewritten as

$$\sum_{c=1}^{\infty} c^{1-2k} \sum_{a|c} \mu\left(\frac{c}{a}\right) N_a(Q) = \zeta(2k-1)^{-1} \sum_{a=1}^{\infty} \frac{N_a(Q)}{a^{2k-1}}, \tag{2.18}$$

where

$$N_a(Q) := \#\{\lambda \pmod{a} \; : \; Q(\lambda) \equiv 0 \pmod{a}\}. \tag{2.19}$$

We seek to explicitly calculate the Dirichlet series in (2.18) in terms of the L-series $L_D(s)$ defined in (2.13). First, for $\gcd(a,m) = 1$, we have that $N_a(Q) = N_a(r^2 - 4nm)$, where

$$N_a\left(r^2 - 4nm\right) := \#\{\lambda \pmod{2a} \; : \; \lambda^2 \equiv D \pmod{4a}\}$$

(where D is the discriminant of Q). Next, for $D = 1$ or D the discriminant of a real quadratic field, it is well known that

$$\sum_{a=1}^{\infty} \frac{N_a(D)}{a^s} = \frac{\zeta(s)}{\zeta(2s)} L_D(s). \tag{2.20}$$

Moreover, in **[509]**, Zagier showed that (2.20) also holds for all integers D with $L_D(s)$ defined in (2.13). Using the functional equations for $\zeta(s)$ and $L_D(s)$, as well as the multiplicativity of N_a, leads to the result. Thanks to the theory of generalized Bernoulli numbers, it is well known that the values $L_D(2-2k)$, where $D < 0$, are rational and non-zero, hence $e_{2k,m}(n,r) \in \mathbb{Q}$ as claimed. □

2.3.3. Weierstrass $\wp$-function. The Weierstrass $\wp$-function (defined in Section 1.2) is also closely related to Jacobi forms. Although it is meromorphic in z (as opposed to holomorphic), it can be expressed as the ratio of two holomorphic Jacobi forms. To explain this, for $\tau \in \mathbb{H}$, define

$$\wp(z;\tau) := \wp_{[\tau,1]}(z).$$

We have the following theorem, which gives $\wp$ as a ratio of Jacobi forms.

THEOREM 2.20. *The Weierstrass $\wp$-function satisfies*

$$\wp(z;\tau) = -\frac{\pi^2}{3} \frac{E_4(\tau)^2 E_{4,1}(z;\tau) - E_6(\tau)E_{6,1}(z;\tau)}{E_6(\tau)E_{4,1}(z;\tau) - E_4(\tau)E_{6,1}(z;\tau)}.$$

REMARK. Theorem 2.20 represents $\wp$ as a weight 12 and index 1 Jacobi form divided by a weight 10 index 1 Jacobi form. In particular, we find that $\wp$ transforms as a meromorphic Jacobi form of weight 2 and index 0. We further discuss meromorphic Jacobi forms (of non-zero index) in Chapter 11.

SKETCH OF PROOF OF THEOREM 2.20. One can show that

$$E_4(\tau)^2E_{4,1}(z;\tau) - E_6(\tau)E_{6,1}(z;\tau) = 1728\Delta(\tau) + O\left(z^2\right), \tag{2.21}$$

$$E_6(\tau)E_{4,1}(z;\tau) - E_4(\tau)E_{6,1}(z;\tau) = -576\pi^2\Delta(\tau)z^2 + O\left(z^4\right), \tag{2.22}$$

where

$$\Delta(\tau) := q\prod_{n=1}^{\infty}(1-q^n)^{24}$$

is the unique normalized weight 12 cusp form on $\mathrm{SL}_2(\mathbb{Z})$. The function on the left hand side of (2.22) is a Jacobi form of index 1 and has a double zero at $z = 0$. By Theorem 2.16, it has no other zeros in the fundamental parallelogram $P_{[\tau,1],0}$. From these facts, we can conclude that the quotient on the right-hand side of the claimed equation is doubly periodic in z with double pole at 0 and no other poles in $P_{[\tau,1],0}$, so must equal $\wp$ up to a constant, which is easily determined to be 1 by considering Fourier expansions. □

2.4. A structure theorem for $J_{k,m}$

We have an analogous decomposition for $J_{k,m}$ into cusp forms and Eisenstein series as holds for general spaces of modular forms. To describe this, for integers $k > 2$, positive integers m, and integers ℓ, we define additional Jacobi-Eisenstein series $E_{k,m,\ell}(z;\tau)$ by

$$E_{k,m,\ell}(z;\tau) := \sum_{(A,\mathbf{n})\in\Gamma_\infty^J\backslash\Gamma^J} q^{a\ell^2}\zeta^{2ab\ell}\big|_{k,m}(A,\mathbf{n}),$$

where $b = b_m$ is the largest integer such that $b^2|m$ and $a = a_m$ is defined by $ab^2 = m$. One can show in a similar manner to the proof of Theorem 2.18 that the functions $E_{k,m,\ell}$ are non-cuspidal Jacobi forms of weight k and index m for each integer ℓ. Their Fourier expansions are of the form

$$E_{k,m,\ell}(z;\tau) = \frac{1}{2}\sum_{\substack{r\in\mathbb{Z}\\ r\equiv 2abs \pmod{2m}}} q^{\frac{r^2}{4m}}\left(\zeta^r + (-1)^k\zeta^{-r}\right) + \dots,$$

with the remainder of the Fourier expansion consisting only of terms $q^n\zeta^r$ with $4nm - r^2 > 0$. It is then clear that $E_{k,m,-\ell} = (-1)^\ell E_{k,m,\ell}$ and that the forms $E_{k,m,\ell}$ with $0 \le \ell \le b/2$ (k even) or $0 < \ell < b/2$ (k odd) are linearly independent. Let $J_{k,m}^{\mathrm{Eis}}$ be the space spanned by the Jacobi-Eisenstein series $E_{k,m,\ell}$. We have the following theorem.

THEOREM 2.21. *For integers $k > 2$, we have that*

$$J_{k,m} = J_{k,m}^{\mathrm{cusp}} \oplus J_{k,m}^{\mathrm{Eis}}.$$

REMARK. For any weight $k > 2$, the proof of Theorem 2.21 shows that the codimension of $J_{k,m}^{\mathrm{cusp}}$ in $J_{k,m}$ is equal to $\lfloor\frac{b}{2}\rfloor + 1$ if k is even, and equal to $\lfloor\frac{b-1}{2}\rfloor$ if k is odd.

SKETCH OF PROOF OF THEOREM 2.21. It suffices to show that the codimension of $J_{k,m}^{\text{cusp}}$ in $J_{k,m}$ is equal to $\lfloor \frac{b}{2} \rfloor + 1$ if k is even, and equal to $\lfloor \frac{b-1}{2} \rfloor$ if k is odd. To show this, first note that to check that the Fourier coefficients $c_\phi(n,r)$ of a Jacobi form ϕ are equal to zero if $4nm = r^2$, one needs only check for a finite set of representatives $r \pmod{2m}$ by Lemma 2.12. There are exactly b classes $r \pmod{2m}$ with $r^2 \equiv 0 \pmod{4m}$. This shows that the codimension of $J_{k,m}^{\text{cusp}}$ in $J_{k,m}$ is at most b. However $c_\phi(n,-r) = (-1)^k c_\phi(n,r)$, from which we deduce the codimension is at most $\lfloor \frac{b}{2} \rfloor + 1$ if k is even, and equal to $\lfloor \frac{b-1}{2} \rfloor$ if k is odd. The theorem now follows from our previous discussion on linear independence among the Eisenstein series $E_{k,m,\ell}$. □

2.5. Relationship with half-integral weight modular forms

Jacobi forms are intimately related to half-integral weight modular forms. We have already alluded to this in the discussion of the Fourier expansions of the Jacobi-Eisenstein series in connection with Cohen's half-integral weight Eisenstein series. Here we make these connections precise.

2.5.1. Theta decompositions. Jacobi theta functions may be used to give an isomorphism between $J_{k,m}$ and a space of ordinary (vector-valued) modular forms. Not only is this result interesting in and of itself, but it is used repeatedly throughout this chapter to relate Jacobi forms to ordinary modular forms. To describe this, we require the notion of half-integer weight Jacobi forms, which mentioned previously has been well defined and developed. For integers a and positive integers m, the Jacobi theta functions

$$\vartheta_{m,a}(z;\tau) := \sum_{\substack{r\in\mathbb{Z} \\ r\equiv a \pmod{2m}}} q^{\frac{r^2}{4m}} \zeta^r \tag{2.23}$$

are examples of weight $1/2$ and index m Jacobi forms (generalizing the Jacobi theta functions discussed in Subsection 2.3.1). More precisely, we have the following.

PROPOSITION 2.22. *For integers a and positive integers m, the Jacobi theta functions $\vartheta_{m,a}$ are weight $1/2$ index m Jacobi forms. In particular, we have that*

$$\vartheta_{m,a}(z;\tau+1) = e\left(\frac{a^2}{4m}\right)\vartheta_{m,a}(z;\tau),$$

$$\vartheta_{m,a}\left(\frac{z}{\tau};-\frac{1}{\tau}\right) = \sqrt{\frac{\tau}{2mi}}\, e\left(\frac{mz^2}{\tau}\right) \sum_{b \pmod{2m}} e\left(-\frac{ab}{2m}\right)\vartheta_{m,b}(z;\tau).$$

REMARK. Strictly speaking, the transformation laws established in Proposition 2.22 do not show that the theta functions $\vartheta_{m,a}$ are Jacobi forms in the sense of the spaces $J_{k,m}$ defined from Definition 2.11. However, they do lead to the fact that these theta functions are Jacobi forms of weight $1/2$ with respect to some congruence subgroup of $\mathrm{SL}_2(\mathbb{Z})$. This transformation is the same as that induced by the Weil representation (cf. Section 18.3).

SKETCH OF PROOF OF PROPOSITION 2.22. By definition, the theta functions $\vartheta_{m,a}$ have the required Fourier expansions. The required elliptic transformation properties, as well as the given transformation $\tau \mapsto \tau + 1$, are easy to establish. The transformation $(z;\tau) \mapsto \left(\frac{z}{\tau}; -\frac{1}{\tau}\right)$ follows from Poisson summation, in a similar manner to the proof of the analogous transformation law for ϑ in Lemma 2.2. □

These theta functions $\vartheta_{m,a}$ may be used to give a *theta decomposition* for Jacobi forms in $J_{k,m}$, as well as an isomorphism to a space of ordinary (vector-valued) modular forms. To make this precise, if $\phi \in J_{k,m}$, then it has a Fourier expansion $\phi(z;\tau) = \sum_{n,r} c_\phi(n,r) q^n \zeta^r$, where dependence among the coefficients $c_\phi(n,r)$ of ϕ is explained in Lemma 2.12. Thus, it makes sense to define for integers $N \geq 0$

$$c_a(N) = c_{a,\phi}(N) := c_\phi\left(\frac{N+r^2}{4m}, r\right)$$

for any integer $r \equiv a \pmod{2m}$. (For ease of notation, we omit the dependence on ϕ if the context is clear.) Here, we set $c_a(N) = 0$ if $N \not\equiv -a^2 \pmod{4m}$. With this, we define for $a \in \mathbb{Z}/2m\mathbb{Z}$ the functions

$$h_a(\tau) := \sum_{N=0}^{\infty} c_a(N) q^{\frac{N}{4m}}. \tag{2.24}$$

As is shown in the proof of Theorem 2.23 below, the functions h_a transform like *vector-valued modular forms*. More precisely, we consider vector-valued modular forms $(f_a)_{a \pmod{2m}}$ of weight $k - 1/2$ on $\mathrm{SL}_2(\mathbb{Z})$ which satisfy

$$f_a(\tau+1) = e\left(\frac{-a^2}{4m}\right) f_a(\tau), \tag{2.25}$$

$$f_a\left(-\frac{1}{\tau}\right) = \frac{\tau^k}{\sqrt{-2mi\tau}} \sum_{b \pmod{2m}} e\left(\frac{ab}{2m}\right) f_b(\tau). \tag{2.26}$$

THEOREM 2.23. *Assume the notation and hypotheses above. Then a Jacobi form $\phi \in J_{k,m}$ has a theta decomposition*

$$\phi(z;\tau) = \sum_{a \pmod{2m}} h_a(\tau) \vartheta_{m,a}(z;\tau).$$

Moreover, this decomposition gives an isomorphism between $J_{k,m}$ and the space of vector-valued modular forms satisfying (2.25) and (2.26) of weight $k - 1/2$ on $\mathrm{SL}_2(\mathbb{Z})$.

SKETCH OF PROOF OF THEOREM 2.23. With notation as above, we first establish the existence of the theta decomposition. We have that

$$\begin{aligned}
\phi(z;\tau) &= \sum_{a \pmod{2m}} \sum_{\substack{r \in \mathbb{Z} \\ r \equiv a \pmod{2m}}} \sum_{n \geq \frac{r^2}{4m}} c_a\left(4nm - r^2\right) q^n \zeta^r \\
&= \sum_{a \pmod{2m}} \sum_{r \equiv a \pmod{2m}} \sum_{n \geq 0} c_a(n) q^{\frac{r^2+n}{4m}} \zeta^r \\
&= \sum_{a \pmod{2m}} h_a(\tau) \vartheta_{m,a}(z;\tau).
\end{aligned}$$

Note that the theta functions $\vartheta_{m,a}$ do not depend on ϕ. With the existence of the claimed decomposition established, a direct calculation using i) from Definition 2.11 and the modular transformation properties of the theta functions $\vartheta_{m,a}$ reveal that the functions h_a form a vector-valued modular form of weight $k-1/2$ as claimed. On the other hand, one can reverse the calculation leading to the decomposition above. Any h_a as defined in (2.24) with $c_a(n) = 0$ for $n \not\equiv -a^2 \pmod{4m}$ defines a function ϕ with Fourier coefficients obeying Lemma 2.12 that satisfies ii) of Definition 2.11

and the correct condition at $i\infty$. The modular transformation properties of the forms $\vartheta_{m,a}$ and h_a establish condition i) in Definition 2.11. □

REMARK. It is convenient to note for the purposes of explicit calculation that the functions h_a exhibit the symmtery property $h_{-a} = (-1)^k h_a$.

2.5.2. An isomorphism to Kohnen's plus space. Holomorphic Jacobi forms are intimately related to half-integral weight modular forms, which in turn are connected to even integer weight forms via the Shimura correspondence **[456]**. This correspondence will be described in Section 19.3.

The next theorem makes this precise in the case of $J_{k,1}$. Let $M^+_{k-\frac{1}{2}}(\Gamma_0(4))$ denote *Kohnen's plus space*, which is the space of modular forms r of weight $k-1/2$ whose Fourier coefficients $c(n)$ vanish unless $(-1)^{k-1}n \equiv 0, 1 \pmod 4$.

THEOREM 2.24. *Let $k > 1$ be a positive integer. We have that*

$$J_{k,1} \cong M^+_{k-\frac{1}{2}}(\Gamma_0(4)) \cong M_{2k-2}.$$

The first isomorphism is described by the identification

$$\sum_{n=0}^{\infty} \sum_{4n \geq r^2} c\left(4n - r^2\right) q^n \zeta^r \longleftrightarrow \sum_{n=0}^{\infty} c(n) q^n.$$

The second isomorphism is via the Shimura correspondence.

Theorem 2.24 may be used to immediately establish the modularity of forms whose Fourier coefficients arise from coefficients of Jacobi forms. For example, Theorem 2.24 combined with Theorem 2.18 and Theorem 2.19 recover the following result, originally due to Cohen, on his half-integral weight Eisenstein series.

COROLLARY 2.25. *If $k \geq 4$ is an even integer, then the function*

$$\sum_{n=0}^{\infty} H(k-1, n) q^n$$

lies in the plus space $M^+_{k-\frac{1}{2}}(\Gamma_0(4))$.

REMARK. In fact, Cohen originally proved Corollary 2.25 for both even and odd k. His discovery of the existence of the half-integral weight Eisenstein series served as motivation for Kohnen's definition of his plus space $M^+_{k-\frac{1}{2}}(\Gamma_0(4))$.

SKETCH OF PROOF OF THEOREM 2.24. That $M^+_{k-\frac{1}{2}}(\Gamma_0(4)) \simeq M_{2k-2}$ is one of the fundamental theorems about Kohnen's plus spaces. We refer the reader to his papers **[322, 323]**. To prove the first isomorphism, let

$$\phi(z;\tau) =: \sum_{n=0}^{\infty} \sum_{4n \geq r^2} c\left(4n - r^2\right) q^n \zeta^r \in J_{k,1},$$

and define

$$h_a(\tau) := \sum_{n=0}^{\infty} c_a(n) q^{\frac{n}{4}},$$

where $c_a(n) := c(n)$ if $n \equiv -a^2 \pmod 4$ and $c_a(n) := 0$ otherwise. Now $c(n) = c_0(n) + c_1(n)$ (for all n), which leads to the decomposition

$$h(\tau) := \sum_{\substack{n\geq 0 \\ n\equiv 0,3 \pmod 4}} c(n)q^n = h_0(4\tau) + h_1(4\tau).$$

By Theorem 2.23, h_0 and h_1 satisfy

(2.27)

$$\begin{aligned} h_0(\tau+1) &= h_0(\tau), & h_1(\tau+1) &= -ih_1(\tau), \\ h_0\left(-\tfrac{1}{\tau}\right) &= \tfrac{1+i}{2}\tau^{k-\frac{1}{2}}(h_0(\tau)+h_1(\tau)), & h_1\left(-\tfrac{1}{\tau}\right) &= \tfrac{1+i}{2}\tau^{k-\frac{1}{2}}(h_0(\tau)-h_1(\tau)). \end{aligned}$$

Together, these imply

$$h(\tau+1) = h(\tau), \quad h\left(\frac{\tau}{4\tau+1}\right) = (4\tau+1)^{k-\frac{1}{2}}h(\tau),$$

from which it follows that $h \in M^+_{k-\frac{1}{2}}(\Gamma_0(4))$.

Conversely, let $h \in M^+_{k-\frac{1}{2}}(\Gamma_0(4))$. Reversing the above calculation shows that h_0 and h_1 satisfy the four transformations given in (2.27). Thus, by Theorem 2.23, we have that the function

$$\sum_{n,r} c\left(4n-r^2\right) q^n \zeta^r$$

is in $J_{k,1}$. □

REMARK. It is natural to ask for a generalization of Theorem 2.24. In other words, are there analogous isomorphisms for spaces of Jacobi forms with higher index $m > 1$? Kohnen's theory of half-integral weight newforms provides a satisfactory theory which provides isomorphisms between suitable spaces of half-integral weight cusp forms and even integer weight cusp forms on $\Gamma_0(N)$ if N is odd and square-free [**323**]. On the Jacobi side, Eichler and Zagier showed an isomorphism between $J_{k,m}$ and a subspace of the plus space of level $4m$ for m prime and k even, making use of the splitting (2.33) discussed in Subsection 2.6.2. Subsequent generalizations have been obtained by Skoruppa [**465**].

2.6. Hecke theory for $J_{k,m}$ and the Jacobi-Petersson inner product

Here we describe the theory of Hecke operators for Jacobi forms, and a generalization for the Petersson inner product.

2.6.1. Hecke theory of $J_{k,m}$. Analogous to the Hecke theory for modular forms, there is a Hecke theory for holomorphic Jacobi forms. In what follows, let $M_2(\mathbb{Z})$ denote the set of all 2×2 matrices with integer entries. There are three types of Hecke operators which are defined as follows.

DEFINITION 2.26. The operators U_ℓ, V_ℓ, and T_ℓ for positive integers ℓ are defined on functions $\phi\colon \mathbb{C}\times\mathbb{H}\to\mathbb{C}$ by

$$\begin{aligned}
(\phi\big|_{k,m}U_\ell)(z;\tau) &:= \phi(\ell z;\tau),\\
(\phi\big|_{k,m}V_\ell)(z;\tau) &:= \ell^{k-1}\sum_{\substack{\left(\begin{smallmatrix}a&b\\c&d\end{smallmatrix}\right)\in \mathrm{SL}_2(\mathbb{Z})\backslash M_2(\mathbb{Z})\\ ad-bc=\ell}} (c\tau+d)^{-k}e\left(\frac{-m\ell cz^2}{c\tau+d}\right)\phi\left(\frac{\ell z}{c\tau+d};\frac{a\tau+b}{c\tau+d}\right),\\
(\phi\big|_{k,m}T_\ell)(z;\tau) &:= \ell^{k-4}\sum_{\substack{M=\left(\begin{smallmatrix}a&b\\c&d\end{smallmatrix}\right)\in \mathrm{SL}_2(\mathbb{Z})\backslash M_2(\mathbb{Z})\\ ad-bc=\ell^2,\ \gcd(a,b,c,d)=\square}}\ \sum_{\boldsymbol{n}\in\mathbb{Z}^2/\ell\mathbb{Z}^2}(\phi\big|_{k,m}(M,\boldsymbol{n}))(z;\tau),
\end{aligned}$$

where $\gcd(a,b,c,d)=\square$ means that $\gcd(a,b,c,d)$ is a perfect square.

REMARK. The slash operator $|_{k,m}(M,\boldsymbol{n})$ is defined as in (2.6); however for $M\in \mathrm{GL}_2^+$, one replaces M by $\det(M)^{-\frac{1}{2}}M\in \mathrm{SL}_2(\mathbb{R})$.

REMARK. The definitions of the operators U_ℓ, V_ℓ, and T_ℓ depend on k and m. This dependence is suppressed from the notation because these quantities are always understood in context.

These operators map holomorphic Jacobi forms to holomorphic Jacobi forms. More precisely, we have the following theorem.

THEOREM 2.27. *For positive integers ℓ, the operators U_ℓ, V_ℓ and T_ℓ are well defined on $J_{k,m}$. Moreover, we have that*

$$\begin{aligned}
U_\ell &: J_{k,m}\to J_{k,m\ell^2},\\
V_\ell &: J_{k,m}\to J_{k,m\ell},\\
T_\ell &: J_{k,m}\to J_{k,m}.
\end{aligned}$$

SKETCH OF PROOF. It is straightforward to verify that U_ℓ, V_ℓ, and T_ℓ are independent of the choice of representatives in their defining sums. Moreover, by definition, it is easy to see for $\phi\in J_{k,m}$ that $\phi|_{k,m}U_\ell$, $\phi|_{k,m}V_\ell$, and $\phi|_{k,m}T_\ell$ transform like Jacobi forms of weight k and indices $m\ell^2, m\ell$, and m, respectively. Theorem 2.28 below establishes their Fourier expansions, so they are indeed Jacobi forms. $\square$

These Hecke operators, in analogy with their modular counterparts, are easily defined in terms of Fourier coefficients. For example, the U_ℓ and V_ℓ operators are defined (with the same notation) on q-series by

$$\begin{aligned}
\sum_n c(n)q^n\big|U_\ell &:= \sum_n c(\ell n)q^n,\\
\sum_n c(n)q^n\big|V_\ell &:= \sum_n c(n)q^{\ell n}.
\end{aligned}\tag{2.28}$$

The image of a Jacobi form under the Hecke operators defined in Definition 2.26 may also be expressed in terms of its expansion, as shown in the following theorem.

THEOREM 2.28. *Let $\phi(z;\tau) = \sum_{n,r} c_\phi(n,r)q^n\zeta^r \in J_{k,m}$. For positive integers ℓ, we have that*

$$(\phi|U_\ell)(z;\tau) = \sum_{n,r} c_\phi\left(n, \frac{r}{\ell}\right) q^n\zeta^r,$$

$$(\phi|V_\ell)(z;\tau) = \sum_{n,r} \sum_{a|\gcd(n,r,\ell)} a^{k-1} c_\phi\left(\frac{n\ell}{a^2}, \frac{r}{a}\right) q^n\zeta^r.$$

Furthermore, for positive integers ℓ such that $\gcd(\ell, m) = 1$, we have

$$(\phi|T_\ell)(z;\tau) = \sum_{n,r} c^*(n,r)q^n\zeta^r,$$

where

$$c^*(n,r) := \sum_{\substack{a|\ell^2,\ a^2|\ell^2(r^2-4nm) \\ a^{-2}\ell^2(r^2-4nm)\ \equiv\ 0,1 \pmod 4}} \varepsilon_{r^2-4nm}(a)a^{k-2}c_\phi(n', r')$$

and

$$r'^2 - 4n'm = \frac{\ell^2}{a^2}\left(r^2 - 4nm\right), \quad ar' \equiv \ell r \pmod{2m}. \tag{2.29}$$

Here, the generalized Legendre symbol ε_D is defined by the Dirichlet series

$$L_D(s) =: \sum_{n=1}^{\infty} \frac{\varepsilon_D(n)}{n^s},$$

where $L_D(s)$ is as in (2.20).

REMARK. The second equation in (2.29) defines r' uniquely $\pmod{2m}$ (resp. $\pmod{m}$) if m is even (resp. odd). If m is odd, then the parity of r' is determined by the first equation in (2.29). Thus $r' \pmod{2m}$ is uniquely defined, and the number n' in the first equation in (2.29) is an integer. Lemma 2.12 then shows that $c_\phi(n', r')$ is independent of further choices of these parameters.

REMARK. Explicitly, the generalized Legendre symbol ε_D is given as follows. For $D \equiv 2, 3 \pmod 4$, $\varepsilon_D(n) = 0$. For $D = 0$, we let

$$\varepsilon_0(n) := \begin{cases} r & \text{if } n = r^2, \text{ with } r \geq 0, \\ 0 & \text{if } n \text{ is not a perfect square.} \end{cases}$$

Otherwise, $D = D_0 f^2$, where $f \geq 1$ and D_0 is the discriminant of the quadratic field $\mathbb{Q}(\sqrt{D})$. In this case, ε_D is given by

$$\varepsilon_D(n) := \begin{cases} \chi(n_0)g & \text{if } n = n_0 g^2,\ g|f, \text{ and } \gcd\left(\frac{f}{g}, n_0\right) = 1, \\ 0 & \text{if } \gcd(n, f^2) \text{ is not a perfect square.} \end{cases}$$

The character χ is the primitive Dirichlet character $\pmod{D_0}$ corresponding to $\mathbb{Q}(\sqrt{D})$.

REMARK. In Theorem 2.28, if $r^2 - 4nm$ is a fundamental discriminant with quadratic character χ^*, then the coefficients c^* can be expressed as

$$c^*(n,r) = \sum_{d|\ell} \chi^*(d)d^{k-2}c_\phi\left(\frac{\ell^2}{d^2}n, \frac{\ell}{d}r\right).$$

SKETCH OF PROOF OF THEOREM 2.28. The expansion given for $\phi|U_\ell$ is clear by definition. The expansion given for $\phi|V_\ell$ follows by choosing a set of representatives and rewriting

$$\phi|V_\ell(z;\tau) = \ell^{k-1}\sum_{ad=\ell}\ \sum_{b \pmod d} d^{-k}\phi\left(az;\frac{a\tau+b}{d}\right).$$

Inserting the Fourier expansion for ϕ leads to the expansion given after some simplifications. The complete proof of the expansion for $\phi|T_\ell$ is given in **[191]**; for brevity we provide a detailed sketch. Begin by rewriting

$$\phi|T_\ell = \ell^{k-2}\sum_{\substack{M=\left(\begin{smallmatrix} a & b\\ c & d\end{smallmatrix}\right)\in \mathrm{SL}_2(\mathbb{Z})\backslash M_2(\mathbb{Z})\\ \det(M)=\ell^2,\ \gcd(a,b,c,d)=\square}} \phi\big|_{k,m}M\,\big|_m\mathcal{A}.$$

(Here, we abbreviate $|M = |(M,\mathbf{1})$.) The averaging operator $|_m\mathcal{A}$ is defined on functions ψ that transform like Jacobi forms with respect to some sublattice $L\subseteq\mathbb{Z}^2$ by

$$\psi\big|_m\mathcal{A} := \frac{1}{[\mathbb{Z}^2:L]}\sum_{\boldsymbol{n}\in\mathbb{Z}^2/L}\psi\big|_m\boldsymbol{n}.$$

(Again, we abbreviate $|\boldsymbol{n} = |(I,\boldsymbol{n})$.) Choosing upper triangular representatives, the action on ϕ by T_ℓ can be rewritten as $\phi|T_\ell = \phi^*|\mathcal{A}$, where

$$\phi^*(z;\tau) := \ell^{-2}\sum_{ad=\ell^2} a^k \sum_{\substack{b \pmod d\\ \gcd(a,b,d)=\square}} \phi\left(\frac{\ell z}{d};\frac{a\tau+b}{d}\right)$$

$$(2.30)\qquad = \sum_{ad=\ell^2} a^{k-1}\sum_{\substack{n\in\frac{a}{\gcd(a,d)}\mathbb{Z}\\ r\in\frac{\ell}{d}\mathbb{Z}}}\Lambda\left(\gcd(a,d),\frac{n\gcd(a,d)}{a}\right)c_\phi\left(\frac{dn}{a},\frac{dr}{\ell}\right)q^n\zeta^r.$$

Here, the numbers $c_\phi(n,r)$ are the Fourier coefficients of ϕ, the function

$$\Lambda(a,b) := \sum_{d|a,\frac{a}{d}|b}\frac{\lambda(d)}{d},$$

and λ is Liouville's multiplicative function ($\lambda(p^k) = (-1)^k$ for primes p). Next, factor the averaging operator $\mathcal{A} = \mathcal{A}_1\circ\mathcal{A}_2$, where $\mathcal{A}_1$ averages with respect to the sublattice $0\times\mathbb{Z}$ and $\mathcal{A}_2$ averages with respect to the sublattice $\mathbb{Z}^2/0\times\mathbb{Z}$. Using (2.30) and the definitions of $\mathcal{A}_1$ and $\mathcal{A}_2$, a lengthy but direct calculation reveals how $\mathcal{A}_2$ acts on (the coefficients of) $\phi^*|\mathcal{A}_1$. Namely, one finds

$$c^*(n,r) = \sum_{ad=\ell^2} a^{k-1}C_a^*\left(r,r^2-4nm\right),$$

where the coefficients $C_a^*(r,r^2-4nm)$ are defined as follows. We let $\Delta = r^2-4nm$ and $\alpha = \gcd(a,d)$ so that $a = X^2\alpha, d = Y^2\alpha, \ell = XY\alpha$ for some $X,Y\in\mathbb{N}$ with $\gcd(X,Y) = 1$. If $X^2\nmid\Delta$ or $X^{-2}\Delta\equiv 2,3 \pmod 4$, then $C_a^*(r,\Delta) = 0$.

If $\Delta = X^2\Delta_0$, then

$$C_a^*(r,\Delta) = \frac{1}{X\alpha} \sum_{\substack{r_0 \pmod{2\alpha m} \\ Xr_0 \equiv r \pmod{2m} \\ r_0^2 \equiv \Delta_0 \pmod{4m}}} \Lambda\left(\alpha, \frac{r_0^2-\Delta_0}{4m}\right) c_\phi\left(Y^2\frac{r_0^2-\Delta_0}{4m}, r_0Y\right).$$

The expansion for $\phi\big|T_\ell$ now follows from the fact that

$$\sum_{\substack{r_0 \pmod{2\alpha m} \\ Xr_0 \equiv r \pmod{2m} \\ r_0^2 \equiv \Delta_0 \pmod{4}}} \Lambda\left(\alpha, \frac{r_0^2-\Delta_0}{4m}\right) = \varepsilon_{\Delta_0}(\alpha),$$

which can be shown by direct calculation. □

2.6.2. The Jacobi-Petersson inner product. Analogous to the Petersson inner product on spaces of modular forms, we have an inner product on spaces of holomorphic Jacobi forms. To describe this, let $\tau = u + iv \in \mathbb{H}$ and $z = x + iy \in \mathbb{C}$. Define the *volume element* dV on $\mathbb{C} \times \mathbb{H}$ by

$$dV := v^{-3} dx dy du dv.$$

One can show that dV is invariant under the action of Γ^J on $\mathbb{C}\times\mathbb{H}$. Using the volume element, we have the following definition of the Jacobi-Petersson inner product.

DEFINITION 2.29. Let $\phi, \psi \in J_{k,m}$ such that $\phi\psi$ is a Jacobi cusp form. The **Jacobi-Petersson inner product** $\langle\phi,\psi\rangle$ of ϕ and ψ is defined by

$$\langle\phi,\psi\rangle := \int_{\Gamma^J\backslash\mathbb{C}\times\mathbb{H}} \phi(z;\tau)\overline{\psi(z;\tau)}v^k \exp\left(\frac{-4\pi m y^2}{v}\right) dV. \tag{2.31}$$

Using the theta decomposition of Theorem 2.23, one can show that the Jacobi-Petersson inner product of two Jacobi forms is nothing but the usual Petersson inner product of the vector-valued modular forms in their theta decomposition.

THEOREM 2.30. *Let $\phi, \psi \in J_{k,m}$ such that $\phi\psi$ is a cusp form, with theta decompositions (as in Theorem 2.23)*

$$\phi(z;\tau) = \sum_{a \pmod{2m}} h_a(\tau)\vartheta_{m,a}(z;\tau), \qquad \psi(z;\tau) = \sum_{a \pmod{2m}} g_a(\tau)\vartheta_{m,a}(z;\tau).$$

Then

$$\langle\phi,\psi\rangle = \frac{1}{\sqrt{4m}} \int_{\mathrm{SL}_2(\mathbb{Z})\backslash\mathbb{H}} \sum_{a \pmod{2m}} h_a(\tau)\overline{g_a(\tau)}v^{k-\frac{5}{2}} du dv.$$

REMARK. Note that there is a slight typo in the leading factor in the formula for $\langle\phi,\psi\rangle$ in Theorem 5.3 of [**191**]. This is also corrected by the authors of [**191**] in their proof of this result on page 63.

Theorem 2.30 immediately reveals the following.

COROLLARY 2.31. *The Jacobi-Petersson inner product is well defined, and finite if $\phi\psi$ is a cusp form.*

SKETCH OF PROOF OF THEOREM 2.30. Firstly, note that a fundamental domain for the action of Γ^J on $\mathbb{C}\times\mathbb{H}$ is given by $(\mathbb{Z}+\mathbb{Z}\tau)\times\mathcal{F}$, where $\mathcal{F}$ is the standard fundamental domain for the action of $\mathrm{SL}_2(\mathbb{Z})$ on $\mathbb{H}$. We insert the theta decompositions of ϕ and ψ into (2.31), and integrate first in z. For any $a, b \pmod{2m}$, we have that

$$\begin{aligned}
&\int_{\mathbb{C}/(\mathbb{Z}+\mathbb{Z}\tau)} \vartheta_{m,a}(z;\tau)\overline{\vartheta_{m,b}(z;\tau)}e^{-\frac{4\pi m y^2}{v}}dxdy \\
&= \int_{\mathbb{C}/(\mathbb{Z}+\mathbb{Z}\tau)} \sum_{\substack{r\equiv a \pmod{2m}\\ s\equiv b \pmod{2m}}} e(rz-s\overline{z})e\left(\frac{r^2\tau-s^2\overline{\tau}}{4m}\right)e^{-\frac{4\pi m y^2}{v}}dxdy \\
&= \delta_{a,b}^{(2m)}\int_{\mathbb{R}/v\mathbb{Z}} \sum_{r\equiv a \pmod{2m}} e^{-\frac{4\pi m}{v}\left(y+\frac{rv}{2m}\right)^2}dy \\
&= \delta_{a,b}^{(2m)}\int_{-\infty}^{\infty} e^{-\frac{4\pi m y^2}{v}}dy = \delta_{a,b}^{(2m)}\sqrt{\frac{v}{4m}}. \qquad (2.32)
\end{aligned}$$

Here, we have used that

$$\int_{\mathbb{R}/\mathbb{Z}} e(rz-s\overline{z})dx = \delta_{r,s}e^{-4\pi r y},$$

where $\delta_{\alpha,\beta}$ is the usual Kronecker δ-symbol. Given our summations $\pmod{2m}$, we have also introduced the notation

$$\delta_{r,s}^{(N)} := \begin{cases} 1 & \text{if } r\equiv s \pmod{N}, \\ 0 & \text{if } r\not\equiv s \pmod{N}. \end{cases}$$

The claim now follows by using (2.32) in (2.31) (using the theta decompositions of ϕ and ψ again), summing over $a, b \pmod{2m}$, and integrating in τ. □

In what follows, for a positive integer m, we let

$$D(m) := \left\{d\in\mathbb{N}:\ d|m,\ \gcd\left(d,\frac{m}{d}\right)=1\right\}.$$

PROPOSITION 2.32. *Let k and m be positive integers. There exist Hermitian operators W_d on $J_{k,m}$, $d\in D(m)$, which commute with all Hecke operators T_ℓ, $\gcd(\ell,m)=1$. Moreover, $J_{k,m}$ has a basis of simultaneous eigenforms for all T_ℓ, $\gcd(\ell,m)=1$, and W_d.*

SKETCH OF PROOF. The operators W_d are defined by permuting the vector-valued modular forms $(h_a)_{a \pmod{2m}}$ appearing in the theta decomposition (cf. Theorem 2.23) of $\phi\in J_{k,m}$. More precisely, for each $d\in D(m)$, let b_d be an integer such that $b_d\equiv 1 \pmod{\frac{2m}{d}}$ and $b_d\equiv -1 \pmod{2d}$. Note that b_d is unique $\pmod{2m}$. We have that

$$\Omega_m := \{b_d\ :\ d\in D(m)\} = \{b \pmod{2m}\ :\ b^2\equiv 1 \pmod{4m}\}\cong(\mathbb{Z}/2\mathbb{Z})^{t(m)},$$

where $t(m)$ is the number of distinct prime divisors of m. Define the operators W_d by

$$(h_a)_{a \pmod{2m}} \mapsto (h_{b_d a})_{a \pmod{2m}}.$$

Because $b_d^2\equiv 1 \pmod{4m}$, $(h_{b_d a})_{a \pmod{2m}}$ is a vector-valued modular form obeying (2.25) and (2.26). The operators W_d are involutions, and their action on

$(h_a)_{a \pmod{2m}}$ extends to an action on $J_{k,m}$ via the theta decomposition of Theorem 2.23. If the coefficient of $q^n\zeta^r$ of $\phi \in J_{k,m}$ is $c_\phi(n,r)$, then one can check that the coefficient of $q^n\zeta^r$ of $\phi|W_d$ is $c_\phi(n',r')$, where $r' \equiv -r \pmod{2d}, r' \equiv r \pmod{\frac{2m}{d}}$, and $4n'm - r'^2 = 4nm - r^2$. Using Theorem 2.28, one sees that the operators W_d commute with the Hecke operators T_ℓ, $\gcd(\ell,m) = 1$. Because they permute the forms $(h_a)_{a \pmod{2m}}$, we have by Theorem 2.30 that the W_d are Hermitian. Now we may realize the space $J_{k,m}$ by the splitting

$$J_{k,m} = \bigoplus_{\chi} J_{k,m}^{\chi}, \tag{2.33}$$

where χ runs over the characters χ of the group Ω_m with $\chi(-1) = (-1)^k$ and $J_{k,m}^{\chi}$ is the eigensubspace of $J_{k,m}$ for χ. That is, if $(h_a)_{a \pmod{2m}}$ are associated via theta decomposition to $\phi \in J_{k,m}^{\chi}$, then $h_{ba} = \chi(b)h_a$ for each $b \in \Omega_m$. From Theorem 2.30, we have that the splitting (2.33) is orthogonal, and each summand has a basis consisting of Hecke eigenforms. □

In Chapter 19, we will recall and make use of deep theorems of Kohnen and Zagier (Theorem 19.15) and Waldspurger (Theorem 19.17) in connection with ranks of elliptic curves. These theorems describe central critical values of twists of modular L-functions in terms of Fourier coefficients of half-integral weight modular forms.

The next result provides an analogous theorem for coefficients of Jacobi eigenforms. We refer the interested reader to **[191]** for a proof.

THEOREM 2.33. *Let $\phi \in J_{k,1}$ (resp. $J_{k,1}^{\mathrm{cusp}}$) be an eigenfunction of all Hecke operators T_ℓ. Then there exists a Hecke eigenform $f(\tau) = \sum_n c_f(n)q^n \in M_{2k-2}$ (resp. S_{2k-2}) with the same eigenvalues. In particular, if $\phi \in J_{k,1}^{\mathrm{cusp}}$, for all n, r with $r^2 < 4n$, then we have*

$$|c_\phi(n,r)|^2 = \frac{(k-2)!}{\pi^{k-1}}\left(4n - r^2\right)^{k-\frac{3}{2}} \frac{\langle \phi,\phi\rangle}{\langle f,f\rangle} L(f, \varepsilon_{r^2-4n}, k-1),$$

where $L(f,\varepsilon_D,s) := \sum_{n=1}^{\infty} \varepsilon_D(n)c_f(n)n^{-s}$.

REMARK. Above, for a modular form f, the expression $\langle f,f\rangle$ is defined in terms of the usual Petersson inner product (given explicitly in (5.8)).

2.7. Taylor expansions

Another way to relate Jacobi forms to ordinary modular forms is via their Taylor expansions. In particular, we see in Theorem 2.34 below how linear combinations of derivatives of Taylor coefficients of holomorphic Jacobi forms, which are functions in the modular variable τ, are ordinary modular forms. This fact, while interesting in and of itself, was used repeatedly by Eichler and Zagier **[191]** in their original development of many aspects of the theory of Jacobi forms. Moreover, we show how to extend this philosophy to the setting of other types of (nonholomorphic) Jacobi forms in later chapters. To make this precise, denote the Taylor coefficients of $\phi \in J_{k,m}$ about $z = 0$ by $\chi_n(\tau)$, that is,

$$\phi(z;\tau) =: \sum_{n=0}^{\infty} \chi_n(\tau)z^n. \tag{2.34}$$

In what follows, we write $f^{(r)}$ to denote the r-th derivative of a function f.

THEOREM 2.34. *With notation as above, for any $\ell \in \mathbb{N}_0$, the function*

$$\xi_\ell := \sum_{\nu=0}^{\lfloor \frac{\ell}{2} \rfloor} \frac{(-2\pi i m)^\nu (k+\ell-\nu-2)!}{(k+\ell-2)!\nu!} \chi_{\ell-2\nu}^{(\nu)}$$

is a (holomorphic) modular form of weight $k+\ell$. Moreover, if

$$\phi(z;\tau) = \sum_{4mn \geq r^2} c_\phi(n,r) q^n \zeta^r \in J_{k,m},$$

then

$$\xi_\ell(\tau) = (2\pi i)^\ell \sum_{4mn \geq r^2} p_\ell(n,r) c_\phi(n,r) q^n,$$

where

$$p_\ell(X,Y) := \sum_{\nu=0}^{\lfloor \frac{\ell}{2} \rfloor} \frac{(k+\ell-\nu-2)!}{(k+\ell-2)!\nu!(\ell-2\nu)!} (-mX)^\nu Y^{\ell-2\nu}.$$

REMARK. The polynomials p_ℓ are essentially the famous Gegenbauer, or ultraspherical, polynomials. They have the important property that their specialization at a certain function of any quadratic form gives a spherical polynomial for that quadratic form (cf. Theorem 7.2 of **[191]**).

SKETCH OF PROOF OF THEOREM 2.34. We offer two ways of thinking about this theorem. The first method is a direct calculation, although our description is incomplete. The second method we offer is a complete sketch of proof, and utilizes the modularity of the functions ξ_ℓ. We begin with the direct calculation. Combining (2.34) with the transformation law i) of Definition 2.11 yields for $\left(\begin{smallmatrix} a & b \\ c & d \end{smallmatrix}\right) \in \mathrm{SL}_2(\mathbb{Z})$ that

(2.35)

$$\chi_n\left(\frac{a\tau+b}{c\tau+d}\right) = (c\tau+d)^{k+n}\left(\chi_n(\tau) + \left(\frac{2\pi i m c}{c\tau+d}\right)\chi_{n-2}(\tau) + \frac{1}{2!}\left(\frac{2\pi i m c}{c\tau+d}\right)^2 \chi_{n-4}(\tau) + \cdots\right).$$

Note that the sum above terminates. In particular, for $n=0$ and $n=1$, we see from (2.35) that χ_0 and χ_1 are modular forms of weight k and $k+1$, respectively:

(2.36)
$$\begin{aligned} \chi_0\left(\frac{a\tau+b}{c\tau+d}\right) &= (c\tau+d)^k \chi_0(\tau), \\ \chi_1\left(\frac{a\tau+b}{c\tau+d}\right) &= (c\tau+d)^{k+1} \chi_1(\tau), \end{aligned}$$

while χ_2 (and χ_n with $n \geq 2$) does not quite transform in the right way:

$$\chi_2\left(\frac{a\tau+b}{c\tau+d}\right) = (c\tau+d)^{k+2}\chi_2(\tau) + 2\pi i m c (c\tau+d)^{k+1}\chi_0(\tau).$$

Fortunately, the error to modularity can be corrected relatively easily via differentiation in τ. Namely, differentiating (2.36) reveals that

$$\xi_2 := \chi_2 - \frac{2\pi i m}{k}\chi_0'$$

is a holomorphic modular form of weight $k+2$. In theory, one can verify similarly, for each $\ell \in \mathbb{N}_0$, that ξ_ℓ defined in Theorem 2.34 transforms as claimed and that the expansions for the functions ξ_ℓ in terms of the polynomials $p_\ell(X,Y)$ are as given by differentiating the Fourier expansion of ϕ.

In practice, this becomes tedious to compute as ℓ increases, so instead we offer a more theoretical method of proof. This method involves utilizing the so-called *heat operator*, which is defined by

$$H = H_m := 8\pi i m \frac{\partial}{\partial \tau} - \frac{\partial^2}{\partial z^2}.$$

The heat operator is a natural object in the theory of Jacobi forms since it is easily seen to preserve the elliptic transformation (but not the modular one) of Jacobi forms ϕ of index m and weight k (cf. Definition 2.11). In fact, this operator plays a key role, especially due to the easily checked fact that it annihilates theta functions (cf. Lemma 2.39 below). In order to correct the erroneous modular transformations of the functions $H(\phi)$, we also define the modified heat operator by

$$H_k^* = H_{k,m}^* := H - \frac{2k-1}{z}\frac{\partial}{\partial z},$$

which then commutes with the slash action for Γ^J. Note in particular that the z in the denominator of this definition does not cause a problem for Jacobi forms without a constant term. In particular, H_k^* acts on *even* Jacobi forms (i.e., Jacobi forms which are even functions in the variable z); it is not difficult to check that it suffices to prove Theorem 2.34 only for even Jacobi forms. The result then follows from a combinatorial proof of the following formula explaining where the functions ξ_ℓ come from:

$$\xi_\ell(\tau) = \left(H_{k+2\ell-1}^* \circ \ldots \circ H_{k+2}^* \circ H_k^*(\phi)\right)(0;\tau).$$

□

Theorem 2.34 has many applications. For example, we can also deduce the modularity properties of the unmodified Taylor coefficients χ_n. To describe this, we require the following definition (see **[298]** by Kaneko and Zagier, and **[458]** by Shimura). Recall that $v = \mathrm{Im}(\tau)$.

DEFINITION 2.35. An **almost holomorphic modular form** F of weight k with respect to the subgroup $\Gamma \subseteq \mathrm{SL}_2(\mathbb{Z})$ satisfies

i) $F(\gamma\tau) = (c\tau + d)^k F(\tau)$, for all $\gamma = \left(\begin{smallmatrix} a & b \\ c & d \end{smallmatrix}\right) \in \Gamma$,

ii) $F(\tau) = \sum_{j=0}^{m} f_j(\tau) v^{-j}$, where each f_j is holomorphic,

iii) F grows at most polynomially in $1/v$ as $v \to 0$.

The holomorphic part f_0 of F is called a **quasimodular form**, and the **depth** of F is the degree of the polynomial in v^{-1} in ii).

COROLLARY 2.36. *Assuming the notation above, for all $n \in \mathbb{N}_0$, the Taylor coefficient χ_n is a quasimodular form of weight $k+n$ and depth $\leq \lfloor \frac{n}{2} \rfloor$.*

PROOF. For $n \in \{0,1\}$, the claim is already established in (2.36). The result in general is a short combinatorial exercise, but here we elaborate in order to promote a different perspective on quasimodular forms due to Nahm. In particular, an alternate characterization of the space of quasimodular forms states that the space of quasimodular forms of weight k and depth $\leq d$ on Γ is exactly the space of holomorphic functions f whose images $f|_k\gamma$ for all $\gamma = \left(\begin{smallmatrix} a & b \\ c & d \end{smallmatrix}\right) \in \Gamma$ are polynomials of degree $\leq d$ in $\frac{c}{c\tau+d}$ over the ring of holomorphic functions on $\mathbb{H}$. The result then follows by induction using (2.35). $\square$

As another application of the ideas used in the proof of Theorem 2.34, we describe a family of bilinear operators on the graded ring of modular forms. These are known as the *Rankin-Cohen operators*, which are related to a general study of polynomials in derivatives which preserve modularity by Rankin in [**433**], and which were defined by Cohen in [**150**]. We return to the question of correcting differential operators to commute with the slash action in Chapter 5, where we see that such constructions lead us to some of the core structures of the theory of harmonic Maass forms.

Specifically, these operators are defined for any natural number n and for modular forms f and g of half-integral weights k and ℓ, respectively, as

$$[f,g]_n := \sum_{\substack{r,s\geq 0\\ r+s=n}} (-1)^r \frac{\Gamma(k+n)\Gamma(\ell+n)}{\Gamma(s+1)\Gamma(k+n-s)\Gamma(r+1)\Gamma(\ell+n-r)} D^r(f)D^s(g), \tag{2.37}$$

where the operator D is defined by

$$D = D_\tau := \frac{1}{2\pi i}\frac{\partial}{\partial\tau}. \tag{2.38}$$

For example, the first two operators are given by

$$[f,g]_0 = fg, \qquad\qquad [f,g]_1 = kfD(g) - \ell gD(f),$$

and are easily checked to be modular forms of weights $k+\ell$ and $k+\ell+2$, respectively. In general, we have the following.

PROPOSITION 2.37. *Let f be a modular form of weight k and g a modular form of weight ℓ. Then $[f,g]_n$ is a modular form of weight $k+\ell+2n$.*

Before proving Proposition 2.37, we establish the following lemma pertaining to the functions ξ_n (cf. Theorem 2.34).

LEMMA 2.38. *Assuming the notation above, we have*

$$\chi_\ell = \sum_{\nu=0}^{\lfloor \frac{\ell}{2}\rfloor} \frac{(2\pi i m)^\nu (k+\ell-2\nu-1)!}{(k+\ell-\nu-1)!\nu!}\xi_{\ell-2\nu}^{(\nu)}.$$

PROOF. Let us suppose for notational convenience that ϕ is an even Jacobi form, so that we must only look at even order Taylor coefficients. The claimed identity is clearly true for $\ell = 0$. Then assume that the claim holds for $\chi_0, \chi_2, \ldots, \chi_{2\ell}$. By induction, it suffices to use this to show that it holds for $\chi_{2\ell+2}$. We then argue as follows. By the definition of $\xi_{2\ell+2}$, we find that

$$\xi_{2\ell+2} = \sum_{\nu=0}^{\ell+1} \frac{(-2\pi i m)^\nu (k+2\ell-\nu)!}{(k+2\ell)!\nu!}\chi_{2\ell-2\nu+2}^{(\nu)},$$

and so

$$\chi_{2\ell+2} = \xi_{2\ell+2} - \sum_{\nu=1}^{\ell+1} \frac{(-2\pi i m)^\nu (k+2\ell-\nu)!}{(k+2\ell)!\nu!}\chi_{2\ell-2\nu+2}^{(\nu)},$$

which by the induction hypothesis is equal to

$$\xi_{2\ell+2} - \sum_{\nu=1}^{\ell+1} \frac{(-2\pi i m)^\nu (k+2\ell-\nu)!}{(k+2\ell)!\nu!} \times \left(\sum_{r=0}^{\ell-\nu+1} \frac{(2\pi i m)^r (k+2\ell-2\nu-2r+1)!}{(k+2\ell-2\nu-r+1)!r!} \xi_{2\ell-2r-2\nu+2}^{(r+\nu)} \right).$$

Rearranging terms and saying $n = r+\nu$, this becomes

$$\xi_{2\ell+2} - \sum_{n=1}^{\ell+1} \frac{(2\pi i m)^n (k+2\ell+1-2n)!}{n(k+2\ell)!} \xi_{2\ell+2-2n}^{(n)} \sum_{\nu=1}^{n} (-1)^\nu \binom{n}{\nu} \binom{k+2\ell-\nu}{n-1}.$$

The proof then follows immediately from the fact that

$$\sum_{\nu=1}^{n} (-1)^\nu \binom{n}{\nu} \binom{k+2\ell-\nu}{n-1} = \frac{k+2\ell}{k+2\ell-n+1} \binom{k+2\ell-1}{n-1}.$$

This can be seen, for example, by first writing

$$\sum_{\nu=1}^{n} (-1)^\nu \binom{n}{\nu} \binom{k+2\ell-\nu}{n-1} = \frac{k+2\ell}{k+2\ell-n-1} \binom{k+2\ell-1}{n-1} \Big({}_2F_1(-n, -k-2\ell+n-1; -k-2\ell; 1) - 1 \Big),$$

where

$$(2.39) \qquad {}_2F_1(a,b;c;z) := \sum_{n=0}^{\infty} \frac{(a)_n (b)_n}{(c)_n n!} z^n.$$

Here the *Pochhammer symbol* $(a)_n$ is the rising factorial $a(a+1)\dots(a+n-1)$. The reader should be warned that in the literature it is also sometimes used to denote the falling factorial. The desired binomial sum identity, and hence the inductive proof, follow from the Chu-Vandermonde identity, valid for natural numbers n (cf. (15.4.24) of [**400**]):

$$(2.40) \qquad {}_2F_1(-n,b;c;1) = \frac{(c-b)_n}{(c)_n}.$$

To see this, note that the numerator of (2.40) becomes $(1-n)_n = 0$. □

Sketch of proof of Proposition 2.37. We begin by forming an auxiliary generating function for each modular form. This generating function, known as the *Cohen-Kuznetsov series*, is a weighted generating function for all the derivatives of a given modular form. To motivate this construction, we begin by noting that any Jacobi form is determined by its Taylor coefficients, and hence by Lemma 2.38, a Jacobi form is determined uniquely by the sequence $\{\xi_\ell\}_{\ell\geq 0}$. Moreover, one can show that if $\{\xi_\ell\}_{\ell\geq 0}$ is an arbitrary sequence of modular forms of the correct weights, then defining a function of z and τ, with Taylor coefficients in z given by Lemma 2.38, yields a form satisfying the modular transformation law of a Jacobi form. That is, one can show that we have an isomorphism

$$M_{k,m}^{\mathrm{even}}(\mathrm{SL}_2(\mathbb{Z})) \xrightarrow{\sim} \prod_{\ell=0}^{\infty} M_{k+2\ell}(\mathrm{SL}_2(\mathbb{Z})),$$

where $M_{k,m}^{\text{even}}(\mathrm{SL}_2(\mathbb{Z}))$ is the space of two variable functions (on $\mathbb{C}\times\mathbb{H}$) whose Taylor coefficients satisfy (2.35), which are even as functions of z (a similar statement holds for Jacobi forms). In particular, we can take the very simple example of $\xi_0 = f$ for a modular form f of weight k, and $\xi_n = 0$ for $n \geq 1$. Then we directly find from the preceding discussion that

$$\Phi_f(z;\tau) := \sum_{r=0}^{\infty} \chi_r(\tau) z^r = \sum_{r=0}^{\infty} \frac{(2\pi i)^r (k-1)!}{r!(k+r-1)!} f^{(r)}(\tau) z^{2r}$$

satisfies the modular Jacobi transformation equation of Definition 2.11 ii) with weight k and index 1. To complete the proof, one considers the function

$$F(z;\tau) := \Phi_f(iz;\tau)\Phi_g(z;\tau),$$

which satisfies the modular Jacobi transformation of weight $k+\ell$ and index 0. Hence, if $\{\xi_\ell\}_{\ell\geq 0}$ is the sequence of modular forms corresponding to F, we see by definition (since $m=0$) that $\xi_\ell = \chi_\ell$ for all ℓ, and hence, by Theorem 2.34, the Taylor coefficients of F are modular. (Alternatively, one can use the transformation equation directly in index $m = 0$ to see the modularity of the Taylor coefficients.) To complete the proof, we note that $[f,g]_n$ is a multiple of the coefficient of z^{2n} in F. If we let $[z^r]F(z;\tau)$ denote the coefficient of z^r when F is expanded as a series in z, then we have

$$\begin{aligned}
&\left[z^{2n}\right] F(z;\tau) \\
&= \left[z^{2n}\right]\left(\sum_{r=0}^{\infty} \frac{(-2\pi i)^r (k-1)!}{r!(k+r-1)!} f^{(r)}(\tau) z^{2r}\right)\left(\sum_{s=0}^{\infty} \frac{(2\pi i)^s (\ell-1)!}{s!(\ell+s-1)!} g^{(s)}(\tau) z^{2s}\right) \\
&= \sum_{\substack{r,s\geq 0\\ r+s=n}} \frac{(-2\pi i)^r (k-1)!}{r!(k+r-1)!} f^{(r)}(\tau) \frac{(2\pi i)^s (\ell-1)!}{s!(\ell+s-1)!} g^{(s)}(\tau) \\
&= \frac{(2\pi i)^{2n}(k-1)!(\ell-1)!}{(k+n-1)!(\ell+n-1)!}[f(\tau), g(\tau)]_n.
\end{aligned}$$

□

We conclude this section by describing the connection between the Taylor and theta expansions of Jacobi forms. Since Jacobi forms are determined by sets of modular forms in these two different manners, it is natural to ask what is the connection between these two expansions. In fact, we can give an explicit answer to this question. For simplicity, we only consider even Jacobi forms. Before stating this result, we first establish an important lemma. Recall that the operator $D = D_\tau$ is defined in (2.38).

LEMMA 2.39. *With notation as above, for any $n \in 2\mathbb{N}_0$, we have*

$$\frac{\partial^n}{\partial z^n}\vartheta_{m,\ell}(z;\tau) = (4\pi i)^n m^{\frac{n}{2}} D_\tau^{\frac{n}{2}}\left(\vartheta_{m,\ell}(z;\tau)\right).$$

In particular, we have

$$H\left(\vartheta_{m,\ell}(z;\tau)\right) = 0.$$

PROOF. We compute that

$$\frac{\partial^n}{\partial z^n}\vartheta_{m,\ell}(z;\tau) = \frac{\partial^n}{\partial z^n} \sum_{\substack{r\in\mathbb{Z}\\ r\equiv a \pmod{2m}}} q^{\frac{r^2}{4m}}\zeta^r = (2\pi i)^n \sum_{\substack{r\in\mathbb{Z}\\ r\equiv a \pmod{2m}}} r^n q^{\frac{r^2}{4m}}\zeta^r,$$

which is equal to

$$(4\pi i)^n m^{\frac{n}{2}} D_\tau^{\frac{n}{2}} \left(\vartheta_{m,\ell}(z;\tau)\right),$$

as claimed. The claim that $\vartheta_{m,\ell}$ is in the kernel of H follows directly by applying this result for $n = 2$ to the definition of H, noting that $\vartheta_{m,\ell}$ is an index m Jacobi form. □

The connection between theta decompositions and Taylor coefficients is given in the following result, where $\vartheta^0_{m,\ell}(\tau)$ is the *Thetanullwert*

$$\vartheta^0_{m,\ell}(\tau) := \vartheta_{m,\ell}(0;\tau).$$

PROPOSITION 2.40. *Assuming the notation above, for any even Jacobi form ϕ of weight k and index m we have for all $n \in 2\mathbb{N}$ that*

$$\xi_n(\tau) = \frac{(2\pi i)^n (4m)^{\frac{n}{2}}}{n! \binom{k+n-2}{\frac{n}{2}}} \sum_{\ell \pmod{2m}} \left[h_\ell(\tau), \vartheta^0_{m,\ell}(\tau)\right]_{\frac{n}{2}}.$$

PROOF. We first write ϕ in its theta expansion

$$\phi(z;\tau) = \sum_{\ell \pmod{2m}} h_\ell(\tau)\vartheta_{m,\ell}(z;\tau)$$

and then compute, using Lemma 2.39 (for n even),

$$\begin{aligned}
\chi_n(\tau) = [z^n]\phi(z;\tau) &= \frac{1}{n!} \sum_{\ell \pmod{2m}} h_\ell(\tau) \frac{\partial^n}{\partial z^n} \left[\vartheta_{m,\ell}(z;\tau)\right]_{z=0} \\
&= \frac{(4\pi i)^n m^{\frac{n}{2}}}{n!} \sum_{\ell \pmod{2m}} h_\ell(\tau) D^{\frac{n}{2}} \left(\vartheta^0_{m,\ell}(\tau)\right).
\end{aligned}$$

Plugging this into the definition of ξ_n, we find, using the product rule, that

$$\begin{aligned}
\xi_n(\tau) &= (2\pi i)^n (4m)^{\frac{n}{2}} \sum_{\ell \pmod{2m}} \sum_{\mu=0}^{\frac{n}{2}} \frac{(-1)^\mu 4^{-\mu}(k+n-\mu-2)!}{(k+n-2)!\mu!(n-2\mu)!} \\
&\quad \times D^\mu \left(h_\ell(\tau) D^{\frac{n}{2}-\mu}\left(\vartheta^0_{m,\ell}(\tau)\right)\right) \\
&= (2\pi i)^n (4m)^{\frac{n}{2}} \sum_{\ell \pmod{2m}} \sum_{\mu=0}^{\frac{n}{2}} \frac{(-1)^\mu 4^{-\mu}(k+n-\mu-2)!}{(k+n-2)!\mu!(n-2\mu)!} \\
&\quad \times \sum_{r=0}^{\mu} \binom{\mu}{r} D^r(h_\ell(\tau)) D^{\frac{n}{2}-r}\left(\vartheta^0_{m,\ell}(\tau)\right) \qquad (2.41) \\
&= (2\pi i)^n (4m)^{\frac{n}{2}} \sum_{\ell \pmod{2m}} \sum_{r=0}^{\frac{n}{2}} D^r(h_\ell(\tau)) D^{\frac{n}{2}-r}\left(\vartheta^0_{m,\ell}(\tau)\right) \\
&\quad \times \sum_{\mu=r}^{\frac{n}{2}} \frac{\binom{\mu}{r}(-1)^\mu 4^{-\mu}(k+n-\mu-2)!}{(k+n-2)!\mu!(n-2\mu)!}.
\end{aligned}$$

To evaluate this last expression, we note that we can express

$$C(k,n,r) := \sum_{\mu=r}^{\frac{n}{2}} \frac{\binom{\mu}{r}(-1)^{\mu}4^{-\mu}(k+n-\mu-2)!}{(k+n-2)!\mu!(n-2\mu)!}$$
$$= \frac{(-1)^r(k+n-r-2)!{}_2F_1\left(r-\frac{n}{2}, r-\frac{n}{2}+\frac{1}{2}; r-k-n+2; 1\right)}{4^r r!(k+n-2)!(n-2r)!},$$

where ${}_2F_1(a,b;c;z)$ is defined in (2.39). Now, given that $r \leq n/2$, we may use the Chu-Vandermonde identity, given above in (2.40). Rewriting, we find that

$$C(k,n,r) = (-1)^r \frac{\binom{(k-\frac{1}{2})+\frac{n}{2}-1}{\frac{n}{2}-r}\binom{\frac{1}{2}+\frac{n}{2}-1}{r}}{\binom{k+n-2}{\frac{n}{2}} n!}.$$

Noting that the weight of h_ℓ is $k-1/2$ and the weight of $\vartheta^0_{m,\ell}$ is $1/2$, and inserting the definition of the Rankin-Cohen brackets into (2.41) directly gives the claimed result. □

This last result is closely related to a well-known result of Cohen on special modular forms built out of class numbers, which is indeed a primary application of his definition of the Rankin-Cohen operators in **[150]**. In particular, if we consider what Proposition 2.40 says for the Jacobi-Eisenstein series $E_{k,1}$ of index 1, then by Theorem 2.19, the associated modular forms ξ_n are certain sums over generalized class numbers. In fact, in this case, we find that the theta coefficient h corresponding to $E_{k,1}$ is just $\mathcal{H}_{k-1}$, where

$$\mathcal{H}_{k-1}(\tau) := \sum_{n=0}^{\infty} H(k-1,n)q^n.$$

Hence Proposition 2.40 implies that if $\theta(\tau) = \sum_{n\in\mathbb{Z}} q^{n^2}$ we have the elegant result that

$$[\theta, \mathcal{H}_{k-1}]_n \left| U_4 \in M_{k+2n}, \right.$$

where the U_ℓ operator is defined on q-series in (2.28). In fact, Cohen proved that his functions are modular forms by defining them in essentially this manner.

REMARK. Similar constructions using Rankin-Cohen brackets were also important in Mertens' proof in **[384]** of a conjecture of Cohen on the modularity of related class number generating functions. In this case, the theory of harmonic Maass forms proved key to proving this conjecture. Special cases of the work of Mertens and further applications of these ideas in the context of mock modular forms will be given in Section 10.3.

2.8. Related topics

In this section, we discuss functions related to holomorphic Jacobi forms, namely Siegel modular forms, and skew-holomorphic Jacobi forms. The former transform not on the upper half-plane $\mathbb{H}$ but on $\mathbb{H}_n$, the Siegel upper half-space of degree n; the latter satisfy modified modular transformation properties as compared to ordinary Jacobi forms, and also satisfy the heat equation.

2.8.1. Siegel modular forms. Analogous to modular forms, *Siegel modular forms* (cf. **[191]** or **[205]**, e.g.) transform on the *Siegel upper half-space* $\mathbb{H}_n$ of degree n, defined by

$$\mathbb{H}_n := \{Z \in M_n(\mathbb{C}) \ : \ Z^T = Z, \ \mathrm{Im}(Z) \text{ is positive definite}\},$$

where we let $M_n(F)$ denote the set of $n \times n$ matrices with entries in F. Analogous to the action of $\mathrm{SL}_2(\mathbb{Z})$ on $\mathbb{H}$, the *symplectic group of degree n over* $\mathbb{R}$, $\mathrm{Sp}_n(\mathbb{R})$, acts on $\mathbb{H}_n$. More precisely, the group $\mathrm{Sp}_n(\mathbb{R})$ is defined by

$$\mathrm{Sp}_n(\mathbb{R}) := \{M \in M_{2n}(\mathbb{R}) \ : \ M^T J_n M = J_n\},$$

where

$$J_n := \begin{pmatrix} 0 & -I_n \\ -I_n & 0 \end{pmatrix},$$

and I_n is the $n \times n$ identity matrix. The action of $\mathrm{Sp}_n(\mathbb{R})$ on the Siegel upper half-space of degree n is given by

$$\begin{pmatrix} A & B \\ C & D \end{pmatrix} \circ Z = (AZ + B)(CZ + D)^{-1},$$

where $\left(\begin{smallmatrix} A & B \\ C & D \end{smallmatrix}\right) \in \mathrm{Sp}_n(\mathbb{R})$, and $Z \in \mathbb{H}_n$. The *Siegel modular group* Γ_n is defined to be $\Gamma_n := \mathrm{Sp}_n(\mathbb{Z})$. We are now equipped to define Siegel modular forms.

DEFINITION 2.41. A **Siegel modular form of degree** $n \in \mathbb{N}$ **and weight** $k \in \mathbb{N}$ with respect to Γ_n is a holomorphic function $F \colon \mathbb{H}_n \to \mathbb{C}$ satisfying

$$F(M \circ Z) = \det(CZ + D)^k F(Z)$$

for all $Z \in \mathbb{H}_n$ and $M = \left(\begin{smallmatrix} A & B \\ C & D \end{smallmatrix}\right) \in \Gamma_n$.

Note that if $n = 1$, holomorphicity at the cusps must be additionally imposed in Definition 2.41. (This condition is automatically satisfied for $n > 1$ by the so-called Köcher principle.) If F is a Siegel modular form of degree $n > 1$, then F has a Fourier expansion of the form

$$F(Z) = \sum_{T \geq 0} A(T) e(\mathrm{tr}(TZ)), \tag{2.42}$$

where the summation is taken over positive semidefinite n-dimensional matrices

$$T = (t_{jk})_{\substack{1 \leq j \leq n \\ 1 \leq k \leq n}}$$

that are semi-integral ($2t_{jk} \in \mathbb{Z}$ for $j \neq k$, and $t_{jj} \in \mathbb{Z}$ for each j). In the case $n = 2$, any matrix $Z \in \mathbb{H}_2$ can be written as $Z = \left(\begin{smallmatrix} \tau & z \\ z & \tau' \end{smallmatrix}\right)$, where $\tau, \tau' \in \mathbb{H}, z \in \mathbb{C}$ satisfy $\mathrm{Im}(z)^2 < \mathrm{Im}(\tau)\mathrm{Im}(\tau')$. In this case, one may write $F(Z) =: F(\tau, z, \tau')$.

Analogous to the theta expansion described in Theorem 2.23 for holomorphic Jacobi forms, the following theorem (see the work of Piatetski-Shapiro **[422]**) shows that Siegel modular forms of weight k and degree 2 have a *Fourier-Jacobi expansion*, with holomorphic Jacobi forms as coefficients.

THEOREM 2.42. *Let F be a Siegel modular form of weight $k \in \mathbb{N}$ and degree 2. Then F has a Fourier-Jacobi expansion of the form*

$$F(\tau, z, \tau') = \sum_{m=0}^{\infty} \phi_m(z; \tau) e(m\tau'),$$

where ϕ_m is a Jacobi form of weight k and index m.

REMARK. Conversely, one can define a Siegel modular form out of a Jacobi form by taking the functions $\phi|V_m$ as Fourier-Jacobi coefficients of a function $F\colon \mathbb{H}_2 \to \mathbb{C}$ which then turns out to be a Siegel modular form. To be more precise, one defines

$$F\left(\begin{pmatrix} \tau & z \\ z & \tau' \end{pmatrix}\right) := \sum_{\substack{n,r,m\in\mathbb{Z} \\ n,m,4nm-r^2\geq 0}} A(n,r,m)e^{2\pi i\left(n\tau+rz+m\tau'\right)}$$

with $(n,r,m) \neq (0,0,0)$

$$A(n,r,m) := \sum_{d|\gcd(n,r,m)} d^{k-1}c\left(\frac{4nm-r^2}{d^2}\right).$$

The subspace of Siegel modular forms defined that way is called the *Maass Spezialschar*. It may be characterized as those Siegel modular forms satisfying the relation

$$A\left(\begin{pmatrix} n & \frac{r}{2} \\ \frac{r}{2} & m \end{pmatrix}\right) = \sum_{d|\gcd(n,r,m)} d^{k-1}A\left(\begin{pmatrix} \frac{nm}{d^2} & \frac{r}{2d} \\ \frac{r}{2d} & 1 \end{pmatrix}\right).$$

SKETCH OF PROOF OF THEOREM 2.42. The positive definite semi-integral 2×2 matrices T in (2.42) may be expressed as $T = \left(\begin{smallmatrix} n & r/2 \\ r/2 & m \end{smallmatrix}\right)$, where n, r, and m are integers satisfying $n, m \in \mathbb{N}_0$ and $r^2 \leq 4nm$. If we let $A(T) := A(n,r,m)$, then we have

$$F(\tau,z,\tau') = \sum_{\substack{n,r,m\in\mathbb{Z} \\ n,m,4nm-r^2\geq 0}} A(n,r,m)e(n\tau+rz+m\tau') = \sum_{m=0}^{\infty} \phi_m(z;\tau)e(m\tau'),$$

where

$$\phi_m(z;\tau) := \sum_{\substack{n,r\in\mathbb{Z} \\ n,4nm-r^2\geq 0}} A(n,r,m)e(n\tau+rz).$$

Thus, the function ϕ_m has the needed Fourier expansion to be a Jacobi form. It remains to be seen that ϕ_m satisfies the required weight k and index m transformation properties. To this end, we have that for any $\left(\begin{smallmatrix} a & b \\ c & d \end{smallmatrix}\right) \in \Gamma_1$ and any $(\lambda,\mu) \in \mathbb{Z}^2$, the matrices

$$M_1 := \begin{pmatrix} a & 0 & b & 0 \\ 0 & 1 & 0 & 0 \\ c & 0 & d & 0 \\ 0 & 0 & 0 & 1 \end{pmatrix}, \quad M_2 := \begin{pmatrix} 1 & 0 & 0 & \mu \\ \lambda & 1 & \mu & 0 \\ 0 & 0 & 1 & -\lambda \\ 0 & 0 & 0 & 1 \end{pmatrix}$$

are in Γ_2. They act on $\mathbb{H}_2$ by

$$M_1 \circ (\tau,z,\tau') = \left(\frac{a\tau+b}{c\tau+d}, \frac{z}{c\tau+d}, \tau' - \frac{cz^2}{c\tau+d}\right),$$
$$M_2 \circ (\tau,z,\tau') = \left(\tau, z+\lambda\tau+\mu, \tau'+2\lambda z+\lambda^2\tau\right).$$

Using this with the transformation law in Definition 2.41 for $F(\tau,z,\tau')$, we see that ϕ_m indeed transforms like a Jacobi form of weight k and index m. $\square$

In another parallel direction, Eichler-Zagier [**191**] expanded the work of Maass on the so-called *Saito-Kurokawa conjecture* [**340**], which purported a certain lifting from weight $2k-2$ ordinary modular forms to Siegel modular forms of weight k. In doing so, they used Theorem 2.42 to give an isomorphism between a certain subspace of Siegel modular forms of weight k and degree 2, and Jacobi forms of weight k and index 1. Theorem 2.24 provides the remainder of their lifting result.

2.8.2. Skew-holomorphic Jacobi forms. Also related to Siegel modular forms and other spaces of modular forms are *skew-holomorphic Jacobi forms*, defined by Skoruppa [**466**]. These functions have modified transformation properties as compared to holomorphic Jacobi forms and also satisfy the heat equation. It is convenient to define skew-holomorphic Jacobi forms using new notation which also encompasses holomorphic Jacobi forms. We follow [**442, 466**] in what follows.

DEFINITION 2.43. Let $k \geq 2$ and $m \geq 1$ be integers, and choose $\varepsilon \in \{+,-\}$. Consider the smooth functions $\phi\colon \mathbb{C}\times\mathbb{H}\to\mathbb{C}$ which have a Fourier expansion of the form $(\tau = u+iv)$

$$\phi(z;\tau) = \sum_{\substack{\Delta, r\in\mathbb{Z}, \varepsilon\Delta\geq 0\\ \Delta\equiv r^2 \pmod{4m}}} c_\phi(\Delta,r) e\left(\frac{r^2-\Delta}{4m}u + \frac{r^2+|\Delta|}{4m}iv + rz\right),$$

where the coefficients $c_\phi(\Delta, r)$ depend only on Δ and on $r \pmod{2m}$, and such that

$$\phi\left(\frac{z}{\tau};-\frac{1}{\tau}\right)e^{-2\pi i m\frac{z^2}{\tau}} = \phi(z;\tau)\times\begin{cases}\tau^k & \text{if } \varepsilon = -,\\ \overline{\tau}^{k-1}|\tau| & \text{if } \varepsilon = +.\end{cases}$$

If $\varepsilon = +$, such forms are called **skew-holomorphic Jacobi forms** of weight k and index m, and we denote the space of such forms by $J^+_{k,m}$. If $\varepsilon = -$, these forms are holomorphic Jacobi forms of weight k and index m as defined in Definition 2.11, and we let $J^-_{k,m} := J_{k,m}$. A form in $J^\varepsilon_{k,m}$ (note the slight abuse of notation with respect to (2.33)) is called **cuspidal** if $c_\phi(0,r) = 0$ for every r.

REMARK. To see that these forms become holomorphic Jacobi forms if $\varepsilon = -$, we need only verify that they exhibit a suitable expansion. (Note that the transformations under $\tau\mapsto\tau+1$ and $z\mapsto z+\lambda\tau+\mu$, where $\lambda,\mu\in\mathbb{Z}$, follow by definition.) We have $\frac{r^2-\Delta}{4m} = \frac{r^2+|\Delta|}{4m}$, and so denoting this number by n yields an expansion as in Definition 2.11.

Skoruppa proved that the space of skew-holomorphic Jacobi forms of weight k, k odd, is isomorphic to the Kohnen plus space of weight $k-1/2$. Later works have subsequently extended Skoruppa's original work, also making connections to spaces of Siegel modular forms. One result of particular interest relates these spaces to spaces of holomorphic cusp forms. To describe this, note that in the case of weight 2 and $\varepsilon = +$, there are certain *trivial cusp forms*; these are Jacobi cusps forms whose coefficients $c(\Delta, r)$ are supported on pairs (Δ, r) for which Δ is equal to a perfect square. We let $S^\varepsilon_{k,m}$ be the subspace of $J^\varepsilon_{k,m}$ consisting of cuspidal forms which are orthogonal to the trivial cusp forms. We also let $S^\varepsilon_k(m)$ denote the space of classical holomorphic cusp forms of weight k on $\Gamma_0(m)$ whose L-functions are such that the sign of their functional equations is ε. The following theorem was proved by Skoruppa and Zagier [**467**] for $\varepsilon = -$. For $\varepsilon = +$, the result was announced in [**466**], and its proof is forthcoming in [**77**].

THEOREM 2.44. *Let $k \geq 2, m \geq 1$ be integers. For any fixed fundamental discriminant Δ_0 and any fixed integer r_0 such that $\Delta_0 \equiv r_0^2 \pmod{4m}$ and $\operatorname{sgn}(\Delta_0) = \varepsilon 1$, there is a Hecke equivariant map*

$$\mathscr{S}_{\Delta_0,r_0} : S_{k,m}^{\varepsilon} \to S_{2k-2}^{\varepsilon}(m)$$

given by

$$\mathscr{S}_{\Delta_0,r_0}(\phi(z,\tau)) := \sum_{n=1}^{\infty} \left(\sum_{d|n} \left(\frac{\Delta_0}{d} \right) c_\phi \left(\frac{n^2}{d^2}\Delta_0, \frac{n}{d} r_0 \right) \right) q^n.$$

Some linear combination of these maps is injective, and its image contains all newforms in $S_{2k-2}^{\varepsilon}(m)$. Moreover, $\mathscr{S}_{\Delta_0,r_0}$ maps newforms to newforms.

CHAPTER 3

Classical Maass Forms

In 1949, Maass introduced nonholomorphic modular functions, his so-called *Maass forms* (see Maass' works **[366, 367, 368, 369, 370]**). This book is primarily about mock modular forms and harmonic Maass forms, functions which are another type of nonholomorphic modular form. For completeness, here we recall some of the main features of the theory of classical Maass forms. We refer the reader to **[134, 227, 283, 284]**, fundamental texts on the subject of Maass forms.

3.1. Definitions

The theory of Maass forms is assembled from special nonholomorphic functions on $\mathbb{H}$ which are eigenfunctions of the hyperbolic Laplacian (or Laplace) operator.

DEFINITION 3.1. Let $\tau = u + iv \in \mathbb{H}$, where $u, v \in \mathbb{R}$. The **hyperbolic Laplacian operator** on $\mathbb{H}$ is defined by

$$\Delta := -v^2 \left(\frac{\partial^2}{\partial u^2} + \frac{\partial^2}{\partial v^2} \right).$$

Of particular interest are eigenfunctions of this operator, that is, those functions f for which

$$\Delta(f) = \lambda f$$

for some $\lambda \in \mathbb{C}$. Typically, eigenvalues are normalized to be of the form $\lambda = r(1-r)$, in terms of a parameter r. If we additionally require that such eigenfunctions f are invariant under the action of some *Fuchsian group* Γ *of the first kind* (discrete, with finite covolume in $\mathrm{SL}_2(\mathbb{Z})$) on $\mathbb{H}$, then we are led to the notion of a *Maass form*. In order to refine this notion even further, we consider Fourier expansions of such functions. For example, for any f which satisfies $f(\tau+1) = f(\tau)$, we have a Fourier expansion at infinity of the form

$$f(\tau) = \sum_{n \in \mathbb{Z}} c_f(v; n) e(nu), \tag{3.1}$$

where

$$c_f(v; n) := \int_0^1 f(u + iv) e(-nu) du.$$

Moreover, such an f has a Fourier expansion at any cusp $\mathfrak{a}$ of Γ, whose coefficients we denote by $c_{f,\mathfrak{a}}(v; n)$.

In what follows, we let $d\mu(\tau) := v^{-2} du dv$ denote the invariant measure in the Poincaré metric and we let $\mathcal{F}(\Gamma \backslash \mathbb{H})$ be a fundamental domain for the action of Γ on $\mathbb{H}$.

DEFINITION 3.2. Let $\Gamma \subseteq \mathrm{SL}_2(\mathbb{Z})$ be a Fuchsian group of the first kind. A **Maass cusp form** f on Γ with eigenvalue $\lambda = r(1-r) \in \mathbb{C}$ is any $f\colon \mathbb{H} \to \mathbb{C}$ satisfying

i) $f(\gamma\tau) = f(\tau)$ for all $\gamma \in \Gamma$,

ii) $c_{f,\mathfrak{a}}(0;0) = 0$ for each cusp $\mathfrak{a}$ of Γ,

iii) $\Delta(f) = r(1-r)f$,

iv) $\displaystyle\int_{\mathcal{F}(\Gamma\backslash\mathbb{H})} |f(\tau)|^2 d\mu(\tau) < \infty.$

We set

$$\mathcal{A}_r^0(\Gamma\backslash\mathbb{H}) := \{\text{Maass cusp forms on } \Gamma \text{ with eigenvalue } \lambda = r(1-r)\}.$$

REMARK. More generally, we call functions satisfying i) in Definition 3.2 *automorphic functions*; functions satisfying i) and iii) comprise the space of *automorphic forms*, denoted by $\mathcal{A}_r(\Gamma\backslash\mathbb{H})$.

3.2. Fourier expansions

Here we consider the Fourier expansions of forms in $\mathcal{A}_r(\Gamma\backslash\mathbb{H})$, which, as discussed above, includes the space of Maass cusp forms. The following key lemma gives the general shape of these expansions.

LEMMA 3.3. *Suppose $f \in \mathcal{A}_r(\Gamma\backslash\mathbb{H})$ satisfies the growth condition $f(\tau) = o(e^{2\pi v})$ as $v \to \infty$. Then there exist $\kappa_1, \kappa_2, c_f(n) \in \mathbb{C}$ such that*

$$f(\tau) = \kappa_1 v^r + \kappa_2 v^{1-r}\delta_r(v) + v^{\frac{1}{2}} \sum_{n\in\mathbb{Z}\backslash\{0\}} c_f(n) K_{r-\frac{1}{2}}(2\pi|n|v)e(nu),$$

where K_ν is the K-Bessel function of order ν, and $\delta_r(v)$ is equal to $\log(v)$ or 1, depending on whether $r = 1/2$ or $r \neq 1/2$, respectively.

SKETCH OF PROOF. Applying Δ to (3.1) shows that the Fourier coefficients of f must satisfy the differential equation

$$\frac{\partial^2}{\partial v^2} c_f(v;n) + \left(\frac{r(1-r)}{v^2} - 4\pi^2 n^2\right) c_f(v;n) = 0. \tag{3.2}$$

If $n = 0$, then (3.2) has the two linearly independent solutions v^r and $v^{1-r}\delta_r(v)$. If $n \geq 1$, then (3.2) has the two linearly independent solutions

$$v^{\frac{1}{2}} K_{r-\frac{1}{2}}(2\pi|n|v), \quad v^{\frac{1}{2}} I_{r-\frac{1}{2}}(2\pi|n|v),$$

in terms of the I- and K-Bessel functions. The I-Bessel function is of exponential growth, while the K-Bessel function is of suitable decay as $v \to \infty$, leading to the claimed expansion. □

The Ramanujan-Petersson Conjecture for Maass forms famously predicts how the Fourier coefficients $c_f(n)$ of Maass cusp forms grow as $n \to \infty$.

CONJECTURE 3.4 (Ramanujan and Petersson). *For $f \in \mathcal{A}_r^0(\Gamma\backslash\mathbb{H})$ with Fourier expansion as in Lemma 3.3, we have that*

$$c_f(n) = O(\sigma_0(n)),$$

where $\sigma_0(n) = \sum_{d|n} 1$ is the divisor function.

3.3. General discussion

For an arbitrary Fuchsian group of the first kind Γ, it is unclear that Maass cusp forms on Γ even exist, rendering their study particularly interesting. In the case that Γ is a congruence subgroup of level $N \geq 1$, however, Selberg used the "trace formula" to show, non-constructively, that Maass cusp forms do indeed exist. (cf. Section 3.6 for more on the existence of Maass cusp forms.) Moreover, Selberg counted (on average) the number of eigenvalues for a given level N. To make this precise (cf. **[284]**), define the eigenvalue counting function

$$\mathcal{N}_N(T) := \#\{j \ : \ |r_j| \leq T\},$$

where the eigenvalue $\lambda_j = r_j(1-r_j)$. This function satisfies *Weyl's law*

$$\mathcal{N}_N(T) = \frac{\mathrm{vol}(\Gamma_0(N)\backslash\mathbb{H})}{4\pi}T^2 + O\left(\sqrt{N}T\log(NT)\right),$$

for $T \geq 2$. Note that as N grows, the first term (in terms of the volume of the fundamental domain) indeed dominates the error term. While this shows that there are on average roughly NT eigenvalues between T and $T+1$, it is non-constructive.

It can be shown that for an eigenvalue $\lambda = r(1-r)$, either $\mathrm{Re}(r) = 1/2$ or $1/2 < r \leq 1$. In the former case we have $\lambda \geq 1/4$. In the latter case we have $\lambda < 1/4$, and such eigenvalues are called *exceptional*. There can only be a finite number of exceptional λ. Selberg conjectured that for congruence subgroups, exceptional eigenvalues do not exist.

CONJECTURE 3.5 (Selberg). *Let Γ be a congruence subgroup of level $N \geq 1$, and let $\lambda_1(\Gamma)$ be the smallest eigenvalue of Δ acting on the space of functions satisfying conditions* i), ii), *and* iv) *in Definition* 3.2. *We have that*

$$\lambda_1(\Gamma) \geq \frac{1}{4}.$$

Selberg already proved that $\lambda_1(\Gamma) \geq 3/16$. At present, the best known result towards this conjecture is the following theorem due to Kim and Sarnak **[309]**.

THEOREM 3.6 (Kim and Sarnak). *Assuming the notation and hypotheses in Conjecture* 3.5, *we have that*

$$\lambda_1(\Gamma) \geq \frac{975}{4096}.$$

Booker and Strömbergsson **[70]** have also recently investigated the problem of numerically computing Maass cusp forms, and they have proved Selberg's conjecture in the case of $\Gamma = \Gamma_1(N)$ for square-free $N < 857$, extending older work of Huxley, who proved Selberg's conjecture for $N < 19$.

Many applications in number theory require the spectral theory of automorphic forms (cf. **[283, 284]**) which arises from the Petersson inner product. These applications ultimately rely on the classical analysis and the theory of Hilbert spaces for the subspace of automorphic cusp forms (on a Fuchsian group Γ of the first kind)

$$\mathcal{C}(\Gamma\backslash\mathbb{H}) := \{\text{smooth, bounded functions satisfying i) and ii) of Definition 3.2}\}.$$

Note that due to ii), the functions in this space are also square-integrable (i.e., satisfy condition iv) of Definition 3.2). Such spaces have a spectral decomposition with respect to the operator Δ (cf. Theorem 4.7 of **[283]**). We have the following Spectral Theorem, where as usual, $\langle\cdot,\cdot\rangle$ denotes the Petersson inner product (defined by the same formula as in (5.8)).

THEOREM 3.7 (Spectral Theorem). *There exists an orthonomal basis of* $\mathcal{C}(\Gamma\backslash\mathbb{H})$ *consisting of Maass cusp forms* $\{u_j\}$ *such that for any* $f \in \mathcal{C}(\Gamma\backslash\mathbb{H})$ *we have a spectral expansion of the form*

$$f(\tau) = \sum_j \langle f, u_j\rangle\, u_j(\tau),$$

which converges in the norm topology. If $\Delta(f)$ *is also a smooth and bounded automorphic function on* Γ, *then this expansion converges absolutely and uniformly on compact sets.*

REMARK. We introduce *Eisenstein series* in Section 3.4 below, which are orthogonal to cusp forms. As in Theorem 3.7, one can obtain a spectral decomposition of a relevant space spanned by Eisenstein series. By considering this space, as well as the cuspidal space above, one may determine the spectral decomposition of the whole space of automorphic functions satisfying i) and iv) of Definition 3.2. We refer the interested reader to **[283]** for details.

3.4. Eisenstein series

A natural way to produce examples of nonholomorphic forms is to use the method of averaging. In light of the difficulty of constructing cusp forms in general, it is important to note that the Eisenstein series constructed in this way below are not cusp forms.

DEFINITION 3.8. Let $s \in \mathbb{C}$ with $\mathrm{Re}(s) > 1$ and $\tau \in \mathbb{H}$. The **Eisenstein series** $E(\tau; s)$ is defined by

$$E(\tau; s) := \sum_{\gamma\in\Gamma_\infty\backslash\mathrm{SL}_2(\mathbb{Z})} \mathrm{Im}(\gamma\tau)^s.$$

Here, $\Gamma_\infty := \{\pm\left(\begin{smallmatrix}1 & n\\ 0 & 1\end{smallmatrix}\right) : n \in \mathbb{Z}\}$ is the stabilizer group of $i\infty$ in $\mathrm{SL}_2(\mathbb{Z})$.

REMARK. One defines Eisenstein series with respect to other groups $\Gamma \subseteq \mathrm{SL}_2(\mathbb{Z})$ and other cusps $\mathfrak{a}$ similarly.

Here we review some essential properties of the Eisenstein series.

THEOREM 3.9. *We have the following:*

i) *The function* $E(\tau; s)$ *satisfies properties* i) *and* iii) *of Definition* 3.2. *In particular, it has eigenvalue* $\lambda = s(1-s)$.
ii) *There is a meromorphic continuation of* $E(\tau; s)$ *(in the s-variable) to* $\mathbb{C}$.
iii) *The Fourier expansion of* $E(\tau; s)$ *is given by*

$$E(\tau; s) = v^s + \phi(s)v^{1-s} + 2v^{\frac12}\sum_{n\in\mathbb{Z}\setminus\{0\}} \phi(n,s)K_{s-\frac12}(2\pi|n|v)e(nu),$$

where

$$\phi(s) := \sqrt{\pi}\frac{\Gamma\left(s-\frac12\right)\zeta(2s-1)}{\Gamma(s)\zeta(2s)},$$

$$\phi(n,s) := \pi^s\Gamma(s)^{-1}\zeta(2s)^{-1}\sum_{ab=|n|}\left(\frac{a}{b}\right)^{s-\frac12}.$$

iv) *The function* $E(\tau; s)$ *has a simple pole at* $s = 1$, *with residue equal to* $1/\mathrm{vol}(\mathrm{SL}_2(\mathbb{Z})\backslash\mathbb{H})$.

REMARK. This Eisenstein series is not square-integrable (i.e., does not satisfy condition iv) in Definition 3.2).

REMARK. Note that E also has a functional equation relating its values at s and $1-s$ (see also (3.31) of **[283]**).

SKETCH OF PROOF OF THEOREM 3.9. It is clear by definition that $E(\tau; s)$ is automorphic (i.e., satisfies i) of Definition 3.2). A short calculation shows that

$$\Delta\left(v^s\right) = s(1-s)v^s.$$

When combined with the fact that Δ is invariant under the action of $\mathrm{SL}_2(\mathbb{Z})$, this leads to the fact that $E(\tau; s)$ satisfies iii) of Definition 3.2 with eigenvalue $\lambda = s(1-s)$.

The Fourier expansion for $E(\tau; s)$ in iii) is found after a technical but straightforward calculation by choosing a set of representatives for $\Gamma_\infty \backslash \mathrm{SL}_2(\mathbb{Z})$, applying Poisson summation, using the (integral) definition of the Bessel functions, and simplifying. See Section 3.4 of **[283]** for details.

The Fourier expansion in part iii) of the theorem can be used to establish meromorphic continuation to $\mathbb{C}$; in the plane $\mathrm{Re}(s) \geq 1/2$ there is just one simple pole at $s = 1$ with constant residue. □

3.5. L-functions of Maass cusp forms

Any $f \in \mathcal{A}_r^0(\mathrm{SL}_2(\mathbb{Z})\backslash\mathbb{H})$ has a Fourier expansion as in Lemma 3.3. The *L-function* for f is defined by

(3.3)
$$L(f, s) := \sum_{n=1}^{\infty} \frac{c_f(n)}{n^s}.$$

By the trivial bound

$$c_f(n) = O\left(n^{\frac{1}{2}}\right),$$

the L-function converges for s with $\mathrm{Re}(s) > 3/2$; however the *Rankin-Selberg* method leads to the non-trivial bound (cf. **[358]**)

$$\sum_{n \leq X} |c_f(n)|^2 = O(X),$$

which shows that $L(f, s)$ actually converges for $\mathrm{Re}(s) > 1$.

These L-functions can be "completed" using the Γ-function into a function $\Lambda(f, s)$ which satisfies a nice functional equation. That is, let $f \in A_r^0(\mathrm{SL}_2(\mathbb{Z})\backslash\mathbb{H})$ with eigenvalue $\lambda = r(1-r)$. Assume f satisfies $f(u+iv) = (-1)^\varepsilon f(-u+iv)$, $\varepsilon \in \{0, 1\}$. We define the *completed L-function* for f as

$$\Lambda(f, s) := \pi^{-s} \Gamma\left(\frac{s+\varepsilon+r-\frac{1}{2}}{2}\right) \Gamma\left(\frac{s+\varepsilon+\frac{1}{2}-r}{2}\right) L(f, s).$$

THEOREM 3.10. *Assume the notation and hypotheses above. Then the L-function $L(f, s)$ satisfies the following:*

i) *The L-function L_f has the Euler product expansion*

$$L(f, s) = \prod_{p \ prime} \left(1 - c_f(p)p^{-s} + p^{-2s}\right)^{-1}.$$

ii) *It has an analytic continuation to all of $\mathbb{C}$.*

iii) *The completed L-function* $\Lambda(f,s)$ *satisfies the functional equation*

$$\Lambda(f,s) = (-1)^{\varepsilon}\Lambda(f,1-s).$$

The value $\varepsilon \in \{0,1\}$ is called the *sign of the functional equation.*

REMARK. Theorem 3.10 also generalizes to Maass cusp forms on congruence subgroups.

REMARK. In the literature, the value $(-1)^{\varepsilon} \in \{1,-1\}$ is also referred to as the sign of the functional equation. In particular, although we find the current notation convenient in this chapter, in Chapter 19 (for example in Theorem 19.6 iii)) we will use this slightly different definition for the sign of the functional equation.

The "simplest" Dirichlet series, in the sense that all of its coefficients are equal to 1, is the Riemann ζ-function (2.14). This seemingly simple L-function is still in many ways very elusive: the famous *Riemann Hypothesis* conjectures that all non-trivial zeros of ζ lie on the line $\mathrm{Re}(s) = 1/2$. The truth of this hypothesis would have many deep implications throughout mathematics.

In analytic number theory, many of these applications can be derived using the weaker *Lindelöf Hypothesis*, which predicts that

$$\zeta\left(\frac{1}{2}+it\right) = O\left(|t|^{\varepsilon}\right) \tag{3.4}$$

($\varepsilon > 0$). The Lindelöf Hypothesis is known to follow from the truth of the Riemann Hypothesis. The standard Phragmén-Lindelöf method establishes the weaker estimate

$$\zeta\left(\frac{1}{2}+it\right) = O\left(|t|^{\frac{1}{4}+\varepsilon}\right),$$

which is commonly referred to as the *convexity bound.*

In general, the Phragmén-Lindelöf method leads to similar convexity bounds for other types of L-functions $L\left(f,1/2+it\right)$ arising from Maass cusp forms, newforms, and so on. (In addition to the references listed in the discussion below, we refer the interested reader to **[280]**.) In practice, breaking convexity (reducing the exponent one obtains by Phragmén-Lindelöf), or establishing a subconvexity bound, is typically very difficult. Nonetheless, a number of authors have broken convexity with respect to various types of L-functions in many different aspects, including Weyl, who first broke convexity for $\zeta\left(1/2+it\right)$, followed by improvements by others including Bombieri-Iwaniec and Huxley more recently. Duke, Friedlander and Iwaniec **[167, 168, 169]** also notably broke convexity for L-functions arising from cusp forms with respect to various parameters. The subconvexity problem has been solved with complete uniformity in the analytic conductor by Michel and Venkatesh **[387]** with an inexplicit savings over the trivial convexity bound; the savings was eventually made explicit by Wu **[501]**.

As another example, consider the *Rankin-Selberg L-function*

$$L(f\otimes g,s) := \sum_{n=1}^{\infty}\frac{c_f(n)\overline{c_g(n)}}{n^s} \tag{3.5}$$

of two cusp forms or Maass cusp forms f and g with Fourier coefficients $c_f(n)$ and $c_g(n)$, respectively. (In the latter case, coefficients are as in Lemma 3.3.) Kowalski, Michel, and VanderKam **[330]** proved a subconvexity bound for the Rankin-Selberg L-function of a fixed holomorphic newform g, as f varies over holomorphic new forms with level $N \to \infty$ of the same weight as g. Sarnak **[444]** established subconvexity for such an L-function as the weight of f goes to ∞. Many other authors including those just mentioned have addressed the subconvexity problem with respect to various types of L-functions in various aspects, including Harcos-Michel **[251]**, Liu-Ye **[359]**, Liu-Masri-Young **[360]**, and many more.

3.6. Maass cusp forms arising from real quadratic fields

Among the most well known examples of Maass cusp forms are those arising from real quadratic fields. In particular, if F is a real quadratic field over $\mathbb{Q}$ with discriminant D, then one can define a Maass cusp form on $\Gamma_0(D)$ with *Hecke character* ψ. We first discuss Grössencharakters, now typically called Hecke characters. These are characters on number fields, first introduced by Hecke, which generalize Dirichlet characters. To define them, in what follows we consider totally real fields F of degree d over $\mathbb{Q}$, before later returning to our construction of Maass cusp forms associated to real quadratic fields.

3.6.1. Hecke characters. Let $\sigma_j, 1 \leq j \leq d$, denote the distinct real embeddings of F, and let $\mathcal{O}_F$ denote the ring of integers of F. For an ideal $A \subset \mathcal{O}_F$, let ψ_A be a primitive character on $(\mathcal{O}_F/A)^\times$. The character ψ_A extends to a function on $\mathcal{O}_F$ as

$$\psi_A(a) := \psi_A(a \,(\mathrm{mod}\ A))$$

if $\gcd(a, A) = 1$; otherwise, $\psi_A(a) := 0$. Next, we seek to use ψ_A to form a character that is trivial on units. Let $\omega_j \in i\mathbb{R}$, $1 \leq j \leq d$, be such that $\sum_{j=1}^d \omega_j = 0$, and let $\varepsilon_j \in \{0, 1\}$, $1 \leq j \leq d$. Using these parameters, one can define characters ψ_j $(1 \leq j \leq d)$ on $\mathbb{R} \setminus \{0\}$ by

$$\psi_j(t) := \mathrm{sgn}(t)^{\varepsilon_j} |t|^{\omega_j}.$$

The characters ψ_j give rise to a character ψ_∞ on $\mathcal{O}_F^\times$ defined by

$$\psi_\infty(t) := \prod_{j=1}^{d} \psi_j(\sigma_j(t)).$$

In fact, it can be shown that the parameters defining ψ_∞ can be chosen in such a way that

$$\psi(t) := \psi_\infty(t)\psi_A(t)$$

is trivial on $\mathcal{O}_F^\times$, yielding a character on principal ideals prime to A.

A *Hecke character of* F on ideals in $\mathcal{O}_F$ is defined to be any character whose restriction to the subgroup of principal ideals arises in the manner described above.

3.6.2. Maass cusp forms from real quadratic fields. Theorem 3.11 below exhibits (weight 0) Maass cusp forms with Nebentypus. To describe this, let F be a real quadratic field and ψ be a Hecke character of F. If $A = \mathcal{O}_F$, then in the notation of Subsection 3.6.1, $\psi = \psi_\infty$, which is defined from parameters $\omega_1 = -\omega_2 =: \omega$ and $\varepsilon_1 = \varepsilon_2 =: \varepsilon$ by

$$\psi_\infty((t_1, t_2)) = \mathrm{sgn}(t_1)^{\varepsilon_1} \mathrm{sgn}(t_2)^{\varepsilon_2} \left| \frac{t_1}{t_2} \right|^\omega,$$

where $(t_1, t_2) \in \mathbb{R}^2$. Using such Hecke characters attached to real quadratic fields, Maass constructed cusp forms on congruence subgroups. These Maass cusp forms are defined from the K-Bessel function; one may wish to compare with Lemma 3.3. Below, $\boldsymbol{N}(A)$ denotes the norm of the ideal A.

THEOREM 3.11 (Maass). *Let F be a real quadratic field F of discriminant D, with ring of integers $\mathcal{O}_F$. Let ψ be a Hecke character for F, with $\omega \neq 0$ and ε defined as above. Then we have that the function*

$$M_\psi(\tau) := \begin{cases} v^{\frac{1}{2}} \sum\limits_{A \subseteq \mathcal{O}_F} \psi(A) K_\omega(2\pi \boldsymbol{N}(A) v) \cos(2\pi \boldsymbol{N}(A) u) & \text{if } \varepsilon = 0, \\ v^{\frac{1}{2}} \sum\limits_{A \subseteq \mathcal{O}_F} \psi(A) K_\omega(2\pi \boldsymbol{N}(A) v) \sin(2\pi \boldsymbol{N}(A) u) & \text{if } \varepsilon = 1 \end{cases}$$

is a Maass cusp form on $\Gamma_0(D)$ with character χ_D.

REMARK. If $\omega = 0$, then $M_\psi(\tau)$ can be defined similarly, up to the addition of the factor $\frac{1}{2} v^{\frac{1}{2}} \operatorname{Log}(\sigma_1(\eta_0))$ involving the regulator of F, where η_0 is the fundamental unit of F.

3.7. Hecke theory on Maass cusp forms

In analogy to the classical Hecke theory for holomorphic modular forms, there is a Hecke theory for Maass cusp forms. We briefly recall some main properties here. To describe this, for $n \in \mathbb{N}$, let

$$\Gamma_n := \left\{ \begin{pmatrix} a & b \\ c & d \end{pmatrix} : a, b, c, d \in \mathbb{Z}, ad - bc = n \right\}.$$

The n-th *Hecke operator* T_n acts on $f \in \mathcal{A}_r^0 (\mathbb{H} \backslash \mathrm{SL}_2(\mathbb{Z}))$ by

$$T_n(f)(\tau) := \frac{1}{\sqrt{n}} \sum_{M \in \mathrm{SL}_2(\mathbb{Z}) \backslash \Gamma_n} f(M\tau).$$

We summarize standard properties of the Hecke operators in the following lemma.

LEMMA 3.12. *The Hecke operators T_n, $n \in \mathbb{N}$, are linear operators on the space of Maass cusp forms. Moreover, for $n, m \in \mathbb{N}$, the following are true:*

i) $T_m T_n = T_n T_m$.
ii) $T_n \Delta = \Delta T_n$.
iii) $T_n(f)$ *has Fourier coefficients*

$$c_{T_n(f)}(m) = \sum_{d \mid \gcd(m,n)} c_f\left(\frac{mn}{d^2}\right),$$

where f is a Maass cusp form with coefficients $c_f(m)$ $(m \neq 0)$ as in Lemma 3.3.

iv) *The Eisenstein series $E(\tau; 1/2 + it)$ $(t \in \mathbb{R})$ are simultaneous eigenfunctions of all Hecke operators T_n, with eigenvalues*

$$\sum_{ad=n} \left(\frac{a}{d}\right)^{it}.$$

3.8. Period functions of Maass cusp forms

It is very difficult to construct non-trivial Maass cusp forms. Inspired by this problem, Lewis and Zagier have developed a program in which Maass cusp forms are in one-to-one correspondence with *period functions*.

The following theorem is due to Lewis and Zagier **[353]**. (The equivalence between i) and iii) below is due to Maass.)

THEOREM 3.13 (Lewis and Zagier). *Let* $r \in \mathbb{C}$ *satisfy* $\mathrm{Re}(r) > 0$. *There is a canonical correspondence between the following functions:*

i) *Maass cusp forms* f *on* $\mathrm{SL}_2(\mathbb{Z})$ *with eigenvalue* $\lambda = r(1-r)$,
ii) *holomorphic functions* $\psi\colon \mathbb{C} \setminus (-\infty, 0] \to \mathbb{C}$ *satisfying*

$$\psi(z) = \psi(z+1) + (z+1)^{-2r}\psi\left(\frac{z}{z+1}\right),$$

and

$$\psi(z) \ll \begin{cases} |\mathrm{Im}(z)|^{-A}(1+|z|^{2A-2\mathrm{Re}(r)}), & \mathrm{Re}(z) \le 0, \\ 1, & \mathrm{Im}(z) \ge 0, |z| \le 1, \\ |z|^{-2\mathrm{Re}(r)}, & \mathrm{Re}(z) \ge 0, |z| \ge 1, \end{cases}$$

for some $A > 0$,

iii) *pairs of L-series* $L_\varepsilon(s)$, $\varepsilon \in \{0,1\}$, *convergent for* s *in some half-plane, such that the functions*

$$L_\varepsilon^*(s) := \frac{1}{4\pi^{s+\varepsilon}}\Gamma\left(\frac{s+\varepsilon-r+\frac{1}{2}}{2}\right)\Gamma\left(\frac{s+\varepsilon+r-\frac{1}{2}}{2}\right)L_\varepsilon(s)$$

are entire, of finite order, and satisfy

$$L_\varepsilon^*(1-s) = (-1)^\varepsilon L_\varepsilon^*(s),$$

iv) *holomorphic functions* $f(z)$, $z \in \mathbb{C} \setminus \mathbb{R}$, *invariant under* $z \mapsto z+1$ *and bounded by* $|\mathrm{Im}(z)|^{-t}$ *for some* $t > 0$ *such that*

$$f(z) - z^{-2r} f\left(-\frac{1}{z}\right)$$

extends holomorphically across the positive real axis and is bounded by a multiple of $\min\{1, |z|^{-2\mathrm{Re}(r)}\}$ *in the right half-plane.*

For such an f, the *period function* ψ is given as

$$\psi(z) := \kappa \int_0^\infty \frac{zt^r}{(z^2+t^2)^{r+1}} f(it)dt, \quad \mathrm{Re}(z) > 0,$$

where κ is some constant.

EXAMPLE 3.14. The generalization of this theorem to congruence subgroups makes contact with some strange q-series studied by Ramanujan. Consider the two q-hypergeometric (see also (9.14)) functions (the first of which appeared in Ramanujan's lost notebook)

$$\sigma(q) := \sum_{n=0}^{\infty} \frac{q^{\frac{n(n+1)}{2}}}{(-q;q)_n} =: \sum_{n=0}^{\infty} T(n) q^{\frac{n-1}{24}}, \tag{3.6}$$

$$\sigma^*(q) := 2\sum_{n=0}^{\infty} \frac{(-1)^n q^{n^2}}{(q;q^2)_n} =: \sum_{n=-\infty}^{-1} T(n) q^{\frac{|n|-1}{24}}, \tag{3.7}$$

famously studied in **[32]** in relation to indefinite theta functions. Here $(a)_n = (a;q)_n := \prod_{j=0}^{n-1}(1-aq^j)$. Note that the coefficients $T(n)$ above equal 0 if $|n| \neq 1$

(mod 24). Cohen [**151**] beautifully showed that the coefficients of these two q-hypergeometric series appear as coefficients in a Maass cusp form with eigenvalue 1/4 on the congruence subgroup $\Gamma_0(2)$:

$$c(\tau) := v^{\frac{1}{2}} \sum_{n \in 1+24\mathbb{Z}} T(n) K_0\left(\frac{2\pi|n|v}{24}\right) e\left(\frac{nu}{24}\right). \tag{3.8}$$

Essentially, one can think of $q^{\frac{1}{24}}\sigma(q)+q^{-\frac{1}{24}}\sigma^*(q)$ as the Lewis-Zagier period function for the Maass cusp form c.

We note that these q-hypergeometric series are also, in fact, related to real quadratic fields. If we let

$$\varphi(q) := q^{\frac{1}{24}}\sigma(q) + q^{-\frac{1}{24}}\sigma^*(q),$$

Cohen proved the identity

$$\varphi(q) = \sum_{\mathfrak{a} \subset \mathbb{Z}[\sqrt{6}]} \chi_1(\mathfrak{a}) q^{\frac{\boldsymbol{N}(\mathfrak{a})}{24}},$$

where $\chi_a(\mathfrak{a})$ is a character of conductor $4(3+\sqrt{6})$ and order 2 on ideals $\mathfrak{a} \subset \mathbb{Z}[\sqrt{6}]$. This gives rise to the *Artin L-function*

$$L(\chi_1, s) := \sum_{\mathfrak{a} \subset \mathbb{Z}[\sqrt{6}]} \frac{\chi_1(\mathfrak{a})}{\boldsymbol{N}(\mathfrak{a})^s},$$

which is completed as

$$\Lambda(\chi_1, s) := (1152)^{\frac{s}{2}} \pi^{-s} \Gamma\left(\frac{s}{2}\right)^2 L(\chi_1, s).$$

Cohen proves that $\Lambda(\chi_1, s)$ can be analytically continued to $\mathbb{C}$ and satisfies the functional equation

$$\Lambda(\chi_1, 1-s) = -\Lambda(\chi_1, s).$$

The connection to Maass cusp forms now follows via an extension of Theorem 3.11. Additional related identities and properties were given in [**151**].

Part 2

Harmonic Maass Forms and Mock Modular Forms

CHAPTER 4

The Basics

Part 1 of this book consists of the background which is necessary to fully grasp the theory of harmonic Maass forms and mock modular forms. This chapter marks the beginning of Part 2, the theory of harmonic Maass forms and mock modular forms. In their paper on geometric theta lifts, Bruinier and Funke **[121]** introduced the notion of harmonic Maass forms. This important paper includes many of the fundamental results in the theory of harmonic Maass forms. The reader is encouraged to read their original work.

In this short chapter, we begin with definitions and then discuss the Fourier expansions of these forms. In Chapter 5 we discuss the action of certain differential operators on such forms. Then in Chapter 6 we give some of the first examples of harmonic Maass forms, and we discuss their properties with respect to the differential operators discussed in Chapter 5.

Harmonic Maass forms are real-analytic functions f on $\mathbb{H}$ which transform like weight k modular forms that are also annihilated by the weight k hyperbolic Laplacian operator. Weight k weakly holomorphic modular forms are also annihilated by the weight k hyperbolic Laplacian operator, and so one can think of the theory of harmonic Maass forms as a natural extension of the classical theory of modular forms.

4.1. Definitions

Throughout, we let $\tau = u + iv \in \mathbb{H}$, where $u, v \in \mathbb{R}$. The theory of harmonic Maass forms begins with the weight k hyperbolic Laplacian operator.

DEFINITION 4.1. For $k \in \mathbb{R}$, the **weight k hyperbolic Laplacian operator** on $\mathbb{H}$ is defined by

$$\Delta_k := -v^2\left(\frac{\partial^2}{\partial u^2} + \frac{\partial^2}{\partial v^2}\right) + ikv\left(\frac{\partial}{\partial u} + i\frac{\partial}{\partial v}\right) = -4v^2\frac{\partial}{\partial \tau}\frac{\partial}{\partial \overline{\tau}} + 2ikv\frac{\partial}{\partial \overline{\tau}}.$$

Weight k harmonic Maass forms are certain real-analytic functions on $\mathbb{H}$ which are annihilated by Δ_k. In this book we focus on forms with weight $k \in \frac{1}{2}\mathbb{Z}$. These functions satisfy transformation laws which are required of ordinary modular forms. Here we briefly recall the set-up, which is slightly more complicated for weights $k \in 1/2 + \mathbb{Z}$. As before, let $\left(\frac{c}{d}\right)$ denote the extended Legendre symbol, let $\sqrt{}$ be the principal branch of the holomorphic square root, and for odd integers d set

$$\varepsilon_d := \begin{cases} 1 & \text{if } d \equiv 1 \pmod 4, \\ i & \text{if } d \equiv 3 \pmod 4. \end{cases} \tag{4.1}$$

Following Bruinier and Funke **[121]**, we define two types of harmonic Maass forms of weights $k \in \frac{1}{2}\mathbb{Z}$.

DEFINITION 4.2. If $k \in \frac{1}{2}\mathbb{Z}$, then a **weight k harmonic Maass form on a subgroup**[1] $\Gamma = \Gamma_0(N)$ for some $N \in \mathbb{N}$, where $4|N$ if $k \in \frac{1}{2}\mathbb{Z} \setminus \mathbb{Z}$ throughout, is any smooth function $f\colon \mathbb{H} \to \mathbb{C}$ satisfying the following properties:

i) For all $\left(\begin{smallmatrix} a & b \\ c & d \end{smallmatrix}\right) \in \Gamma$ and all $\tau \in \mathbb{H}$, we have

$$f\left(\frac{a\tau+b}{c\tau+d}\right) = \begin{cases} (c\tau+d)^k f(\tau) & \text{if } k \in \mathbb{Z}, \\ \left(\frac{c}{d}\right) \varepsilon_d^{-2k} (c\tau+d)^k f(\tau) & \text{if } k \in \frac{1}{2} + \mathbb{Z}. \end{cases}$$

ii) We have that $\Delta_k(f) = 0$.

iii) There exists a polynomial $P_f(\tau) \in \mathbb{C}[q^{-1}]$ such that

$$f(\tau) - P_f(\tau) = O\left(e^{-\varepsilon v}\right)$$

as $v \to \infty$ for some $\varepsilon > 0$. Analogous conditions are required at all cusps.

Note that in particular weakly holomorphic modular forms are harmonic Maass forms. We let $H_k(\Gamma)$ denote the space of weight k harmonic Maass forms on Γ. If we replace the growth condition in iii) by

$$f(\tau) = O\left(e^{\varepsilon v}\right) \text{ as } v \to \infty \text{ for some } \varepsilon > 0,$$

then we call f a **weight k harmonic Maass form of manageable growth on Γ**. We denote the space of such forms by $H_k^!(\Gamma)$. Note that in particular $H_k(\Gamma) \subset H_k^!(\Gamma)$. The motivation behind the notation $H_k^!$ will become apparent in Theorem 5.10, where it will be shown that functions in this space map onto the space of weakly holomorphic modular forms in the "dual weight" under a natural differential operator, whereas the image of the space H_k under this map consists only of cusp forms. An important (and unique) intermediate example when this image is a non-cuspidal, holomorphic modular form is given by Zagier's weight 3/2 Eisenstein series, described in Theorem 6.3.

REMARK. If $k = 0$, then these forms are referred to as *harmonic Maass functions.*

REMARK. It is natural to study eigenfunctions of Δ_k which satisfy i) and iii) in Definition 4.2, whose (complex) eigenvalues are not necessarily equal to 0. Following Bruinier and Funke [**121**], we refer to such functions as *weak Maass forms.*

REMARK. The polynomial $P_f(\tau) \in \mathbb{C}[q^{-1}]$ in Definition 4.2 iii) is called the *principal part of f at $i\infty$.* There are analogous principal parts at all of the cusps.

REMARK. We also consider harmonic Maass forms on $\Gamma_0(N)$ with Nebentypus χ, where χ is a Dirichlet character modulo N. In this case, we replace condition i) in Definition 4.2 by

$$f\left(\frac{a\tau+b}{c\tau+d}\right) = \begin{cases} \chi(d)(c\tau+d)^k f(\tau) & \text{if } k \in \mathbb{Z}, \\ \left(\frac{c}{d}\right) \varepsilon_d^{-2k} \chi(d)(c\tau+d)^k f(\tau) & \text{if } k \in \frac{1}{2} + \mathbb{Z}. \end{cases} \tag{4.2}$$

We let $H_k(\Gamma_0(N),\chi)$ (resp. $H_k^!(\Gamma_0(N),\chi)$) denote the space of weight k harmonic Maass forms (resp. with manageable growth) with Nebentypus χ.

[1] These forms were called *harmonic weak Maass forms* in [**121**].

REMARK. We recall that the transformation laws in Definition 4.2 i) and (4.2) are conveniently described in terms of the *slash operator* as

$$f|_k\gamma(\tau) = f(\tau),$$

where for $\gamma = \left(\begin{smallmatrix} a & b \\ c & d \end{smallmatrix}\right)$ we let

$$f|_k\gamma(\tau) := \begin{cases} (c\tau + d)^{-k} f\left(\frac{a\tau+b}{c\tau+d}\right) & \text{if } k \in \mathbb{Z}, \\ \left(\frac{c}{d}\right)\varepsilon_d^{2k}(c\tau + d)^{-k} f\left(\frac{a\tau+b}{c\tau+d}\right) & \text{if } k \in \frac{1}{2} + \mathbb{Z}. \end{cases} \tag{4.3}$$

REMARK. Bruinier and Funke [**121**] worked out a vector-valued theory of harmonic Maass forms, which of course includes the scalar-valued forms described above. These vector-valued forms will be discussed in Chapter 18. In this setting and others, Definition 4.2 can be generalized to other finite index subgroups $\Gamma \subseteq \mathrm{SL}_2(\mathbb{Z})$ in the obvious way.

REMARK. In [**137**], Candelori developed a geometric framework for (integral weight) harmonic Maass forms. This was built by constructing classes in de Rahm cohomology groups and is further compatible with geometric realizations of raising and lowering operators. As one example of the utility of this perspective, as noted after Theorem 7.7, this directly leads to a (previously conjectured) refinement of the field of definition of lifts of CM modular forms to harmonic Maass forms.

4.2. Fourier expansions

Here we discuss Fourier expansions of harmonic Maass forms. It is not difficult to directly work out the shape of these expansions by using the definition of these forms. These expansions are somewhat similar to the Fourier expansion of ordinary (weight 0) Maass forms (cf. Section 3.1).

In the case of harmonic Maass forms, the Fourier expansions involve the incomplete gamma function. See [**400**] for some basic properties of the incomplete gamma function and other special functions discussed in this section. Recall that the *incomplete gamma function* is initially given by

$$\Gamma(s, z) := \int_z^\infty e^{-t} t^s \frac{dt}{t}, \tag{4.4}$$

for $\mathrm{Re}(s) > 0$ and $z \in \mathbb{C}$ (or any $s \in \mathbb{C}$ and $z \in \mathbb{H}$). It can be analytically continued in s via the functional equation

$$\Gamma(s + 1, z) = s\Gamma(s, z) + z^s e^{-z}.$$

In particular, for $c \in \{0, \frac{1}{2}\}$, continuing $\Gamma(m + c, z)$ for any $m \in \mathbb{Z}$ is equivalent to continuing $\Gamma(\frac{1}{2}, z)$ or $\Gamma(0, z)$. It is well known that

$$\begin{aligned} \Gamma\left(\frac{1}{2}, z\right) &= \sqrt{\pi}\,\mathrm{erfc}\left(\sqrt{z}\right), \\ \Gamma(0, z) &= \mathrm{Ein}(z) - \mathrm{Log}(z) - \gamma, \end{aligned} \tag{4.5}$$

where the *complementary error function* is defined by

$$\mathrm{erfc}(w) := \frac{2}{\sqrt{\pi}} \int_w^\infty e^{-t^2} dt$$

and the entire function Ein, the *complementary error integral*, is defined by

$$\mathrm{Ein}(z) := \int_0^z \left(1 - e^{-t}\right) \frac{dt}{t}.$$

The number γ is the *Euler-Mascheroni constant.* Note that (4.5) requires selecting a branch of the logarithm; throughout, we choose the principal branch. We also require the asymptotic behavior for $x \in \mathbb{R}$,

$$\Gamma(s, x) \sim x^{s-1} e^{-x} \quad \text{as } |x| \to \infty. \tag{4.6}$$

In terms of this special function, we have the following lemma due to Bruinier and Funke [**121**].

LEMMA 4.3. *Let $k \in \frac{1}{2}\mathbb{Z}\backslash\{1\}$ and $\Gamma \in \{\Gamma_0(N), \Gamma_1(N)\}$ for some $N \in \mathbb{N}$. If $f \in H_k^!(\Gamma)$, then f has a Fourier expansion of the shape*

$$f(\tau) = \sum_{n \gg -\infty} c_f^+(n) q^n + c_f^-(0) v^{1-k} + \sum_{\substack{n \ll \infty \\ n \neq 0}} c_f^-(n) \Gamma(1-k, -4\pi n v) q^n.$$

In particular, if $f \in H_k(\Gamma)$, then f has a Fourier expansion of the shape

$$f(\tau) = \sum_{n \gg -\infty} c_f^+(n) q^n + \sum_{n<0} c_f^-(n) \Gamma(1-k, -4\pi n v) q^n.$$

Similar expansions hold at the other cusps.

REMARK. We use the notation $\sum_{n \gg -\infty}$ to mean $\sum_{n=\alpha_f}^{\infty}$ for some $\alpha_f \in \mathbb{Z}$. The definition of $\sum_{n \ll \infty}$ is analogous.

REMARK. If $k = 1$, then the term v^{1-k} in Lemma 4.3 gets replaced by $\log(v)$.

REMARK. We note that for some applications, it can be useful to work with the function

$$\mathrm{W}_k(x) := (-2x)^{1-k} \mathrm{Re}\left(\mathrm{E}_k(-2x)\right),$$

where the *generalized exponential integral*

$$\mathrm{E}_s(z) := \int_1^{\infty} e^{-zt} t^{-s} dt$$

is defined for $s \in \mathbb{C}$ and $z \in \mathbb{C}$ with $\mathrm{Re}(z) > 0$. It can be continued by using its relation to the incomplete gamma function

$$\Gamma(s, z) = z^s \, \mathrm{E}_{1-s}(z).$$

It turns out that for $k \in \frac{1}{2}\mathbb{Z}$, $x \in \mathbb{R} \setminus \{0\}$, we have

$$\mathrm{W}_k(x) = \begin{cases} \Gamma(1-k, -2x) & \text{if } x < 0, \\ \Gamma(1-k, -2x) + \frac{(-1)^{1-k}\pi i}{\Gamma(k)} & \text{if } x > 0. \end{cases}$$

Here $(-1)^{1-k}$ is defined again through the principal branch of the logarithm.

SKETCH OF PROOF OF LEMMA 4.3. The proof follows in a manner similar to that of Lemma 3.3. By condition i) in Definition 4.2, any $f \in H_k^!(\Gamma)$ has an expansion (at $i\infty$) of the form

$$f(\tau) =: \sum_{n \in \mathbb{Z}} a_f(n, v) e\left(nu\right). \tag{4.7}$$

If $n = 0$, then we have that $a_f(0, v)$ is a linear combination of 1 and v^{1-k}. If $n \neq 0$, let $C(2\pi nv) := a_f(n, v)$. Applying Δ_k to (4.7), one finds that the Fourier coefficients $C(w)$ satisfy the differential equation

$$\frac{\partial^2}{\partial w^2}C(w) - C(w) + \frac{k}{w}\left(\frac{\partial}{\partial w}C(w) + C(w)\right) = 0. \tag{4.8}$$

Thus, if $n \neq 0$, equation (4.8) has two linearly independent solutions, namely e^{-w} and $\Gamma(1-k, -2w)e^{-w}$. The restrictions in the summations in Lemma 4.3 follow from (4.6). □

Lemma 4.3 indicates that the Fourier expansions of harmonic Maass forms naturally split into two components. These are defined as follows.

DEFINITION 4.4. Assume the notation and hypotheses above. For $k \in \frac{1}{2}\mathbb{Z} \setminus \{1\}$ we define the **holomorphic part** of the harmonic Maass form $f \in H_k^!(\Gamma)$ as

$$f^+(\tau) := \sum_{n \gg -\infty} c_f^+(n)q^n$$

and the **nonholomorphic part** of f by

$$f^-(\tau) := \sum_{\substack{n \ll \infty \\ n \neq 0}} c_f^-(n)\Gamma(1-k, -4\pi nv)q^n + c_f^-(0)v^{1-k}.$$

In particular, if $f \in H_k(\Gamma)$, then

$$f^+(\tau) = \sum_{n \gg -\infty} c_f^+(n)q^n,$$

and f^- has an expansion of the form

$$f^-(\tau) = \sum_{n<0} c_f^-(n)\Gamma(1-k, -4\pi nv)q^n.$$

REMARK. If $f \in H_k(\Gamma)$ is a harmonic Maass form for which $f^- = 0$, then we have that f is in $M_k^!(\Gamma)$, the space of weight k weakly holomorphic modular forms on Γ. In other words, harmonic Maass forms with trivial nonholomorphic parts are weakly holomorphic modular forms.

A further subspace of $H_k^!(\Gamma)$ that turns out to be of importance is the subset consisting of those $f \in H_k^!(\Gamma)$ whose Fourier expansions have the shape

$$f(\tau) = \sum_{n=0}^{\infty} c_f^+(n)q^n + \sum_{n \ll \infty} c_f^-(n)\Gamma\left(1-k, -4\pi nv\right)q^n$$

at the cusp $i\infty$ (with analogous expansions at all cusps). The holomorphic parts of such forms are bounded near cusps, and so exponential growth at a cusp arises from the nonholomorphic part. We denote this space by $H_k^\sharp(\Gamma)$. Note that if $k < 0$ is an even integer, then

$$H_k^\sharp(\Gamma) = \left\{f \in H_k^!(\Gamma) : D^{1-k}(f) \in S_{2-k}(\Gamma)\right\}, \tag{4.9}$$

where the operator $D = \frac{1}{2\pi i}\frac{\partial}{\partial \tau}$ from (2.38) is discussed in more detail in Chapter 5.

REMARK. Many of the results in this section pertain to harmonic Maass forms of weight $k \neq 1$. We discuss some results in the case of weight $k = 1$ in Section 16.6.

CHAPTER 5

Differential Operators and Mock Modular Forms

In this chapter, we study differential operators and their relations to spaces of harmonic Maass forms. In particular, we see how these operators are used to map from spaces of harmonic Maass forms to spaces of weakly holomorphic modular forms. For simplicity, we state our results in this chapter for forms that transform with trivial character $\chi = 1$, but point out that many of the results below can be easily modified to hold for more general spaces of forms ($\chi \neq 1$). We refer the reader to **[94, 121, 129, 407]** for more on this and the results discussed in this chapter. Throughout, we let $k \in \frac{1}{2}\mathbb{Z}$ unless otherwise specified, and in particular note that if $k \in \mathbb{Z}$, then (because $\chi = 1$) we must have $k \in 2\mathbb{Z}$. Moreover, we restrict to the group $\Gamma_0(N)$.

5.1. Maass operators and harmonic Maass forms

It is natural to study the effect of the differential operator

$$D = D_\tau = \frac{1}{2\pi i}\frac{\partial}{\partial\tau} \tag{5.1}$$

(previously defined in (2.38)) on modular or (harmonic) Maass forms. As is well known, the derivative of a modular form is almost never modular. In particular, if f transforms as

$$f(\gamma\tau) = (c\tau + d)^k f(\tau), \tag{5.2}$$

with $\gamma = \left(\begin{smallmatrix} a & b \\ c & d \end{smallmatrix}\right) \in \mathrm{SL}_2(\mathbb{Z})$, a short calculation reveals that

$$D(f)(\gamma\tau) = (c\tau + d)^{k+2} D(f)(\tau) + \frac{k}{2\pi i} c(c\tau + d)^{k+1} f(\tau). \tag{5.3}$$

Thus, if $k \neq 0$, the expression in (5.3) shows that the operator D does not preserve modularity; however, this deficiency can be corrected. To describe this, we introduce the Maass raising operator R_k and its companion L_k, originally studied by Maass. Recall that $\tau = u + iv \in \mathbb{H}$.

Definition 5.1. The **Maass raising operator** R_k **and Maass lowering operator** L_k are defined by

$$R_k := 2i\frac{\partial}{\partial\tau} + \frac{k}{v} = i\left(\frac{\partial}{\partial u} - i\frac{\partial}{\partial v}\right) + \frac{k}{v},$$

$$L_k := -2iv^2\frac{\partial}{\partial\overline{\tau}} = -iv^2\left(\frac{\partial}{\partial u} + i\frac{\partial}{\partial v}\right).$$

Remark. We alert the reader to the fact that the definitions of the Maass operators in the literature vary by multiplicative constants and also sometimes have a slightly shifted k-dependence as in Chapter 2 of **[134]**.

Note from the definition of R_k that

$$R_k = -4\pi D + \frac{k}{v}.$$

A reason why one may call R_k a "correction" to the D-operator, the reasons for the terms "raising" and "lowering", as well as key properties of these operators in relation to harmonic forms, are made clear in the following lemma.

LEMMA 5.2. *Let f be a smooth function on $\mathbb{H}$, let $k \in \frac{1}{2}\mathbb{Z}$, and let $\gamma \in \mathrm{SL}_2(\mathbb{Z})$, with $\gamma \in \Gamma_0(4)$ if $k \in \frac{1}{2}\mathbb{Z} \setminus \mathbb{Z}$.*

i) *With respect to the slash-operator, we have that*

$$R_k(f|_k\gamma) = R_k(f)|_{k+2}\gamma,$$
$$L_k(f|_k\gamma) = L_k(f)|_{k-2}\gamma.$$

In particular, if f transforms as in (5.2), then we have

$$R_k(f)|_{k+2}\gamma = R_k(f) \quad \text{and} \quad L_k(f)|_{k-2}\gamma = L_k(f).$$

ii) *If f is an eigenfunction of Δ_k with eigenvalue λ, then*

$$\Delta_{k+2}(R_k(f)) = (\lambda + k)R_k(f),$$
$$\Delta_{k-2}(L_k(f)) = (\lambda - k + 2)L_k(f).$$

iii) *The weight k hyperbolic Laplacian operator may be expressed as*

$$-\Delta_k = L_{k+2} \circ R_k + k = R_{k-2} \circ L_k.$$

Combining this fact with Lemma 5.2 reveals that the raising and lowering operators preserve modularity up to a change in weight and preserve eigenfunctions of Δ_k up to a change in eigenvalue. Hence, these operators map weak Maass forms to weak Maass forms, up to a change in weight and eigenvalue.

We discuss further results pertaining to L_k in Section 5.2.

PROOF OF LEMMA 5.2. We let $\gamma = \left(\begin{smallmatrix} a & b \\ c & d \end{smallmatrix}\right) \in \mathrm{SL}_2(\mathbb{Z})$, and for ease of notation, we define

$$\rho_k(c,d) = \rho_k(\gamma) := \begin{cases} 1 & \text{if } k \in \mathbb{Z}, \\ \left(\frac{c}{d}\right)\varepsilon_d^{2k} & \text{if } k \in \frac{1}{2}\mathbb{Z} \setminus \mathbb{Z}. \end{cases} \tag{5.4}$$

i) Using the definition, we find after a short calculation that

$$\left(R_k(f)|_{k+2}\gamma\right)(\tau) = \rho_{k+2}(\gamma)(c\tau+d)^{-(k+2)}\left(2if'(\gamma\tau) + \frac{k}{v}f(\gamma\tau)|c\tau+d|^2\right), \tag{5.5}$$

where we let $f' := \partial f/\partial\tau$, and use the fact that $\mathrm{Im}(\gamma\tau) = v/|c\tau+d|^2$. On the other hand, we calculate that $R_k(f|_k\gamma)(\tau)$ equals

$$\rho_k(\gamma)\left(2i(c\tau+d)^{-(k+2)}f'(\gamma\tau) - 2ick(c\tau+d)^{-k-1}f(\gamma\tau) + \frac{k}{v}(c\tau+d)^{-k}f(\gamma\tau)\right)$$
$$= \rho_k(\gamma)(c\tau+d)^{-(k+2)}\left(2if'(\gamma\tau) + \frac{k}{v}f(\gamma\tau)|c\tau+d|^2\right). \tag{5.6}$$

Because $\rho_{k+2}(\gamma) = \rho_k(\gamma)$, the first claim in part i) of Lemma 5.2 follows from (5.5) and (5.6). The proof of the second claim in part i) pertaining to the lowering operator follows similarly.

ii) Using iii), which is proved below, and the fact that $\Delta_k(f) = \lambda f$, we have that $-(R_{k-2} \circ L_k)(f) = \lambda f$, and hence $-(L_k \circ R_{k-2} \circ L_k) f = \lambda L_k(f)$. We subtract $(k-2)L_k(f)$ from both sides of this expression and find that

$$-(L_k \circ R_{k-2} + k - 2)(L_k(f)) = (\lambda - k + 2) L_k(f).$$

Using iii) again establishes the claim in ii) pertaining to the lowering operator L_k. The result given in ii) pertaining to the raising operator R_k follows similarly.

iii) We have by definition that

$$\begin{aligned} R_{k-2} \circ L_k &= \left(2i\frac{\partial}{\partial \tau} + \frac{k-2}{v}\right)\left(-2iv^2\frac{\partial}{\partial \overline{\tau}}\right) = 4\frac{\partial}{\partial \tau}\left(v^2\frac{\partial}{\partial \overline{\tau}}\right) - 2iv(k-2)\frac{\partial}{\partial \overline{\tau}} \\ &= 4v^2\frac{\partial}{\partial \tau}\frac{\partial}{\partial \overline{\tau}} - 4iv\frac{\partial}{\partial \overline{\tau}} - 2iv(k-2)\frac{\partial}{\partial \overline{\tau}} = 4v^2\frac{\partial}{\partial \tau}\frac{\partial}{\partial \overline{\tau}} - 2ikv\frac{\partial}{\partial \overline{\tau}}, \end{aligned}$$

which by definition equals $-\Delta_k$. The second expression given for $-\Delta_k$ in iii) follows similarly. □

While the raising operator R_k may be used to correct the error to modularity obtained by applying the operator D, by definition, R_k does not in general preserve meromorphicity. As shown above, $R_0 = -4\pi D$, so in this instance, the operator preserving modularity and the original holomorphic differential operator coincide. For general non-positive even weights, there is always a special number of iterations of raising operators which coincides with holomorphic differential operators. In this case, repeated compositions of R_k can be used to define maps from spaces of weakly holomorphic modular forms of non-positive weight to spaces of weakly holomorphic modular forms of positive weight. To describe this, we define for any positive integer n the composition of raising operators, which keep track of the "correct" weights at each step:

$$R_k^n := R_{k+2(n-1)} \circ \cdots \circ R_{k+2} \circ R_k.$$

For $n = 0$, we define $R_k^0 := 1$. The identity relating D^{k-1} and R_{2-k}^{k-1} in Lemma 5.3 below is often referred to as "Bol's identity" (cf. **[353]** and **[519]**), and it turns out to provide a powerful tool for analyzing integer weight harmonic Maass forms.

LEMMA 5.3. *For $k \in \mathbb{N}$, we have that*

$$D^{k-1} = \frac{1}{(-4\pi)^{k-1}} R_{2-k}^{k-1}.$$

In particular, we have that

$$D^{k-1} \colon M_{2-k}^!(\Gamma_0(N)) \to M_k^!(\Gamma_0(N)).$$

We see in Theorem 5.5 below, since D^{k-1} annihilates the non-constant terms of the nonholomorphic parts of harmonic Maass forms and maps the nonholomorphic constant to a constant, that we in fact have

$$D^{k-1} \colon H_{2-k}^!(\Gamma_0(N)) \to M_k^!(\Gamma_0(N)). \tag{5.7}$$

We refer to the image of $f \in H_{2-k}^!(\Gamma_0(N))$ under this map as the *ghost* of f.

We prove Lemma 5.3 from a more general identity connecting iterated raising and holomorphic differential operators. This was considered by Zagier (cf. (56) of **[519]**), rephrased using our notation as in Lemma 5.4.

LEMMA 5.4. *For all $k \in \mathbb{Z}$ and $n \in \mathbb{N}_0$, we have that*

$$R_k^n = \sum_{r=0}^{n} (-1)^r \binom{n}{r} (k+r)_{n-r} v^{r-n} (4\pi D)^r.$$

Here as before, $(a)_n = a(a+1)\dots(a+n-1)$ denotes the rising factorial.

PROOF. We prove the claim by induction. The base case $n = 0$ is trivial. Now suppose that the identity is true for a natural number n. We compute, using the fact that $D(v) = -(4\pi)^{-1}$, that

$$\begin{aligned} R_k^{n+1} &= -4\pi D \circ R_k^n + (k+2n)v^{-1} R_k^n \\ &= \sum_{r=0}^{n} (-1)^r \binom{n}{r} (k+r)_{n-r} (r-n) v^{r-n-1} (4\pi D)^r \\ &\quad + \sum_{r=1}^{n+1} (-1)^r \binom{n}{r-1} (k+r-1)_{n-r+1} v^{r-n-1} (4\pi D)^r \\ &\quad + \sum_{r=0}^{n} (-1)^r \binom{n}{r} (k+r)_{n-r} (k+2n) v^{r-n-1} (4\pi D)^r \\ &= (-1)^{n+1} (4\pi D)^{n+1} \\ &\quad + \sum_{r=0}^{n} (-1)^r \left(\binom{n}{r} (k+r)_{n-r} (r+k+n) + \binom{n}{r-1} (k+r-1)_{n-r+1} \right) v^{r-n-1} (4\pi D)^r \\ &= \sum_{r=0}^{n+1} (-1)^r \binom{n+1}{r} (k+r)_{n+1-r} v^{r-n-1} (4\pi D)^r, \end{aligned}$$

as claimed. In the last line above, we used the following elementary identity, which is valid for $k \in \mathbb{Z}$, $n \in \mathbb{N}$, and $0 \le r \le n$, with the understanding that $\binom{n}{-1} = 0$:

$$\binom{n}{r} (k+r)_{n-r} (r+k+n) + \binom{n}{r-1} (k+r-1)_{n-r+1} = \binom{n+1}{r} (k+r)_{n-r+1}.$$

□

We are now in a position to prove Lemma 5.3.

PROOF OF LEMMA 5.3. For $k \in \mathbb{N}$, in Lemma 5.4, let $k \mapsto 2-k$ and $n \mapsto k-1$. In this case, only one term in the sum survives, and the lemma follows immediately. □

THEOREM 5.5. *For $k \ge 2$ an integer, the following are true:*

i) *We have that*

$$D^{k-1} \colon H_{2-k}(\Gamma_0(N)) \to M_k^!(\Gamma_0(N)).$$

In particular, with notation as in Lemma 4.3, for $f \in H_{2-k}(\Gamma_0(N))$, we have that

$$D^{k-1}(f(\tau)) = D^{k-1}(f^+(\tau)) = \sum_{n \gg -\infty} c_f^+(n) n^{k-1} q^n.$$

ii) *We have that*

$$D^{k-1}\colon H^!_{2-k}(\Gamma_0(N)) \twoheadrightarrow M^!_k(\Gamma_0(N)).$$

For $f \in H^!_{2-k}(\Gamma_0(N))$, *we have that*

$$D^{k-1}(f(\tau)) = -(4\pi)^{1-k}(k-1)!\; c_f^-(0) + \sum_{n \gg -\infty} c_f^+(n) n^{k-1} q^n.$$

REMARK. Theorem 5.5 is closely related to work by Eichler, Knopp, Lehner, and others (for example, see **[190, 312, 313, 315, 347]** and the expository article **[149]**). A standard feature of these works is the relationship to Bol's identity (Lemma 5.3), and the role of obstructions to modularity played by modular integrals such as Eichler integrals (see Chapter 12).

PROOF OF THEOREM 5.5. We first consider i). We begin by showing that the nonholomorphic parts of harmonic Maass forms in $H_{2-k}(\Gamma_0(N))$ are annihilated by D^{k-1}. Specifically, we claim that

$$D^{k-1}\left(\Gamma\left(k-1, 4\pi n v\right) q^{-n}\right) = 0.$$

Indeed, this follows directly, as observed by Zagier, by using the integral representation

$$\Gamma\left(\alpha, w\right) e^{w} = \int_0^\infty \left(w+t\right)^{\alpha-1} e^{-t} dt.$$

Hence, we see that for any $f \in H_{2-k}(\Gamma_0(N))$, the image $D^{k-1}(f)$ is holomorphic on $\mathbb{H}$ and in particular its nonholomorphic part gets annihilated. To show that $D^{k-1}(f)$ is weakly holomorphic, we must prove that it is meromorphic in the cusps of $\Gamma_0(N)$. To see this, note that by Definition 4.2 iii), f exhibits (at worst) linear exponential growth in the cusps of $\Gamma_0(N)$, from which the necessary growth of $D^{k-1}(f)$ at the cusps follows directly using the definition of the D-operator. The claimed Fourier expansion is then obvious since $D^{k-1}(q^n) = n^{k-1}q^n$.

To show ii), we first note that there is an essential "shadow operator" ξ_{2-k} of Bruinier and Funke, which is defined below in (5.11) and which plays a key role in the theory of harmonic Maass forms. There is also an important operator, called the "flipping operator", which is studied in Section 5.3. Parts iii) and iv) of Proposition 5.15 show that this operator interchanges the roles of ξ_{2-k} and D^{k-1} acting on $H^!_{2-k}(\Gamma_0(N))$ (cf. Proposition 5.15 iv) below). In particular, Proposition 5.15 ii) and iv) below show that the desired surjectivity of D^{k-1} follows from a suitable surjectivity statement of ξ_{2-k}. In fact, we will see this shortly in Theorem 5.10 iii), when we show that ξ_{2-k} maps $H^!_{2-k}(\Gamma_0(N))$ surjectively onto $M^!_k(\Gamma_0(N))$. Note that nonholomorphic constant terms get flipped to their negative conjugates. □

While Theorem 5.5 establishes a map from harmonic Maass forms to weakly holomorphic modular forms, D^{k-1} does not map surjectively. The image of this operator in the case was determined in **[129]**. To describe it, we first define the *regularized inner product* of forms $f, g \in M^!_k(\Gamma_0(N))$. We first recall the usual Petersson inner product. For $f, g \in M_k(\Gamma)$ such that $fg \in S_{2k}(\Gamma)$, $\langle f, g\rangle$ denotes *Petersson's inner product*,

$$\langle f, g\rangle := \frac{1}{[\mathrm{SL}_2(\mathbb{Z}) : \Gamma]} \int_{\Gamma\backslash\mathbb{H}} f(\tau)\overline{g(\tau)} v^k \frac{du dv}{v^2}. \tag{5.8}$$

Returning to the regularized inner product, we require a truncated fundamental domain, defined for a positive real number T by

$$\mathcal{F}_T := \{\tau \in \mathbb{H}\colon |u| \leq \tfrac{1}{2}, |\tau| \geq 1, v \leq T\}. \tag{5.9}$$

The domain $\mathcal{F}_T$ may be used to define a truncated fundamental domain for $\Gamma_0(N)$, namely

$$\mathcal{F}_T(\Gamma_0(N)) := \bigcup_{\gamma \in \Gamma_0(N) \backslash \mathrm{SL}_2(\mathbb{Z})} \gamma \mathcal{F}_T.$$

Following Borcherds [**74**], we make the following definition.

DEFINITION 5.6. Let $f, g \in M_k^!(\Gamma_0(N))$. The **regularized inner product** of f and g, denoted by $\langle f, g\rangle^{\mathrm{reg}}$, is defined (whenever it exists) to be the constant term in the Laurent expansion at $s = 0$ of the meromorphic continuation in s of the function

$$\frac{1}{[\mathrm{SL}_2(\mathbb{Z}) : \Gamma_0(N)]} \lim_{T \to \infty} \int_{\mathcal{F}_T(\Gamma_0(N))} f(\tau)\overline{g(\tau)} v^{k-s} \frac{dudv}{v^2}.$$

As explained in [**74**] and [**129**], $\langle f, g\rangle^{\mathrm{reg}}$ exists for $f \in M_k(\Gamma_0(N))$ and $g \in M_k^!(\Gamma_0(N))$ (but not in general for f and g both in $M_k^!(\Gamma_0(N))$). If f and g are both ordinary holomorphic modular forms such that fg is a cusp form, then the regularized inner product reduces to the (unregularized) Petersson inner product.

REMARK. Note that Borcherds' regularized inner product does not exist for all weakly holomorphic modular forms. In particular, one cannot directly integrate weakly holomorphic forms against themselves. This problem was overcome by Diamantis, Ehlen, and one of the authors [**82**]. For simplicity we restrict to even weight forms for $\mathrm{SL}_2(\mathbb{Z})$. The idea is simple. We multiply the integrands with a function that forces convergence and then analytically continue. We start with weakly holomorphic forms. Observe that for $\mathrm{Re}(w) \gg 0$, the integral

$$I(f, g; w, s) := \int_{\mathcal{F}} f(z)\overline{g(z)} y^{k-s} e^{-wy} \frac{dxdy}{y^2}$$

converges and is analytic. For every $\varphi \in (\pi/2, 3\pi/2) \setminus \{\pi\}$ it has an analytic continuation $I_\varphi(f, g; w, s)$ to $U_\varphi \times \mathbb{C}$ with $U_\varphi \subset \mathbb{C}$ a certain open set. Then define

$$\langle f, g\rangle_\varphi := \mathrm{CT}_{s=0}\, I_\varphi(f, g; 0, s) - i \sum_{n>0} c_f(-n)\overline{c_g(-n)}\, \mathrm{Im}\left(E_{2-k,\varphi}(-4\pi n)\right),$$

where $\mathrm{CT}_{s=0}\, F(s)$ denotes the constant term of an analytic function F, c_F stands for the Fourier coefficients of a function F, and $E_{r,\varphi}$ is the generalized exponential integral defined with branch cut on the ray $\{xe^{i\varphi} : x \in \mathbb{R}^+\}$. In [**82**] it is shown that for $f, g \in M_k^!$, $\langle f, g\rangle$ exists and is independent of the choice of φ. It equals Borcherds' regularization if his exists.

The next theorem gives a useful formula expressing the inner product of a modular form and the image of a harmonic Maass form under D^{k-1} in terms of their Fourier expansions. We state Theorem 5.7 in terms of the raising operator, which is related to the D-operator by Lemma 5.3.

THEOREM 5.7. *For $k \geq 2$ an integer, let $g \in M_k(\Gamma_0(N))$ and $f \in H_{2-k}(\Gamma_0(N))$. Denote by $c_{g,\alpha}(0)$ and $c_{f,\alpha}^+(0)$ the respective constant terms of the Fourier expansions of g and f at the cusp α with width w_α. We have that the regularized inner* *product of g and $R_{2-k}^{k-1}(f)$ may be expressed as*

$$\langle g, R_{2-k}^{k-1}(f)\rangle^{\mathrm{reg}} = \frac{(-1)^k}{[\mathrm{SL}_2(\mathbb{Z}) : \Gamma_0(N)]} \sum_{\alpha} w_\alpha c_{g,\alpha}(0)\overline{c_{f,\alpha}^{+}(0)},$$

where the sum runs over the set of cusps of $\Gamma_0(N)$.

In particular, we directly find that cusp forms are orthogonal to weakly holomorphic forms arising from iterated derivatives of Maass forms.

COROLLARY 5.8. *With notation and hypotheses as above, if $g \in S_k(\Gamma_0(N))$, then we have that $\langle g, R_{2-k}^{k-1}(f)\rangle^{\mathrm{reg}} = 0$.*

PROOF OF THEOREM 5.7. For brevity, we provide a proof for $N = 1$, as the general case follows similarly. Let $F := \mathfrak{F}_{2-k}(f)$ be the flip of f (cf. Definition 5.14) and let $h := R_{2-k}^{k-1}(f)$. Using Proposition 5.15 iii), we have that $h = v^{-k}\overline{L_{2-k}(F)}$. The constant term of the function h vanishes by Theorem 5.5 i) and Lemma 5.3, hence $\langle g, R_{2-k}^{k-1}(f)\rangle^{\mathrm{reg}}$ equals

$$\lim_{T\to\infty} \int_{\mathcal{F}_T} g(\tau)\overline{h(\tau)}v^k \frac{dudv}{v^2} = \lim_{T\to\infty} \int_{\mathcal{F}_T} g(\tau) L_{2-k}(F(\tau))\frac{dudv}{v^2}$$

$$(5.10) \qquad = \lim_{T\to\infty} \int_{\mathcal{F}_T} g(\tau)\left(\frac{\partial F(\tau)}{\partial\overline{\tau}}\right) d\tau d\overline{\tau} = -\lim_{T\to\infty} \int_{\mathcal{F}_T} (\overline{\partial} F(\tau)) \wedge g(\tau)d\tau,$$

where $\mathcal{F}_T = \mathcal{F}_T(\mathrm{SL}_2(\mathbb{Z}))$, ∂ and $\overline{\partial}$ are the Dolbeault operators, and $d = \partial + \overline{\partial}$ is the usual exterior derivative. Applying Stokes' theorem (noting that g is holomorphic), (5.10) becomes

$$-\lim_{T\to\infty} \int_{\mathcal{F}_T} d(F(\tau)g(\tau)d\tau) = -\lim_{T\to\infty} \int_{\partial\mathcal{F}_T} F(\tau)g(\tau)d\tau$$

$$= \lim_{T\to\infty} \int_{-\frac{1}{2}}^{\frac{1}{2}} F(u+iT)g(u+iT)du.$$

The integral on the interval $[-1/2, 1/2]$ gives the constant term of the Fourier expansion of the function $F(u+iT)g(u+iT)$. To compute this term explicitly, we use (5.8) and the fact that $\overline{e(\overline{\tau})} = e(-\tau)$, which allows us to easily compute the Fourier expansion of the function $F(\tau)$. (The shape of the Fourier expansion for $g(\tau)$ is obvious.) Letting $T \to \infty$, one finds that the constant term of the function in question is simply $(-1)^k c_g(0)\overline{c_f^{+}(0)}$, as claimed. We conclude by noting that the proof for general congruence subgroups follows along very similar lines, where instead we integrate along the truncated fundamental domains for the corresponding subgroups. □

We are now ready to describe the image of the D^{k-1}-operator on Maass forms, as given in the following theorem from [**129**].

THEOREM 5.9. *If k is an integer satisfying $k \geq 2$, then the image of the map*

$$D^{k-1} \colon H_{2-k}(\Gamma_0(N)) \to M_k^!(\Gamma_0(N))$$

consists of those $h \in M_k^!(\Gamma_0(N))$ which are orthogonal to cusp forms with respect to the regularized inner product and which also have constant term 0 at all cusps of $\Gamma_0(N)$.

REMARK. Theorem 5.9 is essentially contained in work of Petersson (cf. Satz 9 of **[419]**). Note that his series F_{r-2} are the holomorphic parts of Maass-Poincaré series whose associated nonholomorphic parts are annihilated by D^{k-1}. We shall discuss Maass-Poincaré series in Section 6.3.

PROOF OF THEOREM 5.9. Let $h \in M_k^!(\Gamma_0(N))$ have constant term equal to 0 at each cusp of $\Gamma_0(N)$. Further suppose h is orthogonal to cusp forms. It is explained in **[121]**, Lemma 3.11, that there exists some $f \in H_{2-k}(\Gamma_0(N))$ such that the principal parts of $D^{k-1}(f)$ and h at the cusps of $\Gamma_0(N)$ agree, up to the constant terms. Since their constant terms also agree (both are equal to 0), we have that $h - D^{k-1}(f)$ is a cusp form of weight k. By Theorem 5.7, $h - D^{k-1}(f)$ is orthogonal to cusp forms, hence is identically equal to zero. If we let $M^\perp = M_{k,N}^\perp$ denote the set of forms in $M_k^!(\Gamma_0(N))$ which are orthogonal to cusp forms and whose constant terms at cusps are equal to zero, we have shown that $M^\perp \subseteq D^{k-1}(H_{2-k}(\Gamma_0(N)))$. Conversely, $D^{k-1}(H_{2-k}(\Gamma_0(N))) \subseteq M^\perp$ by Theorem 5.5 part i) and Theorem 5.7. □

5.2. The ξ-operator and pairing of Bruinier and Funke

In the previous section, we saw that if we iterate the raising operator R_k just the right number of times we obtain a map D^{k-1} from the space of harmonic Maass forms of weight $2-k$ to the space of modular forms of weight k. It turns out that this operator D^{k-1} has a companion operator which is well defined for general half-integral weight harmonic Maass forms and whose interplay with D^{k-1} plays an important role in our understanding of integral weight mock modular forms. One may think of this operator as an analogous modification of the lowering operator L_{2-k}, which maps between spaces of dual weight, i.e., weights k and $2-k$. To be precise, define the differential operator

$$\xi_\kappa := 2iv^\kappa \overline{\frac{\partial}{\partial \overline{\tau}}}. \tag{5.11}$$

That is, in terms of the lowering operator L_κ, we have that

$$\xi_\kappa = v^{\kappa-2}\overline{L_\kappa}.$$

The following theorem is analogous to Theorem 5.5.

THEOREM 5.10. *Assuming the notation of Lemma 4.3, the following are true:*

i) *We have that*

$$\xi_{2-k}\colon H_{2-k}(\Gamma_0(N)) \twoheadrightarrow S_k(\Gamma_0(N)).$$

ii) *For $f \in H_{2-k}(\Gamma_0(N))$, we have that*

$$\xi_{2-k}(f(\tau)) = \xi_{2-k}(f^-(\tau)) = -(4\pi)^{k-1}\sum_{n=1}^{\infty} \overline{c_f^-(-n)} n^{k-1} q^n.$$

iii) *We have that*

$$\xi_{2-k}\colon H_{2-k}^!(\Gamma_0(N)) \twoheadrightarrow M_k^!(\Gamma_0(N)).$$

iv) *For $f \in H_{2-k}^!(\Gamma_0(N))$, we have that*

$$\xi_{2-k}(f(\tau)) = \xi_{2-k}(f^-(\tau)) = (k-1)\overline{c_f^-(0)} - (4\pi)^{k-1}\sum_{n \gg -\infty} \overline{c_f^-(-n)} n^{k-1} q^n.$$

PROOF OF THEOREM 5.10. Theorem 5.10 is a straightforward modification of Proposition 3.2 of **[121]**. The expansions given in ii) and iv) follow by elementary calculations using Lemma 4.3 and the fact that

$$\frac{\partial}{\partial w}\Gamma(\alpha, w) = w^{\alpha-1}e^{-w}.$$

Additionally, analytic alternatives to Bruinier and Funke's geometric proof in **[121]** for parts i) and iii) follows immediately from the theory of Maass-Poincaré series (cf. Section 6.3). □

It is not difficult to see, using ii) and iv), that the kernel of either map from Theorem 5.10 i) or iii) is

$$\ker\left(\xi_{2-k}\right) = M^!_{2-k}(\Gamma_0(N)). \tag{5.12}$$

Combining with Theorem 5.10, we directly find the following important exact sequence, which was shown by Bruinier and Funke in Corollary 3.8 of **[121]**:

$$0 \longrightarrow M^!_{2-k}(\Gamma_0(N)) \longrightarrow H^!_{2-k}(\Gamma_0(N)) \longrightarrow M^!_k(\Gamma_0(N)) \longrightarrow 0. \tag{5.13}$$

Using the ξ-operator, we can also define an important bilinear pairing, known as the *Bruinier-Funke pairing.* This pairing, mapping

$$\{\cdot,\cdot\}\colon M_k(\Gamma_0(N)) \times H_{2-k}(\Gamma_0(N)) \to \mathbb{C},$$

is given by

$$\{g, f\} := \langle g, \xi_{2-k}(f)\rangle\,. \tag{5.14}$$

Proposition 3.5 of **[121]** gives an alternate description of this pairing in terms of the coefficients of g and f.

PROPOSITION 5.11. *Suppose that $(g, f) \in M_k(\Gamma_0(N)) \times H_{2-k}(\Gamma_0(N))$, and for any cusp α, suppose that the Fourier expansion of g at α is $g(\tau) =: \sum_{n=0}^{\infty} c_{g,\alpha}(n)q^n$ and that $f^+(\tau) =: \sum_{n\gg-\infty} c^+_{f,\alpha}(n)q^n$. Then we have*

$$\{g, f\} = \sum_{\alpha}\sum_{n\leq 0} c_{g,\alpha}(-n)c^+_{f,\alpha}(n),$$

where the first sum runs over the set of cusps of $\Gamma_0(N)$.

REMARK. Proposition 5.11 can be generalized to include weakly holomorphic modular forms (cf. **[82]**).

SKETCH OF PROOF PROPOSITION 5.11. We focus on the case when $N = 1$ for simplicity; the formula can be proven for higher levels by suitably modifying the proof below. We recall the truncated fundamental domain in (5.1). The key ingredient is Stokes' theorem, which shows that

$$\int_{\mathcal{F}_T} g(\tau)L_k\left(f(\tau)\right)\frac{dudv}{v^2} = -\int_{\partial\mathcal{F}_T} g(\tau)f(\tau)d\tau.$$

Now the left-hand side may be rewritten as $\int_{\mathcal{F}_T} g(\tau)\overline{\xi_{2-k}(f(\tau))}v^k\frac{dudv}{v^2}$. As f and g have dual weights, we see that $g(\tau)f(\tau)d\tau$ is an invariant 1-form, so that its integral on all the pieces of $\partial\mathcal{F}_T$ except the top line from $-1/2+iT$ to $1/2+iT$ cancel out. Hence, the right-hand side is equal to

$$\int_{-\frac{1}{2}}^{\frac{1}{2}} g(u+iT)f(u+iT)du.$$

By part iii) of Definition 4.2, and that for any q-series such an integral just picks out the constant term, we see that this integral is equal to

$$\sum_{n\leq 0} c_g(-n)c_f^+(n) + O\left(e^{-\varepsilon T}\right).$$

Taking the limit $T \to \infty$ gives the result. □

Note by definition that $\{g, f\} = 0$ for any $f \in M_k^!(\Gamma_0(N))$, which shows that there is an induced pairing on $H_{2-k}/M_{2-k}^! \times S_k$. The following result of Bruinier and Funke then follows easily from Theorem 5.10 i).

PROPOSITION 5.12. *The pairing between $H_{2-k}/M_{2-k}^!$ and S_k induced by the Bruinier-Funke pairing is non-degenerate.*

REMARK. We note that if one includes harmonic Maass forms with growth at $i\infty$ in the nonholomorphic part, the pairing is no longer non-degenerate (cf. [**82**]).

The formula in Proposition 5.11 has important ramifications which are frequently useful in applications. For example, the following direct corollary was crucial in the proof of [**237**] that Ramanujan's mock theta functions satisfy his definition (cf. Theorem 9.18).

LEMMA 5.13. *Every form $f \in H_{2-k}(\Gamma_1(N)) \setminus M_{2-k}^!(\Gamma_1(N))$ has a non-trivial principal part at some cusp.*

PROOF. By assumption, f has a non-trivial shadow (cf. Definition 5.16, part ii). Pairing f with its shadow, by definition

$$\{\xi_{2-k}(f), f\} = \langle \xi_{2-k}(f), \xi_{2-k}(f)\rangle\,.$$

As the Petersson norm of a non-zero cusp form is always non-zero, it follows from Proposition 5.11 that there is some term $c_{f,\alpha}(n)$ with $n < 0$ which is non-vanishing. That is, at the cusp α, f^+ has a non-trivial principal part. □

REMARK. As an application of the operators investigated in this chapter, we give a classification of certain modules arising in Lie theory. One can associate to every $f \in H_k^!(\Gamma)$ an $\mathrm{SL}_2(\mathbb{Z})$-invariant function $\widetilde{f}$ on $\mathrm{SL}_2(\mathbb{R})$. To motivative this, recall that $\mathrm{SL}_2(\mathbb{R})$ acts on the upper half-plane $\mathbb{H}$, and the stabilizer of i under this action turns out to be the *special orthogonal group*

$$\mathrm{SO}(2) := \left\{k_\vartheta = \begin{pmatrix} \cos(\vartheta) & \sin(\vartheta) \\ -\sin(\vartheta) & \cos(\vartheta)\end{pmatrix} : 0 \leq \vartheta < 2\pi\right\}.$$

It is then not difficult to show that the map sending $g \in \mathrm{SL}_2(\mathbb{R})$ to $g\cdot i$ induces an isomorphism $\mathbb{H} \cong \mathrm{SL}_2(\mathbb{R})/\mathrm{SO}(2)$. Given this realization of $\mathbb{H}$ as a quotient of $\mathrm{SL}_2(\mathbb{R})$, it is natural to try to lift modular objects to the group $G := \mathrm{SL}_2(\mathbb{R})$. Explicitly, for any $g \in G$ and any $f \in H_k^!(\Gamma)$, a suitable function is obtained by defining (for $j((\begin{smallmatrix} a & b \\ c & d\end{smallmatrix}), \tau) := c\tau + d$)

$$\widetilde{f}(g) := j(g, i)^{-k} f(g\cdot i),$$

which inherits the modular transformation properties

$$\widetilde{f}(\gamma g) = \widetilde{f}(g)$$

for all $\gamma \in \Gamma$ and

$$\widetilde{f}(gk_\vartheta) = e^{ik\vartheta}\widetilde{f}(g)$$

for all $k_\vartheta \in \mathrm{SO}(2)$.

The group G, and hence its complexified Lie algebra $\mathfrak{g} := \mathrm{Lie}(G)_\mathbb{C}$, acts on such functions by right translation. Let $M(\widetilde{f})$ be the $(\mathfrak{g}, K)$-module which arises from this action ($K := \mathrm{SO}(2)$ is in fact a maximal compact subgroup of $\mathrm{SL}_2(\mathbb{R})$) associated to $\widetilde{f}$ (cf. Section 2.4 of **[134]** for a detailed exposition of $(\mathfrak{g}, K)$-modules). The full construction of this $(\mathfrak{g}, K)$-module may also be found in Section 3 of **[102]**. In **[102]** Kudla and one of the authors classified the $(\mathfrak{g}, K)$-modules which arise as such an $M(\widetilde{f})$ based on consideration of the structure of a cyclic $(\mathfrak{g}, K)$-module generated by a "harmonic" vector. Note that some of the possible $(\mathfrak{g}, K)$-modules associated to harmonic Maass forms were previously considered in work of Schulze-Pillot **[447]**. As described in **[102]**, the resulting indecomposable modules are built up from the various irreducible $(\mathfrak{g}, K)$-modules occurring as constituents of the principal series for G at points of reducibility. It turns out that there are 9 possibilities for which modules may occur. Here ξ_k is defined in (5.11). Written in classical language, these are

I. Case $k < 1$
   (a) $\xi_k(f) = 0, \quad D_k^{1-k}(f) = 0,$
   (b) $\xi_k(f) = 0, \quad D_k^{1-k}(f) \neq 0,$
   (c) $\xi_k(f) \neq 0, \quad D_k^{1-k}(f) = 0,$
   (d) $\xi_k(f) \neq 0, \quad D_k^{1-k}(f) \neq 0.$

II. Case $k = 1$
   (a) $\xi_1(f) = 0,$
   (b) $\xi_1(f) \neq 0.$

III. Case $k > 1$
   (a) $\xi_k(f) = 0,$
   (b) $\xi_k(f) \neq 0, \quad L_k^k(f) = 0,$
   (c) $\xi_k(f) \neq 0, \quad L_k^k(f) \neq 0,$

where L_k is the lowering operator from Definition 5.1. In **[102]** it was shown that many, but not all, cases can occur. However, if one instead considers the setting of vector-valued forms, then all of the above cases occur.

5.3. The flipping operator

For non-positive integral weight harmonic Maass forms there is another important operator which "switches" the holomorphic and nonholomorphic parts. Throughout this section, since we do not explicitly refer to the dual weight space of cusp forms, for convenience we switch from the convention of Section 5.2 to let $k \leq 0$.

Definition 5.14. For $f \in H_k^!(\Gamma_0(N))$ with $k \in -2\mathbb{N}_0$, define the **flip** of f by

$$\mathfrak{F}_k(f) := -\frac{v^{-k}}{(-k)!}\overline{R_k^{-k}(f)}.$$

The operator $\mathfrak{F}_k$ is called the **flipping operator**.

The flipping operator has the following properties.

PROPOSITION 5.15. *Suppose that $f \in H_k^!(\Gamma_0(N))$ with $k \in -2\mathbb{N}_0$. Then the following are true:*

i) *The image $\mathfrak{F}_k(f)$ lies in $H_k^!(\Gamma_0(N))$.*

ii) *The flipping operator is an involution, that is,*

$$\mathfrak{F}_k(\mathfrak{F}_k(f)) = f.$$

iii) *The shadow of the flip of f is given by*

$$\xi_k\left(\mathfrak{F}_k(f)\right) = \frac{(4\pi)^{1-k}}{(-k)!} D^{1-k}(f).$$

iv) *The image under the Bol operator of the flip of f is given by*

$$D^{1-k}\left(\mathfrak{F}_k(f)\right) = \frac{(-k)!}{(4\pi)^{1-k}} \xi_k(f).$$

v) *Recalling the notation*

$$f(\tau) = c_f^-(0)v^{1-k} + \sum_{n \gg -\infty} c_f^+(n)q^n + \sum_{\substack{n \ll \infty \\ n \neq 0}} c_f^-(n)\Gamma(1-k, -4\pi n v)q^n,$$

we have

$$\mathfrak{F}_k\left(f(\tau)\right) = -\overline{c_f^-(0)}v^{1-k} - (-k)! \sum_{\substack{n \gg -\infty \\ n \neq 0}} \overline{c_f^-(-n)}q^n - \overline{c_f^+(0)}$$
$$- \frac{1}{(-k)!} \sum_{\substack{n \ll \infty \\ n \neq 0}} \overline{c_f^+(-n)}\Gamma\left(1-k, -4\pi n v\right) q^n.$$

REMARK. Proposition 5.15 in particular implies that $\mathfrak{F}_k$ "flips" between the spaces $H_k(\Gamma_0(N))$ and $H_k^\sharp(\Gamma_0(N))$. That is the reason why this operator is called the flipping operator.

REMARK. The flipping operator shares similarities with operators and compositions considered in work of Schulze-Pillot [**447**]. It would be interesting to find a representation-theoretic intepretation of Proposition 5.15 and to determine the relation to Schulze-Pillot's work.

PROOF OF PROPOSITION 5.15. Firstly, using Lemma 5.2 shows that $\mathfrak{F}_k(f)$ transforms like a modular form of weight k. Furthermore, the growth condition of $\mathfrak{F}_k(f)$ is immediate from that of f.

We next show iii), which then yields that f is annihilated by Δ_k, and thus an element of $H_k^!(\Gamma_0(N))$ which also implies i). For any real-analytic function $g\colon \mathbb{H} \to \mathbb{C}$, we have

$$\xi_k\left(v^{-k}\overline{g(\tau)}\right) = R_{-k}(g(\tau)). \tag{5.15}$$

Using this for $g = R_k^{-k}(f)$ we obtain, by Bol's identity (Lemma 5.3),

$$\xi_k\left(v^{-k}\overline{R_k^{-k}(f(\tau))}\right) = (-4\pi)^{1-k}D^{1-k}\left(f(\tau)\right),$$

which directly implies iii).

We next prove ii). For this, we show by induction that, for $0 \leq \ell \leq -k-1$,

$$\mathfrak{F}_k\left(\mathfrak{F}_k\left(f(\tau)\right)\right) = \frac{\ell!}{(-k)!(-k-\ell)!} v^{-k} \overline{R_{k+2\ell}^{-k-\ell}\left(v^{-k-2\ell}\overline{R_k^{-k-\ell}\left(f(\tau)\right)}\right)}. \tag{5.16}$$

The base case $\ell = 0$ follows easily using the definition of $\mathfrak{F}_k$. For the inductive step, we use that for any real-analytic function $g\colon \mathbb{H} \to \mathbb{C}$,

(5.17)

$$R_\kappa\left(v^{-\kappa}\overline{R_{-\kappa-2}(g(\tau))}\right) = v^{-k-2}\overline{L_{-k}\circ R_{-k-2}(g(\tau))} = v^{-\kappa-2}\overline{(-\Delta_{-\kappa-2}+\kappa+2)\,g(\tau)}.$$

Therefore, if g has eigenvalue λ under $\Delta_{-\kappa-2}$, then

$$R_\kappa\left(v^{-\kappa}\overline{R_{-\kappa-2}(g(\tau))}\right) = v^{-\kappa-2}\left(-\overline{\lambda}+\kappa+2\right)\overline{g(\tau)}.$$

We now assume that (5.16) is true for some $0 \le \ell \le -k-1$ and rewrite the right-hand side as

$$\frac{\ell!}{(-k)!(-k-\ell)!}v^{-k}R_{k+2\ell+2}^{-k-\ell-1}\left(\overline{v^{-k-2\ell}R_{-k-2\ell-2}\left(R^{-k-\ell-1}\left(f(\tau)\right)\right)}\right).$$

By induction one can see that $R_k^m(f)$ has eigenvalue $m(k+m-1)$. Thus, using (5.17), we have

$$\begin{aligned}
&R_{k+2\ell}\left(v^{-k-2\ell}\overline{R_{-k-2\ell-2}\left(R_k^{-k-\ell-1}(f(\tau))\right)}\right)\\
&= v^{-k-2\ell-2}\overline{(-\Delta_{-k-2\ell-2}+k+2\ell+2)\,R_k^{-k-\ell-1}(f(\tau))}\\
&= v^{-k-2\ell-2}\left(-(k+\ell+1)(\ell+2)+k+2\ell+2\right)\overline{R_k^{-k-\ell-1}(f(\tau))}\\
&= (-k-\ell)(\ell+1)v^{-k-2\ell-2}\overline{R_k^{-k-\ell-1}(f(\tau))}.
\end{aligned}$$

Thus the right-hand side of (5.16) becomes

$$\frac{(\ell+1)!}{(-k)!(-k-\ell-1)!}v^{-k}R_{k+2\ell+2}^{-k-\ell-1}\left(v^{-k-2\ell-2}\overline{R_k^{-k-\ell-1}(f(\tau))}\right),$$

as claimed. In particular, choosing $\ell = -k$ and simplifying then gives the involution claim of ii).

Using ii) and iii), we obtain by (5.15)

$$\begin{aligned}
R_k^{1-k}\left(\mathfrak{F}_k\left(f(\tau)\right)\right) &= \xi_k\left(v^{-k}\overline{R_k^{-k}\left(\mathfrak{F}_k\left(f(\tau)\right)\right)}\right) = -(-k)!\xi_k\left(\mathfrak{F}_k\circ\mathfrak{F}_k\left(f(\tau)\right)\right)\\
&= -(-k)!\xi_k\left(f(\tau)\right),
\end{aligned}$$

and iv) now follows by Bol's identity.

We are left to show v). Since $\mathfrak{F}_k(f) \in H_k^!\left(\Gamma_0(N)\right)$, it has by Lemma 4.3 a Fourier expansion of the shape

$$\mathfrak{F}_k\left(f(\tau)\right) = d_f^-(0)v^{1-k} + \sum_{n\gg-\infty} d_f^+(n)q^n + \sum_{\substack{n\ll\infty\\ n\neq 0}} d_f^-(n)\Gamma(1-k,-4\pi nv)q^n.$$

We now determine the $d_f^\pm(n)$ explicitly in terms of $c_f^\pm(n)$. By Theorem 5.10 iv), we have

$$\xi_k\left(\mathfrak{F}_k\left(f(\tau)\right)\right) = \overline{d_f^-(0)}(1-k) - (4\pi)^{1-k}\sum_{\substack{n\gg-\infty\\ n\neq 0}}\frac{\overline{d_f^-(-n)}}{n^{k-1}}q^n.$$

On the other hand, by iii) and a direct computation, we find that

$$\xi_k\left(\mathfrak{F}_k\left(f(\tau)\right)\right) = \frac{(4\pi)^{1-k}}{(-k)!}D^{1-k}(f(\tau)) = -(1-k)c_f^-(0) + \frac{(4\pi)^{1-k}}{(-k)!}\sum_{\substack{n\gg-\infty\\ n\neq 0}}\frac{c_f^+(n)}{n^{k-1}}q^n.$$

Comparing coefficients for $n \neq 0$ then gives

$$d_f^-(n) = -\frac{\overline{c_f^+(-n)}}{(-k)!}.$$

Moreover, we have that

$$d_f^-(0) = -\overline{c_f^-(0)}.$$

Next, using Theorem 5.5 we see that

$$D^{1-k}\left(\mathfrak{F}_k\left(f(\tau)\right)\right) = -\frac{(1-k)!}{(4\pi)^{1-k}} d_f^-(0) + \sum_{\substack{n \gg -\infty \\ n \neq 0}} \frac{d_f^+(n)}{n^{k-1}} q^n.$$

On the other hand, using iv), we obtain

$$\begin{aligned} D^{1-k}\left(\mathfrak{F}_k(f(\tau))\right) &= \frac{(-k)!}{(4\pi)^{1-k}} \xi_k\left(f(\tau)\right) \\ &= \frac{(-k)!}{(4\pi)^{1-k}} \left(\overline{c_f^-(0)}(1-k) - (4\pi)^{1-k} \sum_{\substack{n \gg -\infty \\ n \neq 0}} \frac{\overline{c_f^-(-n)}}{n^{k-1}} q^n \right). \end{aligned}$$

Comparing coefficients then for $n \neq 0$, we see that

$$d_f^+(n) = -(-k)!\overline{c_f^-(-n)}.$$

We are left to determine $d_f^+(0)$. One easily sees by induction that, for $\ell \in \mathbb{N}_0$, we have

$$R_k^\ell(1) = \frac{(k)_\ell}{v^\ell}.$$

From this we conclude that

$$\mathfrak{F}_k\left(c_f^+(0)\right) = -\overline{c_f^+(0)},$$

which finishes the proof. □

5.4. Mock modular forms and shadows

In this section, as in Section 5.2, we again emphasize the role of weight k cusp forms. Following Zagier [**520**], we give the following definition, referring to the notation as in Lemma 4.3.

DEFINITION 5.16.

i) A **mock modular form** of weight $2-k$ is the holomorphic part f^+ of a harmonic Maass form of weight $2-k$ for which f^- is nontrivial.
ii) If $f \in H_{2-k}(\Gamma_0(N))$, we refer to the cusp form

$$\xi_{2-k}(f(\tau)) = \xi_{2-k}(f^-(\tau)) = -(4\pi)^{k-1} \sum_{n=1}^{\infty} \overline{c_f^-(-n)} n^{k-1} q^n \in S_k(\Gamma_0(N))$$

from Theorem 5.10 i) as the **shadow** of the mock modular form f^+.
iii) We refer to a mock modular form of weight $1/2$ or $3/2$ whose shadow is a linear combination of unary theta functions as a **mock theta function**.

REMARK. The definition of the shadow of a mock modular form is convenient for our purposes but is also a slight abuse of Zagier's original definition. Zagier's original use of the term in **[520]** highlighted the fact that given a mock modular form, perhaps as a q-hypergeometric series (see (9.14)) or as a natural generating series from combinatorics or physics, there is something "hidden" which is needed to determine a modular completion. That is, as we shall see shortly in Lemma 5.17, the nonholomorphic completion term f^- is a "period integral" of the shadow of f^+. Overall, then, knowledge of the mock modular form f^+ and the shadow, typically a cusp form, allows one to "build" a harmonic Maass form completing f, while here we are taking the full harmonic Maass form f as already given and noting that this shadow may also be obtained by applying the differential operator ξ_{2-k}.

REMARK. More generally, if $f \in H^!_{2-k}(\Gamma)$, then we refer to the modular form $\xi_{2-k}(f)$ as the *shadow* of f.

REMARK. We remark that in the literature, it is often the case that this definition is slightly abused, in that a mock theta function multiplied by a rational power of q, for example, is also referred to as a mock theta function. With this understanding, all of Ramanujan's original examples of mock theta functions as discussed in Chapter 9 satisfy this definition of mock theta function.

We see from (5.12) that given a particular cusp form f there are in fact infinitely many harmonic Maass forms of weight $2-k$ with shadow f. Indeed, shadows are not impacted by the addition of a weakly holomorphic modular form. In view of this phenomenon, it is natural to consider the following general problem.

PROBLEM. *Are there "natural" choices of mock modular forms which correspond to given shadows?*

The reader is referred to Section 6.2 for an important example of such a canonical correspondence in the case of shadows which are cusp forms associated to modular elliptic curves.

We conclude our general discussion of the ξ-operator with one more useful lemma which exhibits a relationship between the shadows of mock modular forms and their corresponding nonholomorphic parts. Loosely speaking, the nonholomorphic part is a *period integral* of the shadow. In this sense, the nonholomorphic part attached to a mock modular form is much simpler than the original mock modular form, as it arises in a very simple way from a classical cusp form.

LEMMA 5.17. *Let $f \in H_{2-k}(\Gamma_0(N))$ and suppose the mock modular form f^+ has shadow $g(\tau) = \sum_{n=1}^{\infty} c_g(n)q^n \in S_k(\Gamma_0(N))$. Then the nonholomorphic part f^- satisfies*

$$f^-(\tau) = 2^{1-k} i \int_{-\overline{\tau}}^{i\infty} \frac{g^c(w)}{(-i(w+\tau))^{2-k}} dw,$$

where $g^c(\tau) := \overline{g(-\overline{\tau})} = \sum_{n=1}^{\infty} \overline{c_g(n)} q^n$.

REMARK. Note that also f^+ can be written as a (regularized) integral. This follows from (12.9) below. Similarly one can also use regularized integrals to define nonholomorphic (regularized) integrals for $f \in H^!_{2-k}(\Gamma_0(N))$.

REMARK. Classically, *periods* of cusp forms $f \in S_{2k}$ are defined as

$$r_n(f) := \int_0^{\infty} f(it)t^n dt, \tag{5.18}$$

where $0 \leq n \leq 2k-2$. They have been studied extensively by Eichler and Shimura (cf. Section 12.1 below), who used them to study the spaces S_{2k} and critical values of modular L-functions. These objects and their connection to the theory of Maass forms are discussed in detail in Chapter 12 below.

PROOF OF LEMMA 5.17. First, using the definition of the incomplete gamma function, we have for $n \in \mathbb{N}$ that

$$i(2\pi n)^{1-k}\Gamma(k-1, 4\pi n v)q^{-n} = \int_{2iv}^{i\infty} \frac{e(n(w-\tau))}{(-iw)^{2-k}} dw = \int_{-\overline{\tau}}^{i\infty} \frac{e(nw)}{(-i(w+\tau))^{2-k}} dw. \tag{5.19}$$

We apply (5.19) and find that

$$\begin{aligned} 2^{1-k} i \int_{-\overline{\tau}}^{i\infty} \frac{\sum_{n=1}^{\infty} \overline{c_g(n)} e(nw)}{(-i(w+\tau))^{2-k}} dw &= 2^{1-k} i \sum_{n=1}^{\infty} \overline{c_g(n)} \int_{-\overline{\tau}}^{i\infty} \frac{e(nw)}{(-i(w+\tau))^{2-k}} dw \\ &= -2^{1-k} \sum_{n=1}^{\infty} \overline{c_g(n)} (2\pi n)^{1-k} \Gamma(k-1, 4\pi n v) q^{-n}. \end{aligned} \tag{5.20}$$

We know from Theorem 5.10 ii) that

$$c_g(n) = -(4\pi)^{k-1} \overline{c_f^-(-n)} n^{k-1}.$$

Thus, (5.20) becomes

$$\sum_{n=1}^{\infty} c_f^-(-n)\Gamma(k-1, 4\pi n v) q^{-n} = \sum_{n<0} c_f^-(n)\Gamma(k-1, 4\pi |n| v) q^n,$$

which is equal to f^- (as in Definition 4.4). □

CHAPTER 6

Examples of Harmonic Maass Forms

Here we discuss some of the first examples of harmonic Maass forms. We describe those forms which arise naturally by generalizing the construction of ordinary modular forms by the method of Eisenstein series and Poincaré series. These include the nonholomorphic Eisenstein series $E_2^*(\tau)$, Zagier's weight 3/2 nonholomorphic Eisenstein series, and Maass-Poincaré series. As we shall see, the harmonic Maass forms which arise from Eisenstein series stem from cases where there are problems with convergence. Hecke's trick, which circumvents these problems by making use of analytic continuation, naturally gives rise to these nonholomorphic modular forms. In an analogous way, we obtain harmonic Maass functions from Eisenstein's nonholomorphic Λ-invariant modification of the Weierstrass ζ-function. We conclude this chapter with recent work of Wagner which offers the first examples of p-adic harmonic Maass forms.

6.1. $E_2^*(\tau)$ and Zagier's weight 3/2 Eisenstein series

Here we discuss two fundamental harmonic Maass forms, the Eisenstein series $E_2^*(\tau)$, which is a modified version of the quasimodular form (cf. (1.4))

$$E_2(\tau) = 1 - 24\sum_{n=1}^{\infty}\sigma_1(n)q^n,$$

and Zagier's weight 3/2 nonholomorphic Eisenstein series.

6.1.1. The Eisenstein series $E_2^*(\tau)$.

The Eisenstein series E_{2k} for integers $k \geq 2$ are some of the first examples of holomorphic modular forms. The function E_2 is a quasimodular form, as defined in Definition 2.35. Although E_2 is not modular, it is not difficult to modify it to obtain a nonholomorphic weight 2 modular form (cf. Chapter 1).

DEFINITION 6.1. For $\tau \in \mathbb{H}$, the **nonholomorphic weight 2 Eisenstein series** $E_2^*(\tau)$ is defined by (cf. (1.3))

$$E_2^*(\tau) = E_2(\tau) - \frac{3}{\pi v},$$

where $\tau = u + iv$ with $u, v \in \mathbb{R}$.

LEMMA 6.2. *We have that* $E_2^* \in H_2^!$.

REMARK. The nonholomorphic part of E_2^* is $-3/\pi v$. By Theorem 5.10 iii), it follows that $\xi_2(E_2^*)$ is a weight 0 weakly holomorphic modular form In particular, by Theorem 5.10 iv), the shadow is the constant modular function $\xi_2(E_2^*) = 3/\pi$.

REMARK. An alternative proof realizes E_2 in terms of the logarithmic derivative of η and then uses the modular transformations of η, namely

$$\eta(\tau+1) = e^{\frac{\pi i}{12}}\eta(\tau),$$
$$\eta\left(-\frac{1}{\tau}\right) = \sqrt{-i\tau}\,\eta(\tau). \tag{6.1}$$

SKETCH OF PROOF OF LEMMA 6.2. The following is a sketch of Zagier's proof (cf. pages 19-20 of [**519**]). It suffices to prove that $E_2^*(\tau+1) = E_2^*(\tau)$ and

$$E_2^*\left(-\frac{1}{\tau}\right) = \tau^2 E_2^*(\tau).$$

The first claim is clear. The proof of the second claim is complicated by the fact that the usual definition of the Eisenstein series is not absolutely convergent for weight 2. To circumvent this problem, one makes use of Hecke's trick. To be more precise, consider

$$G_{2,s}(\tau) := \frac{1}{2}\sum_{(m,n)\in\mathbb{Z}^2\setminus\{(0,0)\}} \frac{1}{(m\tau+n)^2|m\tau+n|^{2s}}, \tag{6.2}$$

where $\tau \in \mathbb{H}$ and $\mathrm{Re}(s) > 0$. Since this series is absolutely convergent, it follows that

$$G_{2,s}\left(\frac{a\tau+b}{c\tau+d}\right) = (c\tau+d)^2|c\tau+d|^{2s}G_{2,s}(\tau)$$

for all $\left(\begin{smallmatrix} a & b \\ c & d\end{smallmatrix}\right) \in \mathrm{SL}_2(\mathbb{Z})$. It suffices to prove that

$$\lim_{s\to 0} G_{2,s}(\tau) = \zeta(2)E_2^*(\tau) = \frac{\pi^2}{6}E_2^*(\tau).$$

To obtain this claim, we let

$$I_s(\tau) := \int_{-\infty}^{\infty} \frac{dt}{(\tau+t)^2|\tau+t|^{2s}}.$$

Then we have that

$$G_{2,s}(\tau) - \sum_{m=1}^{\infty} I_s(m\tau)$$
$$= \sum_{n=1}^{\infty}\frac{1}{n^{2+2s}} + \sum_{m=1}^{\infty}\sum_{n\in\mathbb{Z}}\left(\frac{1}{(m\tau+n)^2|m\tau+n|^{2s}} - \int_n^{n+1}\frac{dt}{(m\tau+t)^2|m\tau+t|^{2s}}\right).$$

One can show that both sums are absolutely and locally uniformly convergent, and so the limit exists as $s \to 0$. Moreover, we can simply replace $s = 0$ term by term in the right-hand side to obtain this limit, which a tedious calculation shows is equal to $\frac{\pi^2}{6}E_2$. Therefore, to complete the proof it suffices to prove that

$$\lim_{s\to 0}\sum_{m=1}^{\infty} I_s(m\tau) = -\frac{\pi}{2v}.$$

A calculation shows that $I_s(\tau) = \frac{I(s)}{v^{1+2s}}$, where

$$I(s) := \int_{-\infty}^{\infty} (t+i)^{-2}\left(t^2+1\right)^{-s} dt,$$

which in turn implies that

$$\sum_{m=1}^{\infty} I_s(m\tau) = \frac{I(s)\zeta(1+2s)}{v^{1+2s}}.$$

One then checks, by a lengthy calculation, that

$$\lim_{s\to 0} \frac{I(s)\zeta(1+2s)}{v^{1+2s}} = -\frac{\pi}{2v},$$

which completes the proof. □

6.1.2. Zagier's weight 3/2 nonholomorphic Eisenstein series. Zagier's weight 3/2 nonholomorphic Eisenstein series is perhaps the first example of a half-integral weight harmonic Maass form to be properly understood. This harmonic Maass form is intimately related to the weight 3/2 modular form

$$\theta(\tau)^3 = 1 + 6q + 12q^2 + 8q^3 + 6q^4 + 24q^5 + 24q^6 + 12q^8 + 30q^9 + \cdots,$$

which we related to Hurwitz class numbers in Example 2.5.

In his 1975 paper **[508]**,[1] Zagier studied the problem of recognizing the generating function for Hurwitz class numbers in terms of weight 3/2 modular forms. He proved the following theorem, which shows that this generating function is the holomorphic part of a weight 3/2 harmonic Maass form.[2] Note that we have slightly rephrased the statement of Zagier's original results in Theorem 6.3 in order to incorporate the newer terminology of harmonic Maass forms.

THEOREM 6.3 (Zagier). *Let $H(n)$ be the usual Hurwitz class number counting the (weighted) number of classes of positive definite binary quadratic forms of discriminant $-n$. Then the function*

$$\mathcal{H}(\tau) := -\frac{1}{12} + \sum_{n=1}^{\infty} H(n)q^n + \frac{1}{4\sqrt{\pi}} \sum_{n=1}^{\infty} n\Gamma\left(-\frac{1}{2}, 4\pi n^2 v\right) q^{-n^2} + \frac{1}{8\pi\sqrt{v}},$$

where $\tau = u + iv$, is a weight 3/2 harmonic Maass form of manageable growth on $\Gamma_0(4)$. Moreover, we have that $\xi_{\frac{3}{2}}(\mathcal{H}) = -\frac{1}{16\pi}\theta$.

REMARK. We refer to

$$\mathcal{H}^+(\tau) = -\frac{1}{12} + \sum_{n=1}^{\infty} H(n)q^n, \tag{6.3}$$

the holomorphic part of $\mathcal{H}$ and the generating function for Hurwitz class numbers, as the *Eisenstein-Hurwitz mock modular form.* Its shadow is $-\frac{1}{16\pi}\theta$.

REMARK. Although $\mathcal{H}$ violates the required existence of a principle part in the definition of a harmonic Maass form, and so properly lies in the space $H^!_{\frac{3}{2}}(\Gamma_0(4))$, it very nearly is an ordinary harmonic Maass form. That is, the presence of the term $1/(8\pi\sqrt{v})$ is the only obstacle to satisfying this condition. This is a reflection of the fact that the shadow of $\mathcal{H}$ is merely non-cuspidal, rather than a weakly holomorphic modular form with a pole at $i\infty$.

[1]This problem is also studied in detail in the famous paper by Hirzebruch and Zagier **[271]** on the intersection numbers of divisors on Hilbert modular surfaces.

[2]This form is often referred to as Zagier's weight 3/2 *nonholomorphic Eisenstein series.*

SKETCH OF PROOF OF THEOREM 6.3. The claim that $\xi_{\frac{3}{2}}(\mathcal{H}) = -\frac{1}{16\pi}\theta$ is a simple calculation using Theorem 5.10. Zagier's construction of $\mathcal{H}$ depends on the theory of half-integral weight Eisenstein series. However, weight 3/2 Eisenstein series do not converge. Therefore, one argues again by using Hecke's trick. To this end, let

$$E_{\frac{3}{2},s}(\tau) := \sum_{\substack{m\geq 1, n\in\mathbb{Z}\\ \gcd(m,2n)=1}} \frac{\left(\frac{n}{m}\right)\left(\frac{-1}{m}\right)^{\frac{1}{2}}}{(m\tau+n)^{\frac{3}{2}}|m\tau+n|^{2s}}.$$

This function converges for $\mathrm{Re}(s) > 1/4$. Moreover, for every $\left(\begin{smallmatrix} a & b\\ c & d\end{smallmatrix}\right) \in \Gamma_0(4)$, the theory of half-integral weight Eisenstein series implies that

$$E_{\frac{3}{2},s}\left(\frac{a\tau+b}{c\tau+d}\right) = \left(\frac{c}{d}\right)\varepsilon_d(c\tau+d)^{\frac{3}{2}}|c\tau+d|^{2s}E_{\frac{3}{2},s}(\tau).$$

If we let

$$F_{\frac{3}{2},s}(\tau) := \tau^{-\frac{3}{2}}|\tau|^{-2s}E_{\frac{3}{2},s}\left(-\frac{1}{4\tau}\right),$$

then we define

$$\mathcal{H}_s(\tau) := -\frac{1}{96}\left((1-i)E_{\frac{3}{2},s}(\tau) - iF_{\frac{3}{2},s}(\tau)\right).$$

Note that this corresponds to the projection into Kohnen's plus space (the plus space was discussed in Subsection 2.5.2, and the projection operator was defined in (6.12)). One easily checks that $\mathcal{H}_s$ transforms like a weight 3/2 modular form on $\Gamma_0(4)$ and has an analytic continuation to the s-plane.

One then computes the Fourier expansion

$$\mathcal{H}_s(\tau) = \sum_{n\in\mathbb{Z}} g_n(s,v)q^n.$$

To complete the proof, it suffices to prove that

$$g_n(0,v) = \begin{cases} H(n) & \text{if } n>0,\\ -\frac{1}{12} + \frac{1}{8\pi}v^{-\frac{1}{2}} & \text{if } n=0,\\ \frac{f}{4\sqrt{\pi}}\Gamma\left(-\frac{1}{2}, 4\pi f^2 v\right) & \text{if } n=-f^2,\\ 0 & \text{otherwise.}\end{cases}$$

To this end, one applies the Poisson summation formula to obtain

$$\sum_{h\in\mathbb{Z}}(\tau+h)^{-\frac{3}{2}}|\tau+h|^{-2s} = \sum_{n\in\mathbb{Z}}\alpha_n(s,v)q^n,$$

where

$$\begin{aligned}\alpha_n(s,v) &:= \int_{iv-\infty}^{iv+\infty}\tau^{-\frac{3}{2}}|\tau|^{-2s}e^{-2\pi i n\tau}d\tau\\ &= v^{-\frac{1}{2}-2s}e^{2\pi nv}\int_{-\infty}^{\infty}(w+i)^{-\frac{3}{2}}\left(w^2+1\right)^{-s}e^{-2\pi i n v w}dw.\end{aligned}$$

The last expression was obtained by letting $\tau = (w+i)v$. One inserts this expression into the definition of $\mathcal{H}_s$. Tedious calculations, combined with some arithmetic involving Gauss sums, yield formulas for $g_n(s,v)$ which can be described in terms of ratios of Dirichlet series in s. For discriminants $-n$, these formulas involve Dirichlet L-functions, and as $s\to 0$ leads to Hurwitz class numbers thanks to

Dirichlet's class number formula. For $n = -f^2$ these formulas involve a simple pole with residue $1/\zeta(2)$. In all other cases the Fourier coefficients visibly vanish.

□

REMARK. Theorem 6.3 plays a central role in recent work of Beckwith **[50]** on the indivisibility of class numbers of imaginary quadratic fields. By combining a method of Kohnen and one of the authors **[327]** with a theorem of Wiles **[497]**, Beckwith proved for odd primes ℓ that

$$\#\{-X < D < 0 \ : \ \ell \nmid h(D) \ \text{ and } \ D \in \Sigma\} \gg_{\Sigma} \frac{\sqrt{X}}{\log(X)},$$

where Σ denotes the set of fundamental discriminants which satisfy a prescribed finite set of local splitting and ramification conditions.

6.2. Weierstrass mock modular forms

Here we revisit the theory of elliptic functions, and we use the modularity of elliptic curves to produce harmonic Maass functions from Eisenstein's modified Weierstrass ζ-function (cf. Section 1.3). Eisenstein's modifications are nonholomorphic doubly periodic functions. Namely, we recall Theorem 1.16.

THEOREM. *Assume the notation in Section* 1.3. *Then* ζ_Λ^* *is a nonholomorphic function for which*

$$\zeta_\Lambda^*(z+\omega) = \zeta_\Lambda^*(z)$$

for all $w \in \Lambda$.

Here we give a proof of this result from **[440]**, which uses the theory of Jacobi forms and the harmonic Maass form E_2^*. For an alternate proof, see Section 1.3.

PROOF. It is enough to rescale the lattice under homothety to write the lattice as $\Lambda_\tau = \mathbb{Z} + \mathbb{Z}\tau$ for $\tau \in \mathbb{H}$. We begin by recalling another classical function of Weierstrass associated to Λ_τ, defined for $z \in \mathbb{C}$ by

$$\sigma_{\Lambda_\tau}(z) := \sigma(z) = z \prod_{w \in \Lambda_\tau \setminus \{0\}} \left(1 - \frac{z}{w}\right) \exp\left(\frac{z}{w} + \frac{z^2}{2w^2}\right).$$

It is a classical fact, which follows directly from the product definition of σ, that the logarithmic derivative of σ is given by

$$\frac{\sigma'_{\Lambda_\tau}(z)}{\sigma_{\Lambda_\tau}(z)} = \zeta_{\Lambda_\tau}(z),$$

where the $'$ denotes differentiation in z. We first rewrite this last equality in terms of Jacobi forms, so that we may apply the standard theory of differential operators for Jacobi forms to obtain our completion. For this, we recall the following classical identity (cf. Theorem 3.9 of **[421]**):

$$\vartheta(z;\tau) = -2\pi\eta(\tau)^3 \exp\left(-\frac{\eta_1 z^2}{2}\right) \sigma_{\Lambda_\tau}(z),$$

where ϑ is the Jacobi theta function (defined in Definition 2.1) and η_1 is the quasi-period defined by

$$\eta_1 = \eta_1(\tau) := \zeta_{\Lambda_\tau}(z+1) - \zeta_{\Lambda_\tau}(z)$$

(note that this is well defined independently of the choice of z as $\zeta'_{\Lambda_\tau}(z) = -\wp_{\Lambda_\tau}(z)$ is elliptic). The interested reader is also invited to prove the preceding equality;

here we simply offer the hint that one should compare the transformations of both sides under shifting z by lattice points. Hence, we find that

$$\frac{\vartheta'(z)}{\vartheta(z)} = \zeta(z) - \eta_1 z. \tag{6.4}$$

Recall by the discussion in Subsection 2.3.1 that ϑ is a Jacobi form of index 1/2. The logarithmic derivative of a Jacobi form is no longer a Jacobi form, but it is nearly one. Analogous to the raising operator for modular forms (cf. Definition 5.1), there is a canonical raising operator on Jacobi forms, which maps the Jacobi theta function to the function

$$Y_+^{\frac{1}{2},\frac{1}{2}}(\vartheta(z;\tau)) = \frac{\partial}{\partial z}\vartheta(z;\tau) + 2\pi i\frac{y}{v}\vartheta(z;\tau),$$

which is then a Jacobi form of weight 3/2 and index 1/2 (cf. **[56]**). Here, for a Jacobi form Φ of weight k and index m, the operator $Y_+^{k,m}$ is defined by

$$Y_+^{k,m}(\Phi(z;\tau)) := \frac{\partial}{\partial z}\Phi(z;\tau) + 4\pi i m\frac{y}{v}\Phi(z;\tau)$$

and yields a Jacobi form of weight $k+1$ and index m. Hence, replacing the logarithmic derivative with a "logarithmic-raising" operator, we find that

$$\frac{Y_+^{\frac{1}{2},\frac{1}{2}}(\vartheta(z;\tau))}{\vartheta(z;\tau)} = \zeta(z;\tau) - \eta_1 z + \frac{2\pi i y}{v}$$

is a real-analytic Jacobi form of index zero, i.e., it transforms as an elliptic function in z. We claim that this last expression is equal to Weierstrass' completion given in Section 1.3. To make this connection, we use the following classical relation (cf. Part 4 of **[342]**):

$$E_2 = \frac{3\eta_1}{\pi^2}. \tag{6.5}$$

Expressed differently, we have $\eta_1 = G_2$. Denoting by G_2^* the nonholomorphic completion

$$2\zeta(2)\left(E_2(\tau) - \frac{3}{\pi v}\right) = G_2(\tau) - \frac{\pi}{v},$$

we have shown that

$$\zeta_{\Lambda_\tau}(z) - zG_2^*(\tau) - \pi\frac{\overline{z}}{v}$$

is a doubly periodic function with respect to Λ_τ. The proof is completed by noting, in the notation of Section 1.3, that $S_{\Lambda_\tau} = G_2^*(\tau)$ (cf. the proof of Lemma 6.2) and that $\mathrm{vol}(\Lambda_\tau) = v$. □

We now use this modified Weierstrass zeta-function to produce harmonic Maass functions. Suppose that $E/\mathbb{Q}$ is an elliptic curve. We first choose a lattice $\Lambda = \Lambda_E$ for which $E \cong \mathbb{C}/\Lambda$. The usual complex analytic parameterization (cf. Section 1.2) of E is given by $z \mapsto (\wp_\Lambda(z), \wp'_\Lambda(z))$, where $\wp_\Lambda$ is the usual Weierstrass $\wp$-function for Λ_E. Here E is represented by the Weierstrass equation

$$E\colon\ Y^2 = 4X^3 - 60G_4(\Lambda)X - 140G_6(\Lambda).$$

Since $E/\mathbb{Q}$ is an elliptic curve, there is another parameterization, the *modular parameterization*. By the celebrated Modularity Theorem (cf. Theorem 19.6), we have the modular parameterization

$$\phi_E \colon X_0(N_E) \to \mathbb{C}/\Lambda_E \cong E, \tag{6.6}$$

where N_E is the conductor of E and $X_0(N_E)$ is the usual modular curve for $\Gamma_0(N_E)$. This deep theorem implies that there is a weight 2 newform

$$F_E(\tau) = \sum_{n=1}^{\infty} a_E(n)q^n \in S_2(\Gamma_0(N_E))$$

with the property that its Hecke L-function $L(F_E, s)$ is the Hasse-Weil L-function for E.

The canonical harmonic Maass function for E arises from this cusp form and Eisenstein's modified Weierstrass ζ-function

$$\zeta_\Lambda^*(z) = \zeta_\Lambda(z) - S_\Lambda z - \frac{\pi}{\mathrm{vol}(\Lambda)}\overline{z}$$

defined in Section 1.3. This function is defined using the *Eichler integral* for F_E given by

$$\mathcal{E}_E(\tau) = -2\pi i \int_\tau^{i\infty} F_E(w)dw = \sum_{n=1}^{\infty} \frac{a_E(n)}{n} q^n. \tag{6.7}$$

Note that by the notation in Chapter 1, this could also be called $\mathcal{E}_{F_E}$, however we have chosen to label the Eichler integral in this way to emphasize the role the elliptic curve plays in the overall construction. Moreover note that there we choose τ_0 instead of $i\infty$ as a bound of integration. Using Eisenstein's modified Weierstrass ζ-function, we then define the nonholomorphic function Z_E on $\mathbb{H}$ by

$$Z_E(\tau) := \zeta_\Lambda^*(\mathcal{E}_E(\tau)) = \zeta_\Lambda(\mathcal{E}_E(\tau)) - S_\Lambda \mathcal{E}_E(\tau) - \frac{\pi}{\mathrm{vol}(\Lambda)}\overline{\mathcal{E}_E(\tau)}. \tag{6.8}$$

As usual, we let Z_E^+ denote the holomorphic part of this function, namely

$$Z_E^+(\tau) := \zeta_\Lambda(\mathcal{E}_E(\tau)) - S_\Lambda \mathcal{E}_E(\tau).$$

We have the following observation of Guerzhoy **[244]**. The present formulation of Theorem 6.4 appears in **[14]**.

THEOREM 6.4. *Assume the notation and hypotheses above. Then the following are true:*

i) *The poles of Z_E^+ are precisely those points τ for which $\mathcal{E}_E(\tau) \in \Lambda$.*
ii) *If Z_E^+ has poles in $\mathbb{H}$, then there is a canonical modular function M_E on $\Gamma_0(N_E)$ which has algebraic coefficients for which $Z_E^+ - M_E$ is holomorphic on $\mathbb{H}$.*
iii) *We have that $Z_E - M_E$ is a weight 0 harmonic Maass function on $\Gamma_0(N_E)$.*
iv) *The shadow of $(Z_E - M_E)^+$ is a non-zero multiple of F_E.*

In Theorem 6.4, we obtain

$$Z_E^+(\tau) - M_E(\tau) = \zeta_\Lambda^*(\mathcal{E}_E(\tau)) - S_\Lambda \mathcal{E}_E(\tau) - M_E(\tau),$$

the *Weierstrass mock modular function* for E. It is not difficult to compute this mock modular form. Using the two Eisenstein numbers $G_4(\Lambda)$ and $G_6(\Lambda)$, one can then easily determine the remaining Eisenstein numbers using the recursion

$$G_{2n}(\Lambda) := \sum_{j=2}^{n-2} \frac{3(2j-1)(2n-2j-1)}{(2n+1)(2n-1)(n-3)} G_{2j}(\Lambda) G_{2n-2j}(\Lambda).$$

To obtain the Weierstrass mock modular function, one then inserts these numbers, the Eichler integral $\mathcal{E}_E$, and S_Λ, into (6.8).

Moreover, if E has complex multiplication, then S_Λ is rational (cf. page 119 of [**378**]). Therefore, it follows that the Weierstrass mock modular function for an elliptic curve with complex multiplication has $\mathbb{Q}$-rational coefficients. For E without complex multiplication, the coefficients of the Weierstrass mock modular function lie in $\overline{\mathbb{Q}}(S_\Lambda)$, where $\overline{\mathbb{Q}}$ is the algebraic closure of $\mathbb{Q}$. Moreover, the coefficients of $Z_E^+(\tau)$ lie in $\mathbb{Q}(S_\Lambda)$, a field with transcendence degree at most one.

EXAMPLE 6.5. For example, consider the elliptic curve

$$E\colon Y^2 + Y = X^3 - 38X + 90$$

of conductor 361 with complex multiplication by $K = \mathbb{Q}(\sqrt{-19})$. We find

$$F_E(\tau) = q - 2q^4 - q^5 + 3q^7 - 3q^9 - 5q^{11} + 4q^{16} - 7q^{17} + \cdots$$

and

$$\zeta_\Lambda(\mathcal{E}_E(\tau)) = q^{-1} + \frac{1}{2}q^2 - \frac{7}{3}q^3 + \frac{12}{5}q^5 + 4q^6 - \frac{6}{7}q^7 - \frac{27}{4}q^8 - \frac{13}{3}q^9 + \frac{17}{2}q^{10} + \cdots.$$

Moreover, $S(\Lambda) = -2$, which in turn gives

$$Z_E^+(\tau) = q^{-1} + 2q + \frac{1}{2}q^2 - \frac{7}{3}q^3 - q^4 + 2q^5 + 4q^6 - \frac{27}{4}q^8 - 5q^9 + \frac{17}{2}q^{10} + 14q^{11} - \cdots.$$

SKETCH OF PROOF OF THEOREM 6.4. Parts i) and iv) are straightforward. The proof of iii) relies on Theorem 1.16, the fact that ζ_Λ^* is Λ-invariant. The Eichler integral $\mathcal{E}_E$ is not modular, however its obstruction to modularity is easily characterized. The map $\Psi_E\colon \Gamma_0(N_E) \to \mathbb{C}$ given by

$$\Psi_E(\gamma) := \mathcal{E}_E(\tau) - \mathcal{E}_E(\gamma\tau) \tag{6.9}$$

is a homomorphism of groups. Its image in $\mathbb{C}$ is the lattice Λ (cf. Theorem 1.17 and [**190, 457**]). Therefore, replacing z by $\mathcal{E}_E(\tau)$ in $\zeta_\Lambda^*(z)$ produces a nonholomorphic modular function. The proof of ii) (i.e., the existence of $M_E(\tau)$) follows by applying standard algebraic facts about modular functions to this nonholomorphic function. It is straightforward to show that $Z_E - M_E$ satisfies a suitable growth condition and is harmonic. □

REMARK. We use the harmonic Maass functions of Theorem 6.4 in Chapter 19 to study the arithmetic of quadratic twists of modular elliptic curves E.

REMARK. In view of Theorem 6.4, it is natural to ask about the distribution of those points $\tau \in \mathbb{H}$ for which $\mathcal{E}_E(\tau) \in \Lambda$. What is the general behavior of such points τ? In her Master's thesis [**317**], Kodgis noted that there are many low conductor examples where these points lie at CM points. Griffin has recently explained to the authors [**234**] that in unpublished work he has shown there to be non-CM examples as well. However, understanding the full nature of these examples, and in particular the distribution of these points among CM points and non-CM points, would be very interesting. As an indication of what explanations one might seek out, we note that Peluse [**415**] has observed that some of the CM points found by Kodgis are fixed points of Atkin-Lehner involutions (or other, related points built out of the properties of the associated modular forms under Atkin-Lehner involutions). One can thus ask, do such considerations "explain" most of the CM point examples, or is there a deeper phenomenon?

6.3. Maass-Poincaré series

In the previous two sections, we constructed harmonic Maass forms by making use of Eisenstein series and the Weierstrass ζ-function. These examples required delicate arguments that dealt with problems related to convergence. In this section, we consider more general Eisenstein series (of integral or half-integral weight) as well as Maass-Poincaré series which constitute a basis of harmonic Maass forms in the case of negative weight. For simplicity we restrict to weights $k < 0$ or $k > 2$ to overcome convergence problems. For weights $0 \leq k \leq 2$ one can analytically continue via Fourier or spectral expansions (the standard technique uses the results of Goldfeld-Sarnak [**225**] or modifications thereof).

A fruitful method for constructing examples of modular forms, as for functions invariant under any group action, is the method of averaging. This idea, although straightforward, gets complicated if delicate convergence issues are taken into consideration. To see the basic idea, let φ be any translation-invariant function, which we call the *seed* of the Poincaré series in question. Then, in the case of absolute convergence, the following function satisfies weight k modularity

$$\mathbb{P}_{k,N}(\varphi;\tau) := \sum_{\gamma\in\Gamma_\infty\backslash\Gamma_0(N)} \varphi|_k\gamma(\tau), \tag{6.10}$$

where $\Gamma_\infty = \{\pm\left(\begin{smallmatrix}1 & n\\ 0 & 1\end{smallmatrix}\right) : n\in\mathbb{Z}\}$ is the group of translations. Convergence is in particular satisfied by functions φ satisfying $\varphi(\tau) = O(v^{2-k+\varepsilon})$ as $v\to 0$. Throughout we assume that $4|N$ if $k\in 1/2+\mathbb{Z}$ and we omit the level N in the notation if it is trivial, i.e., $N=1$ for $k\in\mathbb{Z}$ or $N=4$ for $k\in 1/2+\mathbb{Z}$.

A natural choice for φ is a function which is related to the typical Fourier series uniformizer in the space of automorphic functions in which one is interested. For example, in the case of weakly holomorphic modular forms, for $m\in\mathbb{Z}$, we can choose the function

$$\varphi(\tau) = \varphi_m(\tau) := q^m.$$

DEFINITION 6.6. Define for $k\in\frac{1}{2}\mathbb{N}$ with $k>2$ and $m\in\mathbb{Z}$ the **Poincaré series of exponential type** $\mathcal{P}_{k,m,N}(\tau)$ by

$$\mathcal{P}_{k,m,N}(\tau) := \mathbb{P}_{k,N}(\varphi_m;\tau) = \sum_{\gamma\in\Gamma_\infty\backslash\Gamma_0(N)} \varphi_m|_k\gamma(\tau).$$

REMARK. One can define more general Poincaré series with multipliers, and associated to each cusp (see for example [**282**]).

In the case of the full modular group, we also set

$$\mathcal{P}_{k,m} := \mathcal{P}_{k,m,1}.$$

For $m=0$, we obtain Eisenstein series. To be more precise, with E_{2k} defined in (1.1),

$$\mathcal{P}_{2k,0} = E_{2k}.$$

Corollary 1.7 gives their Fourier expansions.

We further set

$$P_{k,m,N} := \mathcal{P}_{k,m,N}|\,\mathrm{pr}, \tag{6.11}$$

where $|\,\mathrm{pr}$ is the identity for $k \in \mathbb{Z}$ and Kohnen's projection operator for $k \notin \mathbb{Z}$ (see (6.12) and the discussion below). This operator projects a modular form of half-integral weight to Kohnen's plus space. Although this space was already defined in Subsection 2.5.2, for the readers' convenience we recall that the weight $k+1/2$ plus space is the space of modular forms f of weight $k+1/2$ whose Fourier coefficients $c_f(n)$ vanish unless $(-1)^k n \equiv 0, 1 \pmod 4$.

Here we prove an extension of Kohnen's formula for the projection operator into the plus space which holds for harmonic Maass forms. We follow the statement and proof of [**324**, Proposition 3], modified to allow for harmonic Maass forms.

Let $\mathfrak{G}$ denote the set of pairs (γ, ε) where $\gamma = \left(\begin{smallmatrix} a & b \\ c & d \end{smallmatrix}\right) \in \mathrm{GL}_2^+(\mathbb{R})$ and $\varepsilon\colon \mathbb{H} \to \mathbb{C}$ is holomorphic and satisfies $|\varepsilon(\tau)| = \det(\gamma)^{-\frac{1}{4}}\sqrt{|c\tau+d|}$. Then $\mathfrak{G}$ is group with the group operation

$$(\gamma_1, \varepsilon_1) \cdot (\gamma_2, \varepsilon_2) = (\gamma_1\gamma_2, \varepsilon)$$

where $\varepsilon(\tau) := \varepsilon_1(\gamma\tau)\varepsilon_2(\tau)$. Note that $\mathfrak{G}$ contains the metaplectic group $\mathrm{Mp}_2(\mathbb{R})$ defined in Subsection 18.3.1.

If $\gamma = \left(\begin{smallmatrix} a & b \\ c & d \end{smallmatrix}\right) \in \Gamma_0(4)$, then let $\gamma^* := \left(\gamma, \left(\frac{c}{d}\right)\varepsilon_d^{-1}\sqrt{c\tau+d}\right)$. If $\kappa \in \frac{1}{2}\mathbb{Z}$, then an element $(\gamma, \varepsilon) \in \mathfrak{G}$ acts on functions $f\colon \mathbb{H} \to \mathbb{C}$ by the usual weight κ slash operator

$$f|_\kappa(\gamma, \varepsilon)(\tau) := \varepsilon(\tau)^{-2\kappa} f(\gamma\tau).$$

Note that Kohnen moves the dependency on the weight κ onto the group $\mathfrak{G}$ itself, rather than on the slash operator.

Let N be a positive, odd, square-free integer, and $\alpha := \left(\frac{-4}{N}\right)$. We need the following matrices:

$$A_\nu := \begin{pmatrix} 1 & 0 \\ 4N\nu & 1 \end{pmatrix}, \quad B := \left(\begin{pmatrix} 4 & 1 \\ 0 & 4 \end{pmatrix}, i\right), \quad C_\mu := \left(\begin{pmatrix} 1 & 0 \\ \mu N & 1 \end{pmatrix}, \sqrt{\mu N\tau + 1}\right).$$

Suppose $k \in \mathbb{Z}$ and $F \in H_{k+\frac{1}{2}}(\Gamma_0(4N))$ with expansions at the cusps $i\infty$, $\frac{1}{2N}$, and $\frac{-\alpha}{N}$, respectively given by

$$F(\tau) =: \sum_{n \in \mathbb{Z}} c_\infty(n, v) q^n,$$

$$F_{-\alpha}(\tau) := F|_{k+\frac{1}{2}} C_{-\alpha}(\tau) =: \sum_{n \in \mathbb{Z}} c_{-\alpha}(n, v) q^{\frac{n}{4}},$$

$$F_2(\tau) := F|_{k+\frac{1}{2}} C_2(\tau) =: \sum_{\substack{n \in \mathbb{Z} \\ (-1)^k n \equiv 1 \pmod 4}} c_2(n, v) q^{\frac{n}{4}}.$$

Kohnen's projection operator into the plus space is given by

$$F|\,\mathrm{pr} := \frac{1}{3}F + \frac{(-1)^{\left\lfloor \frac{k+1}{2} \right\rfloor}}{3\sqrt{2}} \sum_{\nu \pmod 4} F|_{k+\frac{1}{2}}(BA_\nu^*). \tag{6.12}$$

Then we have the following proposition.

PROPOSITION 6.7. *Suppose $k \in \mathbb{Z}$ and $F \in H_{k+\frac{1}{2}}(\Gamma_0(4N))$ with N as above. The q-expansion of $F|\operatorname{pr}$ is given by*

$$F|\operatorname{pr} = \frac{2}{3} \sum_{\substack{n\in\mathbb{Z} \\ n\equiv 0 \pmod 4}} \left(c_\infty(n,v) + \left(1-(-1)^k i\right) 2^{2k-1} i^{\frac{n}{4}} c_{-\alpha}\left(\frac{n}{4}, 16v\right)\right) q^n$$
$$+ \frac{2}{3} \sum_{\substack{n\in\mathbb{Z} \\ (-1)^k n\equiv 1 \pmod 4}} \left(c_\infty(n,v) + 2^{k-1}\left(\frac{(-1)^k n}{2}\right) c_2(n,4v)\right) q^n.$$

PROOF. This theorem is identical to [**324**, Proposition 3] in statement and method of proof, except that the original only considers holomorphic cusp forms.

Multiplying, we find that

$$BA_2^* B = \left(\begin{pmatrix} 16 & 0 \\ 0 & 16 \end{pmatrix}, 1\right) \begin{pmatrix} 1+2N & \frac{1+N}{2} \\ 8N & 1+2N \end{pmatrix}^* .$$

Since F is modular on $\Gamma_0(4N)$ and A_0^* is the identity, we have that

$$F|_{k+\frac{1}{2}}(BA_0^* + BA_2^*) = F|_{k+\frac{1}{2}}\left(B + B^{-1}\right).$$

Since B and B^{-1} are upper triangular, their action can be computed on powers of q directly. In particular, if $a\colon \mathbb{R} \to \mathbb{C}$ is any function, then

$$\left(a(v)q^n\right)|_{k+\frac{1}{2}}\left(B + B^{-1}\right) = a(v)q^n\left(i^{n-k}e^{\frac{-\pi i}{4}} + i^{k-n}e^{\frac{\pi i}{4}}\right) = a(v)q^n\sqrt{2}(-1)^{\left\lfloor \frac{k-n+1}{2}\right\rfloor}.$$

Thus

$$\frac{1}{3}F + \frac{(-1)^{\left\lfloor\frac{k+1}{2}\right\rfloor}}{3\sqrt{2}} F|(BA_0^* + BA_2^*) = \frac{2}{3} \sum_{\substack{n\in\mathbb{Z} \\ (-1)^k n\equiv 0,1 \pmod 4}} c_\infty(n,v)q^n.$$

We have two more components from the definition of the projection operator. By multiplying we find that

$$BA_{-\alpha}^* = \begin{pmatrix} \left(\frac{1+\alpha N}{2}\right)^2 & \frac{\alpha N+3}{4} \\ N^2+3\alpha N & \alpha N+4 \end{pmatrix}^* C_{-\alpha}\left(\begin{pmatrix} 16 & 1 \\ 0 & 1 \end{pmatrix}, 2^{-1}e^{\frac{2\pi i}{8}}\right).$$

Thus

$$\begin{aligned} \frac{(-1)^{\left\lfloor\frac{k+1}{2}\right\rfloor}}{3\sqrt{2}} F|_{k+\frac{1}{2}} BA_{-\alpha}^* &= \frac{(-1)^{\left\lfloor\frac{k+1}{2}\right\rfloor}}{3\sqrt{2}} F_{-\alpha}|_{k+\frac{1}{2}}\left(\begin{pmatrix} 16 & 1 \\ 0 & 1 \end{pmatrix}, 2^{-1}e^{\frac{2\pi i}{8}}\right) \\ &= \frac{2}{3}\left(1-(-1)^k i\right) 2^{2k-1} \sum_{n\in\mathbb{Z}} i^n c_{-\alpha}(n,16v)q^{4n}. \end{aligned}$$

Lastly, we have that

$$BA_\alpha^* = \begin{pmatrix} \frac{(1+\alpha N)(1+N)}{2} - N & \frac{-\alpha N+1}{4} \\ 2\alpha N(N+1) - 4N & -\alpha N+2 \end{pmatrix}^* C_2\left(\begin{pmatrix} 8 & 1 \\ 0 & 2 \end{pmatrix}, 2^{-\frac{1}{2}}e^{\frac{2\pi i}{8}}\right).$$

Thus

$$\begin{aligned}\frac{(-1)^{\lfloor\frac{k+1}{2}\rfloor}}{3\sqrt{2}}F|_{k+\frac{1}{2}}BA_\alpha^* &= \frac{(-1)^{\lfloor\frac{k+1}{2}\rfloor}}{3\sqrt{2}}F_2|_{k+\frac{1}{2}}\left(\begin{pmatrix}8 & 1\\ 0 & 2\end{pmatrix}, 2^{-\frac{1}{2}}e^{\frac{2\pi i}{8}}\right)\\ &= \frac{(-1)^{\lfloor\frac{k+1}{2}\rfloor}}{3\sqrt{2}}\sum_{\substack{n\in\mathbb{Z}\\ (-1)^k n\equiv 1 \pmod 4}} 2^{k+\frac{1}{2}}e^{2\pi i\left(\frac{n-1}{8}-\frac{k}{4}\right)}c_2(n,4v)q^n\\ &= \frac{2}{3}\sum_{\substack{n\in\mathbb{Z}\\ n\equiv 1 \pmod 4}} 2^{k-1}\left(\frac{(-1)^k n}{2}\right)c_2(n,4v)q^n.\end{aligned}$$

□

We now return to our general discussion of the Poincaré series of Definition 6.6. The case $m > 0$ yields cusp forms. To give their Fourier expansions, we require the following *Kloosterman sums* $K_k(m,n;c)$:

$$K_k(m,n;c) := \begin{cases}\displaystyle\sum_{d \pmod c)^*} e\left(\frac{m\overline{d}+nd}{c}\right) & \text{if } k\in\mathbb{Z},\\ \displaystyle\sum_{d \pmod c)^*} \left(\frac{c}{d}\right)\varepsilon_d^{2k} e\left(\frac{m\overline{d}+nd}{c}\right) & \text{if } k\in\frac{1}{2}\mathbb{Z}\setminus\mathbb{Z}.\end{cases}\tag{6.13}$$

Here, for $d\in\mathbb{Z}$, $\overline{d}$ is such that $d\overline{d}\equiv 1 \pmod c$.

THEOREM 6.8. *Assume the conditions above.*

i) *Denoting by $\delta_{m,n}$ the Kronecker delta-function and J_{k-1} the usual J-Bessel function, we have the Fourier expansion $\mathcal{P}_{k,m,N}(\tau)=\sum_{n=1}^{\infty}b_m(n)q^n$, where*

$$b_m(n) = \left(\frac{n}{m}\right)^{\frac{k-1}{2}}\left(\delta_{m,n} + 2\pi i^{-k}\sum_{\substack{c>0\\ N|c}}\frac{K_k(m,n;c)}{c}J_{k-1}\left(\frac{4\pi\sqrt{nm}}{c}\right)\right).$$

ii) *The Petersson coefficient formula holds. Namely, for $f\in S_k(\Gamma_0(N))$,*

$$\langle f, \mathcal{P}_{k,m,N}\rangle = \frac{\Gamma(k-1)}{(4\pi m)^{k-1}}c_f(m),$$

where $c_f(m)$ denotes the m-th Fourier coefficient of f.

iii) *The set of Poincaré series $\{\mathcal{P}_{k,m,N} : m\in\mathbb{N}\}$ spans $S_k(\Gamma_0(N))$.*

REMARK. Theorem 6.8 ii) also holds for $m = 0$. In particular Eisenstein series and cusp forms are orthogonal.

REMARK. In fact, Theorem 6.8 iii) is a direct corollary of ii), as $S_k(\Gamma_0(N))$ is a finite dimensional space and by ii) any cusp form which is orthogonal to all Poincaré series $P_{k,m,N}$ has identically zero Fourier expansion and is hence zero.

We next turn to the case $m < 0$. In this case we obtain weakly holomorphic modular forms.

THEOREM 6.9. *For $m \in -\mathbb{N}$, the Poincaré series $\mathcal{P}_{k,m,N}$ are elements of $M_k^!(\Gamma_0(N))$. We have the Fourier expansion $\mathcal{P}_{k,m,N}(\tau) = q^m + \sum_{n=1}^{\infty} b_m(n) q^n$, where*

$$b_m(n) = \left|\frac{n}{m}\right|^{\frac{k-1}{2}} 2\pi i^{-k} \sum_{\substack{c>0\\ N|c}} \frac{K_k(m,n;c)}{c} I_{k-1}\left(\frac{4\pi\sqrt{|nm|}}{c}\right).$$

Here I_s denotes the usual I-Bessel function. Moreover, $\mathcal{P}_{k,m,N}$ is holomorphic at cusps of $\Gamma_0(N)$ other than $i\infty$.

REMARK. Note that Theorem 6.9 implies that all principal parts occur. Moreover, the Poincaré series $\mathcal{P}_{k,m,N}$ span $M_k^{!,\infty}$, the space of weakly holomorphic forms with poles supported at $i\infty$.

REMARK. The Poincaré series $\mathcal{P}_{k,m,N}$ and $\mathcal{P}_{k,n,N}$ with $m \in \mathbb{N}$ and $n \in -\mathbb{N}$ are orthogonal to each other with respect to the regularized Petersson inner product defined in Definition 5.6.

We can also compute Fourier expansions for the projections of these Poincaré series to the Kohnen plus space, which is frequently useful in applications. As the method is similar for the various types of Poincaré series in this section, we discuss one example here. Suppose $k > 2$, and $m \in \mathbb{N}$ with $(-1)^k m \equiv 0, 1 \pmod 4$. In order to apply Proposition 6.7 to the Poincaré series $\mathcal{P}_{k+\frac{1}{2},m,N}$, we must compute $\mathcal{P}_{k+\frac{1}{2},m,N}|_{k+\frac{1}{2}} C_\mu$ for $\mu \in \{2, -\alpha\}$.

We have that

$$\mathcal{P}_{k+\frac{1}{2},m,4N}|_{k+\frac{1}{2}} C_\mu = \sum_{\gamma \in \Gamma_\infty \backslash \Gamma_0(4N)} \varphi_m|_{k+\frac{1}{2}} \gamma^* C_\mu.$$

Arguing as above, we can obtain their Fourier expansions. Write

$$\mathcal{P}_{k+\frac{1}{2},m,4N}|_{k+\frac{1}{2}} C_\mu =: \sum_{n=1}^{\infty} b_{\mu,m}(n) q^{\frac{n}{4}}.$$

We find that the action by the matrix primarily affects the Kloosterman sum. We have that for $\mu \in \{2, -\alpha\}$ the coefficients $b_{\mu,m}(n)$ are given by

$$b_{2,m}(n) = \left(\frac{n}{4m}\right)^{\frac{k}{2}-\frac{1}{4}} 2\pi i^{-k-\frac{1}{2}} \sum_{\substack{c\geq 1\\ N|c\\ c \text{ odd}}} \frac{K_{k+\frac{1}{2}}(4m,n;8c)}{8c} J_{k-\frac{1}{2}}\left(\frac{\pi\sqrt{mn}}{c}\right),$$

$$b_{-\alpha,m}(n) = \left(\frac{n}{4m}\right)^{\frac{k}{2}-\frac{1}{4}} 2\pi i^{-k-\frac{1}{2}} \sum_{\substack{c\geq 1\\ N|c\\ c \text{ odd}}} \frac{\mathcal{K}_{-\alpha,k+\frac{1}{2}}(m,n;c)}{c} J_{k-\frac{1}{2}}\left(\frac{2\pi\sqrt{mn}}{c}\right),$$

where

$$\mathcal{K}_{-\alpha,k}(m,n;c) := i^{-n}\frac{(-1)^k i}{4}\left(\frac{4}{-Nc}\right)\left(\frac{-4}{Nc}\right)^{-k-\frac{1}{2}} \sum_{d \pmod{c}^*} e\left(\frac{m\overline{4d}+nd}{c}\right).$$

From here, tedious calculations show that if $m \equiv 0, (-1)^k \pmod 4$, then

$$K_{k+\frac{1}{2}}(m,n;4c) = \begin{cases} (1-(-1)^k i) i^n \mathcal{K}_{-\alpha,k+\frac{1}{2}}(\overline{4}m, \frac{n}{4}; c) & \text{if } n \equiv 0 \pmod 4\\ \sqrt{2}\left(\frac{(-1)^k n}{2}\right) K_{k+\frac{1}{2}}(4m,n;8c) & \text{if } n \equiv (-1)^k \pmod 4. \end{cases}$$

These calculations were given in detail in [**324**].

All together, this gives the following Fourier expansion for $P_{k+\frac{1}{2},m,N}$, as worked out by Kohnen. Specifically, we have that

$$P_{k+\frac{1}{2},m,N}(\tau)=\frac{2}{3}q^m+\sum_{\substack{n=1\\(-1)^k n\equiv 0,1 \pmod 4}}^{\infty} c_m(n)q^n,$$

where

$$c_m(n):=\frac{2}{3}\left(\frac{n}{m}\right)^{\frac{k}{2}-\frac{1}{4}}\sqrt{2}\pi(-1)^{\left\lfloor\frac{k+1}{2}\right\rfloor}\left(1-(-1)^k i\right)$$
$$\times\sum_{\substack{c\geq 1\\4N c}}\left(1+\delta_{\mathrm{odd}}\left(\frac{c}{4}\right)\right)\frac{K_{k+\frac{1}{2}}(m,n;c)}{c}J_{k-\frac{1}{2}}\left(\frac{\pi\sqrt{mn}}{c}\right),$$

and $\delta_{\mathrm{odd}}(n)$ is 1 if n is odd and 0 if n is even.

We next turn to the construction of harmonic Maass forms via Poincaré series. Such series have appeared in many places in the literature, indeed, as in the works of Niebur and Fay in the 1970s, long before the recent advent of harmonic Maass forms. For just a few standard references, the reader is referred to [**108, 111, 119, 192, 261, 397, 398**].

DEFINITION 6.10. Define for $k\in\frac{1}{2}\mathbb{Z}$ with $k<1/2$ and $m\in\mathbb{Z}\setminus\{0\}$ the **Maass-Poincaré series**

$$\mathcal{F}_{k,m,N}:=\sum_{\gamma\in\Gamma_\infty\backslash\Gamma_0(N)}\phi_{k,m}|_k\gamma,$$

where the seed $\phi_{k,m}$ is given by

$$\phi_{k,m}(\tau):=\frac{(-\operatorname{sgn}(m))^{-k+1}}{\Gamma(2-k)}(4\pi|m|v)^{-\frac{k}{2}}M_{\operatorname{sgn}(m)\frac{k}{2},\frac{1-k}{2}}(4\pi|m|v)e(mu).$$

Here the classical M-Whittaker functions $M_{\mu,\nu}(w)$ are solutions to the differential equation

$$\frac{\partial^2 f}{\partial w^2}+\left(-\frac{1}{4}+\frac{\nu}{w}+\frac{\frac{1}{4}-\mu^2}{w^2}\right)f=0,$$

which grow like

$$M_{\mu,\nu}(w)\sim\frac{\Gamma(1+2\nu)}{\Gamma\left(\nu-\mu+\frac{1}{2}\right)}w^{-\mu}e^{\frac{w}{2}}$$

as $w\to\infty$. Explicitly, for $\mu,\nu\in\mathbb{C}$ such that $2\nu\notin-\mathbb{N}$, the M-Whittaker function is defined by

$$M_{\mu,\nu}(w):=e^{-\frac{w}{2}}w^{\frac{1}{2}+\nu}\mathfrak{M}\left(\frac{1}{2}+\nu-\mu,1+2\nu;w\right),$$

where

$$\mathfrak{M}(a,b;t):=\sum_{n=0}^{\infty}\frac{(a)_n}{(b)_n}\frac{t^n}{n!}\tag{6.14}$$

is *Kummer's confluent hypergeometric function* and $(a)_n=a(a+1)\cdots(a+n-1)$ is the rising factorial.

REMARK. We note that the seed may also be written in terms of the incomplete gamma function from Chapter 4. To be more precise, we have

$$\phi_{k,m}(\tau) = (1 - \Gamma^*(1-k, -4\pi mv))\, q^m$$

with

$$\Gamma^*(s,v) := \frac{\Gamma(s,v)}{\Gamma(s)}.$$

This is shown in the proof of Theorem 6.11 below.

REMARK. For weights $0 \leq k \leq 2$, absolute convergence does not hold, however one may introduce a spectral parameter in the summation, compute the Fourier expansion, and analytically continue (the reader is referred to **[201]** for a detailed exposition of this procedure).

REMARK. For $k = 0$, one obtains Niebur's Poincaré series **[400]** defined in (13.4).

We then have the following key relations between Poincaré series, which play a central role in many structural theorems on harmonic Maass forms.

THEOREM 6.11. *Suppose that $k \in \frac{1}{2}\mathbb{Z}$ with $k < 1/2$ and $m \neq 0$. Then the following are true:*

i) *The Poincaré series $\mathcal{F}_{k,m,N}$ lies in $H_k(\Gamma_0(N))$ if $m < 0$ and in $H_k^\sharp(\Gamma_0(N))$ if $m > 0$ (recall that $H_k^\sharp(\Gamma_0(N))$ is as in (4.9)).*
ii) *The shadow of the Poincaré series $\mathcal{F}_{k,m,N}$ is given by*

$$\xi_k\left(\mathcal{F}_{k,m,N}\right) = (-1)^{k+1}\frac{(4\pi m)^{1-k}}{\Gamma(1-k)}\mathcal{P}_{2-k,-m,N}.$$

iii) *For $k \in -2\mathbb{N}_0$, the image under the Bol operator is given by*

$$D^{1-k}\left(\mathcal{F}_{k,m,N}\right) = m^{1-k}\mathcal{P}_{2-k,m,N}.$$

iv) *For any $k \in 2\mathbb{N}$, the image of $\mathcal{F}_{k,m,N}$ under the flipping operator of Definition 5.14 is given by*

$$\mathfrak{F}_k(\mathcal{F}_{k,m,N}) = \mathcal{F}_{k,-m,N}.$$

v) *The Fourier expansion of $\mathcal{F}_{k,m,N}$ at $i\infty$ is given by*

$$\mathcal{F}_{k,m,N}(\tau) = (1-\Gamma^*(1-k,-4\pi mv))\, q^m + \sum_{n=0}^{\infty} b_m^+(n) q^n + \sum_{n=1}^{\infty} b_m^-(n)\Gamma^*(1-k, 4\pi nv) q^{-n},$$

where

$$b_m^+(0) = -\frac{(-2\pi i)^{2-k}(-m)^{1-k}}{\Gamma(2-k)}\sum_{\substack{c>0\\ N|c}}\frac{K_k(m,0;c)}{c^{2-k}},$$

and for $n \geq 1$,

$$b_m^+(n) = 2\pi i^k \left(-\mathrm{sgn}(m)\right)^{1-k} \left|\frac{m}{n}\right|^{\frac{1-k}{2}}$$
$$\times \sum_{\substack{c>0\\ N|c}} \frac{K_k(m,n;c)}{c} \begin{cases} I_{1-k}\left(\frac{4\pi\sqrt{|mn|}}{c}\right) & \textit{if } m<0,\\ J_{1-k}\left(\frac{4\pi\sqrt{|mn|}}{c}\right) & \textit{if } m>0,\end{cases}$$

$$b_m^-(n) = 2\pi i^k \left(-\mathrm{sgn}(m)\right)^{1-k} \left|\frac{m}{n}\right|^{\frac{1-k}{2}}$$
$$\times \sum_{\substack{c>0\\ N|c}} \frac{K_k(m,-n;c)}{c} \begin{cases} J_{1-k}\left(\frac{4\pi\sqrt{|mn|}}{c}\right) & \textit{if } m<0,\\ I_{1-k}\left(\frac{4\pi\sqrt{|mn|}}{c}\right) & \textit{if } m>0.\end{cases}$$

REMARK. In Theorem 6.11, we assumed that $m \neq 0$. Later, in Theorem 6.15, we will address the case when $m = 0$.

REMARK. In Theorem 6.11 v), the Fourier expansion was only given at $i\infty$. However, similar expansions can be described at all other cusps; at cusps inequivalent to $i\infty$, $\mathcal{F}_{k,m,N}$ decays like a cusp form.

In the case of half-integral weight, it is also convenient to have the Kohnen projections of these forms to the plus space, which we require throughout this book. To this end, in analogy with the Poincaré series of exponential type above, we let

$$F_{k,m,N} := \mathcal{F}_{k,m,N}|\mathrm{pr}$$

be the corresponding projections into Kohnen's plus space (cf. (6.12)) for half-integral weight k (while $F_{k,m,N} := \mathcal{F}_{k,m,N}$ for integral weight). Of course, in the case of integer weight k, the Fourier expansion of $F_{k,m,N}$ is given in the previous theorem. In the case of the half-integral weight, a similar application of Proposition 6.7 as was discussed above for the functions $\mathcal{P}_{k+\frac{1}{2},m,N}(\tau)$ where $k>2$, $m \in \mathbb{N}$ with $(-1)^k m \equiv 0, 1 \pmod 4$ can be made (see also Theorem 2.1 of **[108]** for explicit Fourier expansions).

PROOF OF THEOREM 6.11. Here we outline the proof of Theorem 6.11. For simplicity, we also restrict to the case when $k<0$; if $k=0$, an additional analytic continuation must be taken.

That $\mathcal{F}_{k,m,N}$ satisfies the correct transformation law follows by construction. The remaining claims about the spaces follow once we prove Theorem 6.11 ii) and iii). Since ξ_k and D^{1-k} commute with the slash action $|_k$, we only need to compute their actions on the seed $\phi_{k,m}$.

We start by computing the action of the ξ-operator. Using the definition of $\phi_{k,m}$, we have

$$(6.15)\quad \xi_k\left(\phi_{k,m}(\tau)\right)$$
$$= \overline{\frac{(-\mathrm{sgn}(m))^{-k+1}}{\Gamma(2-k)}}\,\overline{q}^m v^k 4\pi|m| \left[\frac{\partial}{\partial w}\left(w^{-\frac{k}{2}} M_{\mathrm{sgn}(m)\frac{k}{2},\frac{1-k}{2}}(w) e^{\frac{\mathrm{sgn}(m)w}{2}}\right)\right]_{w=4\pi|m|v}.$$

Note that

$$\overline{(-\operatorname{sgn}(m))^{-k+1}} = (-\operatorname{sgn}(m))^{k-1}.$$

To compute the derivative of the M-Whittaker function, we use (13.15.15) and (13.15.18) of **[400]** to find, with $\varepsilon \in \{0,1\}$,

$$\frac{\partial^n}{\partial w^n}\left(e^{\frac{\varepsilon w}{2}} w^{\mu-\frac{1}{2}} M_{\kappa,\mu}(w)\right) = (-1)^n(-2\mu)_n e^{\frac{\varepsilon w}{2}} w^{\mu-\frac{n+1}{2}} M_{\kappa-\varepsilon\frac{n}{2},\mu-\frac{n}{2}}(w). \tag{6.16}$$

Thus, in particular,

$$\frac{\partial}{\partial w}\left(w^{-\frac{k}{2}} e^{\frac{\operatorname{sgn}(m)w}{2}} M_{\operatorname{sgn}(m)\frac{k}{2},\frac{1-k}{2}}(w)\right) = -(k-1)e^{\frac{\operatorname{sgn}(m)w}{2}} w^{-\frac{k}{2}-\frac{1}{2}} M_{\operatorname{sgn}(m)\frac{k-1}{2},-\frac{k}{2}}(w). \tag{6.17}$$

Now, by (13.18.2) and (13.18.3) of **[400]**

$$M_{\kappa,\varepsilon\kappa-\frac{1}{2}}(w) = e^{-\frac{\varepsilon w}{2}} w^{\varepsilon\kappa}. \tag{6.18}$$

Thus, we obtain

$$M_{\operatorname{sgn}(m)\frac{k-1}{2},-\frac{k}{2}}(w) = e^{\frac{\operatorname{sgn}(m)w}{2}} w^{\frac{1-k}{2}},$$

while (6.17) equals

$$-(k-1)e^{\operatorname{sgn}(m)w} w^{-k}.$$

Substituting this into (6.15) yields Theorem 6.11 ii).

We next turn to the D-operator and compute

$$\begin{aligned} D^{1-k}\left(\phi_{k,m}(\tau)\right) =& \frac{(-\operatorname{sgn}(m))^{-k+1}}{\Gamma(2-k)}\left(\frac{1}{2\pi i}\right)^{1-k} \overline{q}^{-m}(4\pi|m|)^{1-k}(2i)^{k-1} \\ &\times\left(\frac{\partial}{\partial w}\right)^{1-k}\left[w^{-\frac{k}{2}} M_{\operatorname{sgn}(m)\frac{k}{2},\frac{1-k}{2}}(w)e^{-\frac{\operatorname{sgn}(m)w}{2}}\right]_{w=4\pi|m|v}. \end{aligned} \tag{6.19}$$

We again use (6.16) to obtain

$$\begin{aligned} \left(\frac{\partial}{\partial w}\right)^{1-k}&\left[w^{-\frac{k}{2}} M_{\operatorname{sgn}(m)\frac{k}{2},\frac{1-k}{2}}(w)e^{-\frac{\operatorname{sgn}(m)w}{2}}\right] \\ &= (-1)^{k+1}(k-1)_{1-k} e^{-\frac{\operatorname{sgn}(m)w}{2}} w^{-\frac{1}{2}} M_{\frac{\operatorname{sgn}(m)}{2},0}(w). \end{aligned} \tag{6.20}$$

Note that $(k-1)_{1-k} = (-1)^{k+1}(1-k)!$. Moreover, using (6.18) again gives

$$M_{\frac{\operatorname{sgn}(m)}{2},0}(w) = e^{-\frac{\operatorname{sgn}(m)w}{2}} w^{\frac{1}{2}}.$$

Thus (6.20) becomes

$$(1-k)!e^{-\operatorname{sgn}(m)w}.$$

Inserting in (6.19) then yields Theorem 6.11 iii).

We next turn to computing the Fourier expansions of the Poincaré series. We split off the coset coming from the identity element and in the remaining sums pick $c > 0$. This yields

$$\mathcal{F}_{k,m,N} = \phi_{k,m} + \sum_{\substack{c>0,d\in\mathbb{Z}\\ N|c\\ \gcd(c,d)=1}} \phi_{k,m}\Big|_k \begin{pmatrix} a & b \\ c & d \end{pmatrix},$$

where a, b is an arbitrary integral solution of $ad - bc = 1$.

We first rewrite the principal part. Using (13.18.4) of **[400]** yields, for $m < 0$,

$$\phi_{k,m}(\tau) = \left(1 - \Gamma^*(1-k; 4\pi|m|v)\right) q^m.$$

For $m > 0$, $\phi_{k,m}$ becomes

$$\frac{(-1)^{-k+1}}{\Gamma(2-k)}(4\pi mv)^{-\frac{k}{2}} M_{\frac{k}{2},\frac{1-k}{2}}(4\pi mv)e(mu)$$
$$= \frac{(-1)^{-k+1}}{\Gamma(2-k)}(4\pi mv)^{1-k}\mathfrak{M}(1-k,2-k;4\pi mv)q^m$$

by (13.1.32) of [**1**]. Here $\mathfrak{M}$ is the Kummer function, defined in (6.14). By (13.6.10) of [**1**], this is

$$(1-\Gamma^*(1-k,-4\pi mv))\,q^m.$$

We next consider the non-principal term and compute its Fourier expansion. By definition it equals

$$\frac{(-\operatorname{sgn}(m))^{-k+1}}{\Gamma(2-k)}\sum_{\substack{c>0,d\in\mathbb{Z}\\ N|c\\ \gcd(c,d)=1}}\rho_k(c,d)^{-1}(c\tau+d)^{-k}\left(\frac{4\pi|m|v}{|c\tau+d|^2}\right)^{-\frac{k}{2}}$$
$$\times M_{\operatorname{sgn}(m)\frac{k}{2},\frac{1-k}{2}}\left(\frac{4\pi|m|v}{|c\tau+d|^2}\right)e\left(m\operatorname{Re}\left(\frac{a\tau+b}{c\tau+d}\right)\right),$$

where $\rho_k(c,d)$ is the multiplier defined in (5.4). Now using that

$$\frac{a\tau+b}{c\tau+d}=\frac{a}{c}-\frac{1}{c(c\tau+d)},$$

and letting $d\mapsto d+nc$, the above becomes

$$\text{(6.21)}\quad \frac{(-\operatorname{sgn}(m))^{-k+1}}{\Gamma(2-k)}\sum_{\substack{c>0\\ N|c\\ d\pmod{c}^*}}\rho_k(c,d)^{-1}e\left(m\frac{a}{c}\right)(4\pi|m|v)^{-\frac{k}{2}}$$
$$\sum_{n\in\mathbb{Z}}\left(\frac{c\overline{\tau}+d+nc}{c\tau+d+nc}\right)^{\frac{k}{2}}M_{\operatorname{sgn}(m)\frac{k}{2},\frac{1-k}{2}}\left(\frac{4\pi|m|v}{|c\tau+d+nc|^2}\right)e\left(-m\operatorname{Re}\left(\frac{1}{c(c\tau+d+nc)}\right)\right).$$

We note in passing that since we actually look at $\mathcal{F}(\tau+d/c)$, plugging into (6.21) gives the Kloosterman sums in the Fourier expansion in the theorem. To compute the sum on n, we let

$$F(\tau):=\sum_{n\in\mathbb{Z}}\left(\frac{\overline{\tau}+n}{\tau+n}\right)^{\frac{k}{2}}M_{\operatorname{sgn}(m)\frac{k}{2},\frac{1-k}{2}}\left(\frac{4\pi|m|v}{c^2|\tau+n|^2}\right)e\left(-m\frac{u+n}{c^2|\tau+n|^2}\right).$$

Then F has a Fourier expansion

$$F(\tau)=\sum_{n\in\mathbb{Z}}a_v(n)e^{2\pi inu}$$

where $a_v(n)$ equals

$$(-1)^{\frac{k}{2}}\int_{\mathbb{R}}\left(\frac{v-iu}{v+iu}\right)^{-\frac{k}{2}}M_{\operatorname{sgn}(m)\frac{k}{2},\frac{1-k}{2}}\left(\frac{4\pi|m|v}{c^2(u^2+v^2)}\right)e\left(-\frac{mu}{c^2(u^2+v^2)}-nu\right)du.$$

Making the change of variables $u=-\operatorname{sgn}(m)vt$, we obtain

(6.22)

$$a_v(n)=(-1)^{\frac{k}{2}}v\int_{\mathbb{R}}\left(\frac{1-iu}{1+iu}\right)^{\operatorname{sgn}(m)\frac{k}{2}}M_{\operatorname{sgn}(m)\frac{k}{2},\frac{1-k}{2}}\left(\frac{4\pi B}{u^2+1}\right)e\left(\frac{Bu}{u^2+1}+Au\right)du,$$

where $A := \operatorname{sgn}(m)nv$ and $B := \frac{|m|}{c^2v}$. These integrals were computed by Hejhal on page 357 of [**261**]. To state his results, set for $\operatorname{Re}(z) > 0$

$$G(z) = G_{\ell,s}(z) := \frac{z^s e^{-\frac{z}{2}} \mathfrak{M}(s-\ell; 2s; z)}{\Gamma(2s)}$$

where $\mathfrak{M}$ is defined in (6.14). Using the series representation of the M-Whittaker function, we see

$$M_{\operatorname{sgn}(m)\frac{k}{2}, s-\frac{1}{2}}(z) = \Gamma(2s) G_{\operatorname{sgn}(m)\frac{k}{2}, s}(z).$$

We now distinguish three cases. If $A > 0$, then by [**261**], (6.22) equals

$$\frac{(-1)^{\frac{k}{2}} v\Gamma(2-k)2\pi}{\Gamma\left(1-\frac{k}{2}+\operatorname{sgn}(m)\frac{k}{2}\right)} (4\pi A)^{1-\frac{k}{2}} e^{-2\pi A} \Psi\left(1-\frac{k}{2}-\operatorname{sgn}(m)\frac{k}{2}; 2-k; 4\pi A\right) \times \sqrt{\frac{B}{A}} J_{1-k}\left(4\pi\sqrt{AB}\right), \tag{6.23}$$

where (cf. (4.11) in Chapter 9, page 344 of [**261**])

$$\Psi(a;c;z) := \frac{\Gamma(1-c)}{\Gamma(a-c+1)} \mathfrak{M}(a;c;z) + \frac{\Gamma(c-1)}{\Gamma(a)} z^{1-c} \mathfrak{M}(a-c+1; 2-c; z).$$

Note that for $\operatorname{Re}(a) > 0$ (cf. Lemma 9.3 (vii) in Chapter 9, page 344 of [**261**] for the first identity and (13.2.6) and (13.1.33) of [**1**] for the last identity)

$$\begin{aligned}\Psi(a;c;z) &= \frac{1}{\Gamma(a)} \int_0^\infty e^{-zt} t^{a-1} (1+t)^{c-a-1} dt = \frac{1}{\Gamma(a)} e^z \int_1^\infty e^{-zt} (t-1)^{a-1} t^{c-a-1} dt \\ &= e^{\frac{z}{2}} z^{-\frac{c}{2}} W_{\frac{c}{2}-a, \frac{c-1}{2}}(z).\end{aligned}$$

Thus

$$\Psi\left(1-\frac{k}{2}-\operatorname{sgn}(m)\frac{k}{2}; 2-k; 4\pi A\right) = (4\pi A)^{k-1} \begin{cases} 1 & \text{if } m > 0, \\ \Gamma(1-k, 4\pi A) e^{4\pi A} & \text{if } m < 0, \end{cases}$$

using (13.18.2) and (13.18.5) of [**400**], respectively.

If $A = 0$, then (6.22) equals

$$(-1)^{\frac{k}{2}} 4\pi^{2-\frac{k}{2}} v B^{1-\frac{k}{2}}.$$

If $A < 0$, (6.22) becomes

$$\frac{(-1)^{\frac{k}{2}} v\Gamma(2-k)}{\Gamma\left(1-\frac{k}{2}-\operatorname{sgn}(m)\frac{k}{2}\right)} (4\pi|A|)^{1-\frac{k}{2}} e^{-2\pi|A|} \Psi\left(1-\frac{k}{2}+\operatorname{sgn}(m)\frac{k}{2}; 2-k; 4\pi|A|\right) \times \sqrt{\frac{B}{A}} I_{1-k}\left(4\pi\sqrt{AB}\right). \tag{6.24}$$

As before

$$\Psi\left(1-\frac{k}{2}+\operatorname{sgn}(m)\frac{k}{2}; 2-k; 4\pi|A|\right) = (4\pi A)^{k-1} \begin{cases} \Gamma(1-k, -4\pi|A|) e^{4\pi|A|} & \text{if } m > 0, \\ 1 & \text{if } m < 0. \end{cases}$$

Combining the above gives the claim.

Finally, we turn to the proof of Theorem 6.11 iv). By Theorem 6.11 iii) and Proposition 5.15 iii), we find that

$$\xi_k(\mathfrak{F}_k(\mathcal{F}_{k,m,N})) = -\frac{(-4\pi)^{1-k}}{(-k)!}D^{1-k}(\mathcal{F}_{k,m,N}) = \frac{(4\pi m)^{1-k}}{(-k)!}\mathcal{P}_{2-k,m,N}.$$

Comparing with Theorem 6.11 ii) immediately yields

$$\xi_k\left(\mathcal{F}_{k,-m,N}\right) = \xi_k(\mathfrak{F}_k(\mathcal{F}_{k,m,N})). \tag{6.25}$$

This shows that the two harmonic Maass forms which we claim are equal have the same nonholomorphic parts. To compare holomorphic parts, we compute the action of D^{1-k} on both sides. By ii) and Proposition 5.15 iv),

$$D^{1-k}\left(\mathfrak{F}_k(\mathcal{F}_{k,m,N})\right) = \frac{(-k)!}{(4\pi)^{1-k}}\xi_k(\mathcal{F}_{k,m,N}) = -m^{1-k}\mathcal{P}_{2-k,-m,N}.$$

Thus, by Theorem 6.11 iii),

$$D^{1-k}\left(\mathfrak{F}_k(\mathcal{F}_{k,m,N})\right) = D^{1-k}\left(\mathcal{F}_{k,-m,N}\right).$$

Combining with (6.25) and the fact that $\mathfrak{F}_k(\mathcal{F}_{k,m,N}), \mathcal{F}_{k,-m,N} \in H_k^!(\Gamma_0(N))$ shows that these two functions are equal up to the possible addition of a constant. This constant term could only possibly be non-zero if $k = 0$. In this case,

$$\mathfrak{F}_0(\mathcal{F}_{0,m,N}) = -\overline{\mathcal{F}_{0,m,N}},$$

and so we only require that $c_{\mathcal{F}_{0,m,N}}^-(0) = -\overline{c_{\mathcal{F}_{0,m,N}}^-(0)}$. This follows directly from the Fourier expansion computed earlier in this theorem once one uses the straightforward fact that $K_k(m,n;c)$ is real and independent of k for $k \in \mathbb{Z}$ and the fact that $K_k(m,n;c) = K_k(-m,-n;c)$ for $k \in 2\mathbb{Z}$ (which can be seen by substituting $d \mapsto -d$ in the definition (6.13)). □

REMARK. By construction, the harmonic Maass forms $\mathcal{F}_{k,-m,N}$ with $m \in \mathbb{N}$ have a prescribed non-trivial principal part (namely, just q^{-m}) at the cusp $i\infty$ and have trivial principal parts at all other cusps of $\Gamma_0(N)$. Similarly, for $m \in \mathbb{N}$, $\mathcal{F}_{k,m,N}$ has a prescribed nonholomorphic part.

COROLLARY 6.12. *Assuming the notation and hypotheses from Theorem* 6.11, *we have for* $m > 0$ *that* $\mathcal{F}_{k,-m,N} \in M_k^!(\Gamma_0(N))$ *if and only if* $\mathcal{P}_{2-k,m,N}$ *vanishes identically.*

PROOF. Note that $\mathcal{F}_{k,-m,N}$ is a weakly holomorphic modular form if and only if its nonholomorphic part vanishes. By Theorem 5.10 and Theorem 6.11, this happens if and only if $\mathcal{P}_{k,m,N}$ vanishes identically. □

EXAMPLE 6.13. Here we illustrate Theorem 6.11 for the three Poincaré series for $\Gamma_0(9) : F_{-2,-1,9} \in H_{-2}(\Gamma_0(9))$, $P_{4,1,9} \in S_4(\Gamma_0(9))$, and $P_{4,-1,9} \in M_4^!(\Gamma_0(9))$. Summing the first 150 terms in the Fourier expansion of Theorem 6.11 gives

$$\begin{aligned}-D^3(F_{-2,-1,9}(\tau)) &= P_{4,-1,9}(\tau)\\ &\approx q^{-1} + 1.9999q^2 - 48.9999q^5 + 47.9999q^8 + 770.9999q^{11} + \cdots.\end{aligned}$$

On the other hand, we consider the weight 4 weakly holomorphic modular form

$$g(\tau) := \left(\frac{\eta(\tau)^3}{\eta(9\tau)^3} + 3\right)^2 \eta(3\tau)^8 = q^{-1} + 2q^2 - 49q^5 + 48q^8 + 771q^{11} - \cdots,$$

where η is Dedekind's eta-function. By Theorem 6.9, the Poincaré series $P_{4,-1,9}$ has a principal part only at $i\infty$, and the modular form g is also holomorphic at the other cusps $0, 1/3, 2/3$ of $\Gamma_0(9)$. This may be checked using the well-known formula for the order of vanishing $\mathrm{ord}(\eta(m\tau), r)$ of the function $\eta(m\tau)$ at the cusp $r = -d/c \in \mathbb{Q}$ (cf. Proposition 2.1 of [**319**]):

$$\mathrm{ord}(\eta(m\tau), r) = \frac{1}{24m}\gcd(c, m)^2.$$

Thus, these two modular forms have the same principal parts at all cusps, and so their difference is a weight 4 cusp form on $\Gamma_0(9)$. We claim that

$$P_{4,-1,9} = g. \tag{6.26}$$

The truth of this claim then implies that $F^+_{-2,-1,9}$ has rational coefficients, except for possibly the constant term. The formula for the Fourier expansion of $F^+_{-2,-1,9}$ above, however, shows that the constant term is in fact zero, since each Kloosterman sum of the shape $K_{-2}(-1, 0; 9c)$ vanishes. To prove (6.26), we must show that the cusp form $h := P_{4,-1,9} - g$ is zero. Since the space $S_4(\Gamma_0(9))$ is one dimensional and is spanned by the CM form

$$f(\tau) := \eta(3\tau)^8 = q - 8q^4 + 20q^7 - 70q^{13} + 64q^{16} + 56q^{19} + \cdots,$$

it follows that h must be a multiple of f. However, the non-zero coefficients of f have exponents n supported on the arithmetic progression $n \equiv 1 \pmod 3$. The non-zero coefficients of g are supported on the arithmetic progression of exponents $n \equiv 2 \pmod 3$. Furthermore, a straightforward calculation shows that $K_4(-1, n; 9c) = 0$ for integers $n \equiv 1 \pmod 3$, and so Theorem 6.9 implies that the exponents in the Fourier expansion of $P_{4,-1,9}$ are also supported on the arithmetic progression $n \equiv 2 \pmod 3$. Consequently, we have that h is identically zero. Combining these facts, we find that the coefficients of $F^+_{-2,-1,9}$ are rational, and its first few terms are

$$F^+_{-2,-1,9}(\tau) = q^{-1} - \frac{1}{4}q^2 + \frac{49}{125}q^5 - \frac{3}{32}q^8 - \frac{771}{1331}q^{11} + \cdots.$$

The algebraicity of $F^+_{-2,-1,9}$ is very unusual for holomorphic parts of Maass-Poincaré series. This algebraicity is related to the fact that the corresponding shadow is a CM form and satisfies

$$\xi_{-2}(F_{-2,-1,9}) = 32\pi^3 P_{4,1,9} = \frac{f}{\|f\|^2}.$$

Such algebraicity results are discussed further in Section 7.3 below. The first equality follows from Theorem 6.11 ii). To see the second equality, note that $F_{-2,-1,9}$ has principal part q^{-1} at the cusp $i\infty$ and is holomorphic at the other cusps, and the function f is $q + O(q^2)$. As $S_4(\Gamma_0(9))$ is one dimensional, $\xi_{-2}(F_{-2,-1,9})$ is just a multiple of f, say $\xi_{-2}(F_{-2,-1,9}) = \alpha f$. Then by Proposition 5.11, we find

$$\{f, F_{-2,-1,9}\} = 1 = \langle f, \xi_{-2}(F_{-2,-1,9})\rangle = \langle f, \alpha f\rangle = \overline{\alpha}\|f\|^2,$$

so that $\alpha = \|f\|^{-2}$ as desired.

EXAMPLE 6.14. In contrast with the algebraic coefficients of the previous example, we consider here the case $k = 12$ and $N = 1$, where we observe a very different behavior. In this case S_{12} is spanned by Δ. As in the last case, we have that

$$\xi_{-10}(F_{-10,-1}) = \frac{\Delta}{\|\Delta\|^2}.$$

However, unlike the previous example where the corresponding shadow has complex multiplication, it is conjectured that the holomorphic part of the weight -10 harmonic Maass form $F_{-10,-1}$ has irrational (presumably transcendental) coefficients. Indeed, we find that

$$F^+_{-10,-1}(\tau) = q^{-1} - \frac{65520}{691} - 1842.89472\cdots q - 23274.07545\cdots q^2 - 225028.75877\cdots q^3 - \cdots .$$

Apart from the first two coefficients, the coefficients of this series are presumably all transcendental.

Finally, we consider the $m = 0$ case which was excluded from Definition 6.10. For this, we define

$$\mathcal{Q}_{k,N}(\tau) := \sum_{\gamma \in \Gamma_\infty \backslash \Gamma_0(N)} v^{1-k}\big|_k \gamma.$$

Moreover, let for $k > 2$

$$\mathcal{E}_{k,N}(\tau) := \sum_{\gamma \in \Gamma_\infty \backslash \Gamma_0(N)} 1\big|_k \gamma.$$

We then have the following result, which is analogous to Theorem 6.11. It can be thought of as the "Eisenstein" case of these Poincaré series.

THEOREM 6.15. *For any $n \in \mathbb{N}$, the following are true:*

i) *For any negative $k \in \frac{1}{2}\mathbb{Z}$, the function $\mathcal{Q}_{k,N}$ lies in $H^!_k(\Gamma_0(N))$.*

ii) *For any negative $k \in \frac{1}{2}\mathbb{Z}$, the shadow of $\mathcal{Q}_{k,N}$ is given by*

$$\xi_k\left(\mathcal{Q}_{k,N}\right) = (1-k)\mathcal{E}_{2-k,N}.$$

iii) *For any $k \in -2\mathbb{N}_0$, the image under the Bol operator of $\mathcal{Q}_{k,N}$ is given by*

$$D^{1-k}\left(\mathcal{Q}_{k,N}\right) = -(4\pi)^{k-1}(1-k)!\mathcal{E}_{2-k,N}.$$

iv) *For any $k \in -2\mathbb{N}_0$, the flip of $\mathcal{Q}_{k,N}$ is given by*

$$\mathfrak{F}_k(\mathcal{Q}_{k,N}) = -\mathcal{Q}_{k,N}.$$

v) *For any negative $k \in \frac{1}{2}\mathbb{Z}$, the Fourier expansion of $\mathcal{Q}_{k,N}$ equals*

$$\mathcal{Q}_{k,N}(\tau) = v^{1-k} + (-2i)^k\pi \sum_{n=0}^{\infty} \sum_{\substack{c>0 \\ N|c}} c^{k-2} K_k(0,n;c) q^n + (-2i)^k\pi \sum_{n=1}^{\infty} \sum_{\substack{c>0 \\ N|c}} c^{k-2} K_k(0,-n;c)\Gamma^*(1-k, 4\pi n v) q^{-n}.$$

REMARK. At first glance, it may seem unusual that the flipping operator gives a minus sign when acting on $\mathcal{Q}_{k,N}$ since there is none in Theorem 6.11 iv). However, this becomes less surprising once one notices that the seed $\phi_{k,m}$ of the Maass-Poincaré series incorporates a discontinuous factor depending on $\mathrm{sgn}(m)$, and the flipping operator changes the direction with which we approach this discontinuity when we approach $m = 0$.

As with the functions $\mathcal{F}_{k,m,N}$ described in Theorem 6.11, we also consider the projections of the "Eisenstein versions" $\mathcal{Q}_{k,N}$ to the plus space. Again, for half-integral weight, we let $Q_{k,N}$ be the projection of $Q_{k,N}$ (or $\mathcal{Q}_{k,N}$ itself for integral weight) to Kohnen's plus space (cf. (6.12)). Similarly, we let $E_{k,N}$ be the projection of $\mathcal{E}_{k,N}$ into Kohnen's plus space for k half-integral and $\mathcal{E}_{k,N}$ itself for k integral. As discussed above for the Poincaré series $\mathcal{F}_{k,m,N}$, the corresponding Fourier expansions for the projected forms in half-integral weight can be computed using Proposition 6.7 in the manner described above for $\mathcal{P}_{k+\frac{1}{2},m,N}(\tau)$ with $k > 2$, $m \in \mathbb{N}$, where $(-1)^k m \equiv 0, 1 \pmod 4$.

Proof of Theorem 6.15. We proceed as in the proof of Theorem 6.11.
We have to compute the Fourier expansion of

$$G(\tau) := \sum_{n\in\mathbb{Z}} (\tau+n)^{-1} (\overline{\tau}+n)^{k-1}.$$

We require a result of Shimura, which is given as Lemma 1 in **[457]** (note that Shimura cites Siegel **[460]** for this and related results). Specifically, suppose that α and β are complex numbers with $\mathrm{Re}(\alpha), \mathrm{Re}(\beta) > 0$, and $\mathrm{Re}(\alpha+\beta) > 1$. Then, under the usual requirement that $\tau \in \mathbb{H}$, we have

$$\sum_{m\in\mathbb{Z}} (\tau+m)^{-\alpha}(\overline{\tau}+m)^{-\beta} = \sum_{n\in\mathbb{Z}} c_n(v,\alpha,\beta)e^{2\pi i n u},$$

where $c_n(v,\alpha,\beta)$ is determined by

$$i^{\alpha-\beta}(2\pi)^{\alpha-\beta}\Gamma(\alpha)\Gamma(\beta)c_n(v,\alpha,\beta) = \begin{cases} n^{\alpha+\beta-1}e^{-2\pi n v}\sigma(4\pi n v,\alpha,\beta) & \text{if } n>0, \\ |n|^{\alpha+\beta-1}e^{-2\pi|n|v}\sigma(4\pi|n|v,\beta,\alpha) & \text{if } n<0, \\ \Gamma(\alpha+\beta-1)(4\pi v)^{1-\alpha-\beta} & \text{if } n=0, \end{cases}$$

with

$$\sigma(v,\alpha,\beta) := \int_0^\infty (t+1)^{\alpha-1}t^{\beta-1}e^{-vt}dt.$$

Applying this result with $\alpha = 1$ and $\beta = 1-k$, we obtain

$$G(\tau) = \sum_{n\in\mathbb{Z}} b_v(n)e^{2\pi i n u}$$

with

$$b_v(n) = \frac{i^{-k}(2\pi)^{2-k}}{\Gamma(1-k)} \begin{cases} n^{1-k}e^{-2\pi n v}\sigma(4\pi n v,1,1-k) & \text{if } n>0, \\ |n|^{1-k}e^{-2\pi|n|v}\sigma(4\pi|n|v,1-k,1) & \text{if } n<0, \\ \Gamma(1-k)(4\pi v)^{k-1} & \text{if } n=0. \end{cases}$$

For $n > 0$, we have

$$\sigma(4\pi n v, 1, 1-k) = \int_0^\infty t^{-k+1}e^{-4\pi n v t}\frac{dt}{t} = (4\pi n v)^{k-1}\Gamma(1-k).$$

Moreover, for $n < 0$,

$$\sigma(4\pi|n|v, 1-k, 1) = e^{4\pi|n|v}(4\pi|n|v)^{k-1}\Gamma(1-k)\Gamma^*(1-k, 4\pi|n|v). \tag{6.27}$$

Combining then gives ii) and iii).

To show iv), we argue as in the proof of Theorem 6.11 iii). Using ii), iii), and Proposition 5.15 iii), we find that

$$\xi_k\left(\mathfrak{F}_k(\mathcal{Q}_{k,N})\right) = \frac{(4\pi)^{1-k}}{(-k)!} D^{1-k}(\mathcal{Q}_{k,N}) = (k-1)\mathcal{E}_{2-k,N} = -\xi_k(\mathcal{Q}_{k,n}).$$

Now using ii), iii), and Proposition 5.15 iv), we obtain that

$$D^{1-k}\left(\mathfrak{F}_k(\mathcal{Q}_{k,N})\right) = \frac{(-k)!}{(4\pi)^{1-k}}\xi_k(\mathcal{Q}_{k,N}) = \frac{(1-k)!}{(4\pi)^{1-k}}\mathcal{E}_{2-k,N} = -D^{1-k}(\mathcal{Q}_{k,N}).$$

Thus, $\mathfrak{F}_k(\mathcal{Q}_{k,N})$ and $-\mathcal{Q}_{k,N}$ are equal up to the possible addition of a constant. Since $k < 0$, this constant must be zero. □

We next rewrite the Fourier coefficients in Theorem 6.15 in the case of integral weight. For simplicity we assume that $N = 1$.

COROLLARY 6.16. *For $k \in -2\mathbb{N}$ we have that*

$$\xi_k\left(\mathcal{Q}_k\right) = (1-k)E_{2-k},$$
$$D^{1-k}\left(\mathcal{Q}_k\right) = -(4\pi)^{k-1}(k-1)!E_{2-k}.$$

Moreover, we have that

$$\mathcal{Q}_k(\tau) = \frac{2(2-k)!}{B_{2-k}}\zeta(1-k)(4\pi)^{k-1} + v^{1-k} + \frac{2(2-k)!(4\pi)^{k-1}}{B_{2-k}}\sum_{n=1}^{\infty}\frac{\sigma_{1-k}(n)}{n^{1-k}}q^n$$
$$+\frac{2(2-k)!}{B_{2-k}}(4\pi)^{k-1}\sum_{n=1}^{\infty}\frac{\sigma_{1-k}(n)}{n^{1-k}}\Gamma^*(1-k,4\pi nv)q^{-n}.$$

PROOF. We first consider the (holomorphic) constant term. Note that

$$K_k(0,0;c) = \varphi(c),$$

where φ is Euler's phi-function. Using the identity

$$\sum_{c=1}^{\infty}\frac{\varphi(c)}{c^s} = \frac{\zeta(s-1)}{\zeta(s)}$$

yields the claim. For the remaining coefficients, one uses that

$$K_{k,1}(0,-n;c) = K_{k,1}(0,n;c)$$

and

$$\sum_{c=1}^{\infty}c^{k-2}\sum_{d \pmod{c}^*} e\left(\frac{nd}{c}\right) = \frac{\sigma_{1-k}(n)}{n^{1-k}\zeta(2-k)}.$$

□

REMARK. We have similar looking expansions for $N > 1$.

REMARK. For $k \in 1/2 + \mathbb{Z}$ the Fourier coefficients of the nonholomorphic part may be written in terms of coefficients of Cohen-Eisenstein series. It is an interesting question to investigate the coefficients of the holomorphic part.

REMARK. One can give an alternative proof of Corollary 6.16 using the action of the ξ-operator as well as the D-operator on the seed of $\mathcal{F}_k$.

We end this section by giving examples constructed via Poincaré series which are only annihilated by applying ξ three times. For this, recall that the weight k hyperbolic Laplacian factors as

$$\Delta_k = -\xi_{2-k} \circ \xi_k,$$

where $\xi_k = -2iv^k \overline{\frac{\partial}{\partial \overline{\tau}}}$. One may loosen the harmonicity condition $\Delta_k(F) = 0$ by requiring that the function is annihilated by repeated applications of ξ (in the appropriate weights). Here we restrict to applying ξ three times. For this, we set

$$\Delta_{k,2} := -\xi_k \circ \xi_{2-k} \circ \xi_k.$$

We then make the following definition.

DEFINITION 6.17. A real-analytic function $F\colon \mathbb{H} \to \mathbb{C}$ is called a **sesquiharmonic Maass form of weight k for Γ** if the following conditions are satisfied:

i) We have that $F|_k\gamma = F$ for all $\gamma \in \Gamma$.
ii) We have that $\Delta_{k,2}(F) = 0$.
iii) The function F has at most linear exponential growth at the cusps.

We denote by $H_{k,2}$ the space of sesquiharmonic Maass forms of weight k. The first motivating example of a sesquiharmonic Maass form was given by Duke, Imamoḡlu, and Tóth in **[172]**. There, they described a lift of the Zagier Eisenstein series (defined in Subsection 6.1.2). (See also Theorem 16.11.)

We next discuss how one could construct an infinite family of sesquiharmonic forms. For simplicity we restrict to positive even integral weight and to the space

$$H_{k,2}^{\sharp} := \left\{ f \in H_{k,2}\colon \xi_k(f) \in H_{2-k}^{\sharp} \right\}.$$

The dual space, consisting of forms which map under ξ_k to elements in H_{2-k}, is treated in an analogous way. To produce sesquiharmonic Maass forms, we generalize the Poincaré series considered above. For $k \in \mathbb{N}$ and $m \in -\mathbb{N}$, we define

$$F_{k,m,s}(\tau) := \mathbb{P}_{k,1}\left(\varphi_{k,m,s};\tau\right)$$

with

$$\varphi_{k,m,s}(\tau) := \mathcal{M}_{s,k}(4\pi m v)e(mu).$$

We have that

$$\Delta_k(F_{k,m,s}(\tau)) = \left(s(1-s) - \frac{k}{2}\left(1-\frac{k}{2}\right)\right) F_{k,m,s}(\tau).$$

Note that $F_{k,m,k/2}$ is a weakly holomorphic modular form. The idea is then to differentiate with respect to s and then specialize s in order to obtain the desired sesquiharmonic forms. This idea is not new and has appeared in various contexts (see for example works of Kudla-Rapoport-Yang **[335, 338, 339, 502, 503]**). To be more precise, we set

$$\mathbb{F}_{k,m}(\tau) := \mathbb{P}_{k,1}\left(\psi_{k,m}\right),$$

with

$$\psi_{k,m}(\tau) := \left[\frac{\partial}{\partial s}\mathcal{M}_{k,s}(4\pi m v)\right]_{s=\frac{k}{2}} e(mu).$$

We then have the following result **[83]**.

THEOREM 6.18. *For $m \in -\mathbb{N}$, the function $\mathbb{F}_{k,m}$ are elements of $H^{\sharp}_{k,2}$ and satisfy*

$$\xi_k\left(\mathbb{F}_{k,m}\right) = (4\pi m)^{1-k} F_{k,-m}.$$

In particular, the map

$$\xi_k \colon H^{\sharp}_{k,2} \to H^{\sharp}_{2-k}$$

is surjective.

Although sesquiharmonic forms are relatively new automorphic objects, it seems natural to expect that they will have many applications. As an example, we note that in **[83]** it was shown that they naturally encode non-critical L-values of modular forms.

6.4. p-adic harmonic Maass forms in the sense of Serre

We now briefly consider special cases of the harmonic Maass forms from Theorem 6.15 which are closely related to Serre's theory **[452]** of p-adic modular forms (cf. Subsection 7.4.3). Using Kummer's congruences for Bernoulli numbers, Serre defined the first examples of p-adic modular forms by making use of the classical Eisenstein series.

To make this precise, suppose that p is prime, and let v_p be the p-adic valuation on $\mathbb{Q}_p$. If

$$f(q) = \sum_n a(n)q^n \in \mathbb{Q}[[q]]$$

is a formal power series in q, then define $v_p(f) := \inf_n(v_p(a(n)))$. We then say that f is a *p-adic modular form* in the sense of Serre if there exists a sequence of holomorphic modular forms f_j of weights k_j such that $v_p(f - f_j) \to \infty$ as $j \to \infty$. The weight of a p-adic modular form is given by the limits of weights of the classical (holomorphic) modular forms in $X := \mathbb{Z}_p \times \mathbb{Z}/(p-1)\mathbb{Z}$.

Serre offered examples of such forms by making use of the Eisenstein series

$$G_{2k}(\tau) = \frac{1}{2}\zeta(1-2k) + \sum_{n=1}^{\infty} \sigma_{2k-1}(n)q^n,$$

where $\zeta(s)$ is the Riemann zeta function. For $2k \geq 4$, $G_{2k}(\tau)$ is a weight $2k$ holomorphic modular form on $\mathrm{SL}_2(\mathbb{Z})$. If we let

$$\sigma_k^{(p)} := \sum_{\substack{d|n \\ \gcd(d,p)=1}} d^k$$

and let $\zeta^{(p)}(s)$ be the *p-adic zeta function* **[333]**, then Serre proved that

$$G_{2k}^{(p)}(\tau) = \frac{1}{2}\zeta^{(p)}(1-2k) + \sum_{n=1}^{\infty} \sigma_{2k-1}^{(p)}(n)q^n \tag{6.28}$$

is a p-adic Eisenstein series of weight $2k$. By construction, these p-adic Eisenstein series encode the p-adic continuity of the p-adic zeta function, as well as elementary divisor function congruences. For example, we have that that

$$G_{k_1}^{(p)}(\tau) \equiv G_{k_2}^{(p)}(\tau) \pmod{p^a}$$

whenever $k_1 \equiv k_2 \pmod{(p-1)p^{a-1}}$ with $k_1, k_2 \not\equiv 0 \pmod{p-1}$. For example, $6 \equiv 10 \pmod 4$ and so we have that

$$G_4^{(5)}(\tau) = \frac{781}{126} + q + 33q^2 + 244q^3 + 1057q^4 + q^5 + \cdots$$

and

$$G_{10}^{(5)}(\tau) = \frac{488281}{66} + q + 513q^2 + 19684q^3 + 262657q^4 + q^5 + \cdots$$

are congruent modulo 5.

It is natural to ask whether there is a similar notion for harmonic Maass forms. Here we show that this is indeed the case. Wagner [**485**] has constructed two families, one with integer weight and one with half-integer weight, of harmonic Maass forms which are eigenforms for the Hecke operators which also can be assembled to form p-adic power series. These examples arise as canonical choices of pullbacks under the ξ-operator of the classical Eisenstein series and the Cohen-Eisenstein series.

For positive integers k, we let

$$\begin{aligned} G(\tau, -2k) := {} & \frac{(2k)!\zeta(2k+1)}{(2\pi)^{2k}} + \frac{(-1)^{k+1}v^{1+2k}2^{1+2k}\pi\zeta(-2k-1)}{2k+1} \\ & + (-1)^k(2\pi)^{-2k}(2k)!\sum_{n=1}^{\infty}\frac{\sigma_{2k+1}(n)}{n^{2k+1}}q^n \\ & + (-1)^k(2\pi)^{-2k}\sum_{n=1}^{\infty}\frac{\sigma_{2k+1}(n)}{n^{2k+1}}\Gamma(1+2k, 4\pi nv)q^{-n}. \end{aligned}$$

For half-integral weights we make the following definition. Suppose that r is a positive integer. Then we let

$$T_r^{\chi}(f) := \sum_{a \mid f} \mu(a)\chi(a)a^{r-1}\sigma_{2r-1}\left(\frac{f}{a}\right),$$

where $\mu(a)$ is the Möbius function. Set $(-1)^r N = Df^2$ with D the discriminant of $\mathbb{Q}(\sqrt{D})$. Let

$$c_r(N) := \begin{cases} i^{2r+1}L(1+r,\chi_D)\frac{1}{f^{2r+1}}T_{r+1}^{\chi_D}(f) & \text{if } N > 0, \\ i^{2r-1}\zeta(1+2r) + \frac{2^{2r+4}i\pi^{2r+1}y^{r+\frac{1}{2}}\zeta(-1-2r)}{(2r-3)\Gamma(2r+1)} & \text{if } N = 0, \\ \pi^{\frac{3}{2}}\frac{L(-r,\chi_D)T_{r+1}^{\chi_D}(f)}{N^{r+\frac{1}{2}}}\frac{\Gamma\left(\frac{r+a}{2}\right)}{\Gamma\left(\frac{r+1+a}{2}\right)\Gamma\left(r+\frac{1}{2}\right)}\Gamma\left(r+\frac{1}{2}, -4\pi Nv\right) & \text{if } N < 0, \end{cases}$$

where $a = 0$ if r is odd and $a = 1$ if r is even. Then, for $r \geq 1$, define

$$\mathcal{H}\left(\tau, -r + \frac{1}{2}\right) := \sum_{N \in \mathbb{Z}} c_r(N)q^N.$$

Using the method of Eisenstein series, and recalling the Hecke operators defined in (7.2) and (7.3), we have the following (cf. Theorem 1.1 of [**485**]).

THEOREM 6.19. *Assuming the notation above, the following are true:*

i) *For positive integers k, we have that $G(\tau, -2k)$ is a weight $-2k$ harmonic Maass form on $\mathrm{SL}_2(\mathbb{Z})$. Furthermore, $G(\tau, -2k)$ has eigenvalue $1 + \frac{1}{p^{2k+1}}$ under the Hecke operator T_p.*

ii) *For positive integers* r, *we have that* $\mathcal{H}(\tau, -r+\frac{1}{2})$ *is a weight* $-r+\frac{1}{2}$ *harmonic Maass form on* $\Gamma_0(4)$. *Furthermore,* $\mathcal{H}(\tau, -r+\frac{1}{2})$ *has eigenvalue* $1+\frac{1}{p^{2r+1}}$ *under the Hecke operator* T_{p^2}.

REMARK. The forms in Theorem 6.19 are preimages of the classical Eisenstein series and the Cohen-Eisenstein series under the appropriate ξ-operator.

REMARK. Just as the weight 2 Eisenstein series is not a modular form, the weight 0 form here is not a harmonic Maass form. However, we will see that there is a weight 0 p-adic harmonic Maass form in the same way that there is a weight 2 p-adic Eisenstein series. A more detailed theory of weight 0 examples due to Griffin is described in Section 7.5.

REMARK. An inspection of the Fourier expansions indicates that the forms $G(\tau, -2k)$ are well behaved with respect to the flipping operator discussed in Section 5.3. Namely, they are distinguished by the (near) equality of the Fourier coefficients of their holomorphic and non-holomorphic parts.

In analogy with Serre's notion of a p-adic modular form, we have the following natural generalization.

DEFINITION 6.20. A **weight k p-adic harmonic Maass form** is a formal power series

$$f(\tau) = \sum_{n \gg -\infty} c_f^+(n) q^n + c_f^-(0) v^{1-k} + \sum_{0 \neq n \ll \infty} c_f^-(n) \Gamma(1-k, -4\pi n v)\, q^n$$

with coefficients in $\mathbb{C}_p$ such that there exists a series of harmonic Maass forms $f_j(\tau)$, of weights k_j, such that the following properties are satisfied:

(1) $\lim_{j\to\infty} n^{1-k_j} c_{f_j}^{\pm}(n) = n^{1-k} c_f^{\pm}(n)$ for $n \neq 0$.
(2) $\lim_{j\to\infty} c_{f_j}^{\pm}(0) = c_f^{\pm}(0)$.

REMARK. Here

$$\lim_{j\to\infty} n^{1-k_j} c_{f_i}^{\pm}(n) = n^{1-k} c_f^{\pm}(n)$$

means $v_p(n^{1-k_j} c_{f_j}^{\pm}(n) - n^{1-k} c_f^{\pm}(n))$ tends to ∞ and we have that k is the limit of the k_j in X.

Wagner assembled the forms in Theorem 6.19 to produce the examples of these p-adic harmonic Maass forms. To define them, we let $L_p(s, \chi)$ be the *p-adic L-function* (for example, see **[285]**) and define

$$T_r^{\chi,(p)}(f) := \sum_{\substack{a \mid f \\ \gcd(a,p)=1}} \mu(a)\chi(a) a^{r-1} \sigma_{2r-1}^{(p)}\left(\frac{f}{a}\right).$$

We now have the following theorem (cf. Theorem 1.2 of **[485]**).

THEOREM 6.21. *If p is prime, then the following are true:*

i) *For each $k \in X$, we have that*

$$G^{(p)}(\tau, -2k) = \frac{(2k)!\zeta^{(p)}(2k+1)}{(2\pi)^{2k}} + \frac{(-1)^{k+1}v^{1+2k}2^{1+2k}\pi\zeta^{(p)}(-2k-1)}{2k+1}$$
$$+ (-1)^k(2\pi)^{-2k}(2k)!\sum_{n=1}^{\infty}\frac{\sigma_{2k+1}^{(p)}(n)}{n^{2k+1}}q^n$$
$$+ (-1)^k(2\pi)^{-2k}\sum_{n=1}^{\infty}\frac{\sigma_{2k+1}^{(p)}(n)}{n^{2k+1}}\Gamma(1+2k, 4\pi nv)q^{-n}$$

is a weight $-2k$ p-adic harmonic Maass form.

ii) *For each $r \in X$, let*

$$c_r^{(p)}(N) = \begin{cases} i^{2r+1}L_p(1+r,\chi_D)\frac{1}{f^{2r+1}}T_{r+1}^{\chi_D,(p)}(f) & \text{if } N > 0,\\ i^{2r-1}\zeta^{(p)}(1+2r) + \frac{2^{2r+4}i\pi^{2r+1}y^{r+\frac{1}{2}}\zeta^{(p)}(-1-2r)}{(2r-3)\Gamma(2r+1)} & \text{if } N = 0,\\ \pi^{\frac{3}{2}}\frac{L_p(-r,\chi_D)T_{r+1}^{\chi_D,(p)}(f)}{N^{r+\frac{1}{2}}}\frac{\Gamma\left(\frac{r+a}{2}\right)}{\Gamma\left(\frac{r+1+a}{2}\right)\Gamma\left(r+\frac{1}{2}\right)}\Gamma\left(r+\tfrac{1}{2}, -4\pi Nv\right) & \text{if } N < 0.\end{cases}$$

Then $\mathcal{H}^{(p)}\left(\tau, -r+\frac{1}{2}\right) = \sum_{N\in\mathbb{Z}} c_r^{(p)}q^N$ is a weight $-r+\frac{1}{2}$ p-adic harmonic Maass form.

REMARK. Suppose p is a prime and consider an infinite sequence of even integers which p-adically go to zero (e.g. $\{2p^t\}_{t=1}^{\infty}$). The proof of Theorem 6.21 gives a p-adic harmonic Maass form of weight 0. This is the analogue of the quasimodular form E_2 which (although not modular) is a weight 2 p-adic modular form. In fact, it is the p-adic preimage of E_2 under the ξ-operator.

CHAPTER 7

Hecke Theory

Hecke operators play a central role in the study of modular forms. The classical theory of newforms and modular L-functions relies on the fact that spaces of cusp forms are finite-dimensional and can be diagonalized with respect to the Hecke algebra. Although spaces of harmonic Maass forms are not finite-dimensional, it turns out that Hecke operators continue to play a pivotal role. In fact, an analogous theory of "Hecke eigenforms" for weakly holomorphic modular forms can be described. The fundamental features of the Hecke structure of harmonic Maass forms arise from the intimate relationship between harmonic Maass forms and classical modular forms, facilitated by the ξ_k-operator. Here we discuss the actions of these operators on harmonic Maass forms, and we consider the algebraic and p-adic properties that can be deduced in special cases. To ease notation, throughout we only consider forms with real Nebentypus χ.

7.1. Basic facts

We begin by recalling the definitions of integral and half-integral weight Hecke operators on harmonic Maass forms. For a comprehensive review of the classical Hecke theory, we refer the reader to Chapter 5 of **[162]**. By the results in Section 4.2, any $F \in H_\kappa^!(\Gamma_0(N), \chi)$ has a Fourier expansion of the form ($\tau = u + iv$ as usual)

$$F(\tau) = \sum_{n\in\mathbb{Z}} c_F(n, v)e^{2\pi inu} \tag{7.1}$$

(which is explicitly described in Lemma 4.3). Note that throughout this chapter we make the convention that $k \in -\mathbb{N}_0$, $\kappa \in \frac{1}{2}\mathbb{Z}$.

As in the case of classical modular forms, one can easily describe the action of Hecke operators by their action on Fourier expansions. To this end, suppose throughout that $p \nmid N$ is prime and $4|N$ in the case of half-integral weight. The corresponding integral and half-integral Hecke operators, which generate the Hecke algebra, are defined as follows. If κ is an integer, then we define

$$F(\tau)|T_{p,\kappa,\chi} := \sum_{n\in\mathbb{Z}} \left(c_F\left(pn, \frac{v}{p}\right) + \chi(p)p^{\kappa-1}c_F\left(\frac{n}{p}, pv\right)\right)e^{2\pi inu}. \tag{7.2}$$

As usual, for primes $p|N$, this definition must be slightly modified, and multiplicativity relations among the Hecke operators allow one to describe the Hecke operators $T_{m,\kappa,\chi}$ for general natural numbers m from the prime-indexed ones. We omit these formulas here, as we do not require them in the applications discussed in this book (the interested reader can find equivalent formulas in Section 5.3 of [**162**]). If $\kappa \in \frac{1}{2}\mathbb{Z}\setminus\mathbb{Z}$, then we set

$$F(\tau)|T_{p^2,\kappa,\chi} := \sum_{n\in\mathbb{Z}} \left(c_F\left(p^2 n, \frac{v}{p^2}\right) + \chi^*(p)\left(\frac{n}{p}\right) p^{\kappa-\frac{3}{2}} c_F(n,v) + \chi^*\left(p^2\right) p^{2\kappa-2} c_F\left(\frac{n}{p^2}, p^2 v\right)\right) e^{2\pi i n u}, \tag{7.3}$$

where $\chi^*(n) := \left(\frac{(-1)^{\kappa-\frac{1}{2}}}{n}\right)\chi(n)$. Throughout we write T_p or T_{p^2} for brevity if κ and χ are clear from context.

The first fundamental fact in the Hecke theory of harmonic Maass forms follows as a straightforward generalization of the usual proof that the Hecke operators for modular forms commute with the Petersson slash operator for matrices in $\Gamma_0(N)$, coupled with a standard use of the Fourier expansion of Lemma 4.3.

PROPOSITION 7.1. *Suppose that $F \in H_\kappa^!(\Gamma_0(N),\chi)$ with $\kappa \in \frac{1}{2}\mathbb{Z}$. Then the following are true:*

i) *For $m \in \mathbb{N}$, we have that $F|T_m \in H_\kappa^!(\Gamma_0(N),\chi)$.*

ii) *If $\kappa \in \mathbb{Z}, \varepsilon \in \{\pm\}$, then, unless $n = 0$ and $\varepsilon = -$,*

$$c_{F|T_p}^\varepsilon(n) = c_F^\varepsilon(pn) + \chi(p)p^{\kappa-1}c_F^\varepsilon\left(\frac{n}{p}\right).$$

Moreover, we have that

$$c_{F|T_p}^-(0) = \left(p^{\kappa-1} + \chi(p)\right)c_F^-(0).$$

iii) *If $\kappa \in \frac{1}{2}\mathbb{Z}\setminus\mathbb{Z}$, then, with $\varepsilon \in \{\pm\}$ ($n \neq 0$ for $\varepsilon = -$), we have that*

$$c_{F|T_{p^2}}^\varepsilon(n) = c_F^\varepsilon\left(p^2 n\right) + \chi^*(p)\left(\frac{n}{p}\right)p^{\kappa-\frac{3}{2}}c_F^\varepsilon(n) + \chi^*\left(p^2\right)p^{2\kappa-2}c_F^\varepsilon\left(\frac{n}{p^2}\right).$$

If $n = 0$ and $\varepsilon = -$, then we have that

$$c_{F|T_{p^2}}^-(0) = \left(p^{-2+2\kappa} + \chi^*\left(p^2\right)\right)c_F^-(0).$$

In particular, Proposition 7.1 asserts that the Hecke operators act on the spaces $H_\kappa^!(\Gamma_0(N),\chi)$, and the explicit formulas for the resulting Fourier coefficients are implicitly used throughout this chapter. Since ξ_k maps $H_k(\Gamma_0(N),\chi)$ to classical cusp forms (cf. Theorem 5.10 i)), which come equipped with the Atkin-Lehner-Li theory of newforms, this action inherits some natural extra structure. This interaction is made possible by the following compatibility between harmonic Maass forms and Hecke theory. If $k \in -\mathbb{N}$, then we have

$$p^{1-k}D^{1-k}\left(F|T_{p,k,\chi}\right) = D^{1-k}(F)|T_{p,2-k,\chi}, \tag{7.4}$$

and if $\kappa \in -\frac{1}{2}\mathbb{N}$, then

$$p^{d(1-\kappa)}\xi_\kappa\left(F|T_{p^d,\kappa,\chi}\right) = \xi_\kappa(F)|T_{p^d,2-\kappa,\chi}, \tag{7.5}$$

where

$$d := \begin{cases} 1 & \text{if } \kappa \in \mathbb{Z}, \\ 2 & \text{if } \kappa \in \frac{1}{2}+\mathbb{Z}. \end{cases}$$

The interplay between these structures offers a useful device for relating these harmonic Maass forms to weakly holomorphic modular forms. For example, by using (7.5) to compare nonholomorphic parts, one finds the following weakly holomorphic linear combination.

PROPOSITION 7.2. *Suppose that* $F \in H_\kappa(\Gamma_0(N),\chi)$*, and let* $f := \xi_\kappa(F) \in S_{2-\kappa}(\Gamma_0(N),\chi)$*. If* $f|T_{p^d} = \lambda_{p^d} f$*, then*

$$F|T_{p^d} - p^{d(\kappa-1)}\overline{\lambda_{p^d}}\, F \in M_\kappa^!(\Gamma_0(N),\chi).$$

Proposition 7.2 provides a useful method for studying p-adic properties of mock modular forms, and we make crucial use of it below in the proof of Theorem 7.10 in our study of algebraic properties of mock modular coefficients.

7.2. Weakly holomorphic Hecke eigenforms

The infinite-dimensionality of spaces of harmonic Maass forms, as well as their subspaces of weakly holomorphic modular forms, is a potential roadblock to extending the classical theory of Hecke operators to this more general setting. However, we have already seen that the Hecke operators still act on these spaces and intertwine with the most important differential operators acting on those spaces. Hence, it is natural to search for an analogous theory of eigenforms in this setting.

Recently, Guerzhoy solved this problem for weakly holomorphic forms. He proposed the notion of "weakly holomorphic Hecke eigenforms" in **[243]**. Here we describe this phenomenon. The definition of weakly holomorphic eigenforms uses the iterated derivatives of Bol's identity (Lemma 5.3), which plays a central role in the theory of harmonic Maass forms. Here, we recall the basics of Guerzhoy's theory.

For simplicity, we now restrict to forms on $\mathrm{SL}_2(\mathbb{Z})$. Let $S_k^!$ denote the space of *weakly holomorphic cusp forms*, i.e., those weakly holomorphic forms with vanishing constant term. Furthermore, for a vector space $\mathcal{S}$ and a subspace $\mathcal{J} \subset \mathcal{S}$, we say that $f, g \in \mathcal{S}$ satisfy the congruence

$$f \equiv g \pmod{\mathcal{J}}$$

if $f - g \in \mathcal{J}$.

DEFINITION 7.3. For integral $k < 0$, a weight $2-k$ **weakly holomorphic Hecke eigenform** is any $f \in M_{2-k}^! \setminus D^{1-k}(S_k^!)$ for which there exist $\lambda_p \in \mathbb{C}$ satisfying (for all primes p)

$$f|T_p \equiv \lambda_p f \pmod{D^{1-k}(S_k^!)}. \tag{7.6}$$

REMARK. In new results announced by Funke, a natural cohomological interpretation of this definition is given, in which $D^{1-k}(S_k^!)$ is interpreted as a space of coboundaries.

In view of these remarks, Guerzhoy's main theorem (cf. Theorem 1 of **[243]**) is particularly satisfying. He proved that weakly holomorphic Hecke eigenforms exist. Namely, Guerzhoy and Kent proved the following theorem (cf. Theorem 1 of **[243]** and **[305]**).

THEOREM 7.4. *The quotient*

$$\widehat{M_{2-k}} := M_{k-2}^! / \left(D^{1-k}\left(S_k^!\right) \oplus M_{2-k}\right)$$

has finite dimension, and, in fact,

$$\dim\left(\widehat{M_{2-k}}\right) = \dim(M_{2-k}).$$

Moreover, the space $\widehat{M}_{2-k}$ has a basis consisting of Hecke eigenforms, with eigenvalues equal to those of a corresponding basis of eigenforms of M_k.

Despite Theorem 7.4, the definition of a weakly holomorphic Hecke eigenform might not be very enlightening at first glance. If one simply defines "eigenforms" as elements of some quotient space, then one can replace $D^{1-k}(S_k^!)$ with some other subspace $\mathcal{J}$. The only requirement is that the Hecke operators send $\mathcal{J}$ to itself. The subspace $D^{1-k}(S_k^!)$ must hence be chosen carefully, and it is natural to ask why $D^{1-k}(S_k^!)$ is the correct subspace. There are a few answers to this question, coming from different perspectives. The initial perspective taken in **[243]** was a p-adic one, since the $(p^m n)$-th coefficients of $f \in D^{1-k}(S_k^!)$ become "divisible" by higher powers of p as m gets larger. Moreover, certain regularized critical L-values are zero for elements in this space (cf. Theorem 2.5 of **[90]**, as well as **[208]**). The third justification of this choice comes from the regularized integral of Definition 5.6 and the subsequent remark. If we let

$$M_{2-k}^{!,\perp} := \left\{ f \in M_{2-k}^! : \langle f, g \rangle = 0 \text{ for all } g \in M_{2-k}^! \right\} \tag{7.7}$$

be the space of weakly holomorphic forms which are orthogonal to all of $M_{2-k}^!$, then it turns out that (rewriting Corollary 4.5 of **[82]** using the flipping operator and using Theorem 5.9) $f \in M_{2-k}^{!,\perp}$ if and only if $f \in D^{1-k}(S_k^!)$.

We conclude this section by noting that these results foreshadow the mock modular Eichler-Shimura theory of Section 12.2, where similar quotient spaces play a key role. In particular, the thinking of $D^{1-k}(S_k^!)$ as a sort of "coboundary" space, as Funke has realized it, aligns in a pleasing manner with the vanishing result of Corollary 12.9. In view of this role in Eichler-Shimura theory, and with the recurring theme throughout this book that harmonic Maass forms constitute natural extensions of spaces of weakly holomorphic modular forms, the following question is very natural.

PROBLEM. *Is there a natural theory of Hecke eigenforms for harmonic Maass forms?*

7.3. Harmonic Maass forms and complex multiplication

Returning to the more general setting of harmonic Maass forms, we discuss how Hecke operators can be used to study algebraicity of mock modular coefficients. The case when the shadow has complex multiplication is distinguished and is somewhat simpler. To discuss this, we first fix a normalized (i.e., $c_f(1) = 1$) newform $f(\tau) = \sum_{n=1}^{\infty} c_f(n) q^n \in S_{2-k}(\Gamma_0(N), \chi)$ of weight $2 - k \in \mathbb{N}_{\geq 2}$. Let K_f denote the number field obtained by adjoining the coefficients of f to $\mathbb{Q}$.

DEFINITION 7.5. We say that $F \in H_k(\Gamma_0(N), \chi)$ is **good** for the newform f if it satisfies the following properties:

i) The principal part of F at the cusp $i\infty$ belongs to $K_f[q^{-1}]$.
ii) The principal parts of F at other cusps (if any) are constant.
iii) We have $\xi_k(F) = \frac{f}{\|f\|^2}$, where $\|f\|$ is the Petersson norm of f.

Before explaining the utility of this definition, we first establish a refined version of the crucial surjectivity statement of Theorem 5.10 i). For brevity, we omit the proof (the interested reader should consult Proposition 5.1 of **[129]**).

THEOREM 7.6. *For any* $f \in S_{2-k}(\Gamma_0(N), \chi)$, *there exists a harmonic Maass form* $F \in H_k(\Gamma_0(N), \chi)$ *which is good for* f.

We now discuss the algebraicity of CM-type mock modular form coefficients, i.e., those mock modular forms whose shadows have complex multiplication. For basic facts on classical CM modular forms, we refer the reader to **[412]**. The following result of Bruinier, Rhoades, and one of the authors (cf. Theorem 1.3 of **[129]**) provides a satisfactory answer in the case of CM forms f.

THEOREM 7.7. *Let* $f \in S_{2-k}(\Gamma_0(N), \chi)$ *be a normalized newform with complex multiplication. If* $F \in H_k(\Gamma_0(N), \chi)$ *is good for* f, *then all coefficients of* F^+ *are in* $K_f(\zeta_M)$, *where* $\zeta_M = e^{2\pi i/M}$, *and* $M = Nd$, *where* d *is the discriminant of the field of complex multiplication for* f.

REMARK. In **[137]**, Candelori related his geometric perspectives on such constructions (cf. the last remark in Section 4.1) to a conjecture of Bruinier, Rhoades, and the third author that in fact the coefficients of F^+ in the notation of Theorem 7.7 lie in K_f.

7.4. p-adic properties of integral weight mock modular forms

By Lemma 4.3, a harmonic Maass form naturally consists of two pieces, a nonholomorphic part and a holomorphic part. Recalling Lemma 5.17, if F is a weight k harmonic Maass form, then its nonholomorphic part is a period integral of the shadow of F^+. There is no known simple procedure to "uncover" the shadow of a mock modular form given its q-expansion alone. Due to the numerous applications, it is natural to consider the following general problem.

PROBLEM. *Given the q-expansion of a mock modular form, can one directly "uncover" its shadow, and hence determine its harmonic Maass form completion?*

Here we consider this problem whenever the shadow f is an integer weight newform. In Subsection 7.4.2 we present a p-adic solution to this question which produces the shadow f from the mock modular form of a suitable harmonic Maass form F. It turns out that the corresponding constructions are closely related to p-adic modular forms. This relationship is explained in Subsection 7.4.3.

For the rest of this chapter, we fix a normalized newform in $S_{2-k}(\Gamma_0(N), \chi)$ with weight at least 2 and Nebentypus χ. Define $\mathcal{E}_f$ to be the *(holomorphic) Eichler integral* of f:

$$\mathcal{E}_f(\tau) := \sum_{n=1}^{\infty} c_f(n) n^{k-1} q^n. \tag{7.8}$$

Let $F \in H_k(\Gamma_0(N), \chi)$ be a form which is good for f^c (cf. Lemma 5.17 for the definition of f^c). The objective is to determine the Fourier coefficients of f from those of F^+ . Theorem 7.12 below gives p-adic limit formulae for f in terms of F^+. Theorem 7.18 and the surrounding discussion relate F^+ to p-adic modular forms. Before we can state these results, we must consider the algebraicity of the mock modular form F^+.

7.4.1. Algebraicity. Since the coefficients $c_f(n)$ are algebraic integers (recall that f is a newform), we must first consider the arithmetic properties of the $c_F^+(n)$. By Theorem 7.7, F^+ has algebraic coefficients if f has complex multiplication. Otherwise, we expect the coefficients of F^+ to be transcendental.

EXAMPLE 7.8. For $f = \Delta$, the unique normalized weight 12 cusp form on $\mathrm{SL}_2(\mathbb{Z})$, one may use the first weight -10 Maass-Poincaré series on $\mathrm{SL}_2(\mathbb{Z})$ (cf. Definition 6.10) to find a good form F (cf. [**406**]). We consider the numerical approximations of the first few terms of $11!F^+$:

(7.9)

$$\begin{aligned} 11!F^+(\tau) &:= 11!F^+_{-10,-1,1}(\tau) \\ &\approx \frac{11!}{q} - \frac{2615348736000}{691} - 73562460235.68364q - 929026615019.11308q^2 - \cdots . \end{aligned}$$

After the first two coefficients on the right-hand side of (7.9), the coefficients appear to be transcendental.

In the general non-CM case, it is conjectured that a similar phenomenon holds.

CONJECTURE 7.9. *Assume that a mock modular form F^+ of integral weight has one non-vanishing algebraic Fourier coefficient outside the principal part. Then all of the coefficients of F^+ are algebraic if and only if its shadow has CM.*

REMARK. The transcendence properties of mock modular coefficients are connected to deep problems such as Lehmer's conjecture on the non-vanishing of Fourier coefficients of Δ.

Despite the apparent transcendence of the coefficients of generic mock modular forms F^+, it turns out that F^+ may always be regularized in a simple way to obtain an algebraic q-series. Theorem 1.1 of [**245**] explains this.

THEOREM 7.10. *Assume the hypotheses above. If α is a complex number for which $\alpha - c_F(1) \in K_f$, then the coefficients of*

$$\mathcal{F}_\alpha(\tau) := F^+(\tau) - \alpha\mathcal{E}_f(\tau) = \sum_{n \gg -\infty} c_F^+(n)q^n - \alpha\sum_{n=1}^{\infty} c_f(n)n^{k-1}q^n$$

are in K_f. In particular, the transcendence degree of $K_f(c_F^+(n))$ over K_f is at most 1.

REMARK. One may always take $\alpha = c_F^+(1)$ in Theorem 7.10.

REMARK. It is clear that the Fourier coefficients of $\mathcal{F}_\alpha$ may have unbounded negative p-adic order, due to the presence of unbounded powers of any prime in the denominator of the coefficients of the Eichler integral $\mathcal{E}_f$. One can obtain boundedness by adding a second correction, subtracting multiples of both the Eichler integral and its image under the V_p-operator of (2.28) (which sends $f(\tau)$ to $f(p\tau)$), as seen in Theorem 7.18 below. In the special cases when f has weight two or when $f = \Delta$, Guerzhoy [**244**] showed that the correction term $\alpha = \alpha_p$ appearing in Theorem 7.18 is an adele (i.e., it is a p-adic integer for all but finitely many primes p). This follows a suggestion of Zagier which associates a real number and p-adic numbers for all primes to the weight 12 cusp form Δ.

REMARK. We call the function $\mathcal{F}_\alpha$ of Theorem 7.10 a *regularized mock modular form.*

Let us now reconsider the example of the Δ-function.

EXAMPLE 7.11. For $f = \Delta$, if we take F as in Example 7.8 and let $\alpha = c_F^+(1)$, then Theorem 7.10 implies that $\mathcal{F}_\alpha$ has $\mathbb{Q}$-rational coefficients. Numerically, we indeed find that

(7.10)
$$11!\mathcal{F}_\alpha(\tau) = 11!q^{-1} - \frac{2615348736000}{691} - 929888675100q^2 - \frac{80840909811200}{9}q^3 - \cdots.$$

PROOF OF THEOREM 7.10. We define

$$F_\alpha(\tau) := D^{1-k}\left(\mathcal{F}_\alpha(\tau)\right) =: \sum_{n \gg -\infty} C_\alpha(n)q^n. \tag{7.11}$$

By construction, $C_\alpha(1)$ lies in K_f. We use the action of the Hecke operators on F. Since f is a newform, Proposition 7.2 implies that

$$F|T_m = m^{k-1}c_f(m)F + R_m, \tag{7.12}$$

where $R_m \in M_k^!(\Gamma_0(N), \chi)$ is a weakly holomorphic modular form with coefficients in K_f (by the explicit formulae above which show that $|T_m$ preserves the field of coefficients of a form). Note that F is a Hecke eigenform in the sense of Section 7.2. We now apply the differential operator D^{1-k} to (7.12), using (7.4). This then gives

$$D^{1-k}(F)|T_m = c_f(m)D^{1-k}(F) + m^{1-k}D^{1-k}(R_m).$$

Since $D^{1-k}(F) = D^{1-k}(F^+)$, $D^{1-k}(\mathcal{E}_f) = f$, and $f|T_m = c_f(m)f$, we conclude that

$$\left(D^{1-k}\left(F^+ - \alpha\mathcal{E}_f\right)\right)|T_m = c_f(m)\left(D^{1-k}\left(F^+ - \alpha\mathcal{E}_f\right)\right) + m^{1-k}D^{1-k}(R_m). \tag{7.13}$$

We claim that F_α has its coefficients in K_f. Indeed, we make use of the formula for the action of Hecke operators on Fourier expansions, equate the coefficients of q^n in this formula, and then conclude that for any prime p

$$C_\alpha(pn) + \chi(p)p^{1-k}C_\alpha\left(\frac{n}{p}\right) - c_f(p)C_\alpha(n) \in K_f.$$

Since $C_\alpha(1)$ is in K_f, a simple inductive argument completes the proof. □

7.4.2. p-adic coupling of mock modular forms with newforms. We now return to the original task of this section: to obtain p-adic formulae for the newform f in terms of the mock modular form F^+. We begin by choosing a complex number α for which $\alpha - c_F^+(1) \in K_f$. The regularizations in Theorem 7.10 are the key to the procedure of "coupling" mock modular forms with their shadows described below.

Let p be prime and fix an algebraic closure $\overline{\mathbb{Q}}_p$ of $\mathbb{Q}_p$, along with an embedding of $\overline{\mathbb{Q}}$ into $\overline{\mathbb{Q}}_p$. Furthermore, we embed $\overline{\mathbb{Q}}_p$ into its completion $\mathbb{C}_p$. These embeddings determine an extension of the p-adic valuation to K_f. We denote by $\mathrm{ord}_p\colon \mathbb{C}_p \to \mathbb{Q}$ the p-adic order, normalized so that $\mathrm{ord}_p(p) = 1$.

To relate $\mathcal{F}_\alpha$ to f, if f^c does not have CM, recall that

$$F_\alpha(\tau) = D^{1-k}\left(\mathcal{F}_\alpha(\tau)\right) = \sum_{n \gg -\infty} C_\alpha(n)q^n \tag{7.14}$$

as in the proof of Theorem 7.10. If f^c has CM, then we let

$$G_\alpha(\tau) := D^{1-k}\left(F^+(\tau) - \alpha\mathcal{E}_f(p\tau)\right) =: \sum_{n \gg -\infty} d_\alpha(n)q^n. \tag{7.15}$$

Bol's identity (cf. Theorem 5.3), combined with the fact that $D^{1-k}(\mathcal{E}_f) = f$, implies that such forms are weight $2-k$ weakly holomorphic modular forms.

We iteratively apply Atkin's U_p-operator to these q-series to couple mock modular forms with their shadows. Recall that this operator was defined in (2.28) for $m \in \mathbb{N}$ by

$$\sum_n a(n)q^n | U_m = \sum_n a(mn)q^n. \tag{7.16}$$

Now let β_1 and β_2 be the roots of the equation

$$X^2 - c_f(p)X + \chi(p)p^{1-k} =: (X-\beta_1)(X-\beta_2) \tag{7.17}$$

ordered so that $\mathrm{ord}_p(\beta_1) \le \mathrm{ord}_p(\beta_2)$. Then define the cusp form $\breve{f}$ by

$$\breve{f}(\tau) = \sum_{n=1}^{\infty} \breve{c}_f(n)q^n := f(\tau) - \beta_1^{-1}\chi(p)p^{1-k}f(p\tau). \tag{7.18}$$

Using the properties of the V_p-operator, we have that if $p \nmid N$, then $\breve{f} \in S_{2-k}(\Gamma_0(pN), \chi)$, and if $p|N$, then $\breve{f} = f$ since $\chi(p) = 0$.

The following result, proved by Guerzhoy, Kent, and one of the authors [**245**], relates the Fourier coefficients of f^c to F^+.

THEOREM 7.12. *Assume the hypotheses above.*

i) *Suppose that $p \nmid N$ and $\mathrm{ord}_p(\beta_1) \neq (1-k)/2$, or $p|N$ and $\beta_1 \neq 0$. For all but at most one α with $\alpha - c_F(1) \in K_f$, we have that*

$$\breve{f} = \lim_{r\to\infty} \frac{F_\alpha | U_{p^r}}{C_\alpha(p^r)}.$$

ii) *Suppose that f has CM. If p is inert in the field of complex multiplication, then for all but at most one choice of $\alpha \in K_f$, we have that*

$$f = \lim_{r\to\infty} \frac{G_\alpha | U_{p^{2r+1}}}{d_\alpha(p^{2r+1})}.$$

REMARK. Some of the denominators in the formulae in Theorem 7.12 can vanish. However, the proof shows that there are at most finitely many r for which this is the case.

REMARK. In the case of trivial tame level (i.e., N is a power of p) and trivial Nebentypus χ, the space of p-*ordinary* cusp forms (i.e., those whose p-th Fourier coefficients are not divisible by p) is empty if $p \le 7$ or $k \equiv -2, -4, -6, -8, -12 \pmod{p-1}$. Thus f is necessarily non-p-ordinary, and the proofs of Theorem 7.12 and Proposition 7.15 below give rise to p-adic modular forms. This is discussed in further detail in Subsection 7.4.3.

REMARK. The proof of Theorem 7.12 can break down for one exceptional α. For example, if f has CM, then $\alpha = 0$ can be exceptional if p is a prime which splits in the field of complex multiplication. These exceptional cases are of interest, and they correspond to situations where one directly obtains p-adic modular forms without iteration.

EXAMPLE 7.13. Theorem 7.12 implies infinitely many congruences. For $f = \Delta$ and $p = 3$, we have that $\mathrm{ord}_3(\beta_1) = 2$ and also that $\Delta \equiv \breve{\Delta} \pmod{3^9}$. Using (7.9) and (7.10), we find that the $r = 1$ term in Theorem 7.12 i) numerically gives

$$-\frac{D^{11}\left(F^+ - c_F^+(1)\sum_{n=1}^{\infty}\tau(n)n^{-11}q^n\right)|\,U_3}{39862705122} \equiv \Delta \pmod{27}.$$

The proof of Theorem 7.12 requires three preliminary propositions. The first, whose straightforward proof we omit, is as follows. We remark that the key idea is to multiply by a power of Δ to first obtain a cusp form of positive weight.

PROPOSITION 7.14. *If $R \in M_\kappa^!(\Gamma_0(N), \chi)$ has K_f-rational coefficients, then there is an $A \in \mathbb{Z}$ such that*

$$\mathrm{ord}_p\left(D^{1-\kappa}(R)|U_{p^r}\right) \geq r(1-\kappa) - A.$$

We now prove the existence of the limits which appear in Theorem 7.12.

PROPOSITION 7.15. *Assuming the hypotheses in Theorem 7.12 i), we have that*

$$\lim_{r\to\infty}\left(\beta_1^{-r}F_\alpha|U_{p^r}\right) \in \mathbb{C}_p[[q]].$$

PROOF. We assume that $p \nmid N$; the proof in the case $p \mid N$ is similar. Note that from (7.2), we conclude that

$$F_\alpha|T_p = F_\alpha|U_p + \chi(p)p^{1-k}F_\alpha|V_p. \tag{7.19}$$

Then (7.13) with $m = p$ gives, using (7.19),

$$F_\alpha|U_p + \beta_1\beta_2F_\alpha|V_p = (\beta_1 + \beta_2)\,F_\alpha + r_p,$$

where $r_m = r_{m,k} := m^{1-k}D^{1-k}(R_m)$ is a weakly holomorphic modular form in $M_{2-k}^!(\Gamma_0(N), \chi)$. This makes use of (7.17), which implies that

$$\beta_1 + \beta_2 = c_f(p) \quad \text{and} \quad \beta_1\beta_2 = \chi(p)p^{1-k}.$$

Now we let

$$G := F_\alpha - \beta_2F_\alpha|V_p \quad \text{and} \quad G^* := F_\alpha - \beta_1F_\alpha|V_p. \tag{7.20}$$

A short calculation reveals that

$$G|U_p = \beta_1G + r_p, \qquad\qquad G^*|U_p = \beta_2G^* + r_p,$$

and also that

$$F_\alpha|U_p = \frac{\beta_1}{\beta_1 - \beta_2}(\beta_1G + r_p) - \frac{\beta_2}{\beta_1 - \beta_2}(\beta_2G^* + r_p).$$

By induction on r, we find that

$$\begin{aligned}(\beta_1 - \beta_2)\,\beta_1^{-r}F_\alpha|U_{p^r} = \;&\left(\beta_1G + r_p + \tfrac{1}{\beta_1}r_p|U_p + \cdots + \tfrac{1}{\beta_1^{r-1}}r_p|U_{p^{r-1}}\right)\\ &-\left(\tfrac{\beta_2}{\beta_1}\right)^r\left(\beta_2G^* + r_p + \tfrac{1}{\beta_2}r_p|U_p + \cdots + \tfrac{1}{\beta_2^{r-1}}r_p|U_{p^{r-1}}\right).\end{aligned}$$

Proposition 7.15 now follows from Proposition 7.14 and this formula. □

Next we prove that the limits in Theorem 7.12 ii) are well defined.

PROPOSITION 7.16. *Assuming the hypotheses in Theorem 7.12 ii), we have that*

$$\lim_{r\to\infty}\left(\beta_1^{-2r}G_\alpha|U_{p^{2r+1}}\right) \in \mathbb{C}_p[[q]].$$

PROOF. Observe that

$$\lim_{r\to\infty}\left(\beta_1^{-2r}\mathcal{E}_f|U_{p^{2r}}\right)\in\mathbb{C}_p[[q]].$$

It thus suffices to check the statement of the proposition for $\alpha=0$. Since p is inert in the CM-field, we have $\beta_2=-\beta_1$, and so $\beta_1^2=-\chi(p)p^{1-k}$. As in the proof of Proposition 7.15, we rewrite equation (7.13) with $m=p$:

$$G_0|U_p-\beta_1^2G_0|V_p=r_p.$$

Thus we have that

$$G_0|U_{p^2}=\beta_1^2G_0+r_p|U_p.$$

Acting with the U_p-operator on this identity $2r-1$ times, we obtain

$$\beta_1^{-2r}G_0|U_{p^{2r+1}}=G_0|U_p+\beta_1^{-2}r_p|U_{p^2}+\beta_1^{-4}r_p|U_{p^4}+\cdots+\beta_1^{-2r}r_p|U_{p^{2r}}. \tag{7.21}$$

As in Proposition 7.15, we conclude that the p-adic limit exists. □

We are now in a position to prove the remainder of Theorem 7.12.

PROOF OF THEOREM 7.12. Here we prove Theorem 7.12 i); the proof of Theorem 7.12 ii) is similar, apart from the fact that one applies Proposition 7.16 in place of Proposition 7.15.

We begin by setting

$$L=L(\alpha):=\lim_{r\to\infty}\left(\beta_1^{-r}C_\alpha(p^r)\right).$$

Since $\lim_{r\to\infty}\left(\beta_1^{-r}c_f(p^r)\right)=\beta_1/(\beta_1-\beta_2)$ and $C_\alpha(n)=n^{k-1}c_F^+(n)-\alpha c_f(n)$, there is at most one choice of $\alpha\in K_f$ for which $L(\alpha)=0$. For non-exceptional α, we can then conclude that $C_\alpha(p^n)\neq 0$ for $n\gg 0$.

Let

$$L^{-1}\left(\lim_{r\to\infty}\left(\beta_1^{-r}F_\alpha(\tau)|U_{p^r}\right)\right)=:\sum_{m=1}^{\infty}\breve{c}(m)q^m. \tag{7.22}$$

It follows from the proof of Proposition 7.15 that

$$\left(\lim_{r\to\infty}\left(\beta_1^{-r}F_\alpha|U_{p^r}\right)\right)\Big|U_p=\beta_1\left(\lim_{r\to\infty}\left(\beta_1^{-r}F_\alpha|U_{p^r}\right)\right).$$

Therefore, by (7.18), (7.22), and the recursive formula for $c_f(p^n)$, we inductively find that

$$\frac{\breve{c}(p^n)}{\breve{c}(1)}=\beta_1^n=\breve{c}_f(p^n)$$

for all $n\in\mathbb{N}_0$. For $m\in\mathbb{N}$ such that $p\nmid m$, we then have that

$$F_\alpha|T_m=\breve{c}_f(m)F_\alpha+r_m.$$

Since the operators U_p and T_m commute, we obtain

$$\left(F_\alpha|U_{p^r}\right)|T_m=\breve{c}_f(m)\left(F_\alpha|U_{p^r}\right)+r_m|U_{p^r}.$$

We divide this equation by $C_\alpha(p^r)$, and then take the limit as $r\to\infty$. By Proposition 7.14, the formulas for Hecke operators, and the property that Fourier coefficients of Hecke eigenforms are the eigenvalues, Theorem 7.12 i) follows. □

7.4.3. Relationship with p-adic modular forms. In this subsection, we describe a precise relationship between F^+ and p-adic modular forms following **[91]**.

DEFINITION 7.17. A **p-adic modular form** of level N and Nebentypus χ is a formal power series $H(q) = \sum_{n=-\ell}^{\infty} a(n)q^n$ with coefficients in $\mathbb{C}_p$ satisfying the following condition: for every $m \in \mathbb{N}$ there exist integers k_m and a sequence of weakly holomorphic modular forms $H_m(\tau) = \sum_{n=-\ell}^{\infty} c_m(n)q^n \in M_{k_m}^!(\Gamma_0(N), \chi)$, with algebraic coefficients $c_m(n) \in \overline{\mathbb{Q}}$, which satisfy the congruences

$$H \equiv H_m \pmod{p^m}. \tag{7.23}$$

If $k_m = k$ is constant, then we refer to H as a **p-adic modular form of weight** k, level N, and Nebentypus χ.

REMARK. One can relate p-adic modular forms as defined above to those of Serre **[452]**, where the H_m are required to be holomorphic. For this, let Δ be the unique normalized newform of weight 12 for $\mathrm{SL}_2(\mathbb{Z})$. Consider the case of trivial Nebentypus and level $N = p^j$ ($j \in \mathbb{N}_0$). After multiplication by Δ^ℓ, where $\ell \in \mathbb{Z}$ is the appropriate power as given in the expansion of $H(q)$ above, every p-adic modular form in our sense becomes a p-adic modular form of weight $k + 12$ in the sense of Serre due to Theorem 5.4 of **[453]**.

We now determine the unique correction (up to addition of a p-adic modular form) needed to complete F^+ to a p-adic modular form. If F^+ is not itself a p-adic modular form, this series has an intimate relationship with its shadow. Subsection 7.4.1 shows that $\mathcal{F}_{c_F^+(1)} = F^+ - c_F^+(1)\mathcal{E}_f$, defined in Theorem 7.10, has coefficients in K_f.

The fact that the function $\mathcal{F}_\alpha$ defined in Theorem 7.10 has coefficients in $\mathbb{C}_p$ now permits one to consider its p-adic properties. For the purposes of this book, we consider only the "generic" case that $\mathrm{ord}_p(\beta_1) < \mathrm{ord}_p(\beta_2)$, and for simplicity we restrict to good primes $p \nmid N$. We then define

$$\mathcal{F}_\alpha^* := \mathcal{F}_\alpha - p^{k-1}\beta_2\mathcal{F}_\alpha|V_p = \left(F^+ - \alpha\mathcal{E}_f\right) | \left(1 - p^{k-1}\beta_2 V_p\right). \tag{7.24}$$

To illustrate this, consider Example 7.8 when $f = \Delta$, and denote the coefficients of $11!\mathcal{F}_{c_F^+(1)}$ (resp. $11!\mathcal{F}^*_{c_F(1)}$) by $c(n) \in \mathbb{Q}$ (resp. $c^*(n)$). Writing these coefficients 3-adically, a computer calculation indicates that

$$\begin{aligned} c^*(3) = c(3) &= 3^{-2} + 3^{-1} + 2 + \cdots, \\ c\left(3^6\right) &= 3^{-47} + 3^{-46} + 2\left(3^{-45}\right) + \cdots, \\ c^*\left(3^6\right) &= 3^{-47} + 3^{-46} + 2\left(3^{-45}\right) + \cdots, \\ c\left(3^7\right) &= 3^{-56} + 3^{-55} + 2\left(3^{-54}\right) + \cdots, \\ c^*\left(3^7\right) &= 3^{-56} + 3^{-55} + 2\left(3^{-54}\right) + \cdots. \end{aligned}$$

Having gotten rid of the difficulty arising from transcendental coefficients, our goal of realizing F^+ as a 3-adic modular form now seems to be thwarted by the high powers of 3 appearing in the denominators. However, choosing $\alpha = c_F^+(1) + \gamma/11!$ with

$$\gamma := 154300462955809413372268553898 = 3^7 + 3^8 + \cdots$$

and denoting the coefficients of $11!\mathcal{F}_\alpha$ (resp. $11!\mathcal{F}^*_\alpha$) by $C_\alpha(n)$ (resp. $C^*_\alpha(n)$), one sees that

$$\begin{aligned}
C^*_\alpha(3) &= C_\alpha(3) = 2\left(3^5\right) + 3^7 + 3^8 + \cdots,\\
C_\alpha\left(3^6\right) &= 2\left(3^{-5}\right) + 2\left(3^{-3}\right) + 3^{-2} + \cdots,\\
C^*_\alpha\left(3^6\right) &= 3^5 + 3^6 + 3^8 + \cdots,\\
C_\alpha\left(3^7\right) &= 2\left(3^{-7}\right) + 2\left(3^{-5}\right) + 2\left(3^{-4}\right) + \cdots,\\
C^*_\alpha\left(3^7\right) &= 2 + 2\,(3) + 2\left(3^2\right) + \cdots.
\end{aligned}$$

Hence the denominators of the coefficients of $\mathcal{F}^*_\alpha$ seem to vanish, while the denominators of $\mathcal{F}_\alpha$ still seem to grow.

In order to cancel the denominators of $\mathcal{F}_\alpha$ itself, we require a further refinement of our corrected series. For this purpose, for $\delta \in \mathbb{C}_p$, we define

$$\mathcal{F}_{\alpha,\delta} := \mathcal{F}_\alpha - \delta\left(\mathcal{E}_f - \beta_1 \mathcal{E}_{f|V_p}\right). \tag{7.25}$$

Denoting the coefficients of $11!\mathcal{F}_{\alpha,\delta}$ by $c_{\alpha,\delta}(n)$, we see in our above example (with α chosen as above and δ chosen to be $23974292034 = 2(3^7)+2(3^9)+3^{10}+\cdots$) that

$$\begin{aligned}
c_{\alpha,\delta}\left(3^6\right) &= 3^5 + 3^6 + 3^8 + \cdots,\\
c_{\alpha,\delta}\left(3^7\right) &= 2 + 2\cdot 3 + 2\cdot 3^2 + \cdots.
\end{aligned}$$

The phenomenon we observe here is evidenced by the fact that the correct choice of α and δ results in a p-adic modular form. The choice of α is given in terms of the coefficients of $D^{1-k}(\mathcal{F}_{c_F^+(1)})$ and $D^{1-k}(\mathcal{F}^*_{c_F^+(1)})$. For ease of notation, we define $b_F(n), b^*_F(n) \in \mathbb{C}_p$ by

$$F_{c_F^+(1)}(\tau) := D^{1-k}\left(\mathcal{F}_{c_F^+(1)}(\tau)\right) =: \sum_{n \gg -\infty} b_F(n) q^n, \tag{7.26}$$

$$F^*_{c_F^+(1)}(\tau) := D^{1-k}\left(\mathcal{F}^*_{c_F^+(1)}(\tau)\right) =: \sum_{n \gg -\infty} b^*_F(n) q^n. \tag{7.27}$$

THEOREM 7.18. *Assuming the notation and hypotheses above, then the following are true:*

i) *There exists exactly one $\alpha \in A_F := c_F^+(1) + \mathbb{C}_p$ such that $\mathcal{F}^*_\alpha$ is a p-adic modular form of weight k, level pN, and Nebentypus χ. The unique choice of α is given by the p-adic limit*

$$\alpha = c_F^+(1) + \lim_{m\to\infty} \frac{b^*_F\left(p^m\right)}{\beta_1^m}.$$

ii) *If $\mathrm{ord}_p(\beta_1) < \mathrm{ord}_p(\beta_2) \neq (1-k)$, then there exists exactly one pair $(\alpha,\delta) \in A_F \times \mathbb{C}_p$ such that $\mathcal{F}_{\alpha,\delta}$ is a p-adic modular form of weight k, level pN, and Nebentypus χ. Moreover, α is the unique choice from i) and*

$$\delta = \lim_{m\to\infty} \frac{C_\alpha\left(p^m\right) p^{m(1-k)}}{{\beta_2}^m}.$$

Consider again the mock modular form F^+ in Example 7.8 associated to $f = \Delta$ with principal part q^{-1} which we have computationally investigated above. Since f has trivial level and the principal part is q^{-1}, taking the unique choice of $\alpha \in A_F$

and $\delta \in \mathbb{C}_p$ given by Theorem 7.18 ii), $\Delta\mathcal{F}_{\alpha,\delta}$ is a 3-adic modular form in the sense of Serre. One quickly conjectures numerically that

$$\begin{aligned} 3^7\Delta\mathcal{F}_{\alpha,\delta} &\equiv 1 \pmod 3, \\ 3^7\Delta\mathcal{F}_{\alpha,\delta} &\equiv E_2 \pmod{3^2}, \\ 3^7\Delta\mathcal{F}_{\alpha,\delta} &\equiv E_2 + 9\Delta \pmod{3^3} \\ &\vdots \end{aligned}$$

These were shown as an example of explicit congruence techniques of Kane and Waldherr [**294**] (note that the normalization in the factor of 3^7 on the left-hand sides above corrects a typo in [**91**]). Theorem 7.18 gives a strong interplay between g and the weight k correction of

$$\mathcal{F}_0^* = F^+ - p^{k-1}\beta_2 F^+|V_p$$

in the case that $\alpha \neq 0$. It is of course of particular interest to consider the case that $\alpha = 0$, so that $\mathcal{F}_0^*$ is itself a p-adic modular form. This leads us immediately to consider the case that f has CM by an imaginary quadratic field K.

THEOREM 7.19. *Assume that f has CM by K, $p \nmid N$, and p is split in $\mathcal{O}_K$. Then $\alpha = 0$ is the unique choice from Theorem 7.18 for which $\mathcal{F}_0^*$ is a p-adic modular form of weight k, level pN, and Nebentypus χ.*

REMARK. If we are not in the generic case (or if $p|N$), the situation is somewhat different, but this has been worked out in detail in the original work [**91**].

7.5. p-adic harmonic Maass functions

In [**138**], Candelori and Castella considered certain p-adic modular forms studied by Guerzhoy, Kane, and one of the authors [**91**] associated to certain mock modular forms from a geometric perspective (as described in Subsection 7.4.3). They showed that these are over-convergent p-adic modular forms, and arise from the de Rham cohomology in a way similar to harmonic Maass forms, as shown by Candelori [**137**]. Special examples of p-adic harmonic Maass forms of negative weight due to Wagner were also discussed in detail in Section 6.4.

In recent work, Griffin [**235**] constructed certain p-adic analogs of harmonic Maass functions from an analytic perspective. When the prime in question is p-ordinary (we will often simply say "ordinary" when the prime is clear from context) for a related weight 2 cusp form (that is, if the p-th Fourier coefficient of the form is non-zero modulo all of the primes lying above p in the relevant number field), it turns out these functions align with those studied by Candelori and Castella. However, Griffin's construction extends these functions to the full supersingular locus. This allows us to consider certain generalizations of lifts, such as those described in Section 16.3, which give the coefficients of half-integral weight forms as twisted traces over CM points. For sufficiently large discriminants, such traces will involve supersingular points, as shown by Deuring.

Let N be a square-free positive integer. Throughout this section, we will use the notation $M_k(\Gamma_0(N); K)$, respectively $M_k^!(\Gamma_0(N); K)$, to denote the space of weight k holomorphic and weakly holomorphic modular forms defined on the modular curve $Y_0(N)(K)$ over the field K. We will also consider harmonic Maass functions as functions on $Y_0(N)(\mathbb{C}) \simeq \mathbb{H}/\Gamma_0(N)$. If K is a number field and $\mathfrak{p}$ is a prime of K, we denote the completion of K at $\mathfrak{p}$ by $K_\mathfrak{p}$.

THEOREM 7.20. *Suppose N is a square-free positive integer, K a number field, and $\mathfrak{p}$ a prime of K over a rational prime p, not dividing N. There exists a $K_{\mathfrak{p}}$ vector space $H_0(N; K_{\mathfrak{p}})$ of $\mathfrak{p}$-adically continuous functions $F^{\mathfrak{p}} : Y_0(N)(K_{\mathfrak{p}}) \to K_{\mathfrak{p}}$, satisfying the following properties:*

i) *We have that $M_0^!(\Gamma_0(N; K_{\mathfrak{p}}) \subset H_0(N; K_{\mathfrak{p}})$.*
ii) *The Hecke operators and Atkin–Lehner involutions have well-defined actions on $H_0(N; K_{\mathfrak{p}})$.*
iii) *Each function in $H_0(N; K_{\mathfrak{p}})$ has a well-defined q-expansion in $K_{\mathfrak{p}}((q))$.*
iv) *The usual modular differential operator D acts on $H_0(N; K, \mathfrak{p})$, and the following sequence is exact:*

$$0 \to K_{\mathfrak{p}} \to H_0(N; K_{\mathfrak{p}}) \xrightarrow{D} M_2^!(\Gamma_0(N); K_{\mathfrak{p}}) \to M_2(\Gamma_0(N); K_{\mathfrak{p}}) \to 0.$$

Each function in this space is defined using a sequence of modular functions in $M_0(\Gamma_0(N); K_{\mathfrak{p}})$, whose values converge at all points of the set

$$\{z \in Y_0(N)(K_{\mathfrak{p}}) \,:\, v_{\mathfrak{p}}(j(p)) \geq 0\}.$$

A second, auxiliary sequence of meromorphic modular functions is used to regularize convergence "near the cusps" (i.e., for points $z \in Y_0(N)(K_{\mathfrak{p}})$ with $v_{\mathfrak{p}}(j(p)) < 0$). Griffin constructed these sequences using Hecke operators $T_{p^n,0}$, where p is the norm of the prime $\mathfrak{p}$. The action of the Hecke operators, Atkin–Lehner operators, and the differential operator D are obtained by acting component-wise on the defining sequences.

The spaces $H_0(N; K_{\mathfrak{p}})$ and $H_0(\Gamma_0(N))$ are strongly connected. Suppose $F \in H_0(\Gamma_0(N))$ is a form whose Fourier expansion at all cusps (meaning under each Atkin–Lehner involution) has its principal part in $K[q^{-1}]$. Then there is a corresponding form $F^{\mathfrak{p}} \in H_0(N; K_{\mathfrak{p}})$ with the same principal part in the q-expansion at all cusps. In general, we expect the coefficients of F and its values at algebraic points of $X_0(N)(\mathbb{C})$ to be transcendental. However, this transcendence can be controlled in certain ways. The corresponding $\mathfrak{p}$-adic function $F^{\mathfrak{p}}$ has related properties, and the underlying algebraic structure agrees with that of the complex-valued form F.

Suppose $F \in H_0(\Gamma_0(N))$ is good for a cusp form f^c (see Definition 7.5 and recall again that f^c is obtained from f by applying complex conjugation to the coefficients of f, as in Lemma 5.17) where

$$f(\tau) = \sum_{n=1}^{\infty} c_f(n) q^n \in S_2(\Gamma_0(N))$$

is a multiple of a newform and has coefficients in K. If $\mathcal{E}_f$ is defined as in (7.8), then Theorem 7.10 tells us that there is some fixed $\alpha_{\mathbb{C}} \in \mathbb{C}$, and some q-series $\mathcal{F}_K$ with coefficients coefficients in K, satisfying

$$F^+ = \mathcal{F}_K + \alpha_{\mathbb{C}} \mathcal{E}_f.$$

We can control the transcendence of the CM values of F in a related way. Using a generalization of the theta lifts discussed in Section 16.3, we can obtain weight 3/2 modular forms whose coefficients are given by twisted traces of the singular values of F. Using the notation of Section 18.3, suppose Δ is a fundamental discriminant with some r satisfying $r^2 \equiv \Delta \pmod{4N}$, and define

$$\widetilde{\rho} := \begin{cases} \rho_L & \text{if } \operatorname{sgn}(\Delta) < 0, \\ \widetilde{\rho}_L & \text{if } \operatorname{sgn}(\Delta) > 0. \end{cases}$$

For simplicity, assume that either $\Delta \neq 1$ or that the constant term in the Fourier expansion of F at every cusp is 0. Then there is a vector-valued modular form $\vartheta_{\Delta,r}(F) \in M^!_{\frac{3}{2},\widetilde{\rho}_L}$,

$$\vartheta_{\Delta,r}(F)(\tau) = \sum_{h \in L'/L} \sum_{\substack{n \in Z+Q(h) \\ -\infty \ll n}} c_{F,\Delta,r}(n,h) q^n \mathfrak{e}_h,$$

where $c_{F,\Delta,r}(m,h) \in K$ for all $m \leq 0$, and whose coefficients for $m > 0$ are given by traces of F over twisted Heegner divisors:

$$c_{F,\Delta,r}(m,h) := \frac{1}{\sqrt{\Delta}} \sum_{\lambda \in L_{d\Delta,hr}/\Gamma_0(N)} \chi_\Delta(\lambda) \frac{F(Z(\lambda))}{\omega(\lambda)}.$$

Here d, ω, χ_Δ, and Z are all as in Definition 18.14.

For every $n \in \mathbb{N}$, coprime to N, there is a Hecke operator $T_{n^2,\frac{3}{2},\widetilde{\rho}_L}$ which acts on $M^!_{\frac{3}{2},\widetilde{\rho}_L}$ (cf. (7.1) of [**127**]). If f has eigenvalue λ_f under the Fricke involution, let

$$\widehat{\rho} := \begin{cases} \rho_L & \text{if } \lambda_f = -1, \\ \overline{\rho}_L & \text{if } \lambda_f = 1. \end{cases}$$

The Shimura correspondence for vector-valued forms (cf. Section 7 of [**127**]) tells us that $S_{\frac{3}{2},\rho_L} \oplus S_{\frac{3}{2},\rho_L}$ and $S_2(\Gamma_0(N))$ are isomorphic as Hecke modules. In particular, there is some $g \in S_{\frac{3}{2},\widehat{\rho}}$ such that g has coefficients in K and is an eigenform for each weight 3/2 Hecke operator $T_{n^2,\frac{3}{2},\widetilde{\rho}_L}$, with the same eigenvalue as f under the weight 2 Hecke operator $T_{n,2}$.

Then it turns out there is some $a_{\mathbb{C}}(\Delta) \in \mathbb{C}$ (with, $a_{\mathbb{C}}(\Delta) = 0$ if $\widetilde{\rho} \neq \hat{\rho}$) and some $\mathcal{G}_K \in M^!_{\frac{3}{2},\widetilde{\rho}_L}$ with coefficients in K, so that

$$\vartheta_{\Delta,r}(F) = \mathcal{G}_K + a_{\mathbb{C}}(\Delta)\, g.$$

If F is weakly holomorphic, then the coefficients of $\mathcal{G}_{\Delta,r}$ must all be in K. Moreover, Hecke operators commute with the theta lift, and so the transcendental part must satisfy Hecke relations related to those satisfied by f (see Proposition 7.2).

If $K'_{\mathfrak{p}}$ is any finite extension of $K_{\mathfrak{p}}$, the $\mathfrak{p}$-adic function $F^{\mathfrak{p}}$ extends naturally to a function on $Y_0(N)(K'_{\mathfrak{p}})$. If we interpret $Z(\lambda)$ as a point on this curve, we may define the formal q-series vector

$$\vartheta^{\mathfrak{p}}_{\Delta,r}(F^{\mathfrak{p}})(\tau) := \sum_{h \in L'/L} \sum_{\substack{n \in Z+Q(h) \\ n \gg -\infty}} c_{F^{\mathfrak{p}},\Delta,r}(n,h) q^n \mathfrak{e}_h,$$

where $c_{F^{\mathfrak{p}},\Delta,r}(n,h) := c_{F,\Delta,r}(n,h)$ for $m \leq 0$, and for $m > 0$ we define

$$c_{F^{\mathfrak{p}},\Delta,r}(m,h) := \frac{1}{\sqrt{\Delta}} \sum_{\lambda \in L_{d\Delta,hr}/\Gamma_0(N)} \chi_\Delta(\lambda) \frac{F^{\mathfrak{p}}(Z(\lambda))}{\omega(\lambda)}.$$

With this notation, we have the following theorem.

THEOREM 7.21. *Assume the notation and hypotheses above. There exists some $\alpha_{\mathfrak{p}} \in K_{\mathfrak{p}}$ so that $F^{\mathfrak{p}}$ satisfies the formal q-series identity*

$$F^{\mathfrak{p}} = \mathcal{F}_K + \alpha_{\mathfrak{p}}\, \mathcal{E}_f.$$

Similarly, there exists some $a_{\mathfrak{p}}(\Delta) \in K_{\mathfrak{p}}$ *so that* $\vartheta^{\mathfrak{p}}_{\Delta,r}(F^{\mathfrak{p}})$ *satisfies the formal q-series vector identity*

$$\vartheta^{\mathfrak{p}}_{\Delta,r}(F^{\mathfrak{p}}) = \mathcal{G}_K + a_{\mathfrak{p}}(\Delta)g.$$

Moreover, if $a_{\mathbb{C}}(\Delta) \in K$, *then* $a_{\mathfrak{p}}(\Delta) \in K$ *and*

$$a_{\mathbb{C}}(\Delta) = a_{\mathfrak{p}}(\Delta).$$

REMARK. A theorem closely related to Theorem 19.20 shows that the coefficient $c_{F,\Delta,r}(m,h)$ of $\vartheta_{\Delta,r}(F)$ is in K precisely when

$$L(f_d,1)L'(f_\Delta,1) = 0,$$

where f_D is as in Definition 19.12. When $L(f_d,1) = 0$, Waldspurger's formula (Theorem 19.17) shows that the corresponding coefficient of g vanishes. When $L'(f_\Delta,1) = 0$, we find we can take $a_{\mathbb{C}}(\Delta) = 0$. Thus, under these conditions, Theorem 7.21 shows that the corresponding twisted trace of singular values of $F^{\mathfrak{p}}$ is also in K and is identical to the twisted trace of F.

Given F and f as above, we now briefly sketch the construction of $F^{\mathfrak{p}}$ as a function on $Y_0(N)(K)$, where $\mathfrak{p}$ lies over a rational prime p which is ordinary for f. The function $F^{\mathfrak{p}}$ extends to $Y_0(N)(K_{\mathfrak{p}})$ by continuity. The construction for primes $\mathfrak{p}$ which are non-ordinary is similar but requires a few modifications.

Let d be the inertia degree of $K/\mathbb{Q}$, so that the residue field $K/\mathfrak{p} \simeq \mathbb{F}_{p^d}$, and let β_1 and β_2 be defined as in (7.17). For $n \in \mathbb{N}$, let A_n B_n be the operators of the Hecke algebra

$$B_n := \sum_{j=0}^{n-1} \beta_1^{-j-1} p^j T_{p^j,0} + \sum_{j=n}^{n+d-1} \frac{\beta_1^{-j-1} p^j T_{p^j,0}}{1-\beta_1^{-d}}$$

and

(7.28)

$$\begin{aligned} A_n &:= (c_f(p) - pT_{p,0})B_n \\ &= 1 + \frac{\beta_1^{-n-d}}{1-\beta_1^{-d}} \left(-\beta_2\left(p^{n-1}T_{p^{n-1},0} - p^{n+d-1}T_{p^{n+d-1},0}\right) + \left(p^n T_{p^n,0} - p^{n+d}T_{p^{n+d},0}\right)\right). \end{aligned}$$

Let $\widehat{F} := F|(c_f(p) - pT_{p,0})$, and define the functions $F_n := F|A_n$. By Proposition 7.2, each of these forms is weakly holomorphic. Moreover, we have

$$F_{n+1} - F_n = \frac{\beta_1^{-n-d-1}}{1-\beta_1^{-d}} \widehat{F}|\left(p^{n+d}T_{p^{n+d},0} - p^n T_{p^n,0}\right). \tag{7.29}$$

For points $z \in Y_0(N)(K_{\mathfrak{p}})$ with $v_{\mathfrak{p}}(j(z)) \geq 0$, we define $F^{\mathfrak{p}}(z)$ by the limit of the sequence $(F_n(p))_{n\in\mathbb{N}}$ which (7.29) and the following proposition show converges $\mathfrak{p}$-adically.

PROPOSITION 7.22. *As above, let N be a positive square-free integer and K be a number field with prime $\mathfrak{p}$ over a rational prime p which does not divide N so that the residue field $K/\mathfrak{p} \simeq \mathbb{F}_{p^d}$. If $G \in M^!(\Gamma_0(N);K)$, then the sequence of q-series*

$$\left(G|p^n(T_{p^{n+d},0} - T_{p^n,0})\right)_{n\in\mathbb{N}}$$

converges $\mathfrak{p}$-adically and coefficients-wise to 0.

If $P \in Y_0(N)$ satisfies $v_{\mathfrak{p}}(j(P)) \geq 0$, then the sequence

$$\left(G|p^n(T_{p^{n+d},0} - T_{p^n,0})(P)\right)_{n\in\mathbb{N}}$$

converges $\mathfrak{p}$-adically to 0.

PROOF. Suppose $M \subset M_0^!(\Gamma_0(N);K)$ is the $\mathbb{Z}$-submodule of forms whose q-expansion at every cusp is contained in $\mathbb{Z}((q))$. Then there exists a finite set of generators $\{\varphi_1, ..., \varphi_n\}$ so that

$$M = \mathbb{Z}[\varphi_1, ..., \varphi_n].$$

These generators may be taken to correspond to the coordinates of some model of the smooth affine curve $Y_0(N)(\mathbb{Q})$.

Without loss of generality, assume G has $\mathfrak{p}$-integral coefficients at every cusp, so that $G = R_0(\varphi_1, ..., \varphi_n)$ for some polynomial R_0 with $\mathfrak{p}$-integral coefficients. Standard formulas for the action of the Hecke operators show that

$$G|pT_{p,0} \equiv G|V_p \pmod{\mathfrak{p}}.$$

This implies

$$G|pT_{p,0} = R_0(\varphi_1^p, ..., \varphi_n^p) + pR_1(\varphi_1, ..., \varphi_n),$$

where R_1 is a polynomial with $\mathfrak{p}$-integral coefficients. More generally, we have

$$G|p^nT_{p^n,0} \equiv G|p^{n-1}T_{p^{n-1},0}V_p \pmod{\mathfrak{p}^n}.$$

This, combined with the q-series congruence $\varphi_\ell^{p^{n-1}}|V_p \equiv \varphi_\ell^{p^n} \pmod{\mathfrak{p}^n}$, implies the existence of polynomials R_n for $n \in \mathbb{N}$, each with $\mathfrak{p}$-integral coefficients so that

$$G|p^nT_{p^n,0} = \sum_{m=0}^{n} p^m R_m\left(\varphi_1^{p^{n-m}}, ..., \varphi_n^{p^{n-m}}\right). \tag{7.30}$$

Since each φ_ℓ has integral coefficients, it satisfies a polynomial in $\mathbb{Z}[j, X]$, which is monic in X. In particular, since $v_\mathfrak{p}(j(z)) \geq 0$, we have that $v_\mathfrak{p}(\varphi_\ell(z)) \geq 0$. Moreover, we have that $\varphi_\ell^{p^d}(z) \equiv \varphi_\ell(z) \pmod{\mathfrak{p}}$, and $\varphi_\ell^{p^{dn}}(z)$ converges to some root of the polynomial $X^{p^d} - X$ as n goes to ∞. Then (7.30) implies that the difference given in the proposition must converge to 0. □

In order to extend the function in the region $\{z \in Y_0(K_\mathfrak{p}) : v_\mathfrak{p}(j(z)) < 0\}$, we require a function $\widetilde{F}$, defined so that $E_{p-1}\widetilde{F} \in M_{p-1}^!(\Gamma_0(N);K_\mathfrak{p})$ and so that $\widetilde{F}$ has the same principal part at all cusps as F. Then let $\widetilde{F}_n := \widetilde{F}|(A_n - 1)$. The functions in the sequence $(F_n - \widetilde{F}_n)_{n\in\mathbb{N}}$ share a fixed principal part, and the q-expansions of these functions converge coefficient-wise. Therefore we can evaluate these functions for $\{z \in Y_0(N)(\mathfrak{p}) : v_\mathfrak{p}(j(z)) < 0\}$ using the Tate curve, and their values must converge. The sequence also converges for points with $\mathfrak{p}$-integral j-invariants, which are not supersingular. This can be seen using (7.28) and a slightly modified version of Proposition 7.22, allowing rational functions with poles in the supersingular discs. Not only does the sequence converge, the limit agrees with the value of the original sequence $(F_n)_{n\in\mathbb{N}}$.

SKETCH OF PROOF OF THEOREM 7.20. Since p does not divide the level N, the Atkin–Lehner involutions and Hecke operators commute with the operators A_n, without affecting convergence. Similarly, the operators commute with the differential operator D, up to a change of weight and normalization (cf. (7.4)). Convergence of the derivatives follows from similar arguments as those above. Careful consideration of the convergence also shows that if F is weakly holomorphic to begin with, then $F^\mathfrak{p} = F$ when restricted to $Y_0(N)(K)$.

Linear combinations of forms constructed as above and their images under the Atkin–Lehner involutions span the space $H_0(\Gamma_0(N), K_\mathfrak{p})$. In particular, we may construct a form whose derivative has arbitrarily chosen principal parts at cusps (up to the constant terms which must be 0). Therefore we have the decomposition

$$D\left(H_0(\Gamma_0(N), K_{\mathfrak{p}})\right) \oplus M_2(\Gamma_0(N), K_{\mathfrak{p}}) = M_2^!(\Gamma_0(N), K_{\mathfrak{p}}).$$

□

SKETCH OF PROOF OF THEOREM 7.21. The q-expansions of the forms $\widetilde{F}_n$ converge coefficient-wise to 0, and so we may ignore their contributions to any given coefficient, other than for bounding the order of the principal parts at all cusps.

As noted above, we can write $F^+ = F_K + \alpha_{\mathbb{C}}\mathcal{E}_f$. There is some $\alpha_{\mathfrak{p}} \in K_{\mathfrak{p}}$ so that $F_K + \alpha_{\mathfrak{p}}\mathcal{E}_f$ has bounded $\mathfrak{p}$-denominators. Then some short calculations show that the q-expansions of $(F_K + \alpha_{\mathfrak{p}}\mathcal{E}_f)\,|(A_n - 1)$ converge coefficients-wise to 0. In particular we must have the q-series identity $F^{\mathfrak{p}} = F_K + \alpha_{\mathfrak{p}}\mathcal{E}_f$.

We now consider the lifts to weight 3/2 modular forms. The lift is linear. Moreover, careful consideration of the action of the Hecke operators on CM points shows that

$$\vartheta_{\Delta,r}(F|p^nT_{p^n,0}) = \vartheta_{\Delta,r}(F)|T_{p^{2n},\frac{3}{2},\widetilde{\rho}_L}.$$

In particular, if $\widetilde{A}_n$ is the image of A_n under the implied homomorphism to the weight 3/2 Hecke algebra, we can write $\vartheta_{\Delta,r}(F_n) = \vartheta_{\Delta,r}(F)|\widetilde{A}_n$.

Recall our earlier claim that if the coefficients of $G \in M_0^!(\Gamma_0(N))$ are in K, then the coefficients of $\vartheta_{\Delta,r}(G)$ are also in K. In this case, an involved calculation using the formulas for the action of the Hecke operators on q-series vectors shows that the q-series $\vartheta_{\Delta,r}(G)|(T_{p^{2n+4},\frac{3}{2},\widetilde{\rho}_L} - T_{p^{2n},\frac{3}{2},\widetilde{\rho}_L})$ converges $\mathfrak{p}$-adically to 0. Thus, with these analogs, we can follow the previous argument *mutatis mutandis* to find that the coefficients of $\vartheta_{\Delta,r}(F_n)$ must converge $\mathfrak{p}$-adically to a q-series vector with the same principal part as that of $\vartheta_{\Delta,r}(F)$, and so $\vartheta^{\mathfrak{p}}_{\Delta,r}(F^{\mathfrak{p}}) - \mathcal{G}_K$ is a cusp form. Moreover, this cusp form must satisfy the same Hecke identities as f since for every integer n coprime to N we have

$$\begin{aligned}
&\vartheta^{\mathfrak{p}}_{\Delta,r}(F^{\mathfrak{p}})|(c_f(n) - T_{n^2,\frac{3}{2},\widetilde{\rho}_L}) \\
&= \vartheta^{\mathfrak{p}}_{\Delta,r}(F^{\mathfrak{p}}|(c_f(n) - T_{n,0})) = \vartheta_{\Delta,r}(F|(c_f(n) - T_{n,0})) \\
&= \vartheta_{\Delta,r}(F)|\left(c_f(n) - T_{n^2,\frac{3}{2},\widetilde{\rho}_L}\right) = \mathcal{G}_K|\left(c_f(n) - T_{n^2,\frac{3}{2},\widetilde{\rho}_L}\right).
\end{aligned}$$

In particular, the cusp form $\vartheta^{\mathfrak{p}}_{\Delta,r}(F^{\mathfrak{p}}) - \mathcal{G}_K$ must be a multiple of g. □

EXAMPLE 7.23. Let $N = 43$, $K = \mathbb{Q}$, and $\mathfrak{p} = 3$, and let f be the unique newform of weight 2 for $\Gamma_0(43)$. Let

$$F(\tau) = q^{-1} + 1.707216\ldots q + 1.792783\ldots q^2 + 3.195188\ldots q^3 \cdots + F^-(\tau)$$

be the unique harmonic Maass function for $\Gamma_0(43)$ with the given expansion at the cusp $i\infty$, and which is invariant under the Fricke involution. Then $F^+ = \mathcal{F}_{\mathbb{Q}} + \alpha_{\mathbb{C}}\mathcal{E}_f$, where

$$\mathcal{F}_{\mathbb{Q}}(\tau) = q^{-1} + \frac{5}{3}q + \frac{11}{6}q^2 + \frac{29}{9}q^3 + \frac{23}{6}q^4 + \frac{17}{3}q^5 + \frac{61}{9}q^6 + \frac{72}{7}q^7 + \frac{47}{4}q^8 + \ldots$$

and $\alpha_{\mathbb{C}} = 0.040549914\ldots$.

The prime 3 is ordinary for f, and the form $F^{\mathfrak{p}}$ has the q-expansion

$$F^{\mathfrak{p}}(\tau) = q^{-1} + \cdots 212012_3 q + \cdots 122110_3 q^2 + \cdots 102201_3 q^3 + \cdots = \mathcal{F}_{\mathbb{Q}} + \alpha_{\mathfrak{p}}\mathcal{E}_f,$$

where $\alpha_{\mathfrak{p}} = \ldots 212010.1_3$. Here we have represented each 3-adic number in base 3 format so that, for instance,

$$\alpha_{\mathfrak{p}} = \ldots 212010.1_3 = 1 \cdot 3^{-1} + 0 \cdot 3^0 + 1 \cdot 3^1 + 0 \cdot 3^2 + 2 \cdot 3^3 + 1 \cdot 3^4 + 2 \cdot 3^5 + \ldots.$$

Since the constant term of F is 0, we may take $\Delta = r = 1$. For simplicity in this example we consider a projection of the vector-valued forms indicated in the theorems to scalar-valued forms. This projection is obtained by summing the vector components and multiplying by $-\frac{1}{2}$. Let $\widehat{\vartheta}_{\Delta,r}(F)$ be the given projection of $\vartheta_{\Delta,r}(F)$. Then $\widehat{\vartheta}_{\Delta,r}(F) \in M^!_{\frac{3}{2}}(\Gamma_0(4 \cdot 43))$ and lies in the Kohnen plus space.

The newform

$$\widehat{g}(\tau) = q^3 + q^7 - q^8 - q^{12} + 2q^{19} - q^{20} - 2q^{27} - 3q^{28} + O(q^{30}) \in S^+_{\frac{3}{2}}(\Gamma_0(4 \cdot 43))$$

maps to f under the Shimura correspondence and takes the place of g in Theorem 7.21 under the given projection. Then we find $\widehat{\vartheta}_{\Delta,r}(F) = \widehat{\mathcal{G}}_{\mathbb{Q}} + a_{\mathbb{C}}\,\widehat{g}$, where

$$\mathcal{G}_{\mathbb{Q}}(\tau) = q^{-1} - 1 + q^7 + q^8 + q^{12} + q^{19} + q^{20} + 2q^{27} + q^{28} + 2q^{32} + 4q^{39} + q^{43} + 4q^{48} + \ldots$$

and $a_{\mathbb{C}} = 0.0663160686\ldots$. Finally, we have $\widehat{\vartheta}^{\mathfrak{p}}_{\Delta,r}(F) = \widehat{\mathcal{G}}_{\mathbb{Q}} + a_{\mathfrak{p}}\,\widehat{g}$, where $a_{\mathfrak{p}} = \ldots 00002.1_3$.

CHAPTER 8

Zwegers' Thesis

In this chapter, we discuss some of the key results of Zwegers' thesis which form the basis of the construction of many mock modular forms, and in particular allow one to compute the harmonic Maass forms related to Ramanujan's mock theta functions. In Section 8.1, we consider Appell-Lerch-type series, originally introduced by Appell [**40**], and often also named after Lerch who studied their transformation properties [**351, 352**]. Here, we study Appell-Lerch series of a special type, which, if divided by a theta function, give rise to Zwegers' μ-function. This function can be "completed" giving rise to the function $\widehat{\mu}(z_1, z_2; \tau)$, a prototype for "harmonic Maass-Jacobi forms". This function is an important device for producing mock modular forms and is the first of three constructions used by Zwegers to study the mock theta functions. A second method, that of indefinite theta functions, is then discussed in Section 8.2, and a third method of meromorphic Jacobi forms is considered in a generalized sense in Chapter 11. There are also strong connections between these different viewpoints; for example we will see in Lemma 8.34 that the μ function of Section 8.1 can be seen as an example of the indefinite theta functions of Section 8.2. We study such examples in detail in Chapter 9. Throughout this chapter, our exposition roughly follows Zwegers [**529**], although we offer an alternative proof of the modularity of his indefinite theta functions using work of Vignéras.

8.1. Zwegers' thesis I: Appell-Lerch series

In this section, we describe Zwegers' first method for understanding Ramanujan's mock theta functions. Theorem 8.13, one of the main results in this section, offers the transformation laws satisfied by Zwegers' function $\widehat{\mu}(z_1, z_2; \tau)$.

The construction of this function is based on the properties of the following Appell-Lerch series (recall in the following definition that $\Lambda_\tau = \mathbb{Z} + \mathbb{Z}\tau$).

DEFINITION 8.1. For $\tau \in \mathbb{H}$ and $z_1, z_2 \in \mathbb{C}\backslash\Lambda_\tau$, define **Zwegers' μ-function** by

$$\mu(z_1, z_2) = \mu(z_1, z_2; \tau) := \frac{\zeta_1^{\frac{1}{2}}}{\vartheta(z_2; \tau)} \sum_{n \in \mathbb{Z}} \frac{(-1)^n \zeta_2^n q^{\frac{n(n+1)}{2}}}{1 - \zeta_1 q^n},$$

where $\zeta_j := e^{2\pi i z_j}$ $(j = 1, 2)$ and ϑ is the Jacobi theta function of Definition 2.1.

The function μ is not quite a Jacobi form but exhibits "mock" behavior. In particular, we will see that there is a nonholomorphic correction term which can be added to make it transform as the Jacobi form $\widehat{\mu}$ hinted at above. This nonholomorphic correction must have the same elliptic transformation properties of μ, and hence the following result outlining this behavior is critical.

PROPOSITION 8.2. *We have the following transformation properties:*

i) $\mu(z_1+1,z_2)=\mu(z_1,z_2+1)=-\mu(z_1,z_2)$,
ii) $\mu(z_1+\tau,z_2)=-\zeta_1\zeta_2^{-1}q^{\frac{1}{2}}\mu(z_1,z_2)-i\zeta_1^{\frac{1}{2}}\zeta_2^{-\frac{1}{2}}q^{\frac{3}{8}}$,
iii) $\mu(z_1+\tau,z_2+\tau)=\mu(z_1,z_2)$,
iv) $\mu(-z_1,-z_2)=\mu(z_1,z_2)$.

PROOF. The claims essentially follow by observation or straightforward manipulation. For example, claim i) follows directly from Definition 8.1 and the transformation laws of the ϑ-function given in Lemma 2.2.

To prove ii) we shift $n\mapsto n-1$ in the the sum defining $\mu(z_1+\tau,z_2)$. By combining terms, we obtain

$$\begin{aligned}&-\zeta_2\zeta_1^{-\frac{1}{2}}q^{-\frac{1}{2}}\vartheta(z_2)\mu\left(z_1+\tau,z_2\right)-\zeta_1^{\frac{1}{2}}\vartheta(z_2)\mu\left(z_1,z_2\right)\\&=-\zeta_2\sum_{n\in\mathbb{Z}}\frac{(-1)^n\zeta_2^nq^{\frac{n(n+1)}{2}}}{1-\zeta_1q^{n+1}}-\zeta_1\sum_{n\in\mathbb{Z}}\frac{(-1)^n\zeta_2^nq^{\frac{n(n+1)}{2}}}{1-\zeta_1q^n}\\&=\sum_{n\in\mathbb{Z}}\frac{(-1)^n\zeta_2^nq^{\frac{n(n-1)}{2}}}{1-\zeta_1q^n}\left(1-\zeta_1q^n\right)=\sum_{n\in\mathbb{Z}}(-1)^n\zeta_2^nq^{\frac{n(n-1)}{2}}=iq^{-\frac{1}{8}}\zeta_2^{\frac{1}{2}}\vartheta\left(z_2\right).\end{aligned}$$

Claim iii) follows directly from Lemma 2.2 after a similar shift $n\mapsto n-1$. Finally, to obtain claim iv), replace $n\mapsto -n$ and then use the fact that $\vartheta(-z)=-\vartheta(z)$. □

Once we construct $\widehat{\mu}$, the application to mock modular forms will also require further facts about μ, such as the locations and orders of its poles. Here, we collect several such frequently used facts.

PROPOSITION 8.3. *The following are true:*

i) *The function $z_1\mapsto\mu\left(z_1,z_2;\tau\right)$ is meromorphic with simple poles in Λ_τ, with residue $-\frac{1}{2\pi i}\frac{1}{\vartheta(z_2)}$ at $z_1=0$.*
ii) *We have the following identity of meromorphic functions:*

$$\mu\left(z_1+z,z_2+z\right)=\mu\left(z_1,z_2\right)+\frac{i\eta^3\vartheta\left(z_1+z_2+z\right)\vartheta(z)}{\vartheta\left(z_1\right)\vartheta\left(z_2\right)\vartheta\left(z_1+z\right)\vartheta\left(z_2+z\right)}.$$

iii) *The symmetry property $\mu(z_1,z_2)=\mu(z_2,z_1)$ holds.*

REMARK. The residues at general lattice points $z_1\in\Lambda_\tau$ (which may be useful in various applications or constructions of functions using μ) can be determined from the residue at $z_1=0$ by using the elliptic transformation properties in i) and ii) of Proposition 8.2. Similarly, we see that there are poles if $z_2\in\Lambda_\tau$, and their residues can be computed by the same method.

PROOF OF PROPOSITION 8.3. We begin with the proof of i). The claim on the location and order of the poles is immediate since these can only come from zeros in the denominator $1-\zeta_1q^n$ in the definition of μ. The claimed residue is then computed in a straightforward manner by applying l'Hôpital's rule to the $n=0$ term of μ, which is the only term with a pole:

$$\operatorname{Res}_{z_1=0}\mu(z_1,z_2)=\lim_{z_1\to0}\left(z_1\mu(z_1,z_2)\right)=\frac{1}{\vartheta(z_2)}\lim_{z_1\to0}\frac{z_1}{1-e^{2\pi iz_1}}=-\frac{1}{2\pi i\vartheta(z_2)}.$$

To prove claim ii), we define

$$f(z) := \vartheta\left(z_1+z\right)\vartheta\left(z_2+z\right)\left(\mu\left(z_1+z, z_2+z\right)-\mu\left(z_1, z_2\right)\right).$$

Lemma 2.2, parts i) and iii) of Proposition 8.2, and part i) of this proposition give that f has no poles, has a zero when $z=0$, and satisfies (recall that $\zeta=e^{2\pi i z}$)

$$\begin{aligned} f(z+1) &= f(z), \\ f(z+\tau) &= \zeta_1^{-1}\zeta_2^{-1}\zeta^{-2}q^{-1}f(z). \end{aligned}$$

Therefore, we find that

$$g(z) := \frac{f(z)}{\vartheta(z)\vartheta\left(z+z_1+z_2\right)}$$

is an elliptic function which turns out to be holomorphic except for possible simple poles at points z in $-z_1-z_2+\Lambda_\tau$. Hence, as there is only at most one possible pole in a fundamental parallelogram, by Theorem 1.4 iii), $f(z)$ must in fact be a constant (depending only on τ, z_1, and z_2) multiplied by $\vartheta(z)\vartheta(z_1+z_2+z)$. Denoting this constant by $C=C(z_1,z_2)$, we can determine it by evaluating f at a single point. In fact, we find that

$$\begin{aligned} f(-z_1) &= \vartheta(z_2-z_1)\lim_{z\to 0}\vartheta(z)\mu(z,z_2-z_1) \\ &= \vartheta(z_2-z_1)\lim_{z\to 0}\frac{\vartheta(z)}{z}\lim_{z\to 0}\left(z\mu(z,z_2-z_1)\right), \end{aligned}$$

which, by part i) and (2.1), is equal to $-i\eta^3$. Hence, we find that

$$f(-z_1) = C\vartheta(-z_1)\vartheta(z_2) = -i\eta^3,$$

so that (using the fact that $\vartheta(z)$ is an odd function)

$$C = \frac{i\eta^3}{\vartheta(z_1)\vartheta(z_2)}.$$

This proves ii).

Finally, claim iii) follows by combining Lemma 2.2 and part ii) with $z=-z_1-z_2$. □

Of course, to construct the Jacobi form $\widehat{\mu}$, as well as to understand the modularity of mock theta functions from Zwegers' perspective, we require an understanding of the modularity behavior of μ. Watson **[489]** wrote down examples of such transformations for (e.g. third order) mock theta functions of Ramanujan. These results were early precursors of Zwegers' general theory. In order to describe the modular transformation properties of μ, and subsequently the mock theta functions of Ramanujan (cf. the examples of Chapter 9 and the dictionary of representations for mock theta functions in the Appendix), we need the following auxiliary function. This also arises in the "error to modularity" in the transformation for μ, as we shall shortly see. One of Zwegers' key insights was to construct a nonholomorphic function which corrects the "defect" arising from *Mordell integrals.*

DEFINITION 8.4. For $z\in\mathbb{C}$ and $\tau\in\mathbb{H}$ the **Mordell integral** is given by

$$h(z) = h(z;\tau) := \int_{\mathbb{R}} \frac{e^{\pi i\tau t^2-2\pi z t}}{\cosh(\pi t)}dt.$$

REMARK. Note that $z\mapsto h(z)$ is an even function.

We require the elliptic transformations of h, which we record here for later use.

PROPOSITION 8.5. *The following transformations for h hold:*

i) $h(z+1) = -h(z) + \frac{2}{\sqrt{-i\tau}} e^{\frac{\pi i}{\tau}\left(z+\frac{1}{2}\right)^2}$,

ii) $h(z+\tau) = -\zeta q^{\frac{1}{2}} h(z) + 2\zeta^{\frac{1}{2}} q^{\frac{3}{8}}$.

Moreover, h is the unique holomorphic function which satisfies i) *and* ii).

PROOF. As is the case with many of the proofs in this section, the proofs follow by direct manipulation. To prove i), we compute

$$h(z) + h(z+1) = \int_{\mathbb{R}} \frac{e^{\pi i \tau t^2 - 2\pi z t}}{\cosh(\pi t)} \left(1 + e^{-2\pi t}\right) dt = 2 \int_{\mathbb{R}} e^{\pi i \tau t^2 - 2\pi t\left(z+\frac{1}{2}\right)} dt$$
$$= \frac{2}{\sqrt{-i\tau}} e^{\frac{\pi i}{\tau}\left(z+\frac{1}{2}\right)^2}.$$

We now turn to ii). We have, after shifting $t \mapsto t - i$, the identity

$$\zeta^{-1} q^{-\frac{1}{2}} h(z+\tau) = -\int_{\mathbb{R}} \frac{e^{\pi i \tau (t+i)^2 - 2\pi z (t+i)}}{\cosh(\pi(t+i))} dt = -\int_{i+\mathbb{R}} \frac{e^{\pi i \tau t^2 - 2\pi z t}}{\cosh(\pi t)} dt.$$

The Residue Theorem gives

$$h(z) + \zeta^{-1} q^{-\frac{1}{2}} h(z+\tau) = \left(\int_{\mathbb{R}} - \int_{i+\mathbb{R}}\right) \frac{e^{\pi i \tau t^2 - 2\pi z t}}{\cosh(\pi t)} dt = 2\pi i \operatorname*{Res}_{t=\frac{i}{2}} \frac{e^{\pi i \tau t^2 - 2\pi z t}}{\cosh(\pi t)}$$
$$= 2\zeta^{-\frac{1}{2}} q^{-\frac{1}{8}}.$$

To establish the uniqueness claim, assume that h_1 and h_2 both have the claimed properties. Then $f := h_1 - h_2$ is an entire function which satisfies

$$f(z+1) = -f(z),$$
$$f(z+\tau) = -\zeta q^{\frac{1}{2}} f(z).$$

Iterating these equations yields a bounded function on $\mathbb{C}$, and thus, by Liouville's theorem, f is constant. Letting $y \to \infty$ yields that this constant is zero. □

We now turn to the problem of determining the modular transformation properties of h. The following proposition provides the critical transformation that we shall use later.

PROPOSITION 8.6. *The following modular transformation for h holds:*

$$h\left(\frac{z}{\tau}; -\frac{1}{\tau}\right) = \sqrt{-i\tau} e^{-\frac{\pi i z^2}{\tau}} h(z;\tau).$$

PROOF. The claim follows by showing that

$$g(z) := \frac{1}{\sqrt{-i\tau}} e^{\frac{\pi i z^2}{\tau}} h\left(\frac{z}{\tau}; -\frac{1}{\tau}\right)$$

satisfies i) and ii) of Proposition 8.5 since those properties uniquely determine h. □

Combining the above, we finally obtain the following modularity properties of μ, which are the first key to completing μ to a nonholomorphic Jacobi form; as mentioned above, Zwegers accomplished this by finding a "simple" function with these same modular "defects".

PROPOSITION 8.7. *We have*

i) $\mu(z_1, z_2; \tau + 1) = e^{-\frac{\pi i}{4}} \mu(z_1, z_2; \tau)$.

ii) $\mu\left(\frac{z_1}{\tau}, \frac{z_2}{\tau}; -\frac{1}{\tau}\right)$
$= -\sqrt{-i\tau} e^{-\frac{\pi i}{\tau}(z_1 - z_2)^2} \mu(z_1, z_2; \tau) - \frac{i\sqrt{-i\tau}}{2} e^{-\frac{\pi i}{\tau}(z_1 - z_2)^2} h(z_1 - z_2; \tau)$.

PROOF. Part i) follows directly from the transformation law in Lemma 2.2. In order to show ii), define

$$f(z_1, z_2) := \frac{2i}{\sqrt{i\tau}} e^{\frac{\pi i}{\tau}(z_1 - z_2)^2} \mu\left(\frac{z_1}{\tau}, \frac{z_2}{\tau}; -\frac{1}{\tau}\right) + 2i\mu(z_1, z_2; \tau).$$

Using Proposition 8.3 ii), Lemma 2.2, and the modular transformation properties of the Dedekind η-function, one obtains that f depends only on $z_1 - z_2$. The function $z_1 \mapsto f(z_1, z_2)$ has at most simple poles at $z_1 \in \Lambda_\tau$. But since f is a function of $z_1 - z_2$ there cannot be poles. Using parts i) and ii) of Proposition 8.2 gives that $f(z_1 - z_2)$ satisfies i) and ii) of Proposition 8.5. The claim then follows by the uniqueness property of Proposition 8.5. □

We are now in a position to define the completed function $\widehat{\mu}$, which in particular allows one to find nonholomorphic completions of Ramanujan's mock theta functions.

DEFINITION 8.8. For $z_1, z_2 \in \mathbb{C}$ and $\tau \in \mathbb{H}$, we define the **completed μ-function**

$$\widehat{\mu}(z_1, z_2; \tau) = \widehat{\mu}(z_1, z_2) := \mu(z_1, z_2; \tau) + \frac{i}{2} R(z_1 - z_2; \tau),$$

where $\tau = u + iv$, $z = x + iy$, and

$$R(z; \tau) = R(z) := \sum_{n \in \frac{1}{2} + \mathbb{Z}} \left(\operatorname{sgn}(n) - E\left(\left(n + \frac{y}{v}\right)\sqrt{2v}\right)\right)(-1)^{n - \frac{1}{2}} \zeta^{-n} q^{-\frac{n^2}{2}}. \tag{8.1}$$

Here we have that

$$E(z) := 2\int_0^z e^{-\pi t^2} dt, \tag{8.2}$$

and $\operatorname{sgn}(x)$ denotes the sign function defined in (1.7).

REMARK. For $x \in \mathbb{R}$, we note that

$$E(x) = \operatorname{sgn}(x)\left(1 - \beta\left(x^2\right)\right),$$

where for $y \in \mathbb{R}^+$ we have

$$\beta(y) := \int_y^\infty t^{-\frac{1}{2}} e^{-\pi t} dt.$$

It is also not difficult to check that

$$\beta(y) = \frac{1}{\sqrt{\pi}} \Gamma\left(\frac{1}{2}, \pi y\right).$$

This can be used to show that specializations of $\widehat{\mu}$ to torsion points have Fourier expansions matching the expansions of harmonic Maass forms in $H_k(\Gamma)$ for some Γ in Lemma 4.3. We elaborate on this in Corollary 8.15.

REMARK. One can show that R converges locally uniformly on compact sets (in z and τ). A key step is to split

$$\operatorname{sgn}(n) - E\left(\left(n+\frac{y}{v}\right)\sqrt{2v}\right) = \left(\operatorname{sgn}(n) - \operatorname{sgn}\left(n+\frac{y}{v}\right)\right) + \operatorname{sgn}\left(n+\frac{y}{v}\right)\beta\left(2\left(n+\frac{y}{v}\right)^2 v\right).$$

The first summand is only non-zero for finitely many terms, and the contribution from the second term can be bounded using the estimate $0 \leq \beta(x) \leq e^{-\pi x}$.

The function E will play another critical role in Section 8.2. There we shall follow Zwegers' second realization of the mock theta functions as pieces of harmonic Maass forms, via the theory of *indefinite theta functions*. A central observation underpinning this construction is to build suitable functions satisfying Vignéras' differential equation in (8.21). The fact that Zwegers' functions satisfy that differential equation follows from the following seemingly simple, but powerful, differential equation for E.

LEMMA 8.9. *The special function E satisfies*

$$\left(2\pi z\frac{\partial}{\partial z} + \frac{\partial^2}{\partial z^2}\right)E(z) = 0.$$

PROOF. A direct calculation yields

$$\frac{\partial^2}{\partial z^2}E(z) = 2\frac{\partial}{\partial z}e^{-\pi z^2} = -4\pi z e^{-\pi z^2} = -2\pi z\frac{\partial}{\partial z}E(z),$$

as claimed. □

We next give a proposition which relates the function R at the arguments τ/n and $n\tau$. Its proof just requires the explicit definition of R.

PROPOSITION 8.10. *For $n \in \mathbb{N}, z \in \mathbb{C}$, and $\tau \in \mathbb{H}$, we have*

$$R\left(z;\frac{\tau}{n}\right) = \sum_{\ell=0}^{n-1} q^{-\frac{1}{2n}\left(\ell-\frac{n-1}{2}\right)^2} e^{-2\pi i\left(\ell-\frac{n-1}{2}\right)\left(z+\frac{1}{2}\right)} R\left(nz+\left(\ell-\frac{n-1}{2}\right)\tau+\frac{n-1}{2};n\tau\right).$$

We now aim to study the transformation properties of $\widehat{\mu}$ and show that we have made the "right" choice to yield elliptic and modular behavior. For this, we need to compare the transformation properties of μ with those of the R-function. We begin by determining its behavior under the elliptic transformation laws, which should be compared with the transformation laws of μ in Proposition 8.2.

PROPOSITION 8.11. *We have the following:*

i) $R(z+1) = -R(z)$,
ii) $R(z+\tau) = -\zeta q^{\frac{1}{2}}R(z) + 2\zeta^{\frac{1}{2}}q^{\frac{3}{8}}$,
iii) $R(-z) = R(z)$.

PROOF. Parts i) and iii) follow immediately from the definition of R, since E is an odd function.

For ii) we compute, replacing $n \mapsto n-1$,

$$\zeta^{-1}q^{-\frac{1}{2}}R(z+\tau) = -\sum_{n\in\frac{1}{2}+\mathbb{Z}} \left(\operatorname{sgn}(n-1) - E\left(\left(n+\frac{y}{v}\right)\sqrt{2v}\right)\right)(-1)^{n-\frac{1}{2}}q^{-\frac{n^2}{2}}\zeta^{-n}.$$

Comparing with $R(z)$ then gives the claim, since for $n \in 1/2+\mathbb{Z}$,

$$\operatorname{sgn}(n) - \operatorname{sgn}(n-1) = \begin{cases} 2 & \text{if } n = \frac{1}{2}, \\ 0 & \text{otherwise.} \end{cases}$$

□

We would like to compare the modular transformation properties of μ and R. Just as in Proposition 8.7, R is almost translation-invariant but is modular under inversion only up to a term determined by a Mordell integral.

PROPOSITION 8.12. *We have the following:*

i) $R(z;\tau+1) = e^{-\frac{\pi i}{4}}R(z;\tau)$,

ii) $R\left(\frac{z}{\tau};-\frac{1}{\tau}\right) = -\sqrt{-i\tau}e^{-\frac{\pi i z^2}{\tau}}R(z;\tau) + \sqrt{-i\tau}e^{-\frac{\pi i z^2}{\tau}}h(z;\tau)$.

PROOF. The claim i) follows directly from the definition of R. For ii), we define

$$H(z) := \frac{1}{\sqrt{-i\tau}}e^{\frac{\pi i z^2}{\tau}}R\left(\frac{z}{\tau};-\frac{1}{\tau}\right) + R(z;\tau).$$

From Proposition 8.11, we obtain that H satisfies the transformation laws in Proposition 8.5. So, we are left to show that H is holomorphic. More precisely, it suffices to show that

$$\left(\frac{\partial}{\partial a} + \tau\frac{\partial}{\partial b}\right)H(a\tau - b) = 0, \tag{8.3}$$

where $z = a\tau - b$ for some $a, b \in \mathbb{R}$ (i.e., $a = \operatorname{Im}(z)/\operatorname{Im}(\tau)$). This is because

$$\frac{\partial}{\partial \overline{z}} = \frac{i}{2v}\left(\frac{\partial}{\partial a} + \tau\frac{\partial}{\partial b}\right).$$

Establishing (8.3) follows by slightly lengthy but direct calculations, making use of the inversion transformation in Lemma 2.2. For further details of this calculation, we refer the reader to the proof of Proposition 1.10 in [**529**]. □

Zwegers' key result, which finally solved the mystery of a modular framework for Ramanujan's mock theta functions, and which made many of the applications of the theory of mock modular forms explored in this book possible, is encapsulated in the following result which ties together the sequence of observations above.

THEOREM 8.13. *The following are true:*

i) *For $k, \ell, m, n \in \mathbb{Z}$, we have*

$$\widehat{\mu}(z_1 + k\tau + \ell, z_2 + m\tau + n) = (-1)^{k+\ell+m+n}q^{\frac{1}{2}(k-m)^2}\zeta_1^{k-m}\zeta_2^{m-k}\widehat{\mu}(z_1, z_2).$$

ii) *If ν_η is the multiplier of η, then for all $\gamma = \left(\begin{smallmatrix} a & b \\ c & d \end{smallmatrix}\right) \in \mathrm{SL}_2(\mathbb{Z})$, we have that*

$$\widehat{\mu}\left(\frac{z_1}{c\tau+d}, \frac{z_2}{c\tau+d}; \frac{a\tau+b}{c\tau+d}\right) = \nu_\eta(\gamma)^{-3}(c\tau+d)^{\frac{1}{2}}e^{-\frac{\pi i c}{c\tau+d}(z_1-z_2)^2}\widehat{\mu}(z_1, z_2;\tau).$$

iii) *We have that*

$$\widehat{\mu}(-z_1,-z_2)=\widehat{\mu}(z_1,z_2)=\widehat{\mu}(z_2,z_1).$$

REMARK. In particular, Theorem 8.13 shows that $\widehat{\mu}$ transforms like a two-variable Jacobi form of weight 1/2 and index $\left(\begin{smallmatrix}-1 & 1\\ 1 & -1\end{smallmatrix}\right)$. The broader study of special real-analytic Jacobi forms, called *harmonic Maass-Jacobi forms*, began in [**114**] and extended to include functions with poles like $\widehat{\mu}$ in [**113**]. In [**494**], Westerholt-Raum used this theory to explain why μ "splits" into a sum of a "mock Jacobi form" of one variable and a two-variable Jacobi form. An alternative explanation, due to Zagier, is offered in Theorem 8.18. Roughly speaking, a mock Jacobi form is a function which, like μ, transforms as a holomorphic Jacobi form up to a suitable error term (in this case, a holomorphic Mordell integral).

REMARK. The multiplier of η is explicitly given as follows (where $\left(\begin{smallmatrix}a & b\\ c & d\end{smallmatrix}\right)\in$ $\mathrm{SL}_2(\mathbb{Z}), c>0$):

$$\nu_\eta\begin{pmatrix}a & b\\ c & d\end{pmatrix}=\begin{cases}\left(\frac{d}{c}\right)e^{\frac{\pi i}{12}\left((a+d)c-bd\left(c^2-1\right)-3c\right)} & \text{if } c \text{ is odd,}\\ \left(\frac{c}{d}\right)e^{\frac{\pi i}{12}\left(ac\left(1-d^2\right)+d(b-c+3)-3\right)} & \text{if } d \text{ is odd.}\end{cases}$$

Corollary 8.15 below is an important result, which shows that when specialized at torsion points, the functions $\widehat{\mu}$ are harmonic Maass forms. The shadows of the corresponding mock modular forms, essentially specializations of Zwegers' μ-function, are expressed in terms of the the simple weight 3/2 theta functions. To state this precisely, for $a,b\in\mathbb{R}$, and $\tau\in\mathbb{H}$, we define

$$g_{a,b}(\tau):=\sum_{n\in a+\mathbb{Z}} ne^{2\pi inb}q^{\frac{n^2}{2}}. \tag{8.4}$$

Using the theory from Chapter 2, we obtain the following transformation laws for the weight 3/2 theta functions $g_{a,b}$.

LEMMA 8.14. *We have*

i) $g_{a+1,b}=g_{a,b}$,
ii) $g_{a,b+1}=e^{2\pi ia}g_{a,b}$,
iii) $g_{-a,-b}=-g_{a,b}$,
iv) $g_{a,b}(\tau+1)=e^{-\pi ia(a+1)}g_{a,a+b+\frac{1}{2}}(\tau)$,
v) $g_{a,b}(-\frac{1}{\tau})=ie^{2\pi iab}(-i\tau)^{\frac{3}{2}}g_{b,-a}(\tau)$.

We now state Corollary 8.15. We point out that once one finds representations in terms of μ as given in the Appendix, this corollary directly implies the mock modularity of all of Ramanujan's mock theta functions (see also Theorem 9.4).

COROLLARY 8.15. *Let* $\alpha_1,\alpha_2,\beta_1,\beta_2\in\mathbb{Q}$ *such that* (α_1,β_1), $(\alpha_2,\beta_2)\notin\mathbb{Z}^2$. *Then*

$$\tau\mapsto e^{-\pi i(\alpha_1-\alpha_2)^2\tau}\widehat{\mu}\left(\alpha_1\tau+\beta_1,\alpha_2\tau+\beta_2;\tau\right)$$

is a harmonic Maass form (for some congruence subgroup) of weight 1/2. *Moreover, we have that the shadow of the mock modular form*

$$e^{-\pi i(\alpha_1-\alpha_2)^2\tau}\mu\left(\alpha_1\tau+\beta_1,\alpha_2\tau+\beta_2;\tau\right)$$

is given by

$$i\sqrt{2}e^{-2\pi i(\alpha_1-\alpha_2)\left(\beta_1-\beta_2+\frac{1}{2}\right)}g_{\alpha_1-\alpha_2+\frac{1}{2},\beta_1-\beta_2+\frac{1}{2}}(\tau).$$

PROOF. Using Theorem 8.13, one may show, as in the proof of Theorem 2.15, that the function transforms like a modular form, as claimed. Moreover, using the definition of $\widehat{\mu}$, we can decompose it as

(8.5)

$$e^{-\pi i(\alpha_1-\alpha_2)^2\tau}\mu\left(\alpha_1\tau+\beta_1,\alpha_2\tau+\beta_2\right)+\frac{i}{2}e^{-\pi i(\alpha_1-\alpha_2)^2\tau}R\left((\alpha_1-\alpha_2)\tau+\beta_1-\beta_2\right).$$

To show that $\widehat{\mu}$ is annihilated by the operator $\Delta_{\frac{1}{2}}$, we use the factorization

$$\Delta_{\frac{1}{2}}=-\xi_{\frac{3}{2}}\circ\xi_{\frac{1}{2}}.$$

The first summand in (8.5) is holomorphic and thus clearly annihilated by $\xi_{\frac{1}{2}}$. To investigate the second summand, we set $\alpha:=\alpha_1-\alpha_2$ and $\beta:=\beta_1-\beta_2$ and compute

$$\frac{\partial}{\partial\overline{\tau}}R\left(\alpha\tau+\beta;\tau\right)=-\frac{i}{\sqrt{2v}}\sum_{n\in\frac{1}{2}+\mathbb{Z}}(n+\alpha)(-1)^{n-\frac{1}{2}}e^{-2\pi(n+\alpha)^2v-\pi in^2\overline{\tau}-2\pi in(\alpha\overline{\tau}+\beta)}.$$

Thus we have that

$$\begin{aligned}\xi_{\frac{1}{2}}\left(e^{-\pi i\alpha^2\tau}R(\alpha\tau+\beta;\tau)\right)&=i\sqrt{2}\sum_{n\in\frac{1}{2}+\mathbb{Z}}(n+\alpha)e^{\pi i(n+\alpha)^2\tau+2\pi in\left(\beta+\frac{1}{2}\right)}\\&=i\sqrt{2}e^{-2\pi i\alpha\left(\beta+\frac{1}{2}\right)}g_{\alpha+\frac{1}{2},\beta+\frac{1}{2}}(\tau).\end{aligned}\tag{8.6}$$

That $\widehat{\mu}$ is annihilated by $\Delta_{\frac{1}{2}}$ now follows, since $g_{\alpha+\frac{1}{2},\beta+\frac{1}{2}}$ is holomorphic.

Finally, we briefly discuss the growth condition required of a harmonic Maass form. As $\tau\to i\infty$, it is easy to show that the nonholomorphic part R decays rapidly (and the holomorphic part from μ clearly decays once the negative powers of q are removed). This decay can be shown using the expansion of R in terms of the incomplete-Gamma function, together with the general asymptotic

$$\Gamma(s,x)\sim x^{s-1}e^{-x},$$

which holds as $x\to\infty$. Similar arguments, in particular using the modular transformations established in the first part of this theorem, show that such a principal part exists at all cusps.

The computation of the shadow of the corresponding mock modular forms is given in (8.6). □

It often turns out to be useful to write the nonholomorphic parts of mock theta functions as nonholomorphic Eichler integrals. For example, this has applications to quantum modularity of partial theta functions (cf. Chapter 21). Using an elementary integral evaluation, we find the following expressions for period integrals of the theta functions $g_{a,b}$ (the reader should also compare these results with Lemma 5.17).

THEOREM 8.16. *The following are true:*

i) *For $a\in(-1/2,1/2)$ and $b\in\mathbb{R}$, we have*

$$-e^{-\pi ia^2\tau+2\pi ia\left(b+\frac{1}{2}\right)}R(a\tau-b)=\int_{-\overline{\tau}}^{i\infty}\frac{g_{a+\frac{1}{2},b+\frac{1}{2}}(w)}{\sqrt{-i(\tau+w)}}dw.$$

ii) *For $a,b\in(-1/2,1/2)$, we have*

$$-e^{-\pi ia^2\tau+2\pi ia\left(b+\frac{1}{2}\right)}h(a\tau-b)=\int_{0}^{i\infty}\frac{g_{a+\frac{1}{2},b+\frac{1}{2}}(w)}{\sqrt{-i(\tau+w)}}dw.\tag{8.7}$$

PROOF. We begin with the proof of claim i). Making a change of variables, deforming the path of integration, and then inserting the Fourier expansion of $g_{a+\frac{1}{2},b+\frac{1}{2}}$ gives

$$\begin{aligned}
&\int_{-\overline{\tau}}^{i\infty} \frac{g_{a+\frac{1}{2},b+\frac{1}{2}}(w)}{\sqrt{-i(\tau+w)}}dw \\
&= i\int_{2v}^{\infty} \frac{g_{a+\frac{1}{2},b+\frac{1}{2}}(it-\tau)}{\sqrt{t}}dt = i\sum_{n\in a+\frac{1}{2}+\mathbb{Z}} ne^{-\pi in^2\tau+2\pi in\left(b+\frac{1}{2}\right)}\int_{2v}^{\infty} t^{-\frac{1}{2}}e^{-\pi n^2 t}dt \\
&= -e^{-\pi ia^2\tau+2\pi ia\left(b+\frac{1}{2}\right)}\sum_{n\in\frac{1}{2}+\mathbb{Z}}\left(\operatorname{sgn}(n+a)-E\left((n+a)\sqrt{2v}\right)\right) \\
&\qquad\qquad \times(-1)^{n-\frac{1}{2}}e^{-\pi in^2\tau-2\pi in(a\tau-b)}.
\end{aligned}$$

Using that $\operatorname{sgn}(n+a) = \operatorname{sgn}(n)$ (for $n \in 1/2+\mathbb{Z}$ and $a \in (-1/2, 1/2)$) then gives the claim.

For ii), we first split the integral $\int_0^{i\infty}$ into the sum of $\int_{-\overline{\tau}}^{i\infty}$ and $\int_0^{-\overline{\tau}}$. The first integral is computed in part i). In the second integral, replace w by $-1/w$ to obtain

$$\int_0^{-\overline{\tau}} \frac{g_{a+\frac{1}{2},b+\frac{1}{2}}(w)}{\sqrt{-i(\tau+w)}}dw = \frac{e^{\pi i(2ab+a-b)}}{\sqrt{-i\tau}}\int_{\frac{1}{\overline{\tau}}}^{i\infty} \frac{g_{b+\frac{1}{2},-a+\frac{1}{2}}(w)}{\sqrt{-i\left(-\frac{1}{\tau}+w\right)}}dw,$$

using parts ii) and v) of Lemma 8.14. From part i) with (a,b,τ) replaced by $(b,-a,-1/\tau)$, we find that this last expression is equal to

$$-\frac{1}{\sqrt{-i\tau}}e^{\pi i\left(a+\frac{b^2}{\tau}\right)}R\left(\frac{a\tau-b}{\tau};-\frac{1}{\tau}\right).$$

Combining again with i), we find

$$\begin{aligned}
&\int_0^{i\infty} \frac{g_{a+\frac{1}{2},b+\frac{1}{2}}(w)}{\sqrt{-i(\tau+w)}}dw \\
&= -e^{\pi i\left(-a^2\tau+a(2b+1)\right)}\left(R(a\tau-b)+\frac{1}{\sqrt{-i\tau}}e^{\pi i\frac{(a\tau-b)^2}{\tau}}R\left(\frac{a\tau-b}{\tau};-\frac{1}{\tau}\right)\right) \\
&= -e^{\pi i\left(-a^2\tau+a(2b+1)\right)}h(a\tau-b),
\end{aligned}$$

by Proposition 8.12 ii). □

REMARK. We note that we can also find harmonic Maass forms of weight 3/2 by differentiating Zwegers' μ-function in the Jacobi variable (cf. [**79**]).

It is striking that the error to modularity of μ only depends on $z_1 - z_2$. In an unpublished note, Zagier [**505**] explained this phenomenon by splitting $\mu(z_1, z_2; \tau)$ (non-uniquely) as a sum of a Jacobi form in z_1, z_2 and a mock Jacobi form which depends on $z_1 - z_2$ and τ only. Here, we follow Zagier's proof which he generously shared with the authors. Note that our normalizations differ slightly from his. In what follows, we consider three related functions, namely μ,

$$A(z_1,z_2) := \vartheta(z_2)\mu(z_1,z_2),$$

and

$$M(z_1,z_2) := -i\vartheta(z_1)A(z_1,z_2) = -i\vartheta(z_1)\vartheta(z_2)\mu(z_1,z_2).$$

Although these functions are all related in an elementary way, each has its own useful features. Specifically, $A(z_1, z_2)$ is the simplest and, as shown by Zwegers, a "building block" in the theory of meromorphic Jacobi forms (cf. Chapter 11 and the discussion of the functions A_ℓ at the end of this section). The function μ yields mock modular forms when specialized to torsion points, and $M(z_1, z_2)$ is entire in both z_1, z_2 and symmetric in z_1, z_2 (as is $\mu(z_1, z_2)$).

We first require decompositions of M, which we use to rewrite μ. This function is invariant under $z_j \mapsto z_j + 1$ (for $j = 1, 2$), and so there are unique functions $M_\pm$ such that

$$M(z_1, z_2) = M_+(z_1, z_2) + M_-(z_1, z_2) \tag{8.8}$$

and

$$M_\pm\left(z_1 + \frac{1}{2}, z_2 + \frac{1}{2}\right) = \pm M_\pm(z_1, z_2). \tag{8.9}$$

In particular, these functions are given by

$$M_\pm(z_1, z_2) := \frac{M(z_1, z_2) \pm M\left(z_1 + \frac{1}{2}, z_2 + \frac{1}{2}\right)}{2}.$$

Zagier's first decomposition then states that these two functions split as products involving the theta functions

$$\theta_\pm(z) = \theta_\pm(z;\tau) := \sum_{(-1)^n = \pm 1} q^{\frac{n^2}{4}} \zeta^{\frac{n}{2}}.$$

Note that

$$\vartheta(z;\tau) = \theta_-\left(z + \frac{1}{2}; \tau\right).$$

THEOREM 8.17. *We have that*

$$M_\pm(z_1, z_2) = \theta_\pm(z_1 + z_2) F_\pm(z_1 - z_2), \tag{8.10}$$

where $F_\pm(z) = F_\pm(z;\tau)$ (defined in (8.13)) are even, entire functions of z satisfying

$$\sum_\pm \theta_\pm(z) F_\pm(z) = -\eta^3. \tag{8.11}$$

PROOF. A direct calculation shows that the truncated theta series

$$\gamma_N(\tau) := \sum_{\substack{n \in \frac{1}{2} + \mathbb{Z} \\ n > N}} (-1)^{n - \frac{1}{2}} q^{\frac{n^2}{2}}$$

can be used to expand the Jacobi theta function as

$$(-1)^s i \frac{\zeta^{\frac{1}{2}}}{1 - \zeta q^s} q^{\frac{s(s+1)}{2}} \vartheta(z) = \sum_{r \in \mathbb{Z}} \gamma_{r-s}(\tau) q^{rs} \zeta^r.$$

This is an absolutely convergent expression, since γ_n is supported on powers of q at least $n^2/2$. Moreover, $\gamma_n = \gamma_{-n}$ (to see this, take $s = 0$ in the last formula and use the fact that $\vartheta(z)$ and $\zeta^{\frac{1}{2}}/(1 - \zeta)$ are odd functions of z). Substituting this expansion back into the definition of M gives

$$M(z_1, z_2) = -\sum_{r,s \in \mathbb{Z}} \gamma_{r-s}(\tau) q^{rs} \zeta_1^r \zeta_2^s. \tag{8.12}$$

This shows that $M(z_1, z_2) = M(z_2, z_1)$ (since γ_n is even) and is the key step in computing the desired decomposition of $M_\pm$. Now we claim that the desired functions $F_\pm$ are given by

$$F_\pm(z) := -\sum_{\substack{n\in\mathbb{Z}\\(-1)^n=\pm 1}} \gamma_n(\tau)q^{-\frac{n^2}{4}}\zeta^{\frac{n}{2}}. \tag{8.13}$$

By definition, we have

$$\sum_\pm \theta_\pm(z_1+z_2)F_\pm(z_1-z_2) = -\sum_{\substack{R,S\in\mathbb{Z}\\R\equiv S \pmod 2}} \gamma_S(\tau)q^{\frac{R^2-S^2}{4}}\zeta_1^{\frac{R+S}{2}}\zeta_2^{\frac{R-S}{2}}.$$

Then we change variables by setting $r := (R+S)/2$, $s := (R-S)/2$ and note that as (R,S) runs over pairs of integers with the same parity, (r,s) runs over all pairs of integers, and that $S = r - s$. Hence,

$$\sum_\pm \theta_\pm(z_1+z_2)F_\pm(z_1-z_2) = -\sum_{r,s\in\mathbb{Z}} \gamma_{r-s}(\tau)q^{rs}\zeta_1^r\zeta_2^s = M(z_1,z_2),$$

where we have also used (8.12). To finish verifying that

$$\theta_\pm(z_1+z_2)F_\pm(z_1-z_2) = M_\pm(z_1,z_2),$$

by uniqueness it is enough to show that $\theta_\pm(z_1+z_2)F_\pm(z_1-z_2)$ satisfy (8.9). But this follows directly from the definitions of $\theta_\pm, F_\pm$ and the fact that $\theta_+(z)$ (resp. $\theta_-(z)$) is supported on even (resp. odd) powers of $\zeta^{\frac{1}{2}}$.

The final claim follows by taking $z_1 = z$, $z_2 = 0$ in (8.10) and using (2.1), which then yields $-\eta^3$. □

Given the close relationship between M and μ, it is not surprising that one can use the above decomposition for M to yield interesting formulas for μ. Here we investigate the consequences for μ, explaining why the nonholomorphic completion term only depends on a one-variable function. Firstly, we require the modified Weierstrass zeta-function $\zeta^\times_{\Lambda_{1,\tau}}$, given in (1.6). Renormalizing this function by dividing by $2\pi i$ and using (6.4) and (6.5), we see that

$$Z(z) := \frac{1}{2\pi i}\zeta^\times_{\Lambda_{1,\tau}}(z) = -\frac{1}{2}\sum_{n\in\mathbb{Z}}\left(\frac{1+\zeta q^n}{1-\zeta q^n} - \operatorname{sgn}\left(n+\frac{1}{2}\right)\right) = \frac{\vartheta'(z)}{2\pi i\vartheta(z)}.$$

It is also convenient to define

$$P(z) := \frac{1}{(2\pi i)^2}\wp(z)$$

(recall that $\wp$ was given in Definition 1.8, and that if no lattice Λ is specified, then we take $\Lambda = \Lambda_\tau$). Thus, by the definition of $\zeta^\times_{\Lambda_{1,\tau}}$, and Lemma 1.15, we have

$$P(z) = -\frac{Z'(z)}{2\pi i} + \frac{1}{12}E_2 = -\sum_{n\in\mathbb{Z}}\frac{\zeta q^n}{(1-\zeta q^n)^2} + \frac{1}{12}E_2.$$

The functions Z and P satisfy the following identities:

$$Z(z_1) - Z(z_2) - Z(z_1 - z_2) = -\frac{P'(z_1) + P'(z_2)}{2(P(z_1) - P(z_2))}, \tag{8.14}$$

$$Z(z_1) - Z(z_2) - Z\left(z_1 - z_2 + \frac{1}{2}\right) = -\frac{i\eta^3 \vartheta\left(z_1 + \frac{1}{2}\right)\vartheta\left(z_2 + \frac{1}{2}\right)\vartheta(z_1 - z_2)}{\vartheta\left(\frac{1}{2}\right)\vartheta(z_1)\vartheta(z_2)\vartheta\left(z_1 - z_2 + \frac{1}{2}\right)}, \tag{8.15}$$

$$P(z_1) - P(z_2) = \eta^6 \frac{\vartheta(z_1 - z_2)\vartheta(z_1 + z_2)}{\vartheta(z_1)^2\vartheta(z_2)^2}. \tag{8.16}$$

To check these formulas, one simply notes that P is an elliptic function, while Z is invariant under $z \mapsto z+1$ and satisfies $Z(z+\tau) = Z(z) - 1$ (cf. (1.8)) from which it follows that the left- and right-hand sides of each identity are elliptic functions. It then only remains to verify that the poles and residues of each side of the identities are equal (as Theorem 1.4 then implies that the difference is identically zero), which we leave to the interested reader.

The decomposition for μ may then be written in terms of the following one-variable μ function:

$$\mu(z) := -\frac{\zeta^{\frac{1}{2}}}{\vartheta\left(\frac{1}{2}\right)} \sum_{n\in\mathbb{Z}} \frac{q^{\frac{n(n+1)}{2}}}{1 + \zeta q^n} + \frac{Z(z) - Z\left(z - \frac{1}{2}\right)}{\vartheta(z)}.$$

It is natural to expect that this one-variable function which captures the mock behavior of $\mu(z_1, z_2)$ can essentially be obtained by specializing one of the variables in μ to be a constant. Indeed, it is not difficult to check that

$$\mu(z) = \mu\left(z + \frac{1}{2}, \frac{1}{2}\right) + \frac{Z(z) - Z\left(z - \frac{1}{2}\right)}{\vartheta(z)}. \tag{8.17}$$

THEOREM 8.18. *We have the following splitting of μ as a sum of a one-variable mock Jacobi form and a two-variable Jacobi form:*

$$\begin{aligned}\mu(z_1, z_2) &= \mu(z_1 - z_2) + \frac{Z(z_1) - Z(z_2) - Z(z_1 - z_2)}{\vartheta(z_1 - z_2)} \\ &= \mu(z_1 - z_2) - \frac{1}{4\pi i \vartheta(z_1 - z_2)} \frac{P'(z_1) + P'(z_2)}{P(z_1) - P(z_2)}.\end{aligned}$$

REMARK. Note that by (8.14) the second terms on the right-hand sides of these two decompositions of $\mu(z_1, z_2)$ are Jacobi forms, so that the mock behavior of $\mu(z_1, z_1)$ only arises from the one-variable μ function.

PROOF OF THEOREM 8.18. We begin by considering the function

$$G(z) := \frac{\theta_-(z)F_-(z) - \theta_+(z)F_+(z)}{\theta\left(\frac{1}{2}\right)}.$$

Using Theorem 8.17 and the translation relations $\vartheta_\pm(z+1) = \pm\vartheta_\pm(z)$, we find

$$G(z) = -\frac{M\left(z + \frac{1}{2}, \frac{1}{2}\right)}{\vartheta\left(\frac{1}{2}\right)} = \frac{\zeta^{\frac{1}{2}}\vartheta\left(z + \frac{1}{2}\right)}{\vartheta\left(\frac{1}{2}\right)} \sum_{n\in\mathbb{Z}} \frac{q^{\frac{n(n+1)}{2}}}{1 + \zeta q^n} = -\sum_{n\in\mathbb{Z}} \frac{\zeta^n q^{\frac{n(n+1)}{2}}}{1 + q^n},$$

where in the last equality we use that $M(z_1, z_2) = M(z_2, z_1)$. By employing (8.11), we obtain the identity

$$\theta_\pm(z)F_\pm(z) = \frac{1}{2}\left(-\eta^3 \mp \vartheta\left(\frac{1}{2}\right)G(z)\right),$$

which, combined with the decomposition of M in Theorem 8.17, yields

$$\begin{aligned}M(z_1,z_2) &= \theta_+(z_1+z_2)F_+(z_1-z_2)+\theta_-(z_1+z_2)F_-(z_1-z_2)\\ &= -\frac{\eta^3}{2}\sum_{\pm}\frac{\theta_\pm(z_1+z_2)}{\theta_\pm(z_1-z_2)} - \frac{\theta\left(\frac{1}{2}\right)}{2}G(z_1-z_2)\sum_{\pm}\pm\frac{\theta_\pm(z_1+z_2)}{\theta_\pm(z_1-z_2)}.\end{aligned} \tag{8.18}$$

We now require the following identities, each of which follows from a short calculation:

$$\theta_+(z)\theta_-(z) = \frac{1}{2}\vartheta\left(\frac{1}{2}\right)\vartheta\left(z+\frac{1}{2}\right),$$

$$\theta_+(z_1+z_2)\theta_-(z_1-z_2)+\theta_+(z_1-z_2)\theta_-(z_1+z_2) = \vartheta\left(z_1+\frac{1}{2}\right)\vartheta\left(z_2+\frac{1}{2}\right),$$

$$\theta_+(z_1+z_2)\theta_-(z_1-z_2)-\theta_+(z_1-z_2)\theta_-(z_1+z_2) = \vartheta(z_1)\vartheta(z_2).$$

Directly substituting these identities into (8.18) yields

$$M(z_1,z_2) = -\frac{\eta^3\vartheta\left(z_1+\frac{1}{2}\right)\vartheta\left(z_2+\frac{1}{2}\right)}{\vartheta\left(\frac{1}{2}\right)\vartheta\left(z_1-z_2+\frac{1}{2}\right)} - G(z_1-z_2)\frac{\vartheta(z_1)\vartheta(z_2)}{\vartheta\left(z_1-z_2+\frac{1}{2}\right)}.$$

Dividing by $-i\vartheta(z_1)\vartheta(z_2)$ and using (8.15) shows that

$$\begin{aligned}\mu(z_1,z_2) &= g(z_1-z_2) - \frac{i\eta^3\vartheta\left(z_1+\frac{1}{2}\right)\vartheta\left(z_2+\frac{1}{2}\right)}{\vartheta\left(\frac{1}{2}\right)\vartheta(z_1)\vartheta(z_2)\vartheta\left(z_1-z_2+\frac{1}{2}\right)}\\ &= g(z_1-z_2) + \frac{Z(z_1)-Z(z_2)-Z(z_1-z_2)-Z\left(z_1-z_2+\frac{1}{2}\right)}{\vartheta(z_1-z_2)},\end{aligned}$$

where

$$g(z) := -\frac{iG(z)}{\vartheta\left(z+\frac{1}{2}\right)}.$$

Using (8.17), we find that

$$G(z) = i\vartheta\left(z+\frac{1}{2}\right)\left(\mu(z)-\frac{Z(z)-Z\left(z-\frac{1}{2}\right)}{\vartheta(z)}\right)$$

or

$$g(z) = \mu(z) - \frac{Z(z)-Z\left(z-\frac{1}{2}\right)}{\vartheta(z)},$$

and so finally by (8.14), we find that

$$\begin{aligned}\mu(z_1,z_2) &= \mu(z_1-z_2) + \frac{Z(z_1)-Z(z_2)-Z(z_1-z_2)}{\vartheta(z_1-z_2)}\\ &= \mu(z_1-z_2) - \frac{1}{4\pi i\vartheta(z_1-z_2)}\frac{P'(z_1)+P'(z_2)}{P(z_1)-P(z_2)},\end{aligned}$$

as desired. □

We conclude this section by discussing the so-called higher level Appell functions, which arise frequently in applications and whose modularity properties may be expressed in terms of the above modularity results for μ (cf. [**527**]).

DEFINITION 8.19. For $\ell \in \mathbb{N}$, we define the **level ℓ Appell functions**

$$A_\ell(z_1,z_2) = A_\ell(z_1,z_2;\tau) := \zeta_1^{\frac{\ell}{2}}\sum_{n\in\mathbb{Z}}\frac{(-1)^{\ell n}\zeta_2^n q^{\frac{\ell n(n+1)}{2}}}{1-\zeta_1 q^n}.$$

REMARK. Note that

$$A_1(z_1, z_2) = \vartheta(z_2)\mu(z_1, z_2).$$

The higher level Appell functions are also related to Zwegers' μ-function, as can be read from Lemma 8.20, due to Zwegers.

LEMMA 8.20. *We have*

$$A_\ell\left(z_1, z_2; \tau\right) = \sum_{k=0}^{\ell-1} \zeta_1^k A_1\left(\ell z_1, z_2 + k\tau + \frac{\ell-1}{2}; \ell\tau\right).$$

PROOF. We rewrite the right-hand side as

$$\begin{aligned}\sum_{k=0}^{\ell-1} \zeta_1^{k+\frac{\ell}{2}} \sum_{n\in\mathbb{Z}} \frac{(-1)^{\ell n}\zeta_2^n q^{\frac{\ell n(n+1)}{2}+kn}}{1-\zeta_1^\ell q^{\ell n}} &= \sum_{n\in\mathbb{Z}} \frac{(-1)^{\ell n}\zeta_2^n q^{\frac{\ell n(n+1)}{2}}}{1-\zeta_1^\ell q^{\ell n}} \zeta_1^{\frac{\ell}{2}} \sum_{k=1}^{\ell-1} (\zeta_1 q^n)^k \\ &= \zeta_1^{\frac{\ell}{2}} \sum_{n\in\mathbb{Z}} \frac{(-1)^{\ell n}\zeta_2^n q^{\frac{\ell n(n+1)}{2}}}{1-\zeta_1 q^n} = A_\ell\left(z_1, z_2; \tau\right),\end{aligned}$$

which is the claimed expression. □

As with the μ-functions, we can complete the higher level Appell functions using the R-function defined in (8.1). This is a consequence of the decomposition in Lemma 8.20.

DEFINITION 8.21. The **completed level ℓ Appell functions** are defined as

$$\begin{aligned}&\widehat{A}_\ell(z_1, z_2) = \widehat{A}_\ell(z_1, z_2; \tau) \\ &:= A_\ell(z_1, z_2; \tau) + \frac{i}{2}\sum_{k=0}^{\ell-1} \zeta_1^k \vartheta\left(z_2 + k\tau + \frac{\ell-1}{2}; \ell\tau\right) R\left(\ell z_1 - z_2 - k\tau - \frac{\ell-1}{2}; \ell\tau\right).\end{aligned}$$

The functions $\widehat{A}_\ell$ have the following transformation properties.

THEOREM 8.22. *The following transformation properties hold:*

i) *We have that*

$$\widehat{A}_\ell(-z_1, -z_2) = -\widehat{A}_\ell(z_1, z_2).$$

ii) *For all $n_1, n_2, m_1, m_2 \in \mathbb{Z}$, we have that*

$$\begin{aligned}&\widehat{A}_\ell(z_1 + n_1\tau + m_1, z_2 + n_2\tau + m_2) \\ &\quad = (-1)^{\ell(n_1+m_1)} \zeta_1^{\ell n_1 - n_2} \zeta_2^{-n_1} q^{\frac{\ell n_1^2}{2} - n_1 n_2} \widehat{A}_\ell(z_1, z_2).\end{aligned}$$

iii) *For all $\left(\begin{smallmatrix} a & b \\ c & d\end{smallmatrix}\right) \in \mathrm{SL}_2(\mathbb{Z})$, we have that*

$$\widehat{A}_\ell\left(\frac{z_1}{c\tau+d}, \frac{z_2}{c\tau+d}; \frac{a\tau+b}{c\tau+d}\right) = (c\tau+d) e^{\frac{\pi i c}{c\tau+d}\left(-\ell z_1^2 + 2z_1 z_2\right)} \widehat{A}_\ell(z_1, z_2; \tau).$$

SKETCH OF PROOF. We first note, by Theorem 8.13 and the definition of $\widehat{\mu}$, that the level 1 Appell sum $\widehat{A}_1(z_1, z_2)$ satisfies the transformations of Theorem 8.22. Moreover, by Lemma 8.20, it is clear that $\widehat{A}_\ell$ may be written as

$$\widehat{A}_\ell\left(z_1, z_2; \tau\right) = \sum_{k=0}^{\ell-1} \zeta_1^k \widehat{A}_1\left(\ell z_1, z_2 + k\tau + \frac{\ell-1}{2}; \ell\tau\right).$$

The transformation properties in Theorem 8.22 then follow directly by applying the transformation formulas for $\widehat{A}_1$ and making use of the following formula which follows from a straightforward, but lengthy calculation:

$$\widehat{A}_\ell(z_1, z_2; \tau) = \frac{1}{\ell}\zeta_1^{\frac{\ell-1}{2}} \sum_{k=0}^{\ell-1} \widehat{A}_1\left(z_1, \frac{z_2+k}{\ell} + \frac{(\ell-1)\tau}{2\ell}; \frac{\tau}{\ell}\right).$$

□

Throughout this book, we see the function μ and other Appell sums playing a fundamental role in the construction of examples of mock modular forms. We next switch gears to discuss seemingly different, but closely related objects, namely indefinite theta series.

8.2. Zwegers' thesis II: indefinite theta series

To motivate the definition of indefinite theta functions, we first recall classical theta functions associated to positive definite lattices, and especially to the lattice $\mathbb{Z}^r$. Then, just as with the theta function $\vartheta(z;\tau)$ studied in Chapter 2, there are theta functions attached to any positive definite quadratic form on $\mathbb{R}^r$, which we consider to have integral coefficients (cf. Kapitel V.2 of **[318]** or Chapter 10 of **[488]**).

Given such a quadratic form Q, we recall that its *bilinear form* is defined by $B(X,Y) := Q(X+Y) - Q(X) - Q(Y)$. We also have frequent need of the associated symmetric matrix A, which is given by $Q(X) = \frac{1}{2}\langle X, AX\rangle$ (note that some authors use a different normalization), where $\langle\cdot,\cdot\rangle$ denotes the usual inner product on $\mathbb{R}^r$. Then, for any $x_0 \in \mathbb{Z}^r$, define the theta function

$$\Theta_{Q,x_0}(z;\tau) := \sum_{n\in\mathbb{Z}^r} q^{Q(n)}\zeta^{B(n,x_0)}. \tag{8.19}$$

The following lemma offers the critical feature of these theta functions.

LEMMA 8.23. *Assume the notation and hypotheses above. Then the function* $\Theta_{Q,x_0}(z;\tau)$ *is a Jacobi form of weight* $r/2$ *and index* $Q(x_0)$.

This fact is well known (cf. page 1 of **[191]** for a discussion of the general case, and also **[286]** for a study of these functions if $r = 1$). It can be proved using Poisson summation, exactly as for the ordinary Jacobi theta function.

The situation is more complicated if Q is no longer positive definite. An indication that a theory of modularity of theta functions for more general quadratic forms might exist arose in work of Göttsche and Zagier **[230]**, who studied examples of such series if the quadratic form is not positive definite and the sum defining the theta function is restricted to a proper subset of the lattice $\mathbb{Z}^r$. Related series were also previously investigated by Siegel in **[459]**. However, in general these series have no known nice modularity properties. The general situation was clarified by the results in Zwegers' thesis, which we review in this section.

Further motivation is provided by work of Andrews and Watson, which gave many representations for Ramanujan's mock theta functions (and additional mock theta functions which they discovered) as indefinite theta series. Combining this with the results of Zwegers outlined in this section allows one to complete Ramanujan's mock theta functions to harmonic Maass forms. Examples of such series are discussed in Theorems 9.7 and 9.8 below. We note, in passing, that these indefinite theta functions are in fact closely related to the Appell-Lerch sums in the previous section; we explain this further below.

We first require a few basic definitions from the theory of quadratic forms. All of these can also be found in any standard textbook on quadratic forms; see, for example, **[341]**. We now suppose that A is non-degenerate and has integral entries with off-diagonal entries even. The *type* of the form Q is the pair $(r-s, s)$, where s is the largest dimension of a linear subspace of $\mathbb{R}^r$ on which Q is negative definite. We warn the reader that some authors switch the order of this pair of integers, and so care must be taken if consulting the literature.

Zwegers defined indefinite theta functions with the proper convergence and modularity properties attached to any Q with $s = 1$, and we assume henceforth that the type of Q is $(r-1, 1)$ unless stated otherwise. It is also important to note that one must be careful when defining theta series which are indefinite to obtain absolutely convergent series, but it turns out that for quadratic forms of type $(r-1, 1)$, suitably defined theta series converge. The main goal of this section is to study the mock modularity of such series. This is explicitly given below in Theorem 8.30. In what follows, we roughly follow Zwegers' exposition in Chapter 2 of **[529]**.

In order to define the indefinite theta functions we are interested in, we need a few more definitions. The set of vectors $c \in \mathbb{R}^r$ with $Q(c) < 0$ splits into two connected components. If two vectors c_1 and c_2 with $Q(c_j) < 0$ are given, they lie in the same component if and only if $B(c_1, c_2) < 0$. We fix one of the components and denote it by C_Q. Picking any vector $c_0 \in C_Q$, we then have

$$C_Q = \{c \in \mathbb{R}^r : Q(c) < 0,\ B(c, c_0) < 0\}.$$

Zwegers' indefinite theta functions may naturally be considered as automorphic forms on the $(r-1)$-dimensional hyperbolic space $C_Q/\mathbb{R}^+$ with respect to an orthogonal group (although we focus on the case that the functions are simply functions of a variable $\tau \in \mathbb{H}$).

We resume our general discussion for quadratic forms of type $(r-1, 1)$. In analogy with the situation for cusps of the usual upper half-plane, we define (the possibly empty) set S_Q by

$$S_Q := \{c = (c_1, c_2, \ldots, c_r) \in \mathbb{Z}^r : \gcd(c_1, c_2, \ldots, c_r) = 1,\ Q(c) = 0,\ B(c, c_0) < 0\},$$

and the set of *cusps*, needed to compactify the hyperbolic space $C_Q/\mathbb{R}^+$, by

$$\{c \in \mathbb{Q}^r : Q(c) = 0,\ B(c, c_0) < 0\}/\mathbb{Q}^+.$$

We then set $\overline{C}_Q := C_Q \cup S_Q$, which serves as the compactification of C_Q. Defining for any $c \in \overline{C}_Q$

$$\mathcal{R}(c) := \begin{cases} \mathbb{R}^r & \text{if } c \in C_Q, \\ \{a \in \mathbb{R}^r : B(c, a) \notin \mathbb{Z}\} & \text{if } c \in S_Q, \end{cases}$$

and recalling that $\operatorname{Im}(z) = y$, $\operatorname{Im}(\tau) = v$, we set

$$D(c) := \left\{(z, \tau) \in \mathbb{C}^r \times \mathbb{H} : \frac{y}{v} \in \mathcal{R}(c)\right\}.$$

Zwegers' indefinite theta functions, which transform as modular forms and which are (generically) nonholomorphic, are then defined as follows.

DEFINITION 8.24. Assume the notation above and let $c_1, c_2 \in \overline{C}_Q$. Then the **theta function** associated to Q with **characteristic** $a \in \mathcal{R}(c_1) \cap \mathcal{R}(c_2)$ and $b \in \mathbb{R}^r$, with respect to (c_1, c_2), is the series

$$\Theta_{a,b}(\tau) = \Theta_{a,b,c_1,c_2}(\tau) = \Theta_{A,a,b,c_1,c_2}(\tau) := \sum_{n \in a+\mathbb{Z}^r} \rho(n;\tau) e^{2\pi i B(n,b)} q^{Q(n)}.$$

Here, we have let

$$\rho(n;\tau) = \rho_Q^{c_1,c_2}(n;\tau) := \rho^{c_1}(n;\tau) - \rho^{c_2}(n;\tau), \tag{8.20}$$

with

$$\rho^c(n;\tau) := \begin{cases} E\left(\frac{B(c,n)v^{\frac{1}{2}}}{\sqrt{-Q(c)}}\right) & \text{if } c \in C_Q, \\ \operatorname{sgn}(B(c,n)) & \text{if } c \in S_Q, \end{cases}$$

where we recall that the special function E is defined in (8.2).

We also require Zwegers' Jacobi form version of $\Theta_{a,b}$ which neatly ties together the modularity properties of the indefinite theta functions with characteristic.

DEFINITION 8.25. For $(z,\tau) \in D(c_1) \cap D(c_2)$, we define the **theta function** of A with respect to (c_1, c_2) by $(z = a\tau + b)$

$$\begin{aligned}\Theta(z;\tau) = \Theta_{A,c_1,c_2}(z;\tau) &:= e^{-2\pi i B(a,b)} q^{-Q(a)} \Theta_{a,b}(\tau) \\ &= \sum_{n \in \mathbb{Z}^r} \rho(n+a;\tau) e^{2\pi i B(n,z)} q^{Q(n)}.\end{aligned}$$

We next state that these indefinite theta functions indeed converge. Note that this is far from obvious, since in particular $q^{Q(n)}$ is not bounded as n ranges over $\mathbb{Z}^r$, due to the indefiniteness of Q. The proof may be found in [**529**].

THEOREM 8.26. *The series $\Theta(z;\tau)$, and hence also $\Theta_{a,b}(\tau)$, converges absolutely.*

We next state the main results of this section, which allows us to use indefinite theta series to build examples of mock modular forms. We begin by stating the Jacobi transformation properties of Θ, where in the following we require

$$\begin{aligned}\mathcal{D}(c) : &= \left\{(z,\tau) \in D(c) : \left(\frac{z}{\tau}, -\frac{1}{\tau}\right) \in D(c)\right\} \\ &= \left\{(a\tau + b, \tau) : \tau \in \mathbb{H}, a, b \in \mathbb{R}^r, B(c,a), B(c,b) \notin \mathbb{Z}\right\}.\end{aligned}$$

Instead of the original proof that can be found in Zwegers' thesis, we use a framework of Vignéras [**484**] which is well suited for producing many examples of theta series. We present it here in its general form since it can also be employed for generalizations to arbitrary type $(r-s, s)$.

THEOREM 8.27. *Let Q be a quadratic form of type $(r-s, s)$, $L \subset \mathbb{R}^r$ a lattice on which Q takes integral values, $L' := \{x \in \mathbb{R}^r : B(x,\ell) \in \mathbb{Z}\ \forall \ell \in L\}$ its dual lattice, and $p\colon \mathbb{R}^r \to \mathbb{C}$ a Schwartz function satisfying Vignéras' differential equation for some $\lambda \in \mathbb{Z}$:*

$$\left(\mathcal{E} - \frac{\Delta_Q}{4\pi}\right) p = \lambda p, \tag{8.21}$$

where $\mathcal{E}$ is the Euler operator

$$\mathcal{E} := \sum_{j=1}^{r} x_j \frac{\partial}{\partial x_j}$$

and Δ_Q is the Laplacian associated to Q, given in terms of the usual inner product $\langle \cdot, \cdot \rangle$ on $\mathbb{R}^r$ by

$$\Delta_A := \left\langle \frac{\partial}{\partial x}, A^{-1}\frac{\partial}{\partial x} \right\rangle .$$

Then the theta function

$$\Theta_{A,\mu}(z;\tau) = \Theta_\mu(z;\tau) := v^{-\frac{\lambda}{2}} \sum_{n\in\mu+L} p\left(\sqrt{v}\left(n+\frac{y}{v}\right)\right) e^{2\pi i B(n,z)} q^{Q(n)}$$

satisfies the modular transformations

$$\Theta_\mu\left(\frac{z}{\tau};-\frac{1}{\tau}\right) = \frac{(-i\tau)^{\lambda+\frac{r}{2}}}{\sqrt{|L'/L|}} e^{\pi i Q(A^{-1}A^*)} \sum_{\nu\in L'/L} e^{-2\pi i B(\mu,\nu)+2\pi i \frac{Q(z)}{\tau}} \Theta_\nu(z;\tau),$$

$$\Theta_\mu(z;\tau+1) = e^{2\pi i Q(\mu+\frac{1}{2}A^{-1}A^*)}\Theta_\mu(z;\tau),$$

where $A^* := (A_{1,1},\ldots,A_{r,r})^T$.

The main difficulty lies in finding functions p which satisfy (8.21), "approximate" fixed discontinuous functions defined by sgn functions (in a specific sense which becomes apparent), and which yield convergent theta series. The special case of signature $(r-1,1)$ was treated by Zwegers in his thesis [**529**], and an alternative perspective on why his indefinite theta functions "work" is that his differences built from error functions have all the desired properties from the perspective of Theorem 8.27. Zwegers' main achievement on indefinite theta series can then be described as follows.

THEOREM 8.28. *Assuming the notation above, the function Θ satisfies the following transformations:*

i) *For all $\lambda \in \mathbb{Z}^r$ and $\mu \in A^{-1}\mathbb{Z}^r$, we have*

$$\Theta(z+\lambda\tau+\mu;\tau) = e^{-2\pi i B(z,\lambda)} q^{-Q(\lambda)}\Theta(z;\tau).$$

ii) *If $A^* := (A_{11},\ldots,A_{rr})^T$, then*

$$\Theta(z;\tau+1) = \Theta\left(z+\frac{1}{2}A^{-1}A^*;\tau\right).$$

In particular, $\Theta(z;\tau+2) - \Theta(z;\tau)$, and $\Theta(z;\tau \mid 1) = \Theta(z;\tau)$ if A has even entries.

iii) *If $(z,\tau) \in \mathcal{D}(c_1)\cap\mathcal{D}(c_2)$, then*

$$\Theta\left(\frac{z}{\tau};-\frac{1}{\tau}\right) = \frac{i(-i\tau)^{\frac{r}{2}}}{\sqrt{-\det(A)}} \sum_{n\in A^{-1}\mathbb{Z}^r/\mathbb{Z}^r} e^{\frac{2\pi i}{\tau}Q(z+n\tau)}\Theta(z+n\tau;\tau).$$

In particular, for the one-variable theta functions, we directly obtain the following.

COROLLARY 8.29. *The function $\Theta_{a,b}$ satisfies the following transformation laws:*

i) *For all $\lambda \in \mathbb{Z}^r$, we have*
$$\Theta_{a+\lambda,b} = \Theta_{a,b}.$$

ii) *For all $\mu \in A^{-1}\mathbb{Z}^r$, we have*
$$\Theta_{a,b+\mu} = e^{2\pi i B(a,\mu)}\Theta_{a,b}.$$

iii) *As $\tau \mapsto \tau+1$, we have*
$$\Theta_{a,b}(\tau+1) = e^{-2\pi i Q(a) - \pi i B\left(A^{-1}A^*,a\right)}\Theta_{a,a+b+\frac{1}{2}A^{-1}A^*}(\tau).$$

iv) *If $a, b \in \mathcal{R}(c_1) \cap \mathcal{R}(c_2)$, then as $\tau \mapsto -1/\tau$, we have*
$$\Theta_{a,b}\left(-\frac{1}{\tau}\right) = \frac{i(-i\tau)^{\frac{r}{2}} e^{2\pi i B(a,b)}}{\sqrt{-\det(A)}} \sum_{n \in A^{-1}\mathbb{Z}^r/\mathbb{Z}^r} \Theta_{b+n,-a}(\tau).$$

SKETCH OF PROOF OF THEOREM 8.28. For the elliptic transformation property i), we note that the identity
$$\Theta(z+\mu;\tau) = \Theta(z;\tau)$$
for $\mu \in A^{-1}\mathbb{Z}^r$ is straightforward, while the transformation property for $z \mapsto z + \lambda\tau$ follows by replacing n by $n+\lambda$ in the sum defining Θ.

The modular transformation properties can be concluded from Theorem 8.27. Indeed, combining Lemma 8.9 with the chain rule shows that $\rho^c(n;\tau)$ satisfies the differential equation, and the transformations for $\mu = 0$ give ii) and iii). □

We next discuss the relation between indefinite theta functions and mock modular forms. To do so, we require the notion of *mixed harmonic Maass forms,* which are essentially linear combinations of products of weakly holomorphic modular forms with harmonic Maass forms. These are formally defined and studied in Section 13.2. In anticipation of the results below on the mock modularity of series built from indefinite theta series, we now give a convenient decomposition of ρ (cf. (8.20)). Recalling the special function E and its relation to the beta function (see the remark following Definition 8.8), we find that for arbitrary $c_1, c_2 \in C_Q$ we may write $\rho(n;\tau)$ as a sum of the expressions
$$\rho_j(n;\tau) = \rho_j^{c_j}(n;\tau) := (-1)^j \operatorname{sgn}\left(B(c_j,n)\right)\beta\left(-\frac{B(c_j,n)^2}{Q(c_j)}v\right) \quad (j \in \{1,2\}),$$
$$\rho_3(n;\tau) = \rho_3^{c_1,c_2}(n;\tau) := \operatorname{sgn}\left(B(c_1,n)\right) - \operatorname{sgn}\left(B(c_2,n)\right).$$

Hence we may decompose ρ as
$$\rho(n;\tau) = \begin{cases} \rho_1(n;\tau) + \rho_2(n;\tau) + \rho_3(n;\tau) & \text{if } c_1, c_2 \in C_Q, \\ \rho_1(n;\tau) + \rho_3(n;\tau) & \text{if } c_1 \in C_Q\ c_2 \in S_Q, \\ \rho_2(n;\tau) + \rho_3(n;\tau) & \text{if } c_1 \in S_Q\ c_2 \in C_Q, \\ \rho_3(n;\tau) & \text{if } c_1, c_2 \in S_Q. \end{cases} \tag{8.22}$$

Then we define for any $c_1, c_2 \in \overline{C}_Q$ the series

$$(8.23)\quad \begin{aligned} \Theta^+_{a,b,c_1,c_2}(\tau) = \Theta^+_{A,a,b,c_1,c_2}(\tau) &:= \sum_{n\in a+\mathbb{Z}^r} \rho_3(n;\tau) e^{2\pi i B(n,b)} q^{Q(n)}, \\ \Theta^-_{a,b,c_1,c_2}(\tau) = \Theta^-_{A,a,b,c_1,c_2}(\tau) &:= \sum_{n\in a+\mathbb{Z}^r} \left(\rho(n;\tau) - \rho_3(n;\tau)\right) e^{2\pi i B(n,b)} q^{Q(n)}. \end{aligned}$$

We set

$$(8.24)\qquad \Theta^+_{c_1,c_2}(z,\tau) = \Theta^+_{A,c_1,c_2}(z;\tau) := \sum_{n\in\mathbb{Z}^r} \rho_3\left(n + \frac{y}{v};\tau\right) e^{2\pi i B(n,z)} q^{Q(n)}.$$

Using Theorem 8.28, we show the following, which easily gives many examples of mock and mixed mock modular forms.

THEOREM 8.30. *Suppose that $c_1, c_2 \in \overline{C}_Q$ and that $a, b \in \mathcal{R}(c_1) \cap \mathcal{R}(c_2)$. Moreover, we require that each c_j lies in $\mathbb{Z}^r$ and has relatively prime coordinates. Then the theta function Θ_{a,b,c_1,c_2} is a component of a vector-valued, mixed harmonic Maass form of weight $r/2$ with holomorphic part Θ^+_{a,b,c_1,c_2} and with image under $\xi_{r/2}$ given in Proposition* 8.33.

REMARK. As we shall see in Theorem 13.3 (see also the discussion before that theorem), much recent work has been done on generalizations of Zwegers' theory for indefinite quadratic forms of general types, and a new picture has emerged involving "higher depth" harmonic modular forms. The crucial breakthrough relies on finding the correct generalization of the function E that satisfies an analogue of Lemma 8.9.

We now give an example to illustrate how Theorem 8.30 can be used to establish the mock modularity of a fifth order mock theta function of Ramanujan.

EXAMPLE 8.31. In (1.4) of **[23]**, Andrews provided a Hecke-type double sum representation of Ramanujan's fifth order mock theta function f_0, which can be expressed as

$$f_0(q) = \frac{1}{(q)_\infty}\left(\sum_{\substack{n+j\geq 0\\ n-j\geq 0}} - \sum_{\substack{n+j<0\\ n-j<0}}\right)(-1)^j q^{\frac{5n^2}{2}+\frac{n}{2}-j^2}.$$

A short calculation then shows that we may write this in terms of the type $(1,1)$ indefinite theta function

$$\frac{q^{\frac{1}{60}}}{2\eta(\tau)} \sum_{n\in a+\mathbb{Z}^2} \left(\operatorname{sgn}(B(n,c_1)) - \operatorname{sgn}(B(n,c_2))\right) e^{2\pi i B(n,b)} q^{Q(n)},$$

where

$$Q(j,k) := \frac{1}{2}\left(5j^2 - 2k^2\right),\ a := \begin{pmatrix} \frac{1}{10} \\ 0 \end{pmatrix},\ b := \begin{pmatrix} 0 \\ \frac{1}{4} \end{pmatrix}, c_1 := \begin{pmatrix} 2 \\ 5 \end{pmatrix},\ c_2 := \begin{pmatrix} -2 \\ 5 \end{pmatrix}.$$

Hence, as $c_1, c_2 \in C_Q$, we find that

$$f_0(q) = \frac{q^{\frac{1}{60}}}{2\eta(\tau)} \Theta^+_{a,b,c_1,c_2}(\tau).$$

The mock modularity and shadow of f_0 may then be deduced from Theorem 8.30.

We next prove the Jacobi transformation properties of Θ. The elliptic translations are straightforward, but the modular transformation properties require an application of Theorem 8.27.

We now consider special nonholomorphic functions which allow us to specify the shadows of the mock modular forms considered in Theorem 8.30. To this end, we define, for any $a, b \in \mathbb{R}$ and $\tau \in \mathbb{H}$, the function

$$R_{a,b}(\tau) := \sum_{n \in a+\mathbb{Z}} \operatorname{sgn}(n) \beta\left(2n^2 v\right) e^{-2\pi i n b} q^{-\frac{n^2}{2}},$$

which is closely related to the R-function that we used to complete the μ-function in Section 8.1. Recall that $g_{a,b}$ denotes a special unary theta function (cf. (8.4)). One may then show by the same method of proof presented in Section 8.1 that we have the following.

LEMMA 8.32. *Assuming the notation above, we have*

$$\xi_{\frac{r}{2}}\left(R_{a,b}(\tau)\right) = -v^{\frac{r-1}{2}} \sqrt{2} g_{a,b}(\tau).$$

Our next goal is to rewrite Θ^{-}_{a,b,c_1,c_2} in terms of the functions $R_{a,b}$. We assume from now on that $c_j \in \overline{C}_Q \cap \mathbb{Z}^r$ has relatively prime coordinates. Then we write $n = \ell + N c_j$ with $\ell \in a + \mathbb{Z}^r$, $N \in \mathbb{Z}$, such that $\frac{B(c_j, \ell)}{2Q(c_j)} \in [0,1)$. As $c_j \in \mathbb{Z}^r$, it is easy to see that we may write the set of such ℓ as a disjoint union

$$\left\{\ell \in a + \mathbb{Z}^r : \frac{B(c_j, \ell)}{2Q(c_j)} \in [0,1)\right\} = \dot{\bigcup}_{\ell_0 \in P_{0,j}} \left(\ell_0 + \langle c_j \rangle_{\mathbb{Z}}^{\perp}\right), \tag{8.25}$$

where $P_{0,j}$ is a suitable finite set and where $\langle c_j \rangle_{\mathbb{Z}}^{\perp} := \{\lambda \in \mathbb{Z}^r : B(c_j, \lambda) = 0\}$. We then show the following, where $\ell_0^{\perp} := \ell_0 - \frac{B(c_j, \ell_0)}{2Q(c_j)} c_j$ and

$$\mathcal{J} := \{j \in \{1,2\} : c_j \in C_Q\}.$$

PROPOSITION 8.33. *Suppose that $c_1, c_2 \in \overline{C}_Q$ and that $c_j \in \mathbb{Z}^r$ have relatively prime coefficients. Then we have*

$$\begin{aligned}\xi_{\frac{r}{2}}\left(\Theta_{a,b,c_1,c_2}(\tau)\right) &= \sqrt{2v^{r-1}} \\ &\times \sum_{j \in \mathcal{I}} (-1)^j \sum_{\ell_0 \in P_{0,j}} g^{c}_{\frac{B(c_j,\, \ell_0)}{2Q(c_j)}, B(c_j, b)}\left(-2Q\left(c_j\right)\tau\right) \overline{\sum_{\nu \in \ell_0^{\perp} + \langle c_j \rangle_{\mathbb{Z}}^{\perp}} e^{2\pi i B(v, b^{\perp})} q^{Q(\nu)}}.\end{aligned}$$

REMARK. The rightmost sum in the last equation is a theta function on a shifted $(r-1)$-dimensional lattice, and so in particular is a modular form of weight $(r-1)/2$.

REMARK. In many examples, such as those considered in **[529]** and those which we investigate in Chapter 9, the theta functions in the right-most sum of the formula in Proposition 8.33 are essentially equal to each other for $c = c_1$ and $c = c_2$. If these theta functions also have no zeros on $\mathbb{H}$, one obtains an ordinary harmonic Maass form by dividing $\Theta_{a,b}$ by this weight $(r-1)/2$ theta series.

PROOF OF PROPOSITION 8.33. As $\tau \mapsto \rho_3(n;\tau)$ is a holomorphic function, its image under $\xi_{\frac{r}{2}}$ is zero, and hence the contribution coming from any term in S_Q is zero. Suppose now that $c \in C_Q \cap \mathbb{Z}^r$ and has relatively prime coordinates.

By using Lemma 8.32, it is enough to show that a generic piece of $\Theta^-_{a,b}$ may be written as follows:

$$\begin{aligned}&\sum_{n\in a+\mathbb{Z}^r}\operatorname{sgn}(B(c,n))\beta\left(-\frac{B(c,n)^2}{Q(c)}v\right)e^{2\pi iB(n,b)}q^{Q(n)}\\ &=-\sum_{\ell_0\in P_0}R_{\frac{B(c,\ell_0)}{2Q(c)},-B(c,b)}(-2Q(c)\tau)\sum_{\nu\in\ell_0^\perp+\langle c\rangle_\mathbb{Z}^\perp}e^{2\pi iB\left(\nu,b^\perp\right)}q^{Q(\nu)}.\end{aligned}\tag{8.26}$$

To see (8.26), write $n=\ell_0+\nu+Nc$, with $\ell_0\in P_0$ (with P_0 as in (8.25)), $\nu\in\langle c\rangle_\mathbb{Z}^\perp$, and $N\in\mathbb{Z}$. Then we have that

$$n=\left(N+\frac{B(c,\ell_0)}{2Q(c)}\right)c+\ell_0^\perp+\nu,$$

and thus

$$\begin{aligned}&\sum_{n\in a+\mathbb{Z}^r}\operatorname{sgn}(B(c,n))\beta\left(-\frac{B(c,n)^2}{Q(c)}v\right)e(B(n,b))q^{Q(n)}\\ &=-\sum_{\ell_0\in P_0}\sum_{\nu\in\langle c\rangle_\mathbb{Z}^\perp}\sum_{N\in\mathbb{Z}}\operatorname{sgn}\left(N+\frac{B(c,\ell_0)}{2Q(c)}\right)\beta\left(-4Q(c)\left(N+\frac{B(c,\ell_0)}{2Q(c)}\right)^2v\right)\\ &\quad\times e^{2\pi i\left(B(c,b)\left(N+\frac{B(c,\ell_0)}{2Q(c)}\right)+B\left(\ell_0^\perp+\nu,b\right)\right)}q^{Q(c)\left(N+\frac{B(c,\ell_0)}{2Q(c)}\right)^2+Q\left(\ell_0^\perp+\nu\right)}\\ &=-\sum_{\ell_0\in P_0}R_{\frac{B(c,\ell_0)}{2Q(c)},-B(c,b)}(-2Q(c)\tau)\sum_{\alpha\in\ell_0^\perp+\langle c\rangle_\mathbb{Z}^\perp}e^{2\pi iB(\alpha,b)}q^{Q(\alpha)}.\end{aligned}$$

The claim then follows by using the fact that $B(\alpha,b)=B(\alpha,b^\perp)$ for all $\alpha\in\ell_0^\perp+\langle c\rangle_\mathbb{Z}^\perp$. □

We are now in a position to prove Theorem 8.30.

PROOF OF THEOREM 8.30. The proof follows directly by combining Corollary 8.29 and Proposition 8.33. □

We conclude with an example which ties together this section with Section 8.1, and in particular which shows that we can write Zwegers' $\widehat{\mu}$-function as an indefinite theta function of signature $(1,1)$.

LEMMA 8.34. *For* $0<\frac{y_1}{v},\frac{y_2-y_1}{v}+\frac{1}{2}<1$, *we have that*

$$\widehat{\mu}(z_1,z_2;\tau)=\frac{\zeta_1^{\frac{1}{2}}}{2\vartheta(z_2;\tau)}\Theta_{A,c_1,c_2}\left(z_1,z_2-z_1+\frac{\tau+1}{2};\tau\right),$$

where $A=\left(\begin{smallmatrix}1&1\\1&0\end{smallmatrix}\right)$, $c_1=\left(\begin{smallmatrix}0\\1\end{smallmatrix}\right)$, $c_2=\left(\begin{smallmatrix}-1\\1\end{smallmatrix}\right)$.

PROOF. To compare the holomorphic parts, we directly verify that (recalling the notation of (8.24))

$$2\sum_{n\in\mathbb{Z}} \frac{(-1)^n \zeta_2^n q^{\frac{n(n+1)}{2}}}{1-\zeta_1 q^n} = 2\left(\sum_{n,\ell\geq 0} - \sum_{n,\ell<0}\right) q^{\frac{n(n+1)}{2}+n\ell}(-1)^n\zeta_2^n\zeta_1^\ell$$

$$= \sum_{N=\binom{n}{\ell}\in\mathbb{Z}^2} \left(\operatorname{sgn}\left(n+\frac{y_1}{v}\right)+\operatorname{sgn}\left(\ell+\frac{1}{2}+\frac{y_2-y_1}{v}\right)\right)e^{2\pi i(\ell z_1+nz_2)}(-1)^n q^{\frac{n^2}{2}+n\ell+\frac{n}{2}}$$

$$= \sum_{N\in\mathbb{Z}^2}\left[\operatorname{sgn}\left(B\left(c_1, N+\frac{1}{v}\begin{pmatrix} y_1 \\ y_2-y_1+\frac{v}{2}\end{pmatrix}\right)\right)\right.$$

$$\left. -\operatorname{sgn}\left(B\left(c_2, N+\frac{1}{v}\begin{pmatrix} y_1 \\ y_2-y_1+\frac{v}{2}\end{pmatrix}\right)\right)\right] e^{2\pi i B\left(N,\left(z_1,z_2-z_1+\frac{\tau+1}{2}\right)^T\right)} q^{Q(N)}$$

$$= \Theta^+_{A,c_1,c_2}\left(z_1, z_2-z_1+\frac{\tau+1}{2};\tau\right),$$

where in the first step we use a geometric series expansion.

The completion term $\widehat{\mu}-\mu$ is given by

$$\frac{i}{2}R(z_1-z_2;\tau) = \frac{i}{2}R(z_2-z_1;\tau)$$

$$= \frac{i\zeta_1^{\frac{1}{2}}}{2}\sum_{j\in\frac{1}{2}+\mathbb{Z}}\left(\operatorname{sgn}(j) - E\left(\left(j+\frac{y_2-y_1}{v}\right)\sqrt{2v}\right)\right)\zeta_1^{j-\frac{1}{2}}\zeta_2^{-j}(-1)^{j-\frac{1}{2}}q^{-\frac{j^2}{2}}.$$

Shifting in the series expansion of ϑ by $n\mapsto n+j$ ($j\in 1/2+\mathbb{Z}$) yields

$$\vartheta(z_2;\tau) = q^{\frac{j^2}{2}}\zeta_2^j i(-1)^{j-\frac{1}{2}}\sum_{n\in\mathbb{Z}}(-1)^n\zeta_2^n q^{\frac{n^2}{2}+nj},$$

which in turn gives

$$\frac{i}{2}\vartheta(z_2;\tau)R(z_1-z_2;\tau)$$

$$= \frac{\zeta_1^{\frac{1}{2}}}{2}\sum_{\substack{n\in\mathbb{Z}\\ j\in\frac{1}{2}+\mathbb{Z}}}\left(E\left(\left(j+\frac{y_2-y_1}{v}\right)\sqrt{2v}\right)-\operatorname{sgn}(j)\right)(-1)^n\zeta_1^{j-\frac{1}{2}}\zeta_2^n q^{\frac{n(n+1)}{2}+n\left(j-\frac{1}{2}\right)}.$$

Using that $0<y_2-y_1+v/2<v$ and $j\in 1/2+\mathbb{Z}$, we find that

$$\operatorname{sgn}(j) = \operatorname{sgn}\left(j+\frac{y_2-y_1}{v}\right).$$

Setting $\ell := j - 1/2$, we then obtain

$$\begin{aligned}
&\frac{i}{2}\vartheta(z_2;\tau)R(z_1-z_2;\tau)\\
&= \frac{\zeta_1^{\frac{1}{2}}}{2}\sum_{n,\ell\in\mathbb{Z}}\left(E\left(\left(\ell+\frac{1}{2}+\frac{y_2-y_1}{v}\right)\sqrt{2v}\right)\right.\\
&\qquad\qquad \left.-\operatorname{sgn}\left(\ell+\frac{1}{2}+\frac{y_2-y_1}{v}\right)\right)(-1)^n\,\zeta_1^{\ell}\zeta_2^{n}q^{\frac{n(n+1)}{2}+n\ell}\\
&= \frac{\zeta_1^{\frac{1}{2}}}{2}\sum_{N=\binom{n}{\ell}\in\mathbb{Z}^2}\left[E\left(B\left(c_2,N+\frac{1}{v}\begin{pmatrix} y_1\\ y_2-y_1+\frac{v}{2}\end{pmatrix}\right)v^{\frac{1}{2}}\right)\right.\\
&\qquad\qquad \left.-\operatorname{sgn}\left(B\left(c_2,N+\frac{1}{v}\begin{pmatrix} y_1\\ y_2-y_1+\frac{v}{2}\end{pmatrix}\right)\right)\right]e^{2\pi iB\left(N,A^{-1}\left(z_2+\frac{\tau+1}{2},z_1\right)^T\right)}q^{Q(N)}\\
&= \frac{\zeta_1^{\frac{1}{2}}}{2}\left(\Theta_{A,c_1,c_2}\left(z_1,z_2-z_1+\frac{\tau+1}{2};\tau\right)-\Theta^{+}_{A,c_1,c_2}\left(z_1,z_2-z_1+\frac{\tau+1}{2};\tau\right)\right).
\end{aligned}$$

□

CHAPTER 9

Ramanujan's Mock Theta Functions

The Bruinier-Funke [**121**] theory of harmonic Maass forms came to fruition at about the same time that Zwegers wrote his celebrated Ph.D. thesis [**529**] on Ramanujan's enigmatic "mock theta functions." These functions were defined in Ramanujan's famous 1920 deathbed letter to Hardy. Thanks to Zwegers' thesis, Ramanujan's q-series are now, nearly 100 years later, understood as examples of weight $1/2$ mock modular forms. Ramanujan coined the term "mock theta function," and it is for this reason that holomorphic parts of harmonic Maass forms are now referred to as mock modular forms. In this chapter we explain Ramanujan's examples as mock modular forms, and we discuss the resolution of some well-known problems about these strange q-series which have now been solved by making use of the theory of harmonic Maass forms.

9.1. Ramanujan's last letter to Hardy

In his deathbed letter to Hardy (cf. pages 220–224 of [**54**]), Ramanujan introduced his so-called *mock theta functions.*

"I am extremely sorry for not writing you a single letter up to now... I discovered very interesting functions recently which I call 'Mock' ϑ-functions. Unlike the 'False' θ-functions (studied partially by Prof. Rogers in his interesting paper) they enter into mathematics as beautifully as the ordinary[1] theta functions. I am sending you with this letter some examples."

Ramanujan offered 17 examples[2] of mock theta functions in this letter, which he grouped by *order*, a notion that he did not define. His third order mock theta functions included the examples

$$f(q) := \sum_{n=0}^{\infty} \frac{q^{n^2}}{(-q;q)_n^2} =: \sum_{n=0}^{\infty} \alpha_f(n) q^n, \tag{9.1}$$

$$\omega(q) := \sum_{n=0}^{\infty} \frac{q^{2n(n+1)}}{(q;q^2)_{n+1}^2}. \tag{9.2}$$

Ramanujan's fifth order mock theta function

$$f_0(q) := \sum_{n=0}^{\infty} \frac{q^{n^2}}{(-q;q)_n} \tag{9.3}$$

[1]To Ramanujan, ordinary theta functions are weakly holomorphic modular forms.

[2]These mock theta functions are listed in the Appendix, where we also give all of their known representations.

is another prominent example of these functions. In these expressions we have used the *q-Pochhammer symbol*

$$(9.4)\qquad (a)_n = (a;q)_n := \begin{cases} \prod_{j=0}^{n-1} (1-aq^j) & \text{if } n \in \mathbb{N}_0, \\ \prod_{j=0}^{\infty} (1-aq^j) & \text{if } n = \infty. \end{cases}$$

REMARK. The infinite product $(q)_\infty = (q;q)_\infty$ occurs frequently in the q-series literature. This infinite product is essentially Dedekind's eta-function. Indeed, by definition, we have that

$$(q)_\infty = q^{-\frac{1}{24}}\eta(\tau).$$

What did Ramanujan mean by a "mock theta function"? Here is his own description (cf. page 221 of [**54**]):

"*... Suppose there is a function in the Eulerian form*[3] *and suppose that all or an infinity of points are exponential singularities, and also suppose that at these points the asymptotic form of the function closes neatly... The question is: Is the function taken the sum of two functions one of which is an ordinary theta function and the other a (trivial) function which is* $O(1)$ *at* all *the points* $e^{2m\pi i/n}$*? The answer is* it is not necessarily so. *When it is not so I call the function Mock* ϑ*-function. I have not proved rigorously that* it is not necessarily so. *But I have constructed a number of examples in which it is inconceivable to construct a* ϑ*-function to cut out the singularities of the original function.*"

To understand the context of the letter, it is important to recall that some of Ramanujan's earlier work involved Eulerian modular forms such as

$$\frac{1}{\eta(\tau)} = q^{-\frac{1}{24}} \sum_{n=0}^{\infty} \frac{q^{n^2}}{(q;q)_n^2},$$

and the so-called *Rogers-Ramanujan* identities

$$(9.5)\qquad \frac{q^{-\frac{1}{60}}}{(q;q^5)_\infty (q^4;q^5)_\infty} = q^{-\frac{1}{60}} \sum_{n=0}^{\infty} \frac{q^{n^2}}{(q;q)_n},$$

$$(9.6)\qquad \frac{q^{\frac{11}{60}}}{(q^2;q^5)_\infty (q^3;q^5)_\infty} = q^{\frac{11}{60}} \sum_{n=0}^{\infty} \frac{q^{n^2+n}}{(q;q)_n}.$$

The Rogers-Ramanujan q-series on the right-hand sides of (9.5) and (9.6) are q-hypergeometric functions, which may be viewed as modular forms using the infinite product expansions given on the left-hand sides of (9.5) and (9.6).

Ramanujan's letter was about Eulerian series which he did not believe could be modular. He made certain claims about the behavior of his peculiar Eulerian series near roots of unity. He claimed that they "behave like" weakly holomorphic modular forms near roots of unity without being modular. These claims then suggested that the coefficients of these functions should possess asymptotics which are analogous

[3]Eulerian forms are q-series similar in "shape" to (9.1), (9.2), and (9.3), which can be expressed using *basic hypergeometric*, also known as *q-hypergeometric*, series. See the proof of Theorem 9.11 for more detail.

to the famous asymptotic formula for $p(n)$ (cf. (14.2) below) that he obtained with Hardy. This is what he meant by "an asymptotic form which closes neatly". In particular, for the third order mock theta function f defined by (9.1), Ramanujan claimed that

$$\alpha_f(n) = (-1)^{n+1}\frac{\sqrt{6}\exp\left(\frac{\pi}{12}\sqrt{24n-1}\right)}{\sqrt{24n-1}} + O\left(\frac{\exp\left(\frac{\pi}{24}\sqrt{24n-1}\right)}{\sqrt{24n-1}}\right). \tag{9.7}$$

Ramanujan's letter had frustrated generations of mathematicians. It offered only few clues as to how he obtained his examples. Moreover, as mentioned above, he did not define his notion of *order*. Despite these challenges, the following definition captures the spirit of Ramanujan's notion of a mock theta function.

DEFINITION 9.1. A **mock theta function** is a function F, defined on $\mathbb{H}$, satisfying the following conditions:

i) There are infinitely many roots of unity ζ for which $F(\tau)$ grows exponentially as $q = e^{2\pi i\tau}$ approaches ζ radially from inside the unit disk.
ii) For every root of unity ζ, there exists a (weakly holomorphic) modular form M_ζ and a rational number α_ζ such that
$$F(\tau) - q^{\alpha_\zeta}M_\zeta(\tau)$$
is bounded as $q \to \zeta$ radially from within the unit disk.
iii) There does not exist a single (weakly holomorphic) modular form M that satisfies ii).

REMARK. In this definition we abuse notation and view the q-series as functions of $\tau \in \mathbb{H}$. In this chapter we view Ramanujan's examples as both q-series and functions of τ. If we think of them as formal power series we use q, and if we think of them in terms of mock modular forms we use τ with the usual understanding that $q = e^{2\pi i\tau}$.

9.2. Work of Watson and Andrews

Watson proved transformation laws [**489**] for Ramanujan's functions which resemble those of weight 1/2 modular forms. For example, using Watson's classical notation, the mock theta functions f and ω, defined in (9.1) and (9.2), respectively, satisfy the following transformation law.

THEOREM 9.2 (Watson). *We have that*

$$q^{-\frac{1}{24}}f(q) = 2\sqrt{\frac{2\pi}{\alpha}}q_1^{\frac{4}{3}}\omega\left(q_1^2\right) + 4\sqrt{\frac{3\alpha}{2\pi}}\int_0^\infty \frac{\sinh(\alpha t)}{\sinh\left(\frac{3\alpha t}{2}\right)}e^{-\frac{3\alpha t^2}{2}}\,dt,$$

where $q := e^{-\alpha}, \beta := \pi^2/\alpha, q_1 := e^{-\beta}$ ($\alpha \in \mathbb{C}$ *with* $\mathrm{Re}(\alpha) > 0$).

REMARK. Note that we use the notation in Theorem 9.2 for historical reasons and q should not be confused with $q = e^{2\pi i\tau}$ from before. Using more modern terminology, Theorem 9.2 can be interpreted as exhibiting a transformation under $\tau \mapsto -1/(2\tau)$, ($\tau \in \mathbb{H}$).

REMARK. The strange integral in Theorem 9.2 is related to the *Mordell integrals* defined by Definition 8.4. The quantity can be viewed as an obstruction to modularity. If the integral was not present, then it would be straightforward to assemble a weight 1/2 vector-valued modular form using f and ω. We revisit this observation shortly. Namely, we explain a result of Zwegers which reformulates this theorem as an example of a weight $1/2$ vector-valued harmonic Maass form. The strange integral turns out to be related to period integrals of unary weight $3/2$ theta functions.

SKETCH OF PROOF OF THEOREM 9.2. Watson's proof relies on the representation of the mock theta functions as *Appell-Lerch-type sums* (cf. Section 8.1). In the case of f and ω, we have

$$f(q) = \frac{2}{(q)_\infty} \sum_{n\in\mathbb{Z}} \frac{(-1)^n q^{\frac{n(3n+1)}{2}}}{1+q^n}, \tag{9.8}$$

$$\omega(q) = \frac{2}{(q)_\infty} \sum_{n\in\mathbb{Z}} \frac{(-1)^n q^{\frac{n(3n+1)}{2}}}{1+q^{2n}}. \tag{9.9}$$

Therefore, one may apply the method of Poisson summation together with the known transformation properties of $(q)_\infty$ to obtain a modular-like transformation law (as one does for theta functions). In this case, one obtains the desired transformation law by applying the Residue Theorem after shifting the associated path of integration. The q-series arises from the calculation of the residues of the integrand, and the strange integral is obtained by factoring the function on the new path of integration in a clever way. □

Theorem 9.11 below gives infinite families of q-series identities which are directly analogous to (9.8) and (9.9). Such formulas express mock theta functions as full lattice sums (i.e., over $\mathbb{Z}$). Therefore, one expects to obtain infinite families of functions with similar modular transformation laws which arise in a similar way. In Section 9.5 we revisit results such as Theorem 9.2. We discuss infinite families of q-hypergeometric series which arise from the so-called "universal mock theta functions". Basic hypergeometric or q-hypergeometric series are similar in shape to (9.1), (9.2), and (9.3) (see the proof of Theorem 9.11 below). It turns out that the universal mock theta functions additionally have representations in terms of Appell-Lerch-type sums similar to (9.8) and (9.9). Using the results in Section 8.1, we determine the modular transformation properties of infinite families of q-hypergeometric series.

The work of Watson placed Ramanujan's mock theta functions in the vicinity of the theory of modular forms. Unfortunately, the task of assembling a theory of such functions was complicated by the technical nature of the formulas involved. Further identities for mock theta functions added another element of mystery to Ramanujan's q-series. Many of them possessed "Hecke-type" expansions which resembled those of certain modular forms. For example, we have the following identity of Andrews [**23**].

THEOREM 9.3 (Andrews). *Ramanujan's fifth order mock theta function $f_0(q)$ (9.3) satisfies the identity*

$$f_0(q) = \frac{1}{(q;q)_\infty} \sum_{n=0}^{\infty} \sum_{|j|\le n} (-1)^j q^{\frac{5n^2}{2}+\frac{n}{2}-j^2} \left(1-q^{4n+2}\right).$$

REMARK. The right-hand side of this expression is analogous to the functions which appear in Hecke's theory of indefinite theta functions. For example, it is interesting to compare the formula of Andrews with the following well-known Hecke-type expansion for the weight 1 cusp form:

$$\eta(\tau)^2 = \left(\sum_{n,m \geq 0} - \sum_{n,m<0} \right) (-1)^{n+m} q^{\frac{n^2}{2}+2nm+\frac{m^2}{2}+\frac{n}{2}+\frac{m}{2}+\frac{1}{12}}.$$

Similar expressions are given in Theorem 9.7 below.

Inspired by such identities, Zwegers developed a theory of indefinite binary theta series in his thesis [**529**]. We described many of his key results in the previous chapter, and in particular we saw how Theorem 9.3 allows one to use Zwegers' work to realize f_0 as a mock modular form in Example 8.31. In view of the works of Watson and Andrews and others, Zwegers recognized that Ramanujan's theory must turn out to be a theory of modular forms which involves Mordell integrals, Appell-Lerch-type series, and an extended theory of indefinite theta functions. In his Ph.D. thesis (cf. Sections 8.1 and 8.2), Zwegers developed these theories, and he finally determined the modularity properties of Ramanujan's mock theta functions. Using these results, we now have the following satisfactory resolution to Ramanujan's enigmatic letter, rephrasing Zwegers' original results [**528, 529**] using the language of harmonic Maass forms (see also Corollary 8.15).

THEOREM 9.4 (Zwegers). *Ramanujan's mock theta functions are (up to multiplication by a power of q and addition of a constant) weight 1/2 mock modular forms. More precisely, if F is one of Ramanujan's mock theta functions, then*

$$F(\tau) = q^{\alpha} G^{+}(\tau) + c,$$

for some $\alpha \in \mathbb{Q}$ and $c \in \mathbb{C}$, where G^+ is the holomorphic part of a weight 1/2 harmonic Maass form whose shadow is a weight 3/2 unary theta function.

REMARK. Ramanujan's notion of order for his mock theta functions has been related to standard invariants for modular forms. In some cases they coincide with the notion of Nebentypus. For example, Ramanujan's third order mock theta function f (as we shall see in the next section) is essentially the mock modular form associated to a weight 1/2 harmonic Maass form with Nebentypus $\chi_{12}(\cdot) = \left(\frac{12}{\cdot}\right)$, which is the Kronecker character for the real quadratic field $\mathbb{Q}(\sqrt{3})$. An alternative interpretation is given in [**158**]. There the order is related to the level of the associated harmonic Maass form as well as to the index of the corresponding mock Jacobi form (cf. pages 90-91 of [**158**]).

In the next two sections we give explicit examples of Theorem 9.4. The remainder of this chapter is devoted to the following general topics: universal mock theta functions, the Mock Theta Conjectures, the Andrews-Dragonette Conjecture, and the resolution of Ramanujan's original claim (cf. Definition 9.1 iii)) that for any mock theta function, there is not a single weakly holomorphic modular form which exactly cuts out its singularities.

9.3. Third order mock theta functions revisited

Here we give an important example of a vector-valued weight 1/2 harmonic Maass form whose mock modular forms arise from Ramanujan's mock theta functions. This example [**528**] was worked out by Zwegers, and it corresponds to the third order mock theta functions f and ω, whose modular transformation laws are given by Theorem 9.2.

Define the vector-valued function F by setting

$$(9.10)\quad F(\tau) = (F_0(\tau), F_1(\tau), F_2(\tau))^T := \left(q^{-\frac{1}{24}} f(q), 2q^{\frac{1}{3}}\omega\left(q^{\frac{1}{2}}\right), 2q^{\frac{1}{3}}\omega\left(-q^{\frac{1}{2}}\right)\right)^T.$$

Furthermore, define the vector-valued nonholomorphic function G by

$$(9.11)\qquad G(\tau) := 2i\sqrt{3}\int_{-\overline{\tau}}^{i\infty} \frac{(\mathcal{G}_1(w),\ \mathcal{G}_0(w),\ -\mathcal{G}_2(w))^T}{\sqrt{-i(w+\tau)}}\, dw.$$

Here, the functions $\mathcal{G}_j$ are the cuspidal weight 3/2 theta functions (cf. Section 2.1) defined by

$$\mathcal{G}_0(\tau) := \sum_{n\in\mathbb{Z}} (-1)^n \left(n+\frac{1}{3}\right) e^{3\pi i\left(n+\frac{1}{3}\right)^2\tau},$$

$$\mathcal{G}_1(\tau) := -\sum_{n\in\mathbb{Z}} \left(n+\frac{1}{6}\right) e^{3\pi i\left(n+\frac{1}{6}\right)^2\tau},$$

$$\mathcal{G}_2(\tau) := \sum_{n\in\mathbb{Z}} \left(n+\frac{1}{3}\right) e^{3\pi i\left(n+\frac{1}{3}\right)^2\tau}.$$

Therefore, G is a nonholomorphic vector-valued Eichler integral.

Using these functions, Zwegers defined the vector-valued function h by

$$(9.12)\qquad h(\tau) := F(\tau) - G(\tau),$$

and he proved the following theorem **[528]**.

THEOREM 9.5 (Zwegers). *The function h is a vector-valued real-analytic modular form of weight $1/2$ that satisfies the following transformation laws:*

$$h(\tau+1) = \begin{pmatrix} \zeta_{24}^{-1} & 0 & 0 \\ 0 & 0 & \zeta_3 \\ 0 & \zeta_3 & 0 \end{pmatrix} h(\tau),$$

$$h\left(-\frac{1}{\tau}\right) = \sqrt{-i\tau}\begin{pmatrix} 0 & 1 & 0 \\ 1 & 0 & 0 \\ 0 & 0 & -1 \end{pmatrix} h(\tau).$$

In particular, h is a vector-valued weight 1/2 harmonic Maass form whose shadow is a vector-valued weight 3/2 cuspidal unary theta function.

REMARK. Theorem 9.5, combined with similar results in Zwegers' thesis (like Theorem 9.8 below) for the other mock theta functions, collectively imply Theorem 9.4 above.

This theorem gives the modularity properties of Ramanujan's original mock theta functions f and ω. In particular, we have the following for f.

COROLLARY 9.6. *The first component of h, which we denote by h_0, is an element of $H_{\frac{1}{2}}(\Gamma_0(144), \chi_{12})$.*

SKETCH OF PROOF OF THEOREM 9.5. That h is annihilated by $\Delta_{\frac{1}{2}}$ is a straightforward calculation. Therefore, it suffices to check the claimed modular transformation laws. The first transformation law (under the map $\tau \mapsto \tau+1$) is confirmed by inspection. Using Theorem 9.2, one can determine the image of the holomorphic function F under $\tau \mapsto -1/\tau$. Using the modular transformation properties of the theta functions $\mathcal{G}_j$ one can directly compute the image of the nonholomorphic function G under $\tau \mapsto -1/\tau$. One finds that F and G share the same obstruction to modularity. Therefore, their difference must satisfy the second claimed modular transformation law.

To be more precise, to show that the holomorphic functions F and the nonholomorphic function G obey the same transformation laws, Zwegers used the Mittag-Leffler partial fraction decomposition, and the integral identity

$$\int_{-\infty}^{\infty} \frac{e^{-\pi t w^2}}{w - ir} dw = \pi i r \int_0^{\infty} \frac{e^{-\pi r^2 w}}{\sqrt{w+t}} dw$$

which holds for $r \in \mathbb{R}\backslash\{0\}$. This can then be used to rewrite the error of modularity of F, which we denote by R, as follows:

$$\frac{1}{\sqrt{-i\tau}} F\left(-\frac{1}{\tau}\right) = \begin{pmatrix} 0 & 1 & 0 \\ 1 & 0 & 0 \\ 0 & 0 & -1 \end{pmatrix} F(\tau) + R(\tau),$$

where

$$R(\tau) := 4\sqrt{3}\sqrt{-i\tau}\left(r_2(\tau), -r_1(\tau), r_3(\tau)\right)^T,$$

with

$$r_1(\tau) := \int_0^{\infty} \frac{\sin(2\pi\tau w)}{\sin(3\pi\tau w)} e^{3\pi i \tau w^2} dw,$$
$$r_2(\tau) := \int_0^{\infty} \frac{\cos(\pi\tau w)}{\cos(3\pi\tau w)} e^{3\pi i \tau w^2} dw,$$
$$r_3(\tau) := \int_0^{\infty} \frac{\sin(\pi\tau w)}{\sin(3\pi\tau w)} e^{3\pi i \tau w^2} dw.$$

It turns out that

$$R(\tau) = -2i\sqrt{3} \int_0^{i\infty} \frac{\mathcal{G}(w)}{\sqrt{-i(\tau + w)}} dw,$$

where $\mathcal{G}$ is the vector $\mathcal{G} := (\mathcal{G}_0, \mathcal{G}_1, \mathcal{G}_2)^T$. This completes the sketch of the proof. □

9.4. Mock theta functions as indefinite theta series

Many of Ramanujan's mock theta functions have representations as indefinite binary theta series. Using work of Zwegers on indefinite theta functions described in Section 8.2, we have a second method of proving that (some of) Ramanujan's mock theta functions are mock modular forms. We illustrate this with an example. We consider Ramanujan's seventh order mock theta functions

$$\mathcal{F}_0(q) := \sum_{n=0}^{\infty} \frac{q^{n^2}}{\left(q^{n+1}\right)_n},$$
$$\mathcal{F}_1(q) := \sum_{n=1}^{\infty} \frac{q^{n^2}}{\left(q^{n}\right)_n},$$
$$\mathcal{F}_2(q) := \sum_{n=0}^{\infty} \frac{q^{n^2+n}}{\left(q^{n+1}\right)_{n+1}}.$$

The main observation we use is the following theorem of Hickerson [**263**] which essentially describes these mock theta functions as products of indefinite binary theta series with the reciprocal of Dedekind's eta-function.

THEOREM 9.7 (Hickerson). *The following q-series identities are true:*

$$(q)_\infty \mathcal{F}_0(q) = \left(\sum_{r,s\geq 0} - \sum_{r,s<0}\right)(-1)^{r+s} q^{\frac{3r^2}{2}+4rs+\frac{3s^2}{2}+\frac{r}{2}+\frac{s}{2}},$$

$$(q)_\infty \mathcal{F}_1(q) = \left(\sum_{r,s\geq 0} - \sum_{r,s<0}\right)(-1)^{r+s} q^{\frac{3r^2}{2}+4rs+\frac{3s^2}{2}+\frac{5r}{2}+\frac{5s}{2}+1},$$

$$(q)_\infty \mathcal{F}_2(q) = \left(\sum_{r,s\geq 0} - \sum_{r,s<0}\right)(-1)^{r+s} q^{\frac{3r^2}{2}+4rs+\frac{3s^2}{2}+\frac{3r}{2}+\frac{3s}{2}}.$$

Following [**529**], to build a vector-valued harmonic Maass form analogous to (9.12), we combine these functions in a single vector as

$$F_7(\tau) := \begin{pmatrix} q^{-\frac{1}{168}}\mathcal{F}_0(q) \\ q^{\frac{47}{168}}\mathcal{F}_2(q) \\ q^{-\frac{25}{168}}\mathcal{F}_1(q) \end{pmatrix}.$$

Using the notation from Section 8.1, we then define the nonholomorphic function

$$G_7(\tau) := \begin{pmatrix} -\zeta_{84}^{-13} R_{\frac{13}{43},-\frac{1}{2}} + \zeta_{84} R_{\frac{41}{42},-\frac{1}{2}} \\ \zeta_{84}^{29} R_{\frac{11}{42},-\frac{5}{2}} + \zeta_{84}^{-41} R_{\frac{25}{42},-\frac{5}{2}} \\ \zeta_{28}^{5} R_{\frac{23}{42},-\frac{3}{2}} + \zeta_{28}^{-9} R_{\frac{37}{42},-\frac{3}{2}} \end{pmatrix} (21\tau).$$

With these two functions, we then define the vector-valued function

$$h_7(\tau) := F_7(\tau) - G_7(\tau). \tag{9.13}$$

By the results on indefinite binary theta functions described in Section 8.2, we have the following theorem.

THEOREM 9.8. *The function h_7 is a vector-valued weight 1/2 harmonic Maass form. In particular, h_7 satisfies the following modular transformation laws:*

$$h_7(\tau+1) = \begin{pmatrix} \zeta_{168}^{-1} & 0 & 0 \\ 0 & \zeta_{168}^{47} & 0 \\ 0 & 0 & \zeta_{168}^{-25} \end{pmatrix} h_7(\tau),$$

$$h_7\left(-\frac{1}{\tau}\right) = \frac{2}{\sqrt{7}}\sqrt{-i\tau} \begin{pmatrix} \sin\left(\frac{\pi}{7}\right) & \sin\left(\frac{3\pi}{7}\right) & \sin\left(\frac{2\pi}{7}\right) \\ \sin\left(\frac{3\pi}{7}\right) & -\sin\left(\frac{2\pi}{7}\right) & \sin\left(\frac{\pi}{7}\right) \\ \sin\left(\frac{2\pi}{7}\right) & \sin\left(\frac{\pi}{7}\right) & -\sin\left(\frac{3\pi}{7}\right) \end{pmatrix} h_7(\tau).$$

PROOF. It is straightforward to show that h_7 is annihilated by $\Delta_{\frac{1}{2}}$ and satisfies the correct growth condition to be a vector-valued harmonic Maass form. The first transformation (under $\tau \mapsto \tau+1$) follows by inspection. To prove the second transformation equation (under $\tau \mapsto -1/\tau$), we make use of Theorem 9.7. In terms of the notation in Section 8.2 on indefinite theta series, we find that

$$h_7(\tau) = \frac{\zeta_{14}^{-1}}{2\eta(\tau)} \begin{pmatrix} \vartheta_{\frac{1}{14}e,\frac{1}{14}e}(\tau) \\ \vartheta_{\frac{3}{14}e,\frac{5}{14}e}(\tau) \\ \vartheta_{\frac{5}{14}e,\frac{3}{14}e}(\tau) \end{pmatrix}.$$

The claimed transformation law under $\tau \mapsto -1/\tau$ now follows from Corollary 8.29, combined with the modular transformation laws (6.1) for Dedekind's η-function.

□

9.5. Universal mock theta functions

It is natural to ask whether Ramanujan's strange q-series arise from a more general family which gives rise to mock modular forms. It turns out that this is indeed the case. They are specializations of the so-called *universal mock-theta functions* $g_2(\zeta; q)$ and $g_3(\zeta; q)$ of Gordon and McIntosh [**228**].

DEFINITION 9.9. The **universal mock theta functions** $g_2(\zeta; q)$ and $g_3(\zeta; q)$ are defined by the following q-series:

$$g_2(\zeta; q) := \sum_{n=0}^{\infty} \frac{(-q)_n \, q^{\frac{n(n+1)}{2}}}{(\zeta)_{n+1} \left(\zeta^{-1} q\right)_{n+1}},$$

$$g_3(\zeta; q) := \sum_{n=0}^{\infty} \frac{q^{n(n+1)}}{(\zeta)_{n+1} \left(\zeta^{-1} q\right)_{n+1}}.$$

REMARK. The notation g_2 and g_3 used to define the universal mock theta functions should not be confused with similar notation sometimes used in the literature to define Eisenstein series.

These functions are called "universal" because all of Ramanujan's mock theta functions can essentially be written as a linear combination of specializations of g_2 and g_3 and classical modular forms. To be more precise, for the mock theta functions of even order (resp. odd order) we employ g_2 (resp. g_3). For example, the order 2 mock theta function $B(q)$ satisfies

$$B(q) := \sum_{n=0}^{\infty} \frac{\left(-q; q^2\right)_n}{\left(q; q^2\right)_{n+1}} = g_2\left(q; q^2\right).$$

Moreover, for the third order mock theta functions f and ω, we have

$$\omega(q) = g_3\left(q; q^2\right),$$

$$f(q) = 2 - 2g_3(-1; q).$$

REMARK. Note that in the Appendix we write all mock theta functions in terms of g_2 and g_3.

It turns out that one can express g_3 in terms of g_2 and specific modular and Jacobi forms. Namely, we have the following identity of Gordon and McIntosh [**228, 381**].

PROPOSITION 9.10 (Gordon and McIntosh). *We have that*

$$g_3\left(\zeta^4; q^4\right) = q\zeta^{-2} g_2\left(\zeta^6 q; q^6\right) + q^{-1}\zeta^2 g_2\left(\zeta^6 q^{-1}; q^6\right) - \frac{\zeta^4 q^{\frac{5}{3}} \eta(2\tau)^3 \eta(12\tau) \vartheta(2z+\tau; 2\tau) \vartheta(12z + 6\tau; 12\tau)}{\eta(4\tau)\eta(6\tau)^2 \vartheta(4z; 2\tau)\vartheta(6z-\tau; 2\tau)}.$$

The importance of the universal mock theta functions arises from the fact that one readily obtains hypergeometric mock modular forms by specializing their parameters. This follows from identities for the universal mock theta functions which involve Appell-Lerch-type sums and the μ-function of Zwegers. The following theorem, due to Gordon and McIntosh, and Hickerson, expresses g_2 and g_3 as Appell-Lerch-type sums (cf. [**228, 264**]).

THEOREM 9.11 (Gordon and McIntosh). *We have that*

$$g_2(\zeta; q) = \frac{(-q)_\infty}{(q)_\infty} \sum_{n\in\mathbb{Z}} \frac{(-1)^n q^{n(n+1)}}{1-\zeta q^n},$$

$$g_3(\zeta; q) = \frac{1}{(q)_\infty} \sum_{n\in\mathbb{Z}} \frac{(-1)^n q^{\frac{3n(n+1)}{2}}}{1-\zeta q^n}.$$

SKETCH OF PROOF. The universal mock theta functions are defined in Definition 9.9 as so-called q-hypergeometric series. More precisely, the *q-hypergeometric series*, also known as *basic hypergeometric series*, are defined for non-negative integers r, s by

(9.14)

$$ {}_r\Phi_s \left[\begin{matrix} \alpha_1, \alpha_2, & \dots & \alpha_r \\ \beta_1, \beta_2, & \dots & \beta_s \end{matrix} ; q, w \right] := \sum_{n=0}^{\infty} \frac{(\alpha_1, \alpha_2, \dots, \alpha_r; q)_n}{(\beta_1, \beta_2, \dots, \beta_s, q; q)_n} \left((-1)^n q^{\frac{n}{2}}\right)^{1+s-r} w^n,$$

with

$$(a_1, a_2, \dots, a_\ell)_n = (a_1, a_2, \dots, a_\ell; q)_n := (a_1)_n \cdot \dots \cdot (a_\ell)_n.$$

The main idea of the proof is to first rewrite the universal mock theta functions using (9.14) and then to rewrite these series as lattice sums. To write g_2 and g_3 as lattice sums, one uses the celebrated Watson-Whipple transformation law (cf. page 360 of [**224**]), which states that

$$(9.15)\quad {}_8\Phi_7 \left[\begin{matrix} a, & \sqrt{a}q, & -\sqrt{a}q, & b, & c, & d, & e, & f \\ & \sqrt{a}, & -\sqrt{a}, & \frac{aq}{b}, & \frac{aq}{c}, & \frac{aq}{d}, & \frac{aq}{e}, & \frac{aq}{f} \end{matrix} ; q, \frac{a^2q^2}{bcdef} \right]$$
$$= \frac{\left(aq, \frac{aq}{de}, \frac{aq}{df}, \frac{aq}{ef}\right)_\infty}{\left(\frac{aq}{d}, \frac{aq}{e}, \frac{aq}{d}, \frac{aq}{def}\right)_\infty} {}_4\Phi_3 \left[\begin{matrix} \frac{aq}{bc}, & d, & e, & f \\ & \frac{def}{a}, & \frac{aq}{b}, & \frac{aq}{c} \end{matrix} ; q, q \right]$$

if d, e, or f is of the form q^{-N} with $N \in \mathbb{N}$ (so that the ${}_4\Phi_3$-series terminates).

To prove the claimed equation for g_2, we let $a = q,\ b = \zeta,\ c = \zeta^{-1}q,\ d, e \to \infty$, and $f = -q$. We use that

$$\lim_{d\to\infty} \frac{(d)_n}{d^n} = (-1)^n q^{\frac{n(n-1)}{2}}$$

and multiply by

$$\frac{(-q)_\infty}{(q^2)_\infty (1-\zeta)(1-\zeta^{-1}q)}$$

to obtain that

$$g_2(\zeta;q) = \frac{(-q)_\infty}{(q)_\infty} \sum_{n=0}^{\infty} \frac{\left(1-q^{2n+1}\right)(-1)^n q^{n^2+n}}{(1-\zeta q^n)\left(1-\zeta^{-1}q^{n+1}\right)}$$

$$= \frac{(-q)_\infty}{(q)_\infty} \left(\sum_{n=0}^{\infty} \frac{(-1)^n q^{n^2+n}}{1-\zeta q^n} + \zeta^{-1} \sum_{n=0}^{\infty} \frac{(-1)^n q^{n^2+2n+1}}{1-\zeta^{-1}q^{n+1}} \right). \tag{9.16}$$

Here we used that

$$\frac{1-q^{2n+1}}{(1-\zeta q^n)\left(1-\zeta^{-1}q^{n+1}\right)} = \frac{1}{1-\zeta q^n} + \frac{\zeta^{-1}q^{n+1}}{1-\zeta^{-1}q^{n+1}}. \tag{9.17}$$

Now shifting $n \mapsto -n-1$ in the second sum in (9.16) gives the claim.

In the case of g_3, we let $a = q$, $b = \zeta$, $c = \zeta^{-1}q$, and $d, e, f \to \infty$ to similarly give

$$g_3(\zeta;q) = \frac{1}{(q)_\infty} \sum_{n=0}^{\infty} \frac{\left(1-q^{2n+1}\right)(-1)^n q^{\frac{3n(n+1)}{2}}}{(1-\zeta q^n)\left(1-\zeta^{-1}q^{n+1}\right)}.$$

The claim now follows as before using (9.17) and a shift. □

Along similar lines, we have the following theorem of Kang **[299]** which expresses g_2 in terms of the Dedekind's η-function and the μ-function of Zwegers.

THEOREM 9.12 (Kang). *We have*

$$\zeta g_2(\zeta;q) = -i\zeta q^{-\frac{1}{4}} \mu(2z, \tau; 2\tau) - \frac{i\eta(2\tau)^4}{\eta(\tau)^2 \vartheta(2z;2\tau)}.$$

REMARK. Theorem 9.12 is a minor reformulation of Kang's original result, where she used a slightly different normalization of the theta function and the μ-function. Moreover, we offer a different proof of this result, making use of Zwegers' theory of Appell functions.

REMARK. Kang offered a companion theorem to Theorem 9.12 in **[299]**, in which g_3 is similarly expressed in terms of two μ-functions of Zwegers, and the functions η and ϑ.

PROOF OF THEOREM 9.12. In terms of Appell functions, we can write

$$\zeta g_2(\zeta;q) = \frac{(-q)_\infty}{(q)_\infty} A_2\left(z, \frac{1}{2}; \tau\right).$$

Using Lemma 8.20, we can reduce to a sum of level 1 Appell functions, giving

$$A_2\left(z, \frac{1}{2}; \tau\right) = A_1(2z, 0; 2\tau) + \zeta A_1(2z, \tau; 2\tau).$$

We can now rewrite

$$A_1(2z, 0; 2\tau) = \zeta \sum_{n\in\mathbb{Z}} \frac{(-1)^n q^{n(n+1)}}{1-\zeta^2 q^{2n}} = \frac{-i\eta^3(2\tau)}{\vartheta(2z;2\tau)},$$

where in the final equality we use a special case of the Mittag-Leffler Theorem (cf. (14.8)). Moreover, we have the elementary identity

$$\vartheta(\tau; 2\tau) = -iq^{-\frac{1}{4}} \frac{\eta^2(\tau)}{\eta(2\tau)}$$

and we may also write

$$\zeta A_1(2z,\tau;2\tau)=\zeta\vartheta(\tau;2\tau)\mu(2z,\tau;2\tau),$$

which together imply the claimed identity. □

By combining the results in Section 8.1 with Theorem 9.12, we obtain the following infinite family of basic hypergeometric mock modular forms.

COROLLARY 9.13. *If $\zeta\neq 1$ is a root of unity, the function $\zeta g_2(\zeta;q)+1/2$ is a weight $1/2$ mock modular form whose shadow is proportional to*

$$\sum_{n\in\mathbb{Z}}(-1)^n n\zeta^{-2n}q^{n^2}.$$

REMARK. A similar result holds for g_3 (cf. **[299]** and **[520]**).

REMARK. Instead of specializing ζ to be a root of unity, one may view the functions $g_2(\zeta;q)$ and $g_3(\zeta;q)$ as mock Jacobi forms in the variables z and τ.

9.6. The Mock Theta Conjectures

The "Mock Theta Conjectures" were a hot topic long before Zwegers wrote his thesis on the mock theta functions. The conjectures are a collection of ten formal power series identities for the mock theta functions

$$f_0(q):=\sum_{n=0}^{\infty}\frac{q^{n^2}}{(-q)_n},$$

$$\Phi(q):=-1+\sum_{n=0}^{\infty}\frac{q^{5n^2}}{(q;q^5)_{n+1}(q^4;q^5)_n},$$

$$f_1(q):=\sum_{n=0}^{\infty}\frac{q^{n^2+n}}{(-q)_n},$$

$$\Psi(q):=-1+\sum_{n=0}^{\infty}\frac{q^{5n^2}}{(q^2;q^5)_{n+1}(q^3;q^5)_n}.$$

Andrews and Garvan **[34]** proved that these ten identities are implied from the truth of a pair of the identities.

CONJECTURE 9.14 (Andrews-Garvan **[34]**). *The following identities are true:*

$$f_0(q)+2\Phi\left(q^2\right)=\frac{(q^5;q^5)_\infty(q^5;q^{10})_\infty}{(q;q^5)_\infty(q^4;q^5)_\infty},$$

$$f_1(q)+2q^{-1}\Psi\left(q^2\right)=\frac{(q^5;q^5)_\infty(q^5;q^{10})_\infty}{(q^2;q^5)_\infty(q^3;q^5)_\infty}.$$

Hickerson **[264]** famously proved this pair of identities by making use of ingenious and intricate q-series manipulations.

THEOREM 9.15 (Hickerson). *The Mock Theta Conjectures are true.*

REMARK. Recent work **[265]** by Hickerson and Mortenson, in the spirit of Hickerson's proof of the Mock Theta Conjectures, gives many further identities where sums of mock modular forms are shown to be modular. This important paper gives a general framework of identities relating q-hypergeometric series, such as the universal mock theta functions, to weakly holomorphic modular forms. The methods in their work do not rely on the theory of harmonic Maass forms.

It is now worthwhile to revisit this conjecture with the understanding that Ramanujan's mock theta functions are mock modular forms. The infinite products in Conjecture 9.14 are (up to powers of q) weight $1/2$ weakly holomorphic modular forms. These identities are explicit examples of the fact that the difference of two harmonic weight k Maass forms with the same nonholomorphic part is a weight k weakly holomorphic modular form. Armed with this understanding, one can now prove such identities by means of the standard theory of modular forms. Namely, one applies the principle that finitely many coefficients determine a modular form of given weight, group, and divisor. In the case of Conjecture 9.14, this idea was suggested by two of the authors in **[112]** and was carried out by one of the authors in **[194]**. Note however that the number of coefficients to check is very large. This problem has been circumvented by using vector-valued forms by Andersen **[19]**.

9.7. The Andrews-Dragonette Conjecture

In the last two sections of this chapter we will revisit the contents of Ramanujan's last letter. We recall his claimed asymptotic formula (9.7) for the coefficients of his third order mock theta function f, namely that

$$\alpha_f(n) = (-1)^{n+1}\frac{\sqrt{6}\exp\left(\frac{\pi}{12}\sqrt{24n-1}\right)}{\sqrt{24n-1}} + O\left(\frac{\exp\left(\frac{\pi}{24}\sqrt{24n-1}\right)}{\sqrt{24n-1}}\right).$$

As is typical of his writings, Ramanujan did not offer a proof of this claim. These asymptotics remained open until Dragonette proved the claim (9.7) in her 1951 Ph.D. thesis **[164]** written under the direction of Rademacher. In his 1964 Ph.D. thesis, also written under Rademacher, Andrews **[20]** improved upon Dragonette's work, and proved, for any $\varepsilon > 0$, that $\alpha_f(n)$ equals

(9.18)

$$\frac{\pi}{(24n-1)^{\frac{1}{4}}}\sum_{k=1}^{\lfloor\sqrt{n}\rfloor}\frac{(-1)^{\left\lfloor\frac{k+1}{2}\right\rfloor}A_{2k}\left(n-\frac{k\left(1+(-1)^k\right)}{4}\right)}{k}I_{\frac{1}{2}}\left(\frac{\pi\sqrt{24n-1}}{12k}\right)+O\left(n^{\varepsilon}\right).$$

Here, for positive integers k and n, the *Kloosterman sum* $A_k(n)$ is defined by

$$A_k(n) := \frac{1}{2}\sqrt{\frac{k}{12}}\sum_{\substack{d \pmod{24k} \\ d^2\equiv -24n+1 \pmod{24k}}}\left(\frac{12}{d}\right)e\left(\frac{d}{12k}\right). \tag{9.19}$$

REMARK. These are the same Kloosterman sums which appear in Rademacher's exact formula for partition numbers (cf. Theorem 14.3 below)

Based on these works (which use the Circle Method), Andrews and Dragonette conjectured **[20, 26]** the following exact formula for the coefficients $\alpha_f(n)$.

CONJECTURE 9.16 (Andrews and Dragonette). *If $n \in \mathbb{N}$, then*

$$\alpha_f(n) = \frac{\pi}{(24n-1)^{\frac{1}{4}}}\sum_{k=1}^{\infty}\frac{(-1)^{\left\lfloor\frac{k+1}{2}\right\rfloor}A_{2k}\left(n-\frac{k\left(1+(-1)^k\right)}{4}\right)}{k}I_{\frac{1}{2}}\left(\frac{\pi\sqrt{24n-1}}{12k}\right).$$

Two of the authors proved this conjecture in [**107**] using the theory of Maass-Poincaré series.

THEOREM 9.17. *The Andrews-Dragonette Conjecture is true.*

REMARK. Garthwaite [**215, 216**] generalized the proof in [**107**], and she obtained exact formulas for the coefficients for several of Ramanujan's other mock theta functions.

SKETCH OF PROOF OF THEOREM 9.17. From Theorem 9.5, we conclude that the h_0 component of the vector-valued harmonic Maass form has the property that its holomorphic part is $q^{-\frac{1}{24}}f(\tau)$. Therefore, the idea is to use the method of Maass-Poincaré series (cf. Section 6.3) to construct a harmonic Maass form with the same modular transformation properties and the same principal parts at cusps. This harmonic Maass form must then equal h_0 up to a holomorphic weight 1/2 modular form, one which turns out to be identically zero. One then finds (after some manipulation) that the conjectured exact formula of Andrews and Dragonette is indeed the formula that one obtains by the method of Poincaré series.

To successfully carry out this strategy requires care because the relevant series in this case is not absolutely convergent. To this end, we use Hecke's trick, and we apply analytic continuation to the Fourier expansion to prove the conjecture. Note that similar Poincaré series are studied in Section 6.3.

More precisely, let

$$\varphi_{s,k}(\tau) := \left(\frac{6}{\pi v}\right)^{\frac{k}{2}} M_{-\frac{k}{2},s-\frac{1}{2}}\left(\frac{\pi v}{6}\right) e\left(-\frac{u}{24}\right),$$

where M denotes the standard M-Whittaker function. For matrices $\left(\begin{smallmatrix} a & b \\ c & d \end{smallmatrix}\right) \in \Gamma_0(2)$ with $c \in \mathbb{N}_0$, set

$$\chi\left(\begin{pmatrix} a & b \\ c & d \end{pmatrix}\right) := \begin{cases} e\left(-\frac{b}{24}\right) & \text{if } c = 0, \\ i^{-\frac{1}{2}}(-1)^{\frac{1}{2}(c+ad+1)} e\left(-\frac{a+d}{24c} - \frac{a}{4} + \frac{3dc}{8}\right)\omega_{-d,c}^{-1} & \text{if } c > 0. \end{cases}$$

Here we let

$$\omega_{a,b} := e\left(\frac{1}{2}s(a,b)\right),$$

where we have the usual *Dedekind sum*

$$s(a,b) := \sum_{\mu \pmod{b}} \left(\left(\frac{\mu}{b}\right)\right)\left(\left(\frac{a\mu}{b}\right)\right),$$

which uses

$$((x)) := \begin{cases} x - \lfloor x \rfloor - \frac{1}{2} & \text{if } x \in \mathbb{R} \setminus \mathbb{Z}, \\ 0 & \text{if } x \in \mathbb{Z}. \end{cases}$$

Using this notation, we define the Poincaré series $P_k(s;\tau)$ by

$$P_k(s;\tau) := \frac{2}{\sqrt{\pi}} \sum_{M \in \Gamma_\infty \setminus \Gamma_0(2)} \chi(M)^{-1}(c\tau+d)^{-k}\varphi_{s,k}(M\tau). \tag{9.20}$$

The series $P_k(1-k/2;\tau)$ is absolutely convergent for $k < 1/2$ and annihilated by $\Delta_{\frac{k}{2}}$.

The function $P_{\frac{1}{2}}(s;\tau)$ can be analytically continued by its Fourier expansion to $s = 3/4$. This continuation requires an interpretation of the Kloosterman sums

$A_{2k}(n)$ as sums of quadratic forms of fixed discriminant. The continuation then follows by a generalization of an argument of Hooley [**274**] on the "equidistribution" of CM points. This calculation is lengthy and is carried out in detail in [**107**].

One then checks that the Maass-Poincaré series $P_{\frac{1}{2}}(\frac{3}{4};\tau)$ has the same principal parts at cusps and the same modular transformation properties as h_0. Therefore, these harmonic Maass forms agree up to a weight $1/2$ holomorphic modular form.

A direct calculation of the Fourier expansion of $P_{\frac{1}{2}}(\frac{3}{4};\tau)$ (after some manipulation) gives the conjectured formula for $\alpha_f(n)$. Therefore, the difference between h_0 and this Maass-Poincaré series is a weight $1/2$ harmonic Maass form whose holomorphic part is identically zero and whose nonholomorphic part is the period integral of a weight $3/2$ cusp form. Since such period integrals are not harmonic Maass forms by themselves, it follows that this difference is identically zero. This completes the sketch of the proof. □

9.8. Ramanujan's original claim revisited

Here we revisit Ramanujan's original notion of a mock theta function. Throughout this section, the limit "$q \to \zeta$" (ζ a root of unity) means a radial limit from within the unit disk. For Ramanujan, mock theta functions are holomorphic functions on $\mathbb{H}$ which have singularities at some cusps and which also have asymptotics which closely mirror those of modular forms. However, he insisted that the singularities of a mock theta function cannot be cut out by a single weakly holomorphic modular form. This condition is Definition 9.1 iii).

Here we confirm that Ramanujan's examples do indeed satisfy his own definition. By making use of the realization that Ramanujan's examples are essentially mock modular forms of weight $1/2$, we can prove his claim. The key result is the following theorem which was proved by Griffin in joint work with two of the authors (cf. Theorem 1.1 of [**237**]).

THEOREM 9.18. *Suppose that $f = f^+ + f^- \in H_k(\Gamma_1(N))$, where $k \in \frac{1}{2}\mathbb{Z}$. If f^- is non-trivial and g is a weight k weakly holomorphic modular form on $\Gamma_1(N')$ for some $N' \in \mathbb{N}$, then there are infinitely many roots of unity ζ for which $f^+ - g$ has exponential growth as q approaches ζ.*

This theorem confirmed (cf. Corollary 1.2 of [**237**]) Ramanujan's original claim about his examples.

COROLLARY 9.19. *Suppose that M is one of Ramanujan's mock theta functions and $a \in \mathbb{N}, b \in \mathbb{Z}, c \in \mathbb{C}$ such that $q^b M(a\tau) + c$ is the holomorphic part of a weight $1/2$ harmonic Maass form. Then there does not exist a weakly holomorphic modular form g of any half-integral weight on any group $\Gamma_1(N)$ such that at every root of unity ζ we have*

$$\lim_{q \to \zeta} \left(q^b M(a\tau) + c - g(\tau)\right) = O(1).$$

REMARK. Although this corollary asserts that a weakly holomorphic modular form and a mock theta function cannot cut out each other's singularities, Ramanujan discovered a surprising near miss in his last letter to Hardy. For his third order mock theta function f, he finds the q-series

$$b(q) := (1-q)\left(1-q^2\right)\left(1-q^3\right)\left(1-q^5\right)\cdots\left(1-2q+2q^4-2q^9+2q^{16}-\cdots\right),$$

which is essentially a weight $1/2$ modular form that nearly cuts out its singularities. By inspection, one shows that the singularities of f occur at the even order roots of unity. For half of these roots of unity, b cuts out the singularities, while for the other half it turns out that $-b$ cuts out the singularities. More precisely, Ramanujan claimed that as q approaches an even order $2k$ primitive root of unity ζ, we have

$$\lim_{q\to\zeta}\left(f(q)-(-1)^k b(q)\right)=O(1). \tag{9.21}$$

Watson investigated this in **[489]**, and his work indicates how one could prove the boundedness of the radial limit in (9.21). Rhoades, and two of the authors went further **[203, 204]** and derived exact formulas for the $O(1)$ numbers. These numbers turn out to be explicit numbers in $\mathbb{Z}[\zeta]$. They are special values of L-functions, and they are values of "quantum modular forms", a subject which we shall discuss in Chapter 21. Namely, we have that **[203, 204]**

$$\lim_{q\to\zeta}\left(f(q)-(-1)^k b(q)\right)=-4\sum_{n=0}^{k-1}(1+\zeta)^2\left(1+\zeta^2\right)^2\cdots\left(1+\zeta^n\right)^2\zeta^{n+1}.$$

This is a special case of a general theory. In particular, this result is an instance of Theorem 14.25, which we discuss in Chapter 14 and revisit in Chapter 21. Moreover, such formulas are now known for the universal mock theta function $g_2(\zeta;q)$. Proving a conjecture of Rhoades **[435]**, two of the authors **[115]** determined associated modular forms $f_{a,b,A,B,h,k}$ such that

$$g_2\left(\zeta_b^a q^A;q^B\right)-f_{a,b,A,B,h,k}(q)$$

is bounded as $q\to\zeta_k^h$. Moreover, they obtained explicit finite formulas for the constants

$$Q_{a,b,A,B,h,k}:=\lim_{q\to\zeta_k^h}\left(g_2\left(\zeta_b^a q^A;q^B\right)-f_{a,b,A,B,h,k}(q)\right).$$

Analogous results in **[203]** and **[287]** pertain to the universal mock theta function g_3.

REMARK. We point out that Rhoades **[435]** recently showed that Ramanujan's definition of the mock theta function is not equivalent to the modern definition.

Next we sketch the proof of Theorem 9.18 and Corollary 9.19. These results rely on the Bruinier-Funke pairing described in Section 5.2. Namely, we require the important fact that every harmonic Maass form on $H_k(\Gamma_1(N))$ with a non-trivial nonholomorphic part has a non-trivial principal part at some cusp.

SKETCH OF PROOF OF THEOREM 9.18. Suppose that g is a weakly holomorphic modular form on $\Gamma_1(N')$, for some N', which cuts out the exponential singularities of f as q approaches roots of unity. Then $h:=f-g$ is a harmonic Maass form of weight k on $\Gamma_1(\mathrm{lcm}(N,N'))$ with a non-constant nonholomorphic part. By Lemma 5.13, h has a non-constant principal part at some cusp. Since the nonholomorphic part f^- is $O(1)$ at cusps, it follows that h^- is also $O(1)$ at cusps. Suppose that h has a non-constant principal part at infinity (a similar argument applies at other cusps). By choosing matrices $\left(\begin{smallmatrix}a&b\\c&d\end{smallmatrix}\right)\in\Gamma_1(\mathrm{lcm}(N,N'))$, combined with the fact that

$$\lim_{\tau\to i\infty}h\left(\frac{a\tau+b}{c\tau+d}\right)=\lim_{\tau\to i\infty}(c\tau+d)^k h(\tau),$$

we find that infinitely many roots of unity are exponential singularities for h. $\square$

To obtain Corollary 9.19, we require the following simple lemma.

LEMMA 9.20. *Suppose that f and g are weakly holomorphic modular forms of weights k_1 and k_2, respectively. If both have exponential singularities at some cusps and $f-g$ has no exponential singularities at cusps, then $k_1 = k_2$.*

PROOF. Arguing as in the proof of Theorem 9.18, both f and g must have the same principal parts at all cusps for $f-g$ to be bounded as one approaches all cusps. Moreover, we know that at least one principal part at a cusp is non-constant. Without loss of generality, suppose that the principal part at the cusp infinity is non-constant. Then consider the function $h := f-g$. By hypothesis, h has bounded radial limits as q approaches every root of unity. Now f and g are modular on some common subgroup $\Gamma_1(N')$. Thus, if we take $\left(\begin{smallmatrix} a & b \\ c & d \end{smallmatrix}\right) \in \Gamma_1(N')$ with $cd \neq 0$, then we have

$$h\left(\frac{a\tau+b}{c\tau+d}\right) = f\left(\frac{a\tau+b}{c\tau+d}\right) - g\left(\frac{a\tau+b}{c\tau+d}\right) = (c\tau+d)^{k_1} f(\tau) - (c\tau+d)^{k_2} g(\tau).$$

Letting $\tau \to i\infty$, we find that f and g cannot cut out the same exponential singularities at roots of unity because of the difference between the weights. Therefore, it follows that $k_1 = k_2$. ☐

SKETCH OF PROOF OF COROLLARY 9.19. Suppose that M is one of Ramanujan's mock theta functions. Then by Theorem 9.4, there are integers a and b for which $q^b M(a\tau) + c =: f^+(\tau)$ is the holomorphic part of a weight $1/2$ harmonic Maass form. Moreover, the associated shadow is a weight $3/2$ unary theta function. Using the theory of quadratic twists, we can obtain a non-trivial weakly holomorphic modular form $\widetilde{f}$ whose non-zero coefficients agree with those of the harmonic Maass form f in certain arithmetic progressions. One uses quadratic and trivial twists to assemble[4] $\widetilde{f}$ by annihilating the nonholomorphic part, the period integral of the unary theta function.

Suppose that g is a weakly holomorphic modular form of some weight k which cuts out the exponential singularities of f. Applying the same procedure to g gives a weakly holomorphic modular form $\widetilde{g}$. We then have that $\widetilde{f}$ and $\widetilde{g}$ cut out exactly the same exponential singularities at all roots of unity. By Lemma 9.20, this implies that g must also have weight $1/2$. Therefore, if there is such a g, then $f-g$ is a weight $1/2$ harmonic Maass form which has a non-vanishing nonholomorphic part and which also has the property that $f^+ - g$ has no exponential singularities at any roots of unity. This contradicts Theorem 9.18. ☐

[4]This is analogous to the proof of Theorem 1.4 in **[110]**.

CHAPTER 10

Holomorphic Projection

We have already related harmonic Maass forms to modular forms using differential operators. For example, in Chapter 5 we discussed the differential operators ξ_{2-k} and D^{k-1}; they map integer weight $2-k$ harmonic Maass forms to weight k weakly holomorphic modular forms. Here we consider another method for relating modular forms to harmonic Maass forms. It is the principle of *holomorphic projection*, a procedure which canonically associates (weakly) holomorphic modular forms to suitable nonholomorphic modular forms. Holomorphic projection operators were first introduced by Sturm **[475]** in 1980, and they were further developed in the famous paper by Gross and Zagier **[242]** on Heegner points and derivatives of L-functions.

As first observed by Zwegers in unpublished work, holomorphic projection is very useful for studying the coefficients of mixed mock modular forms, natural components of nonholomorphic modular forms which are obtained by multiplying harmonic Maass forms with holomorphic modular forms. These ideas play important roles in the recent proof of the Umbral Moonshine Conjecture (cf. Section 20.3), as well as in the study of shifted convolution L-functions (cf. Chapter 17).

In this chapter we recall the essential features of holomorphic projection, and we give examples of some elegant recurrence relations related to Ramanujan's original mock theta functions and Hurwitz class numbers in Section 10.3.

10.1. Principle of holomorphic projection

Here we describe the essential features of the theory following unpublished notes of Zagier.[1] A description of holomorphic projection may also be found in **[514]**. The idea underlying holomorphic projection is simple. Suppose that f is a weight k real-analytic modular form of moderate growth at cusps. Then a linear functional on the space of weight k cusp forms is given by

$$g \mapsto \langle g, f \rangle,$$

where $\langle \cdot, \cdot \rangle$ is the usual Petersson inner product. Since the Petersson inner product is non-degenerate, this functional must be given by $\langle \cdot, \widetilde{f} \rangle$ for some weight k cusp form $\widetilde{f}$. This $\widetilde{f}$ is essentially the holomorphic projection of f. Here we discuss the case of real-analytic modular forms on $\Gamma_0(N)$. It is not difficult to modify the statement for multiplier systems and other groups.

[1]This treatment is based on Zagier's lecture "Mock modular forms: Theory and Examples" at the Max Planck Institute in Bonn in May 2009.

DEFINITION 10.1. Suppose that $f\colon \mathbb{H} \to \mathbb{C}$ is a (not necessarily holomorphic) modular form with integer weight $k \geq 2$ on $\Gamma_0(N)$ with Fourier expansion

$$f(\tau) = \sum_{n\in\mathbb{Z}} c_f(n,v)q^n,$$

where, as before, $\tau = u + iv$. For a cusp κ_j, $j = 1, ..., M$, and $\kappa_1 := i\infty$, of $\Gamma_0(N)$, fix $\gamma_j \in \mathrm{SL}_2(\mathbb{Z})$ with $\gamma_j\kappa_j = i\infty$. Suppose that the following are true:

i) There is an $\varepsilon > 0$ and a $c_0^{(j)} \in \mathbb{C}$ for which

$$f\left(\gamma_j^{-1}w\right)\left(\frac{d}{dw}\tau\right)^{\frac{k}{2}} = c_0^{(j)} + O(\mathrm{Im}(w)^{-\varepsilon})$$

for all $j = 1, ..., M$ and $w := \gamma_j\tau$,

ii) For all $n > 0$ we have that $c_f(n,v) = O(v^{2-k})$ as $v \to 0$.

Then the **holomorphic projection** of f is defined by

$$\pi_{\mathrm{hol}}(f)(\tau) := c_0 + \sum_{n=1}^{\infty} c(n)q^n,$$

where $c_0 = c_0^{(1)}$, and for $n \in \mathbb{N}$,

$$c(n) := \frac{(4\pi n)^{k-1}}{(k-2)!}\int_0^\infty c_f(n,v)e^{-4\pi nv}v^{k-2}dv.$$

The holomorphic projection operator enjoys some important properties. For weights $k \geq 4$, holomorphic projections are holomorphic weight k modular forms. If the weight is $k = 2$, then the holomorphic projection is a quasimodular form. The essential properties are summarized in the next proposition.

PROPOSITION 10.2. *Assume the notation and hypotheses in Definition* 10.1. *Then the following are true:*

i) *If f is holomorphic, then $\pi_{\mathrm{hol}}(f) = f$.*
ii) *We have that*

$$\pi_{\mathrm{hol}}(f) \in \begin{cases} M_k(\Gamma_0(N)) & \text{if } k \geq 4,\\ \mathbb{C}E_2 \oplus M_2(\Gamma_0(N)) & \text{if } k = 2.\end{cases}$$

SKETCH OF PROOF. Suppose that $k \geq 4$. For $m \in \mathbb{N}$, let $P_{k,m,N} \in S_k(\Gamma_0(N))$ be the usual absolutely convergent m-th weight k cuspidal Poincaré series (cf. (6.11)). We have that

$$P_{k,m,N}(\tau) = \sum_{\left(\begin{smallmatrix} a & b\\ c & d\end{smallmatrix}\right)\in\Gamma_\infty\backslash\Gamma_0(N)} (c\tau+d)^{-k}e^{2\pi i m\frac{a\tau+b}{c\tau+d}}. \tag{10.1}$$

The idea is to show that the Petersson inner product $\langle f, P_{k,m,N}\rangle$ can be expressed as

$$\langle f, P_{k,m,N}\rangle = \int_{\Gamma_\infty\backslash\mathbb{H}} \overline{e^{2\pi i m\tau}}f(\tau)v^{k-2}dv = \int_0^\infty e^{-4\pi mv}c_f(m,v)v^{k-2}dv. \tag{10.2}$$

This amounts to the legality of replacing $P_{k,m,N}$ by (10.1), and then interchanging summation with integration in the definition of the Petersson inner product. Then one uses the standard unfolding argument to obtain (10.2). This interchange of summation and integration is justified by bounds for

$$P^*_{k,m,N}(\tau) := \sum_{\left(\begin{smallmatrix} a & b \\ c & d \end{smallmatrix}\right) \in \Gamma_\infty \backslash \Gamma_0(N)} \left| (c\tau + d)^{-k} e^{2\pi i m \frac{a\tau+b}{c\tau+d}} \right|$$

which can be proven by comparing against a level 1 Eisenstein series.

To complete the proof, we note that as mentioned above the map $g \mapsto \langle g, f \rangle$ is a linear functional on $S_k(\Gamma_0(N))$, and so it must be of the form $g \mapsto \langle g, \widetilde{f} \rangle$ for some fixed $\widetilde{f} \in S_k(\Gamma_0(N))$. The claimed formula for the coefficients follows easily.

The proof above does not quite work if $k = 2$ because of the divergence of the series definition of the corresponding $P^*_{k,m,N}$. In this case one applies Hecke's trick (as we have done several times before), which amounts to studying, for $\mathrm{Re}(s) > 0$, the Poincaré series

$$P^*_{k,m,N,s}(\tau) := \sum_{\left(\begin{smallmatrix} a & b \\ c & d \end{smallmatrix}\right) \in \Gamma_\infty \backslash \Gamma_0(N)} \frac{1}{(c\tau + d)^2} \frac{v^s}{|c\tau + d|^{2s}} e^{2\pi i m \frac{a\tau+b}{c\tau+d}}.$$

One proceeds as above but argues with analytic continuation where $s \to 0$. □

REMARK. Under suitable conditions of convergence, one may express $\pi_{\mathrm{hol}}(f)$ alternatively as

$$\pi_{\mathrm{hol}}(f)(\tau) = \frac{(k-1)(2i)^k}{4\pi} \int_{\mathbb{H}} \frac{f(w) w_2^k}{(\tau - \overline{w})^k} \frac{dw_1 dw_2}{w_2^2}, \tag{10.3}$$

where $w =: w_1 + iw_2$. This uses the invariance $f(\tau + 1) = f(\tau)$ and the fact that the integral may be rewritten as

$$\int_0^\infty w_2^{k-2} \int_0^1 f(w_1 + iw_2) \sum_{n \in \mathbb{Z}} \frac{1}{(\tau - w_1 + iw_2 + n)^k} dw_1 dw_2.$$

One then inserts the Fourier expansion of f and employs the classical Lipschitz summation formula **[357]**

$$\sum_{n \in \mathbb{Z}} \frac{1}{(w+n)^k} = \frac{(-2\pi i)^k}{(k-1)!} \sum_{n=1}^\infty n^{k-1} e^{2\pi i n w}.$$

In view of this calculation, one can define π_{hol} on arbitrary translation invariant functions.

10.2. Regularized holomorphic projection

The holomorphic projection operator defined in the previous section requires that the functions f have moderate growth at cusps. The construction of holomorphic projections can be suitably modified in some cases to relax this growth condition at cusps. To this end, we make use of the regularized Petersson inner product (cf. Definition 5.6). We begin by defining *regularized holomorphic projections*, which were introduced in **[386]** in connection with the study of shifted convolution L-functions.

DEFINITION 10.3. Let $f\colon \mathbb{H} \to \mathbb{C}$ be a real-analytic modular form with integer weight $k \geq 2$ on $\Gamma_0(N)$ with Fourier expansion

$$f(\tau) = \sum_{n \in \mathbb{Z}} c_f(n, v) q^n.$$

Suppose that for each cusp ρ of $\Gamma_0(N)$ there is a polynomial $H_\kappa(X) \in \mathbb{C}[X]$ and $\varepsilon > 0$ for which

$$f|_k \gamma_\rho^{-1}(\tau) - H_\rho\left(q^{-1}\right) = v^{-\varepsilon}.$$

Furthermore, suppose that $c_f(n,v) = O(v^{2-k})$ as $v \to 0$ for all $n > 0$. Then the **regularized holomorphic projection** of f is defined by

$$\pi_{\mathrm{hol}}^{\mathrm{reg}}(f)(\tau) = H_{i\infty}\left(q^{-1}\right) + \sum_{n=1}^{\infty} c(n) q^n,$$

where

$$c(n) = \lim_{s \to 0} \frac{(4\pi n)^{k-1}}{(k-2)!} \int_0^\infty c_f(n,v) e^{-4\pi n v} v^{k-2-s} dv.$$

This operator enjoys the following generalization of Proposition 10.2 (cf. Proposition 3.4 of **[386]**), which is obtained by employing the regularized Petersson inner product (cf. Definition 5.6) in place of the ordinary Petersson inner product in the proof of Proposition 10.2.

PROPOSITION 10.4. *Assume the notation and hypotheses in Definition* 10.3. *Then the following are true:*

i) *If f is holomorphic on $\mathbb{H}$, then $\pi_{\mathrm{hol}}^{\mathrm{reg}}(f) = f$.*
ii) *We have that*

$$\pi_{\mathrm{hol}}^{\mathrm{reg}}(f) \in \begin{cases} M_k^!(\Gamma_0(N)) & \text{if } k \geq 4, \\ E_2 M_0^!(N) \oplus M_2^!(\Gamma_0(N)) & \text{if } k = 2. \end{cases}$$

iii) *If f satisfies the conditions in Definition* 10.1, *then $\pi_{\mathrm{hol}}^{\mathrm{reg}}(f) = \pi_{\mathrm{hol}}(f)$.*

10.3. Kronecker-type relations for mock modular forms

Here we illustrate how holomorphic projections give rise to elegant recurrence relations for the coefficients of mock modular forms such as Ramanujan's original mock theta functions and the generating function for Hurwitz class numbers. Such applications were also studied in unpublished work of Zwegers. We are content with presenting two simple examples.

Recall Ramanujan's third order mock theta function (cf. (9.1)):

$$f(q) = 1 + \sum_{n=1}^{\infty} \alpha_f(n) q^n = 1 + \sum_{n=1}^{\infty} \frac{q^{n^2}}{(-q;q)_n^2}.$$

Imamoğlu, Raum, and Richter proved (cf. Theorem 1 of **[278]**) the following elegant recurrence relation for the coefficients $\alpha_f(n)$.

THEOREM 10.5. *If $n \in \mathbb{Q}$, then let*

$$\sigma(n) := \begin{cases} \sum_{0<d|n} d & \text{if } n \in \mathbb{Z}, \\ 0 & \text{otherwise}, \end{cases}$$

$$\operatorname{sgn}^+(n) := \begin{cases} \operatorname{sgn}(n) & \text{if } n \neq 0, \\ 1 & \text{if } n = 0, \end{cases}$$

and then define

$$d(N_1, N_2, t_1, t_2) := \operatorname{sgn}^+(N_1)\operatorname{sgn}^+(N_2)\left(|N_1 + t_1| - |N_2 + t_2|\right).$$

Then for positive integers n, we have that

$$\sum_{\substack{m\in\mathbb{Z} \\ 3m^2+m\le 2n}} \left(m + \frac{1}{6}\right)\alpha_f\left(n - \frac{3m^2}{2} - \frac{m}{2}\right)$$
$$= \frac{4}{3}\sigma(n) - \frac{16}{3}\sigma\left(\frac{n}{2}\right) - 2\sum_{\substack{a,b\in\mathbb{Z} \\ 2n=ab}} d\left(N_1, N_2, \frac{1}{6}, \frac{1}{6}\right),$$

where $N_1 := \frac{1}{6}(-3a+b-1)$ and $N_2 := \frac{1}{6}(3a+b-1)$, and the sum is over integers a, b for which $N_1, N_2 \in \mathbb{Z}$.

REMARK. For each positive integer n, it is obvious that the sums in Theorem 10.5 are finite. Therefore, these formulas give a recurrence for the coefficients $\alpha_f(n)$. Indeed, the $m = 0$ term in the left-hand side of the expression in the theorem leads to an expression for $\alpha_f(n)$ in terms of earlier coefficients and the expression on the right-hand side. The work of Imamoğlu, Raum, and Richter is actually formulated in terms of vector-valued harmonic Maass forms. This is an important feature of their work, and so their paper can be viewed as a prototype for holomorphic projections in this setting.

SKETCH OF PROOF OF THEOREM 10.5. Using Corollary 9.6, we can deduce that

$$F(\tau) := q^{-1}f\left(q^{24}\right) - 2i\sqrt{3}\int_{-24\bar{\tau}}^{i\infty} \frac{\mathcal{G}_1(w)}{\sqrt{-i(24\tau + w)}}dw$$

is a weight $1/2$ harmonic Maass form on $\Gamma_0(144)$ with Nebentypus $\chi(\cdot) = \left(\frac{12}{\cdot}\right)$. Here, the function

$$\mathcal{G}_1(\tau) = -\sum_{n\in\mathbb{Z}} \left(n + \frac{1}{6}\right) e^{3\pi i\left(n+\frac{1}{6}\right)^2\tau}$$

is as defined in Section 9.3. The result follows by applying a modification of Proposition 10.2 to the weight 2 nonholomorphic modular form $F(\tau)\mathcal{G}_1(24\tau)$. The holomorphic projection is a linear combination of weight 2 Eisenstein series, and the terms with $d(N_1, N_2, 1/6, 1/6)$ arise from integrals involving the nonholomorphic part of F. □

In **[385]**, Mertens generalized the work of Imamoğlu, Raum, and Richter (see also **[384]**). He obtained such recurrence relations for general mixed mock modular forms and their generalizations constructed by Rankin-Cohen brackets. This work has many applications. For example, he proved Eichler-Selberg[2] type class number relations. To make this precise, for $k \in \mathbb{N}$ we let

$$\lambda_k(n) := \frac{1}{2}\sum_{d|n} \min\left(d, \frac{n}{d}\right)^k. \tag{10.4}$$

---

[2]We refer to such relations as Eichler-Selberg relations because of their resemblance to the class number identities, which follow from the Eichler-Selberg trace formula for the Hecke operators.

Classic works of Hurwitz **[276]** and Kronecker **[332]** give relations such as

$$\sum_{s\in\mathbb{Z}} H\left(4n - s^2\right) - 2\lambda_1(n) = 2\sigma_1(n).$$

Such identities inspired Cohen **[150]** to conjecture further identities. Using the non-holomorphic weight 3/2 Eisenstein series (cf. Subsection 6.1.2), whose holomorphic part is the generating function for Hurwitz class numbers, Mertens studied **[385]** the holomorphic projections of mixed mock modular forms which are the product of this Eisenstein series with unary theta functions. Up to "error" q-series which are described in terms of the $\lambda_k(n)$-numbers, these holomorphic projections are linear combinations of holomorphic modular forms. We highlight two identities (cf. (1.1), (1.2) of **[384]**), the first of which was previously obtained by Eichler **[189]** and the second of which was conjectured by Cohen and proved in **[384, 385]**.

THEOREM 10.6. *For positive odd integers n, we have*

$$\sum_{s\in\mathbb{Z}} H\left(n - s^2\right) + \lambda_1(n) = \frac{1}{3}\sigma_1(n),$$

$$\sum_{s\in\mathbb{Z}} \left(4s^2 - n\right) H\left(n - s^2\right) + \lambda_3(n) = 0.$$

CHAPTER 11

Meromorphic Jacobi Forms

In Chapter 2 we discussed the theory of (classical) Jacobi forms. It turns out that there is an extension and generalization of this theory which we describe here. In analogy with the classical theory, the theory of "meromorphic Jacobi forms" offers convenient packaging of functions such as some of Ramanujan's mock theta functions. After discussing the applications of meromorphic Jacobi forms to the theory of mock theta functions, we will discuss the cases of both positive and negative index, which have somewhat different flavors. We also remark that in Subsection 20.2.4, a special family of index zero forms are considered.

11.1. Mock theta functions as coefficients of meromorphic forms

Theorem 2.23 shows how holomorphic Jacobi forms (in the elliptic variable z) admit a theta decomposition in which the Fourier coefficients h_a inherit true vector-valued modular transformation properties. In this chapter, we consider the analogous situation for meromorphic Jacobi forms, which is somewhat parallel to the holomorphic setting but inherently more complicated.

To best illustrate this, we begin with a beautiful example due to Andrews, who in [**24**] realized all of Ramanujan's fifth order mock theta functions as constant terms of Jacobi forms. For example, the fifth order mock theta function f_0 defined in (9.3) satisfies

$$f_0(q) = [\zeta^0]\left(-\zeta q^{\frac{5}{8}}\frac{\eta(5\tau)^3\vartheta(4\tau;5\tau)\,\vartheta\left(\frac{1}{2}+z+4\tau;3\tau\right)\vartheta(z;5\tau)}{\vartheta\left(\frac{1}{2}+2\tau;5\tau\right)\vartheta\left(\frac{1}{2}+2\tau-z;5\tau\right)\vartheta\left(\frac{1}{2}+2\tau+z;5\tau\right)}\right), \tag{11.1}$$

where here and throughout, $q = e(\tau), \zeta = e(z)$ (where $e(\alpha) = e^{2\pi i\alpha}$), with $\tau \in \mathbb{H}, z \in \mathbb{C}$, and $[\zeta^r]\,f$ denotes the coefficient of ζ^r of the function f. The Jacobi form appearing on the right-hand side of (11.1) is meromorphic, with positive weight and index.

Among other things, this section describes how this realization gives the mock modularity of functions like f_0, which arise as constant terms of meromorphic Jacobi forms with positive index. We also discuss the case of meromorphic Jacobi forms of negative index in Section 11.3, where related objects involving partial theta functions arise.

11.2. Positive index Jacobi forms

In this section, we consider meromorphic Jacobi forms of positive index, i.e., functions on $\mathbb{C}\times\mathbb{H}$ that transform like in Definition 2.11 i) and ii) but may have poles at the elliptic z-variable. Note that for fixed $z \in \mathbb{C}$, poles in τ lying at the cusps are allowed. Explicit examples are considered in Section 20.2 below. We assume throughout that such a form has only poles at points of the form $z = z_s = \alpha\tau + \beta$

where $s = (\alpha, \beta)$ belongs to some subset $S = S_\tau$ of $\mathbb{Q}^2$. The elliptic transformation law of Definition 2.11 ii) implies that S is invariant under shifts by $\mathbb{Z}^2$ and that $S/\mathbb{Z}^2$ is finite.

The goal of this section is to find a theta-type decomposition, resembling the one for holomorphic Jacobi forms as given in Theorem 2.23. Zwegers showed in **[529]** that meromorphic Jacobi forms also have a theta-type decomposition, however the coefficients are not modular forms. Instead, they are related to mock modular forms.

A different approach was used by Dabholkar-Murthy-Zagier **[158]** in the case of poles of order 1 and 2 and by two of the authors **[86]** and Olivetto **[403]** in the case of higher order poles. The advantage to this approach, as opposed to the one of Zwegers, is that the forms decompose canonically into two pieces, one of which is constructed directly from the poles, and the other of which is a so-called mock Jacobi form. We explain this in the remainder of this section and provide a specific application to representation theory in Chapter 20.

It turns out that if we have poles of order higher than 2, we require new automorphic objects. To describe this, we start by recalling *almost holomorphic modular forms*, which, as originally defined by Kaneko-Zagier **[298]**, transform like ordinary modular forms, but are polynomials in $1/v$ ($v = \mathrm{Im}(\tau)$) with holomorphic coefficients (cf. Definition 2.35). Here we use a slightly modified definition, allowing weakly holomorphic coefficients. Standard examples of almost holomorphic modular forms include derivatives of weakly holomorphic modular forms, as well as the nonholomorphic Eisenstein series E_2^*, discussed in Section 6.1. In general, the holomorphic part of an almost holomorphic modular form is called a *quasimodular form* (see the proof of Corollary 2.36 for a discussion of another characterization of quasimodular forms, as well as for another connection between quasimodular forms and Jacobi forms).

We are now ready to describe the new automorphic objects mentioned previously. The definitions and results given in the remainder of this section may be found in **[86, 158, 403]**. To begin, we recall the classical Maass raising operator of weight κ introduced in Chapter 5,

$$R_\kappa = 2i\frac{\partial}{\partial\tau} + \frac{\kappa}{v}. \tag{11.2}$$

If F is of weight κ and satisfies $\Delta_\kappa(F) = sF$ for some $s \in \mathbb{C}$, then we recall from Lemma 5.2 that $R_\kappa(F)$ is of weight $\kappa + 2$ and that

$$\Delta_{\kappa+2}(R_\kappa(F)) = (s + \kappa)R_\kappa(F).$$

Moreover, we have for positive integers n the iteration

$$R_\kappa^n = R_{\kappa+2(n-1)} \circ \cdots \circ R_{\kappa+2} \circ R_\kappa.$$

Abusing notation, we furthermore set for $F\colon \mathbb{H} \to \mathbb{C}$

$$R^{-1}(F) = 1.$$

If F is a modular form of weight κ, then it is obvious from the above facts that $R_\kappa^n(F)$ is an almost holomorphic modular form of weight $\kappa + 2n$.

DEFINITION 11.1. An **almost harmonic Maass form of weight** $\kappa \in \frac{1}{2}\mathbb{Z}$ **for a subgroup** $\Gamma \subseteq \mathrm{SL}_2(\mathbb{Z})$ **with Nebentypus character** χ is a smooth function $f\colon \mathbb{H} \to \mathbb{C}$ satisfying

i) For all $\gamma = \left(\begin{smallmatrix} a & b \\ c & d \end{smallmatrix}\right) \in \Gamma$ and all $\tau \in \mathbb{H}$, we have
$$f\big|_\kappa \gamma(\tau) = \chi(d) f(\tau).$$

ii) The function f may be written as finite linear combination (indexed by ℓ) of objects of the form
$$\sum_{j=0}^{r_\ell} g_{j,\ell} R_{\kappa+2-n_\ell}^{j-1}(g_\ell),$$
where the g_ℓ are harmonic Maass forms of weight $\kappa+2-n_\ell$ ($n_\ell \in \frac{1}{2}\mathbb{Z}$ fixed) and the $g_{j,\ell}$ are almost holomorphic modular forms of weight $n_\ell - 2j$. The **depth** of f is defined as the maximal j occurring.

In particular, harmonic Maass forms are almost harmonic Maass forms of depth 1 and almost holomorphic modular forms may be viewed as almost harmonic Maass forms of depth 0 or 1. The holomorphic parts of almost harmonic Maass forms are called *almost mock modular forms*.

We further require certain higher index Jacobi forms, but first recall the following "mock Jacobi forms" of **[529]**, defined for $z, w \in \mathbb{C}$ with $z \neq w$ and $M \in \mathbb{N}$ by

$$f_{M,w}(z;\tau) := \sum_{n\in\mathbb{Z}} \frac{e^{4\pi i M n z} q^{Mn^2}}{1 - e^{2\pi i(z-w)} q^n}. \tag{11.3}$$

The forms $f_{M,w}$ are completed in **[529]** as follows:

$$\widehat{f}_{M,w}(z;\tau) := f_{M,w}(z;\tau) - \frac{1}{2} \sum_{\ell \pmod{2M}} R_{M,\ell}(w;\tau) \vartheta_{M,\ell}(z;\tau), \tag{11.4}$$

where

$$\begin{aligned} &R_{M,\ell}(w;\tau) \\ := &\sum_{\substack{\lambda\in\mathbb{Z} \\ \lambda\equiv\ell \pmod{2M}}} \left\{ \mathrm{sgn}\left(\lambda + \frac{1}{2}\right) - E\left(\left(\lambda + \frac{2M\mathrm{Im}(w)}{v}\right)\sqrt{\frac{v}{M}}\right)\right\} e^{-\frac{\pi i \lambda^2 \tau}{2M} - 2\pi i \lambda w}. \end{aligned} \tag{11.5}$$

The function E is as in Definition 8.8, and the functions $\vartheta_{M,\ell}$ are given in (2.23). The completed form $\widehat{f}_{M,w}$ transforms like a 2-variable Jacobi form of weight 1 and matrix index $\left(\begin{smallmatrix} 2M & 0 \\ 0 & -2M \end{smallmatrix}\right)$, as stated in the following lemma **[529]**.

LEMMA 11.2 (Zwegers). *With notation and hypotheses as above, for all* $\gamma = \left(\begin{smallmatrix} a & b \\ c & d \end{smallmatrix}\right) \in \mathrm{SL}_2(\mathbb{Z})$ *and* $\lambda, \mu \in \mathbb{Z}$, *we have that*
$$\widehat{f}_{M,\frac{w}{c\tau+d}}\left(\frac{z}{c\tau+d}; \frac{a\tau+b}{c\tau+d}\right) = (c\tau+d) e^{\frac{2\pi i M c\left(z^2-w^2\right)}{c\tau+d}} \widehat{f}_{M,w}(z;\tau),$$
$$\widehat{f}_{M,w}(z+\lambda\tau+\mu;\tau) = e^{-2\pi i M\left(\lambda^2\tau + 2\lambda z\right)} \widehat{f}_{M,w}(z;\tau),$$
$$\widehat{f}_{M,w+\lambda\tau+\mu}(z;\tau) = e^{2\pi i M\left(\lambda^2\tau+2\lambda w\right)} \widehat{f}_{M,w}(z;\tau).$$
Moreover, the function $z \mapsto f_w(z;\tau)$ *is meromorphic with simple poles in* $w + \mathbb{Z} + \mathbb{Z}\tau$ *of residue* $-\frac{1}{2\pi i}$ *in* $z = w$ *and no other poles.*

REMARK. Restricting w and z in $\widehat{f}_{M,w}(z;\tau)$ to torsion points gives linear combinations of modular forms multiplied by harmonic Maass forms (directly from definition (11.4)).

We now return to the question of determining a theta-type decomposition for meromorphic Jacobi forms of positive index. To mimic the analogous construction for holomorphic Jacobi forms, we consider a function ϕ on $\mathbb{C}\times\mathbb{H}$ that satisfies the elliptic transformation law of Definition 2.11 ii) (for $\lambda,\mu\in\mathbb{Z}$):

$$\phi(z+\lambda\tau+\mu)=e\left(-m\left(\lambda^2\tau+2\lambda z\right)\right)\phi(z). \tag{11.6}$$

As for holomorphic Jacobi forms, we define functions which look like coefficients in the theta decomposition of Theorem 2.23, namely

$$h_{\ell,z_0}(\tau):=q^{-\frac{\ell^2}{4m}}\int_{z_0}^{z_0+1}\phi(z;\tau)e(-\ell z)dz, \tag{11.7}$$

where $\ell\in\mathbb{Z}$ and $z_0\in\mathbb{C}$. Here the path of integration is the straight line connecting z_0 and z_0+1 if there are no poles on the path of integration. Otherwise, if there is a pole which is not an endpoint, we define the path to be the average of the paths deformed to pass above or below the pole. If there is a pole at an endpoint, we first remark that the integral defined in (11.7) depends only on the height (i.e., y_0) of the path of integration, and not on the initial point of the line. In this case, therefore, we replace the path $[z_0,z_0+1]$ in (11.7) with the path $[z_0-\delta,z_0+1-\delta]$ for some real $\delta>0$ chosen sufficiently small to ensure the endpoints $z_0-\delta$ and $z_0+1-\delta$ are not poles of the integrand. Note that in the holomorphic case (11.7) is independent of the choice of z_0, the path connecting z_0 and z_0+1, and the coefficients only depend on $\ell \pmod{2m}$. For general meromorphic forms, using Definition 2.11 ii), we have that

$$h_{\ell,z_0+\tau}(\tau)=h_{\ell+2m,z_0}(\tau).$$

In particular, if we define the *canonical Fourier coefficients*

$$h_\ell(\tau):=h_{\ell,-\frac{\ell\tau}{2m}}(\tau),$$

then we have that

$$h_{\ell+2m}(\tau)=h_\ell(\tau). \tag{11.8}$$

We now use these coefficients to mimic the theta-decomposition of Theorem 2.23. In contrast to the holomorphic setting, this piece, which we call the *finite part*, does not represent the whole function and an additional *polar part* is required. To describe this in more detail, set

$$\phi^F(z;\tau):=\sum_{\ell\in\mathbb{Z}}h_\ell(\tau)q^{\frac{\ell^2}{4m}}\zeta^\ell=\sum_{\ell \pmod{2m}}h_\ell(\tau)\vartheta_{m,\ell}(z;\tau), \tag{11.9}$$

where we used (11.8). We then define the *polar part* of ϕ as

$$\phi^P:=\phi-\phi^F.$$

As we see below in (11.12), ϕ^P only depends on the principal part of ϕ at its singularities.

We now suppose that ϕ is a meromorphic Jacobi form, i.e., that it also satisfies the Jacobi modular transformation law. The goal now is to determine modular properties of the functions h_ℓ – which, interestingly, are defined only using a function ϕ satisfying an elliptic transformation as in (11.6). For this, assume that

$z_s = \alpha\tau + \beta$ ($s = (\alpha, \beta) \in \mathbb{Q}^2$) is a pole of order n_s of ϕ ($n_s \in \mathbb{N}$). Then we define the Laurent coefficients $\widetilde{D}_{z_s,j}$ as

$$\phi(z;\tau) =: \sum_{j=1}^{n_s} \frac{\widetilde{D}_{z_s,j}(\tau)}{(2\pi i(z-z_s))^j} + O(1). \tag{11.10}$$

It turns out to be more convenient to rather consider renormalized Laurent coefficients, namely, setting $\varepsilon := z - z_s$,

$$e^{\frac{\pi m\varepsilon^2}{v} + 2\pi i m\alpha(2\varepsilon+\beta)} q^{m\alpha^2} \phi(z;\tau) =: \sum_{j=1}^{n_s} \frac{D_{z_s,j}(\tau)}{(2\pi i\varepsilon)^j} + O(1). \tag{11.11}$$

A short calculation shows the following properties of the $D_{z_s,j}$'s, which will be used to establish the modularity properties of the functions h_ℓ.

LEMMA 11.3. *The functions $D_{z_s,j}$ are almost holomorphic modular forms of weight $k-j$.*

With notation and hypotheses as above, we define

$$\xi_{m,\ell,(\alpha,\beta)}(z;\tau) := \sum_{\substack{r\in\mathbb{Z} \\ r\equiv \ell \pmod{2m}}} \frac{\operatorname{sgn}\left(r+\frac{1}{2}\right) - \operatorname{sgn}(r+2m\alpha)}{2} q^{-\frac{r^2}{4m}} \zeta^{-r}.$$

Note that the right-hand side is independent of β. Let $S_0(\tau)$ be a set of representatives of poles of $z \mapsto \phi(z;\tau)$ modulo $\mathbb{Z} + \mathbb{Z}\tau$.

We next turn to completing the mock-type objects. First define the completions $\widehat{h}_\ell$ of h_ℓ by

$$\widehat{h}_\ell(\tau) := h_\ell(\tau) - \sum_{z_s\in S_0(\tau)} \sum_{j=1}^{n_s} \frac{D_{z_s,j}(\tau)}{(j-1)!}$$
$$\times D_\varepsilon^{j-1}\left[\left(\frac{1}{2}R_{m,\ell}(z_s+\varepsilon;\tau) - \xi_{m,\ell,s}(z_s+\varepsilon;\tau)\right) e^{-\frac{\pi m\varepsilon^2}{v} - 2\pi i m\alpha(2\varepsilon+\beta)} q^{-m\alpha^2}\right]_{\varepsilon=0}.$$

We have the following modularity properties of the functions $\widehat{h}_\ell$.

THEOREM 11.4. *The functions $\widehat{h}_\ell$ are almost harmonic Maass forms of weight $k-1/2$ and depth* $\max_{z_s\in S_0(\tau)}\{\lfloor (n_s-1)/2\rfloor\}$.

SKETCH OF PROOF. We first define completions of the polar and the finite part by

$$\widehat{\phi}^F(z;\tau) := \sum_{\ell \pmod{2m}} \widehat{h}_\ell(\tau)\vartheta_{m,\ell}(z;\tau)$$

and

$$\widehat{\phi}^P(z;\tau) := \phi^P(z;\tau) + \sum_{z_s\in S_0(\tau)} \sum_{j=1}^{n_s} \frac{D_{z_s,j}(\tau)}{(j-1)!}$$
$$\times \sum_{\ell \pmod{2m}} D_\varepsilon^{j-1}\left[\left(\frac{1}{2}R_{m,\ell}(z_s+\varepsilon;\tau) - \xi_{m,\ell,s}(z_s+\varepsilon;\tau)\right)\right.$$
$$\left.\times\, e^{-\frac{\pi m\varepsilon^2}{v} - 2\pi i m\alpha(2\varepsilon+\beta)} q^{-m\alpha^2}\right]_{\varepsilon=0} \vartheta_{m,\ell}(z;\tau).$$

We next rewrite the polar part. Using the Residue Theorem, a direct calculation gives

(11.12)

$$\phi^P(z;\tau) = - \sum_{z_s\in S_0(\tau)} \sum_{j=1}^{n_s} \frac{\widetilde{D}_{z_s,j}}{(j-1)!} \times D_\varepsilon^{j-1}\left[f_{m,z_s+\varepsilon}(z;\tau) - \sum_{\ell \pmod{2m}} \xi_{m,\ell,s}\left(z_s+\varepsilon;\tau\right)\vartheta_{m,\ell}(z;\tau)\right]_{\varepsilon=0},$$

with $f_{M,w}$ defined in (11.3). Another direct calculation then shows that we may rewrite

$$\widehat{\phi}^P(z;\tau) = \sum_{z_s\in S_0(\tau)} \sum_{j=1}^{n_s} \frac{D_{z_s,j}(\tau)}{(j-1)!} D_\varepsilon^{j-1}\left[\widehat{f}_{z_s+\varepsilon,m}(z;\tau) e^{-\frac{\pi m\varepsilon^2}{v}-2\pi i m\alpha(2\varepsilon+\beta)} q^{-m\alpha^2}\right]_{\varepsilon=0}.$$

By Lemma 11.2, it thus follows that $\widehat{\phi}^F$ and $\widehat{\phi}^P$ transform like Jacobi forms. The modularity for $(\widehat{h}_\ell)_{\ell \pmod{2m}}$ can then be concluded as in the holomorphic case (cf. Theorem 2.23). Thus we are left to show that $\widehat{h}_\ell$ can be written as the approriate linear combination of forms as given in Definition 11.1. A key step here is the identity

$$D_\varepsilon^{2j}\left[\left(\frac{1}{2}R_{m,\ell}\left(z_s+\varepsilon;\tau\right) - \xi_{m,\ell,s}\left(z_s+\varepsilon;\tau\right)\right) e^{-\frac{\pi m\varepsilon^2}{v}-2\pi i m\alpha\beta-4\pi i m\alpha\varepsilon} q^{-m\alpha^2}\right]_{\varepsilon=0}$$
$$= \left(\frac{m}{\pi}\right)^j R_{\frac{1}{2}}^j\left(e^{-2\pi i m\alpha\beta} q^{-m\alpha^2}\left(\frac{1}{2}R_{m,\ell}\left(z_s;\tau\right) - \xi_{m,\ell,s}\left(z_s;\tau\right)\right)\right)$$

(and a similar identity for an odd number of derivatives). The rest then follows by a lengthy calculation. □

REMARK. An interesting question concerns the structure of Fourier coefficients of multivariable meromorphic Jacobi forms. Such an example occurs in the context of vector bundles on complex (ruled) surfaces in work of Manschot **[376]** (see also **[229]** and **[389]**).

11.3. Negative index Jacobi forms

In this section we describe the situation of negative index Jacobi forms, following **[117]** (see also **[81]**). We give an explicit example in Subsection 20.2.3 below. Here, in contrast to the case of positive index, only a polar part occurs. We are mainly interested in the explicit elliptic transformation property and thus consider meromorphic functions which transform in the complex z-variable as (for $\varepsilon \in \{0,1\}$)

$$\phi\left(z+\lambda\tau+\mu\right) := (-1)^{2m\mu+\lambda\varepsilon} e^{-2\pi i m\left(\lambda^2\tau+2\lambda z\right)}\phi(z) \tag{11.13}$$

for all $\lambda, \mu \in \mathbb{Z}$, where $m \in \frac{1}{2}\mathbb{Z}$. These modified transformations, when compared to Definition 2.11 ii), which have half-integral indices and multipliers, also frequently occur in applications. Here, and in what follows, we may suppress dependence on the variable τ, and let $\phi(z) = \phi(z;\tau)$.

To describe the results, we require some notation. We use slightly different notation than in the previous subsection, since this slightly improves the required formulas. We consider the level $2M$ Appell sum given for $M \in \frac{1}{2}\mathbb{N}$ by

$$F_{M,\varepsilon}(\rho_1,\rho_2) = F_{M,\varepsilon}(\rho_1,\rho_2;\tau) := \left(\zeta_1\zeta_2^{-1}\right)^M \sum_{n\in\mathbb{Z}} \frac{(-1)^{n\varepsilon}\zeta_2^{-2Mn}q^{Mn(n+1)}}{1-\zeta_1\zeta_2^{-1}q^n}, \tag{11.14}$$

where $\zeta_j := e(\rho_j), j \in \{1,2\}$. So we have that

$$F_{M,\varepsilon}(\rho_1,\rho_2) = \left(\zeta_1\zeta_2^{-1}\right)^M f_{M,-\rho_1+\frac{\varepsilon}{4M}+\frac{\tau}{2}}\left(\frac{\varepsilon}{4M}-\rho_2+\frac{\tau}{2};\tau\right).$$

As in (11.10), let $\widetilde{D}_{w,j}$ be the j-th Laurent coefficient of ϕ around $z = w$. Recalling Definition 1.3 and with $\Lambda_\tau = \mathbb{Z}+\mathbb{Z}\tau$, for $z_0 \in \mathbb{C}$ we let

$$P_{z_0} = P_{\Lambda,z_0} = z_0 + [0,1) + \tau\,[0,1)$$

be a fundamental parallelogram for the lattice Λ_τ, and we denote by $S_{z_0}(\tau)$ the complete set of poles of ϕ in P_{z_0}. In the following theorem, although the dependence on τ is suppressed, both sides of the decomposition given there depend on τ.

THEOREM 11.5. *Let $m \in -\frac{1}{2}\mathbb{N}$ and $\varepsilon \in \{0,1\}$, and suppose that z_0 is chosen so that ϕ has no poles on the boundary of P_{z_0}. If ϕ is a meromorphic function satisfying* (11.13), *then*

$$\phi(z) = -\sum_{z_s\in S_{z_0}(\tau)}\sum_{j=1}^{n_s}\frac{\widetilde{D}_{z_s,j}}{(j-1)!}D_\varepsilon^{j-1}\left[F_{-m,\varepsilon}(z,z_s+\varepsilon)\right]_{\varepsilon=0}.$$

REMARK. As ϕ is a meromorphic function, there are only finitely many non-zero terms in the sum over n in the right-hand side of the expansion in Theorem 11.5.

We begin by giving the key properties of the functions $F_{M,\varepsilon}$ needed for the proof of Theorem 11.5, which follow from direct calculations generalizing Lemma 11.2.

LEMMA 11.6. *Let $M \in \frac{1}{2}\mathbb{N}$ and $\tau \in \mathbb{H}$. We have the elliptic transformation property*

$$F_{M,\varepsilon}(\rho_1,\rho_2+\lambda\tau+\mu) = (-1)^{2M\mu+\lambda\varepsilon}e^{-2\pi i M\left(\lambda^2\tau+2\lambda\rho_2\right)}F_{M,\varepsilon}(\rho_1,\rho_2),$$

for all $\lambda,\mu\in\mathbb{Z}$. Furthermore, $\rho_2 \mapsto F_{M,\varepsilon}(\rho_1,\rho_2)$ is meromorphic, with only simple poles in $\rho_1+\mathbb{Z}+\mathbb{Z}\tau$ and residue $\frac{1}{2\pi i}$ in $\rho_2=\rho_1$.

We can now prove Theorem 11.5.

PROOF OF THEOREM 11.5. Let $z \in \mathbb{C}$ be such that ϕ is holomorphic at z and that ϕ has no poles on the boundary of P_{z_0}. We consider the integral

$$\int_{\partial P_{z_0}} \phi(\rho)F_{-m,\varepsilon}(z,\rho)d\rho,$$

which we now compute in two different ways. On the one hand, we find from equation (11.13) and Lemma 11.6 that the integrand is both 1- and τ-periodic. Hence, we immediately see that the integral vanishes. On the other hand, we can use the Residue Theorem to give another evaluation of the integral: the poles of

$\rho \mapsto \phi(\rho)F_{-m,\varepsilon}(z,\rho)$ in P_{z_0} are the poles of ϕ in P_{z_0} together with z. In $\rho = z$ the function has residue $\frac{1}{2\pi i}\phi(z)$, so we have

$$\int_{\partial P_{z_0}} \phi(\rho)F_{-m,\varepsilon}(z,\rho)d\rho = \phi(z) + 2\pi i \sum_{z_s \in S_{z_0}(\tau)} \operatorname*{Res}_{\rho=z_s} \left[\phi(\rho)F_{-m,\varepsilon}(z,\rho)\right],$$

and thus, we obtain

$$\phi(z) = -2\pi i \sum_{z_s \in S_{z_0}(\tau)} \operatorname*{Res}_{\rho=z_s} \left[\phi(\rho)F_{-m,\varepsilon}(z,\rho)\right].$$

Since the function $\rho \mapsto \phi(\rho)F_{-m,\varepsilon}(z,\rho)$ is invariant under translation by a lattice point, so is $w \mapsto \mathrm{Res}_{\rho=w}[\phi(\rho)F_{-m,\varepsilon}(z,\rho)]$. Hence we can drop the condition that z_0 is such that $z \in P_{z_0}$ and take z_0 to be arbitrary (as long as there are no poles of ϕ on the boundary of P_{z_0}). The theorem then follows immediately by inserting the definition of the Laurent coefficients $\widetilde{D}_{w,n}$. □

We next turn to the Fourier coefficients of ϕ. We first require partial theta functions defined for $z \in \mathbb{C}$, $\tau \in \mathbb{H}$, $M \in \frac{1}{2}\mathbb{N}$, and $\ell \in M + \mathbb{Z}$ by

$$\theta^+_{\ell,\varepsilon,M}(z) = \theta^+_{\ell,\varepsilon,M}(z;\tau) := \sum_{n=0}^{\infty} (-1)^{n\varepsilon} q^{\frac{(2Mn-\ell)^2}{4M}} \zeta^{2Mn-\ell}. \tag{11.15}$$

Moreover, we let $h_{\ell,z_0}(\tau)$ be the ℓ-th Fourier coefficient of ϕ with respect to z_0 as defined in (11.7).

THEOREM 11.7. *Let $m \in -\frac{1}{2}\mathbb{N}$, $\tau \in \mathbb{H}$, and let ϕ be a meromorphic function satisfying* (11.13) *with $\varepsilon \in \{0,1\}$. If $z_0 \in \mathbb{C}$ is chosen so that ϕ has no poles on the boundary ∂P_{z_0}, then we have for any $\ell \in m + \mathbb{Z}$ that*

$$h_{\ell,z_0}(\tau) = \sum_{z_s \in S_{z_0}(\tau)} \sum_{j=1}^{n_s} \frac{\widetilde{D}_{z_s,j}(\tau)}{(j-1)!} D_\varepsilon^{j-1} \left[\theta^+_{\ell,\varepsilon,-m}(z_s + \varepsilon;\tau)\right]_{\varepsilon=0}.$$

REMARK. As Zagier has pointed out to the authors, it is curious that this expansion in terms of partial theta functions mirrors that of the polar part of meromorphic Jacobi forms of positive index, given the (still mysterious) relationship between mock theta functions and partial theta functions which is mentioned for example in Zwegers' thesis **[529]**. It would be extremely interesting to make sense of this strange "coincidence".

Before giving the proof of Theorem 11.7, we require the following properties of the partial theta functions under consideration, which follow from a direct calculation.

LEMMA 11.8. *For all $\lambda, \mu \in \mathbb{Z}$, $M \in \frac{1}{2}\mathbb{N}$ and $\ell \in M + \mathbb{Z}$, we have*

$$(-1)^{2\ell\mu} q^{M\lambda^2} \zeta^{2M\lambda} \theta^+_{\ell,\varepsilon,M}(z + \lambda\tau + \mu) = \theta^+_{\ell-2M\lambda,\varepsilon,M}(z).$$

Furthermore, we have that

$$\theta^+_{\ell,\varepsilon,M}(z) - (-1)^\varepsilon q^M \zeta^{2M} \theta^+_{\ell,\varepsilon,M}(z+\tau) = q^{\frac{\ell^2}{4M}} \zeta^{-\ell}.$$

PROOF OF THEOREM 11.7. By the Residue Theorem we have

$$\int_{\partial P_{z_0}} \phi(\rho)\theta^+_{\ell,\varepsilon,-m}(\rho)d\rho = 2\pi i \sum_{z_s \in S_{z_0}(\tau)} \operatorname*{Res}_{\rho=z_s} \left[\phi(\rho)\theta^+_{\ell,\varepsilon,-m}(\rho)\right].$$

On the other hand, we can compute the integral directly. Since $\phi\,\theta^+_{\ell,\varepsilon,-m}$ is 1-periodic (and using the fact that $\ell \in m+\mathbb{Z}$), we find, using (11.13) and both transformations formulas from Lemma 11.8, that

$$\begin{aligned}\int_{\partial P_{z_0}} \phi(\rho)\theta^+_{\ell,\varepsilon,-m}(\rho)d\rho &= \int_{z_0}^{z_0+1} \phi(\rho)\theta^+_{\ell,\varepsilon,-m}(\rho)d\rho - \int_{z_0+\tau}^{z_0+\tau+1} \phi(\rho)\theta^+_{\ell,\varepsilon,-m}(\rho)d\rho\\
&= \int_{z_0}^{z_0+1} \left(\phi(\rho)\theta^+_{\ell,\varepsilon,-m}(\rho) - \phi(\rho+\tau)\theta^+_{\ell,\varepsilon,-m}(\rho+\tau)\right) d\rho\\
&= \int_{z_0}^{z_0+1} \phi(\rho) \left(\theta^+_{\ell,\varepsilon,-m}(\rho) - (-1)^{\varepsilon} e^{-2\pi i m(\tau+2\rho)}\theta^+_{\ell,\varepsilon,-m}(\rho+\tau)\right) d\rho\\
&= e^{-\frac{\pi i \ell^2 \tau}{2m}} \int_{z_0}^{z_0+1} \phi(\rho)e^{-2\pi i \ell\rho}d\rho = h_{\ell,z_0}(\tau).\end{aligned}$$

Comparing the two evaluations of the integral implies that

$$h_{\ell,z_0}(\tau) = 2\pi i \sum_{z_s \in S_{z_0}(\tau)} \operatorname*{Res}_{\rho=w} \left[\phi(\rho)\theta^+_{\ell,\varepsilon,-m}(\rho)\right].$$

The result then follows directly by inserting the definition of the Laurent coefficients into the last formula. ◻

It is natural to ask in this case of negative index whether the forms h_{ℓ,z_0} exhibit modular properties, as they do in the case of meromorphic forms of positive index, as shown in the previous section. As explained in **[117]**, it is not difficult to deduce using Theorem 11.7 that the Fourier coefficients of a general negative index Jacobi form are expressible as derivatives of quantum modular forms multiplied by quasimodular forms. This is explained further in Chapter 20 and Chapter 21.

CHAPTER 12

Mock Modular Eichler-Shimura Theory

In Section 6.2, we constructed canonical harmonic Maass functions associated to elliptic curves $E/\mathbb{Q}$. This construction makes use of the Weierstrass zeta-function for the lattice Λ_E which realizes the complexification of E as a torus, i.e., the lattice such that $E \simeq_{\mathbb{C}} \mathbb{C}/\Lambda_E$. As discussed in Section 1.4, Λ_E is generated by the numbers

$$\mathcal{E}_E(\gamma\tau) - \mathcal{E}_E(\tau)$$

as one varies the matrices $\gamma \in \Gamma_0(N_E)$, where N_E is the conductor of E,

$$F_E(\tau) = \sum_{n=1}^{\infty} a_E(n) q^n \in S_2(\Gamma_0(N_E))$$

is the weight 2 newform associated to E via the Modularity Theorem, and $\mathcal{E}_E$ is the Eichler integral of F_E given by

$$\mathcal{E}_E(\tau) := \sum_{n=1}^{\infty} \frac{a_E(n)}{n} q^n.$$

This fact plays a critical role in the construction of the harmonic Maass function associated to E. As this case demonstrates, Eichler integrals have a deep connection to the arithmetic theory of modular forms, as well as the general structure of harmonic Maass forms. Further connections to arithmetic are given by Eichler-Shimura theory, which explains the relationship between periods of integer weight modular forms $f \in S_k$ and their Eichler integrals $\mathcal{E}_f$, the formal inverse images under the operator D^{k-1} of Bol's identity (cf. Lemma 5.3). Loosely speaking, the obstruction to modularity for such Eichler integrals are polynomials, the so-called *period polynomials*. The coefficients of these polynomials are the periods of f and are essentially critical L-values of f. As mentioned in Chapter 5, works by Knopp, Lehner, and others are closely related (for example, see **[312, 313, 315, 347]** and the expository article **[149]**).

In this chapter we explain the role of harmonic Maass forms and mock modular forms in the Eichler-Shimura theory for the full modular group $\mathrm{SL}_2(\mathbb{Z})$, which in particular gives alternative explanations for many classical facts. Throughout this chapter, we only consider modular forms of even weight, and so throughout k denotes a positive even integer.

12.1. Classical Eichler-Shimura theory

We begin by recalling the classical Eichler-Shimura isomorphism. We follow the treatment as presented by Paşol and Popa in Section 2 of **[412]**. This excellent summary is highly recommended to the reader who is interested in the details and proofs of the results presented below.

Suppose that Γ is a finite index subgroup of $\mathrm{SL}_2(\mathbb{Z})$ and that $k \geq 2$ is an even integer. Then we denote by V_{k-2} the vector space of complex polynomials of degree at most $k-2$. This vector space is equipped with the structure of a right Γ-module by defining the action using the Petersson slash-operator. That is, for any $P \in V_{k-2}$ and any $\gamma = \left(\begin{smallmatrix} a & b \\ c & d \end{smallmatrix}\right) \in \Gamma$ we have the action

$$P(X)|_{2-k}\gamma = (cX+d)^{k-2}P(\gamma X).$$

We require a natural extension of V_{k-2} to an $\mathrm{SL}_2(\mathbb{Z})$-module, which we denote by $\widetilde{V}_{k-2}^{\Gamma}$. As a module, this is simply the set of maps

$$P\colon \Gamma\backslash\mathrm{SL}_2(\mathbb{Z}) \to V_{k-2}$$

together with the $\mathrm{SL}_2(\mathbb{Z})$-action defined for $P \in \widetilde{V}_{k-2}^{\Gamma}$ and $\gamma \in \mathrm{SL}_2(\mathbb{Z})$ as the function which sends P to a function whose value on a coset A is given by

$$P|\gamma(A) = P\left(A\gamma^{-1}\right)|_{2-k}\gamma.$$

We note in passing that each element P of $\widetilde{V}_{k-2}^{\Gamma}$ can be thought of as a collection of $[\mathrm{SL}_2(\mathbb{Z})\colon \Gamma]$ many complex polynomials. We then require the following simple submodule:

$$V_{k-2}^{\Gamma} := \left\{P \in \widetilde{V}_{k-2}^{\Gamma}\colon P(A) = P(-A)\right\}.$$

REMARK. This submodule is natural from a cohomological point of view as the polynomial $\frac{1}{2}(P(A)+P(-A))$ lies in the same "cohomology class" as P.

The content of Eichler and Shimura's theory is the construction of an explicit isomorphism between a (quotient of a) subspace of V_{k-2}^{Γ} satisfying a special set of relations and the space of cusp forms $S_k(\Gamma)$. We first describe how cusp forms on Γ naturally yield special elements of V_{k-2}^{Γ}. Supposing that $f \in S_k(\Gamma)$, we obtain the *period polynomial* of f by defining for any $A \in \Gamma\backslash\mathrm{SL}_2(\mathbb{Z})$ the polynomial

$$r(f;X)(A) := \int_0^{i\infty} f|_k A(w)(X-w)^{k-2}dw.$$

One can check that this definition is independent of the choice of coset representative, using the Γ-modularity of f. This polynomial automatically satisfies many simple relations, as can be seen using the explicit integral representation and changes of variables. Moreover, such relations actually "cut out" the space of cusp forms. To describe these, we first let J, S, T, and U be the matrices

$$J := \begin{pmatrix} -1 & 0 \\ 0 & -1 \end{pmatrix}, \quad S := \begin{pmatrix} 0 & -1 \\ 1 & 0 \end{pmatrix}, \quad T := \begin{pmatrix} 1 & 1 \\ 0 & 1 \end{pmatrix}, \quad U := \begin{pmatrix} 1 & -1 \\ 1 & 0 \end{pmatrix}.$$

The space of *period polynomials* for Γ is then defined as the subspace of elements of V_{k-2}^{Γ} which satisfy the so-called *period relations.* That is, we define

$$W_{k-2}^{\Gamma} := \left\{P \in V_{k-2}^{\Gamma}\colon P + P|_{2-k}S = P + P|_{2-k}U + P|_{2-k}U^2 = P - P|_{2-k}J = 0\right\}.$$

The Eichler-Shimura isomorphism is naturally interpreted as a cohomological statement, and the period polynomials, as we have defined them, play the role of cocycles in this theory (more precisely, they are specializations of cocycles associated to modular forms which uniquely determine the cohomology class of these cycles). The corresponding coboundaries are defined by the space

$$C_{k-2}^{\Gamma} := \left\{P - P|_{2-k}S\colon P \in V_{k-2}^{\Gamma}, \quad P|_{2-k}T = P\right\}.$$

REMARK. In relation to an alternate formulation of the Eichler-Shimura Theorem which we see in Theorem 12.2 below, in the case of $\Gamma = \mathrm{SL}_2(\mathbb{Z})$, we clearly have that $V_{k-2}^{\Gamma} = \widetilde{V}_{k-2}^{\Gamma}$ and that only constant values are fixed by the slash-action of T. Hence, in this case, we find that

$$C_{k-2}^{\mathrm{SL}_2(\mathbb{Z})} = \{c\left(X^{k-2} - 1\right) : c \in \mathbb{C}\}.$$

More generally, it turns out that the dimension of C_{k-2}^{Γ} is the same as the dimension of the Eisenstein subspace of $M_k(\Gamma)$, which in this case is reflected by the fact that the even part of a suitably defined "period polynomial" of the weight k Eisenstein series E_k is precisely of this shape (cf. Section 4 of [**321**]).

We now have all of the ingredients to state the general Eichler-Shimura theorem. From now on, we assume that Γ is normalized by $\varepsilon := \left(\begin{smallmatrix} -1 & 0 \\ 0 & 1 \end{smallmatrix}\right)$. This matrix acts on elements P of V_{k-2}^{Γ} by

$$P|_{2-k}\varepsilon(A) = P\left(\varepsilon A \varepsilon\right)|_{2-k}\varepsilon.$$

Essentially, this operation controls the action of complex conjugation on the Fourier coefficients of our modular form, as one easily checks that

$$\overline{r\left(f^c; X\right)} = -r(f;X)|_{2-k}\varepsilon,$$

where the action of complex conjugation on an element P of V_{k-2}^{Γ} is naturally interpreted as conjugating the coefficients of each polynomial corresponding to P and where f^c is the modular form obtained by taking the complex conjugate of the coefficients of f (cf. Lemma 5.17). This action splits V_{k-2}^{Γ} into ± 1-eigenspaces, which we denote by $\left(V_{k-2}^{\Gamma}\right)^{\pm}$. Of course, an element $P \in V_{k-2}^{\Gamma}$ may be split into its $+1$ and -1 pieces P^+ and P^- by setting

$$P^{\pm} := \frac{1}{2}\left(P \pm P|\varepsilon\right) \in \left(V_{k-2}^{\Gamma}\right)^{\pm}.$$

Since $P(I)^+$ (resp. $P(I)^-$) is an even (resp. odd) polynomial, we refer to P^+ (resp. P^-) as the *even part* of P (resp. *odd part* of P). It follows from a straightforward calculation that W_{k-2}^{Γ} is stable under the action of ε, which implies that W_{k-2}^{Γ} also decomposes into eigenspaces $\left(W_{k-2}^{\Gamma}\right)^{\pm}$, and a similar statement holds for C_{k-2}^{Γ}. The main theorem we have been seeking may be elegantly stated as follows.

THEOREM 12.1 (Eichler-Shimura isomorphism). *The two maps*

$$r^{\pm} \colon S_k(\Gamma) \to \left(W_{k-2}^{\Gamma}\right)^{\pm}$$

defined by sending $f \mapsto r(f;\tau)^{\pm}$ give rise to isomorphisms

$$r^{\pm} \colon S_k(\Gamma) \to \left(W_{k-2}^{\Gamma}\right)^{\pm} / \left(C_{k-2}^{\Gamma}\right)^{\pm}.$$

Thus, we see that the space of cusp forms may be defined very explicitly via a finite set of relations on spaces of polynomials of bounded degree. This connection, besides being elegant, is very useful. In particular, Eichler-Shimura theory is in a sense "dual" to the theory of modular symbols which is used to compute modular forms in MAGMA and SAGE. One can also compute spaces of modular forms using period polynomials, as explained in Section 5.5 of [**412**] (for a direct derivation of the standard algorithms for computing modular forms on congruence subgroups using the Eichler-Shimura isomorphism, see also [**472**] and [**495**]). These explicit realizations are further helped along by the important facts that the Eichler-Shimura period maps commute with natural Hecke operators defined on period polynomials, so that the isomorphism works on a deeper level than simply an isomorphism of abstract vector spaces (which really only says that they have equal dimensions). In particular, they provide isomorphisms of Hecke modules. Finally, we note that the proof of Theorem 12.1 follows from just such a dimension count (which itself is proven using the classical Eichler-Shimura isomorphism given in e.g., Chapter 8 of [**455**]) combined with a generalized Haberland's formula. This formula was proven in Theorem 3.3 of [**412**], and it expresses the Petersson inner product of two cusp forms in terms of the coefficients of their period polynomials. We also note that such Haberland-type identities are very analogous (and indeed proven in a similar manner, namely, using Stokes' theorem) to the identities for the Bruinier-Funke pairing between harmonic Maass forms and classical cusp forms described in Section 5.2.

For the remainder of this section, we further expound on the simplest case when $\Gamma = \mathrm{SL}_2(\mathbb{Z})$, and the exposition which follows is largely inspired by the beautiful paper of Kohnen and Zagier [**321**]. Hence, we assume throughout that $\Gamma = \mathrm{SL}_2(\mathbb{Z})$ and all superscripts above which involve the congruence subgroup Γ will now be suppressed. For example, elements of $\widetilde{V}_{k-2}^{\Gamma}$ are now determined by their values at the identity matrix I, so we may think of them simply as ordinary polynomials. These assumptions being made, we now describe the importance of these polynomials in the theory of periods and special L-values. To be more precise, define for each integer $0 \leq n \leq k-2$ the n-th *period* of an integer weight cusp form f by

$$r_n(f) := \int_0^\infty f(it)t^n dt. \tag{12.1}$$

Then the generating function

$$r(f;X) := \sum_{n=0}^{k-2} i^{-n+1}\binom{k-2}{n} r_n(f)X^{k-2-n} = \int_0^{i\infty} f(w)(X-w)^{k-2}dw \tag{12.2}$$

is an element of W_{k-2} (cf. [**321**]). Decomposing $r(f;X) = r^-(f;X) + ir^+(f;X)$, we explicitly find that

$$r^+(f;X) = \sum_{\substack{0\leq n\leq k-2\\ n \text{ even}}} (-1)^{\frac{n}{2}}\binom{k-2}{n} r_n(f)X^{k-2-n}, \tag{12.3}$$

$$r^-(f;X) = \sum_{\substack{0\leq n\leq k-2\\ n \text{ odd}}} (-1)^{\frac{n-1}{2}}\binom{k-2}{n} r_n(f)X^{k-2-n}. \tag{12.4}$$

The Eichler-Shimura isomorphism above may then be restated in this case as follows (cf. [**321**] and the references therein).

THEOREM 12.2 (Eichler-Shimura isomorphism for $\mathrm{SL}_2(\mathbb{Z})$)). *Let k be an even positive integer. Then for each $P \in W_{k-2}$ there exists a unique pair $(f,g) \in S_k \times S_k$ and $c \in \mathbb{C}$ such that*

$$P(X) = r(f;X) + r(g;-X) + c\left(X^{k-2} - 1\right).$$

REMARK. Usually, the second term is written as $\overline{r(g;\overline{X})}$, which is the polynomial obtained by replacing each coefficient of the polynomial $r(g;X)$ with its conjugate. However, this may be rewritten as

$$\overline{r(g;\overline{X})} = -r\left(g^c;-X\right).$$

REMARK. Theorem 12.2 may also be stated as saying that r^- is an isomorphism onto W_{k-2}^- and that r^+ is an isomorphism onto a codimension one subspace $W_{k-2,0}$ of W_{k-2}^+ not containing $X^{k-2}-1$. This implies that the even periods of f satisfy an "extra relation", which was explicitly written down in terms of Bernoulli numbers in Section 4 of [**321**]. A generalized version of this relation is explored in the context of weakly holomorphic cusp forms below.

The period polynomials encode the L-values of cusp forms inside the critical strip. More specifically, for $f(\tau) = \sum_{m=1}^{\infty} c_f(m)q^m \in S_k$, one defines for $\mathrm{Re}(s) \gg 0$ the *Hecke L-series* associated to f by

$$L(f,s) := \sum_{m=1}^{\infty} \frac{c_f(m)}{m^s}. \tag{12.5}$$

It is a classical fact going back to Hecke that $L(f,s)$ satisfies a functional equation under $s \leftrightarrow k-s$ and has an analytic continuation to the complex plane (for example, see Theorem 19.6 iii) in the case of newforms associated to elliptic curves $E/\mathbb{Q}$). The *special values* of $L(f,s)$ are those values at integral arguments. In particular, the *critical values*, those special values of $L(f,s)$ at integers within the critical strip $0 < \mathrm{Re}(s) < k$, may be written as periods of f as follows:

$$L(f,n) = \frac{(2\pi)^n}{(n-1)!} r_{n-1}(f). \tag{12.6}$$

To illustrate the above results, we conclude with a canonical example. It may easily be checked that for $k \in \{4,6,8,10,14\}$, the space W_{k-2}^- is empty, and that $W_{k-2}^+ = \mathbb{C}(X^{k-2}-1)$. This is consistent with Theorem 12.2, as $S_k = 0$ for these k. Of course, if $k=12$ there is a unique (up to multiplication by a constant) cusp form, namely Δ. In this case, W_{10}^+ is the $\mathbb{C}$-span of

$$\begin{aligned} p_0(X) &:= X^{10} - 1, \\ p_1(X) &:= X^8 - 3X^6 + 3X^4 - X^2. \end{aligned}$$

Furthermore, W_{10}^- is generated by

$$p_2(X) := 4X^9 - 25X^7 + 42X^5 - 25X^3 + 4X.$$

The periods of Δ may then be computed to be those values listed in the following table, where $\omega_\pm$ are certain real constants

$$\begin{aligned} \omega_+ &= 0.0214460667\ldots, \\ \omega_- &= 0.0000482774800\ldots. \end{aligned}$$

Moreover, it turns out that

$$\omega_+^{-1} r^+(\Delta; X) = -\frac{192}{691} p_0(X) + \frac{16}{3} p_1(X),$$
$$\omega_-^{-1} r^-(\Delta; X) = 192 p_2(X).$$

We conclude by noting that the fact that such constants $\omega_\pm$ exist which normalize the even and odd period polynomials to have rational coefficients is not a coincidence, and follows for general Hecke eigenforms as a corollary of Manin's proof that the maps in Theorem 12.2 commute with a certain Hecke operator acting on period polynomials **[374]**.

12.2. Period polynomials for weakly holomorphic modular forms

Our next goal is to generalize the classical picture sketched above to the case of weakly holomorphic modular forms. For simplicity, we consider only weakly holomorphic modular forms on $\mathrm{SL}_2(\mathbb{Z})$. We give two perspectives, one which utilizes canonical "regularization" procedures of certain integrals **[90]** and one which arises from viewing period polynomials as errors of modularity for harmonic Maass forms **[94]**. We begin with the first perspective. We first define a notion of regularization for divergent integrals via a special process of analytic continuation.

DEFINITION 12.3. Consider a continuous function $f \colon \mathbb{H} \to \mathbb{C}$ and assume that there is a constant $c \in \mathbb{R}^+$ such that $f(\tau) = O(e^{cv})$ uniformly in u as $v \to \infty$. Then, for each $w_0 \in \mathbb{H}$, the integral

$$\int_{w_0}^{i\infty} e^{izw} f(w)\, dw$$

(where the path of integration lies within a fixed vertical strip) is convergent for $z \in \mathbb{C}$ with $\mathrm{Re}(z) \gg 0$. If this integral has an analytic continuation to $z = 0$, then we define the **regularized integral** of f by

$$R.\int_{w_0}^{i\infty} f(w)\, dw := \left[\int_{w_0}^{i\infty} e^{izw} f(w)\, dw\right]_{z=0},$$

where the right-hand side denotes the analytic continuation to $z = 0$ of the integral.

REMARK. It is not hard to see that this integral is independent of $w_0 \in \mathbb{H}$.

REMARK. Regularized integrals for cusps other than $i\infty$ may be defined similarly.

Using these regularized integrals, we may naturally define period integrals for weakly holomorphic modular forms by regularizing the analogous period integrals in (12.2).

DEFINITION 12.4. For $f \in S_k^!$, the space of weakly holomorphic modular forms of weight k with vanishing constant term, define the **regularized period integral**,

$$r(f; X) := R.\int_0^{i\infty} f(w)(X - w)^{k-2} dw.$$

n	0 or 10	1 or 9	2 or 8	3 or 7	4 or 6	5
$r_n(\Delta)$	$\frac{192}{691}\omega_+$	$\frac{384}{5}\omega_-$	$\frac{16}{135}\omega_+$	$40\omega_-$	$\frac{8}{105}\omega_+$	$40\omega_-$

REMARK. In **[513]**, for $f(\tau) = \sum_{m=0}^{\infty} c_f(m)q^m \in M_k$, Zagier defined the regularized period integral

$$r(f;X) := \int_{t_0}^{\infty} (f(t) - c_f(0))(t-X)^{k-2}dt + \int_0^{t_0} \left(f(t) - \frac{c_f(0)}{t^k}\right)(t-X)^{k-2}dt$$
$$- \frac{c_f(0)}{k-1}\left((X-t_0)^{k-1} + \frac{1}{X}\left(1 - \frac{X}{t_0}\right)^{k-1}\right),$$

where $t_0 \in \mathbb{R}^+$.

REMARK. Regularized periods for weight 0 forms were also considered in **[123]**.

We will shortly see that period polynomials for weakly holomorphic modular forms are closely related to L-values, as in the case for cusp forms. To realize this general connection, we must carefully define a notion of L-values for weakly holomorphic modular forms. For this, we define a regularized L-function as in **[90]**.

DEFINITION 12.5. For $f(\tau) = \sum_{m=m_0}^{\infty} c_f(m)q^m \in S_k^!$ and $t_0 \in \mathbb{R}^+$, define the **regularized L-series** associated to f by

$$L(f,s) := \frac{(2\pi)^s}{\Gamma(s)} L^*(f,s)$$

$$L^*(f,s) := \sum_{m \geq m_0} \frac{c_f(m)\Gamma(s, 2\pi m t_0)}{(2\pi m)^s} + i^k \sum_{m \geq m_0} \frac{c_f(m)\Gamma\left(k-s, \frac{2\pi m}{t_0}\right)}{(2\pi m)^{k-s}}.$$

REMARK. For cusp forms, the L-series $L^*(f,s)$ has also been expressed as in Definition 12.5 by Kohnen and Zagier **[321]** in order to accelerate convergence for numerical calculations.

REMARK. The defining series of $L^*(f,s)$ is absolutely convergent since $\Gamma(s,x) \sim x^{s-1}e^{-x}$ as $x \to \infty$, and the Circle Method (described in Section 14.1) then implies that

$$c_f(m) \ll m^{\frac{1}{2}\left(k-\frac{3}{2}\right)} e^{4\pi\sqrt{m_0 m}}. \tag{12.7}$$

REMARK. Using the modularity of f, it is not hard to see that $L_f(s)$ is independent of t_0.

REMARK. We note there is also a twisted version of these L-series (cf. equation (1.7) of **[90]**).

As one should hope for any satisfying theory of L-functions, these weakly holomorphic L-functions satisfy nice functional equations, and the connection between L-values and periods in the case of cusp forms naturally extends to this setting as well.

THEOREM 12.6. *For $f \in S_k^!$, we can also express $L^*(f,s)$ as a regularized Mellin transform:*

$$L^*(f,s) = R.\int_0^{\infty} f(it)t^{s-1}dt.$$

Furthermore, $L^*(f,s)$ satisfies the functional equation

$$L^*(f,k-s) = i^k L^*(f,s).$$

SKETCH OF PROOF. The proof of the integral representation follows by a direct calculation inserting the Fourier expansion of f and splitting the integral at t_0. The functional equation is obvious by construction of L. □

We next show that the period polynomials can also be viewed as the "errors of modularity" of Eichler integrals. Recall that for $f(\tau) = \sum_{m=m_0}^{\infty} c_f(m)q^m \in M_k^!$, the *Eichler integral* of f is defined by

$$\mathcal{E}_f(\tau) := \sum_{m\neq 0} \frac{c_f(m)}{m^{k-1}} q^m. \tag{12.8}$$

In analogy with the cusp form case, the failures of such Eichler integrals to transform as modular forms give weakly holomorphic period integrals.

THEOREM 12.7. *For any $f \in S_k^!$, we may express*

$$r(f;\tau) = c_k \left(\mathcal{E}_f(\tau) - \mathcal{E}_f(\tau)|_{2-k}S\right),$$

where $c_k := -\frac{(k-2)!}{(2\pi i)^{k-1}}$.

REMARK. It turns out that, up to the constant term, Eichler integrals are mock modular forms (whose harmonic completions have growth in the nonholomorphic part). This understanding yields a mock-picture of period polynomials. More generally, for any mock modular form F^+, $P_{F^+} := F^+|_{2-k}(1-S)$ is a period polynomial. This follows since P_{F^+} is holomorphic on $\mathbb{H}$, $D^{k-1}(P_{F^+}) = 0$, and the period relations follow by construction.

SKETCH OF PROOF OF THEOREM 12.7. Firstly we can prove, inserting the Fourier expansion of f, that

$$\mathcal{E}_f(\tau) = c_k^{-1} R. \int_\tau^{i\infty} f(w)(\tau - w)^{k-2} dw. \tag{12.9}$$

The claim then follows after a short calculation using the modularity of f. □

Continuing the analogy with the classical cusp form case, these regularized L-values are also essentially the coefficients of the period polynomial of a weakly holomorphic form. More generally, non-critical L-values are also encoded in modular-type objects; in this case, one requires sesqui-harmonic Maass forms, which are briefly discussed in Section 6.3.

THEOREM 12.8. *For any $f \in S_k^!$, we have the identity*

$$r(f;X) = \sum_{n=0}^{k-2} i^{n-1} \binom{k-2}{n} L^*(f, n+1) X^n.$$

SKETCH OF PROOF. The claim directly follows from Definition 12.5, using the binomial expansion and the regularized Mellin transform representation of $L^*(f,s)$ in Theorem 12.6. □

This theorem directly implies the following vanishing result for critical L-values.

COROLLARY 12.9. *If $f \in D^{k-1}(M_{2-k}^!)$, then, for $2 \leq n \leq k-2$, we have $L(f,n) = 0$. Moreover $L(f,1) = L(f,k-1) = 0$ if and only if $f \in D^{k-1}(S_{2-k}^!)$.*

PROOF. If $f = D^{k-1}(F)$ with $F \in M^!_{2-k}$, then

$$\mathcal{E}_f = F + c_F^+(0).$$

Using that $F \in M^!_{2-k}$ yields

$$\mathcal{E}_f(\tau)|_{2-k}(1-S) = c\left(1-\tau^{k-2}\right),$$

which directly gives the claim by Theorem 12.8. □

Fitting in the above results with the standard operators between harmonic Maass forms and classical modular forms, we obtain the following period relations between the shadow and the ghost (defined directly after (5.7)) of a mock modular form (we would like to point out here the analogous role to the polynomial $X^{k-2}-1$ below that the polynomial $X^{k-2}-1$ played in the Eichler-Shimura isomorphism in Theorem 12.2 and the extra relation among the even periods of cusp forms).

THEOREM 12.10. *If $\mathcal{F} \in H_{2-k}$, then we have that*

$$r\left(\xi_{2-k}(\mathcal{F}); X\right) \equiv -\frac{(4\pi)^{k-1}}{(k-2)!}\overline{r\left(D^{k-1}(\mathcal{F}); \overline{X}\right)} \pmod{X^{k-2}-1},$$

where equivalence modulo $X^{k-2}-1$ means that the difference of the two functions is a constant multiple of $X^{k-2}-1$. Moreover, there is a function $\widehat{\mathcal{F}} \in H_{2-k}$ for which $\xi_{2-k}(\widehat{\mathcal{F}}) = \xi_{2-k}(\mathcal{F})$ and

$$r\left(\xi_{2-k}(\mathcal{F}); X\right) = -\frac{(4\pi)^{k-1}}{(k-2)!}\overline{r\left(D^{k-1}\left(\widehat{\mathcal{F}}\right); \overline{X}\right)}.$$

REMARK. If $\mathcal{F} \in H^!_{2-k}$ has constant term zero, then we have

$$r(D^{k-1}(\mathcal{F}); X) = r(D^{k-1}(\mathcal{F}^+); X). \tag{12.10}$$

Moreover, note that there are always weakly holomorphic forms $F \in M^!_{2-k}$ for which $\mathcal{F} + F \in H^!_{2-k}$ has constant term zero.

REMARK. Most of the theorems in this chapter also hold for forms in $H^\sharp_{2-k}$ (defined in (4.9)) via the flipping operator defined in Section 5.3.

SKETCH OF PROOF OF THEOREM 12.10. Define

$$\mathcal{G}(\tau) := \mathcal{G}_1(\tau) + \mathcal{G}_2(\tau)$$

with

$$\mathcal{G}_1(\tau) := -(2i)^{1-k} R. \int_{-\overline{\tau}}^{i\infty} \xi_{2-k}(\mathcal{F})^c(w)(\tau+w)^{k-2} dw,$$

$$\mathcal{G}_2(\tau) := c_k^{-1} R. \int_{\tau}^{i\infty} D^{k-1}(\mathcal{F})(w)(\tau-w)^{k-2} dw.$$

By applying ξ_{2-k} and D^{k-1} one easily obtains that $\mathcal{G}$ and $\mathcal{F}$ are identical up to a constant term. Modularity of $\mathcal{F}$, as well as of $\xi_{2-k}(\mathcal{F})^c$ and $D^{k-1}(\mathcal{F})$, easily implies the claim using the representation (24.6). □

COROLLARY 12.11. *If $\mathcal{F} \in H_{2-k}$ and $0 < n < k-2$, then*

$$L(\xi_{2-k}(\mathcal{F}), n+1) = (-1)^n \frac{(4\pi)^{k-1}}{(k-2)!}\overline{L(D^{k-1}(\mathcal{F}), n+1)}.$$

PROOF. The corollary follows easily from the integral representation in Theorem 12.6 using the Binomial Theorem. □

Let us now mention one application of the mock Eichler-Shimura theory, namely a Haberland-type formula which expresses the Petersson inner product and a related antisymmetric inner product on $M_k^!$ in terms of periods. To describe it, we introduce an inner product on $M_k^!$ as follows (cf. [**121**]). For $f(\tau) = \sum_{m \gg -\infty} c_f(m)q^m$ and $g(\tau) = \sum_{m \gg -\infty} c_g(m)q^m$ in $M_k^!$, we define $\{f, g\}$ by the (finite) sum

$$\{f, g\} := \sum_{m \in \mathbb{Z} \setminus \{0\}} \frac{c_f(-m)c_g(m)}{m^{k-1}}. \tag{12.11}$$

REMARK. A similar pairing, defined by Bruinier and Funke, is considered in Section 5.2 (cf. (5.14)).

This pairing is antisymmetric, bilinear, and Hecke equivariant. Generalizing an argument of Kohnen and Zagier [**321**], we obtain the following closed formula for these products, which is an analogue of the classical Haberland identity [**247**].

THEOREM 12.12. *For $f, g \in M_k^!$ we have*

$$\{f, g\} = \frac{(2\pi)^{k-1}}{3(k-2)!} \sum_{\substack{0 \le m < n \le k-2 \\ m \not\equiv n \pmod 2}} i^{n+1+m} \binom{k-2}{n} \binom{n}{m} r_n(f) r_{k-2-m}(g)$$
$$+ \frac{2(2\pi)^{k-1}}{3(k-2)!} \sum_{\substack{0 \le n \le k-2 \\ n \equiv 0 \pmod 2}} i^{k-n} \binom{k-1}{n+1} \left(r_n(g)\frac{c_f(0)}{k-1} - r_n(f)\frac{c_g(0)}{k-1} \right).$$

We next turn to an Eichler-Shimura isomorphism theorem for weakly holomorphic forms. To state it, we first restrict our extended period function r (cf. Definition 12.4) to $S_k^!$, defining the maps

$$r\colon S_k^! \to W_{k-2} \qquad \text{and} \qquad \widetilde{r}\colon S_k^! \to W_{k-2,0},$$

where the second map is the composition of r and the projection from W_{k-2} to $W_{k-2,0}$ (cf. the second remark following Theorem 12.2). Furthermore, there are maps $r^{\pm}\colon S_k^! \to W_{k-2}^{\pm}$ which extend the classical even and odd period maps on S_k. The extended Eichler-Shimura isomorphisms for these two maps are then stated as follows (cf. [**94**]).

THEOREM 12.13. *The following sequences are exact:*

$$0 \longrightarrow D^{k-1}\left(M_{2-k}^!\right) \longrightarrow S_k^! \xrightarrow{\widetilde{r}} W_{k-2,0} \longrightarrow 0,$$
$$0 \longrightarrow D^{k-1}\left(S_{2-k}^!\right) \longrightarrow S_k^! \xrightarrow{r} W_{k-2} \longrightarrow 0.$$

Note that the first exact sequence from Theorem 12.13 tells us that

$$S_k^!/D^{k-1}\left(M_{2-k}^!\right) \cong W_{k-2,0}, \tag{12.12}$$

and the second exact sequence explains the role of the codimension one subspace $W_{k-2,0}$ in the classical setting (cf. the second remark following Theorem 12.2). We also note that such quotients echo the notion of weakly holomorphic Hecke eigenforms, which were discussed in Section 7.2. The presence of non-zero constant terms of weight $2-k$ weakly holomorphic modular forms gives

$$r\colon D^{k-1}\left(M_{2-k}^!\right)/D^{k-1}\left(S_{2-k}^!\right) \xrightarrow{\sim} W_{k-2}/W_{k-2,0} \cong \left\langle \tau^{k-2} - 1 \right\rangle.$$

Theorem 12.13 sheds further light on the classical Eichler-Shimura isomorphism, where the maps

$$r^{\pm}\colon S_k \longrightarrow W_{k-2,0} \tag{12.13}$$

each give one copy of S_k inside $W_{k-2,0}$ so that $W_{k-2,0} \cong S_k \oplus S_k$. Equations (12.12) and (12.13) tell us that $S_k^!/D^{k-1}(M_{2-k}^!) \cong S_k \oplus S_k$. We have that D^{k-1} only sees the holomorphic parts M^+ of harmonic Maass forms $M \in H_{2-k}$, and it is shown in **[94]** that the two copies of S_k arise from the quotient space $H_{2-k}/M_{2-k}^!$ and the inclusion of $S_k \subseteq S_k^!$. In particular, we find that

$$W_{k-2,0} \cong \widetilde{r}\left(D^{k-1}\left(H_{2-k}\right)\right) \oplus \widetilde{r}\left(\xi_{2-k}\left(H_{2-k}\right)\right).$$

12.3. Cycle integrals of weakly holomorphic modular forms

We conclude this chapter by sketching the extension of one more classical object connected with modular L-functions to the mock modular setting. Namely, we recall that *cycle integrals* are special integrals along geodesiscs of modular forms multiplied with polynomials, and they play a critical role in the explicit understanding of the Shintani lift between modular forms of integral and half-integral weight. This itself is the key to a deep connection between central L-values of twisted modular forms of integral weight and Fourier coefficients of modular forms of half-integral weight.

Before discussing a recent result pertaining to mock modular forms, we first recall a related result by Kohnen **[324]** which holds for modular forms. (See also Waldspurger's paper **[487]** as well as an explicit formulation due to Kohnen and Zagier **[320]**.) These results will be reformulated in Section 19.4.

Here, we discuss results for general level N (odd and square-free) as they are presented in **[324]**. To state them, let $S_\kappa(\Gamma_0(N))$ be the space of cusp forms on $\Gamma_0(N)$ with weight κ; if $\kappa \in 1/2 + \mathbb{Z}$, then we denote by $S_\kappa^+(\Gamma_0(4N))$ the space of cusp forms on $\Gamma_0(4N)$ whose Fourier coefficients are supported on positive integers n such that $(-1)^{\kappa-\frac{1}{2}}n \equiv 0, 1 \pmod 4$. If $f(\tau) = \sum_{n=1}^\infty c_f(n)q^n \in S_{2k}^{\text{new}}(\Gamma_0(N))$ is a normalized Hecke eigenform in the space of newforms of weight $2k$ ($k \in \mathbb{N}$) and level N, an earlier celebrated result of Kohnen in **[323]** shows that there is a corresponding form $g(\tau) = \sum_{n=1}^\infty c_g(n)q^n \in S_{k+\frac{1}{2}}^{\text{new}}(\Gamma_0(4N))$ of half-integral weight which arises from the "Shimura correspondence" (cf. Section 19.3) such that

$$c_g\left(n^2|\delta|\right) = c_g(|\delta|) \sum_{d|n, \gcd(d,N)=1} \mu(d) \left(\frac{\delta}{d}\right) d^{k-1} c_f\left(\tfrac{n}{d}\right)$$

for every fundamental discriminant δ satisfying $(-1)^k\delta > 0$.

The coefficients $c_g(n)$ are related to central values of modular L-functions. This connection relies on the theory of cycle integrals. Before stating this result, we introduce some additional notation. For more on the technical aspects of these terms, see **[324]**. For $f \in S_{2k}^{\text{new}}(\Gamma_0(N))$, and primes $p|N$, define the numbers $w_p \in \{\pm 1\}$ by $f_{2k}|W_p = w_p f$, where W_p is the Atkin-Lehner involution. Next, for $\Delta > 0$, let $\mathcal{Q}_{\Delta,N}$ denote the set of integral binary quadratic forms $Q = [a, b, c]$ of discriminant Δ on $\Gamma_0(N)$ and with $N|a$.For fixed k and N, we let δ be a fundamental discriminant with $(-1)^k\delta > 0$, and let $m \in \mathbb{N}$ satisfy $(-1)^k m \equiv 0, 1 \pmod 4$. Kohnen defined the following sum of cycle integrals of cusp forms $f \in S_{2k}(\Gamma_0(N))$:

$$r_{k,N}\left(f; \delta, (-1)^k m\right) := \sum_{\substack{Q \in \mathcal{Q}_{|\delta| m, N}/\Gamma_0(N) \\ Q(1,0) \equiv 0 \pmod N}} \chi_\delta(Q) \int_{C_Q} f(\tau) Q(\tau, 1)^{k-1} d\tau, \tag{12.14}$$

where for $Q = [a, b, c]$, the *genus character* χ_δ is defined by

(12.15)

$$\chi_\delta(Q) := \begin{cases} \chi_\delta(\ell) & \text{if } \gcd(a, b, c, \delta) = 1 \text{ and } Q \text{ represents } \ell \text{ with } \gcd(\ell, \delta) = 1, \\ 0 & \text{if } \gcd(a, b, c, \delta) > 1 \end{cases}$$

(cf. pages 59–62 of [**461**]). Note that χ_δ is independent of the choice of the integer ℓ with $\gcd(\ell, \delta) = 1$ that Q represents. The integrals defining $r_{k,N}$ are taken over geodesics C_Q. Precisely, for $Q = [a, b, c]$, C_Q is defined to be the image of the semicircle $a|\tau|^2 + bu + c = 0$ in $\Gamma_0(N)\backslash\mathbb{H}$, oriented from left to right if $a > 0$, from right to left if $a < 0$, and from $-c/b$ to $i\infty$ if $a = 0$. Kohnen related the cycle integrals (12.14) to certain L-values, defined for $f \in S_{2k}^{\mathrm{new}}(\Gamma_0(N))$, and fundamental discriminants δ satisfying $\gcd(\delta, N) = 1$, as follows:

$$L(f_\delta, s) := \sum_{n=1}^{\infty} \frac{\left(\frac{\delta}{n}\right) c_f(n)}{n^s},$$

for $\mathrm{Re}(s) \gg 0$. These L-functions are also similarly defined for more arbitrary non-zero discriminants δ in [**324**]. The following results are proved in [**324**]. See also [**320**] for related results.

THEOREM 12.14 (Kohnen). *Let $k \in \mathbb{N}$ and $f(\tau) = \sum_{n=1}^{\infty} c_f(n) q^n \in S_{2k}(\Gamma_0(N))$ (further recall that we assume throughout that $N \in \mathbb{N}$ is odd and square-free). Let $m, n \in \mathbb{N}$ satisfy $(-1)^k m, (-1)^k n \equiv 0, 1 \pmod 4$, and suppose that $(-1)^k n$ is a fundamental discriminant. Let δ, f, g, c_f, c_g, and $r_{k,N}$ be as above, and further assume that for all primes $p|N$, $\left(\frac{\delta}{p}\right) = w_p$. The following are true:*

$$r_{k,N}(f; (-1)^k n, (-1)^k m) = \frac{(-1)^{[\frac{k}{2}]}}{2^k} \frac{\langle f, f\rangle}{\langle g, g\rangle} c_g(m) \overline{c_g(n)},$$

$$L(f_\delta, k) = \frac{\pi^k}{2^{t(N)} |\delta|^{k-\frac{1}{2}} (k-1)!} \frac{\langle f, f\rangle}{\langle g, g\rangle} |c_g(|\delta|)|^2,$$

where $t(N)$ denotes the number of distinct prime divisors of N.

Using the cycle integrals $r_{k,N}$, define *the D-th Shintani lift of $f \in S_{2k}(\Gamma_0(N))$* by

(12.16)

$$\mathscr{S}^*_{k,N,D}(f)(\tau) := \sum_{\substack{m=1 \\ (-1)^k m \equiv 0,1 \pmod 4}}^{\infty} \left(\sum_{t|N} \mu(t) \left(\frac{D}{t}\right) t^{k-1} r_{k,Nt}\left(f; D, (-1)^k m t^2\right) \right) q^m.$$

Kohnen proved [**324**] that the Shintani lifts $\mathscr{S}^*_{k,N,D}$ map f into $S^+_{k+\frac{1}{2}}(\Gamma_0(4N))$. We can also view the Shintani lift as a theta lift using functions defined by Zagier in the context of the Doi-Naganuma lifting, following Kohnen [**324**] (this is discussed in detail in Section 13.4).

We now turn to a more recent result on mock modular forms related to Theorem 12.14, and introduce some additional notation. For the remainder of this section we restrict our attention to level $N = 1$. We first define cycle integrals as in (12.14), but more explicitly describe the path of integration. For each non-square discriminant $\delta > 0$, denote the set of integral binary quadratic forms on $\mathrm{SL}_2(\mathbb{Z})$ of discriminant δ by $\mathcal{Q}_\delta$ ($= \mathcal{Q}_{\delta,1}$). If δ is not a square, then the (infinite cyclic) group of automorphs of an indefinite binary quadratic form $Q = [a, b, c] \in \mathcal{Q}_\delta$ is generated by

$$g_Q := \begin{pmatrix} \frac{t+bX}{2} & cX \\ -aX & \frac{t-bX}{2} \end{pmatrix},$$

where (t, X) is the smallest positive solution to the *Pell equation* $t^2 - \delta X^2 = 4$ (the automorphs are trivial if δ is a square, and we let g_Q be the identity in this case). For $a \neq 0$ (resp. $a = 0$), let S_Q be the oriented semicircle (resp. vertical line) given by $a\left|\tau\right|^2 + bu + c = 0$ directed counterclockwise if $a > 0$, clockwise if $a < 0$, and up from the real axis if $a = 0$. Note that $\frac{d\tau}{Q(\tau,1)}$ is an invariant measure on S_Q. Since every $Q \in \mathcal{Q}_\delta$ is modular of weight -2 for $\Gamma_Q := \langle g_Q \rangle$, for $\kappa \in \mathbb{Z}$ it is natural to integrate the product of $Q(\tau, 1)^{\kappa-1}$ multiplied with a function F satisfying weight $2\kappa \in 2\mathbb{Z}$ modularity on Γ_Q along $\Gamma_Q \backslash S_Q$. Hence, for $\tau \in S_Q$, we let C_Q be the directed arc from τ to $g_Q\tau$ along S_Q if D is non-square and $C_Q := S_Q$ if δ is a square.

DEFINITION 12.15. Assuming the notation above (in particular, recall Definition 12.3), we define the **regularized cycle integral** by

$$\mathcal{C}\left(F; Q\right) := R. \int_{C_Q} F(\tau) Q(\tau, 1)^{k-1} d\tau.$$

REMARK. We note that the above regularization is in particular necessary if δ is a square and F does not decay towards the cusps.

REMARK. Regularized cycle integrals for weight 0 forms were also considered in [**123**].

We begin by stating a relation between the period integrals of the shadow and the ghost of a mock modular form (cf. [**93**]).

THEOREM 12.16. *For every harmonic Maass form $M \in H_{2-2k}$, every discriminant $\delta > 0$, and every $Q \in \mathcal{Q}_\delta$, the regularized cycle integrals satisfy*

$$\mathcal{C}\left(\xi_{2-2k}(M); Q\right) = -\frac{(4\pi)^{2k-1}}{(2k-2)!}\overline{\mathcal{C}\left(D^{2k-1}(M); Q\right)}.$$

REMARK. Note that there is a related result by Funke [**209**] which was independently developed. In his forthcoming work, he will show that the two forms $\xi_{2-2k}(M)$ and $-\frac{(4\pi)^{2k-1}}{(2k-2)!}\overline{D^{2k-1}(M)}$ give rise to the same cohomology classes for the modular curve. Hence, their cycle integrals coincide.

REMARK. Related objects, known as sesquiharmonic Maass forms (cf. Section 6.3), were also shown in [**83**] to play a role in encoding *non-critical* modular L-values.

SKETCH OF PROOF OF THEOREM 12.16. We first assume that δ is not a square. A key step is to represent $\mathcal{C}(f;Q)$ in terms of the periods of f. To be more precise, one can show that

$$\mathcal{C}\left(f;Q\right) = \sum_{n=0}^{2k-2} i^{-n+1} q_{k,\delta,\mathcal{A}}^{(n)} r_n(f),$$

where $\mathcal{A}$ is an equivalence class of binary quadratic forms containing Q and $q_{k,\delta,\mathcal{A}}^{(n)}$ denotes the coefficient of X^n in

$$\sum_{\substack{Q\in\mathcal{A} \\ Q \text{ reduced}}} Q(X,-1)^{k-1}.$$

Here we recall that a binary quadratic form $Q = [a,b,c] \in \mathcal{Q}_D$ $(D > 0)$ is called *reduced* if $a > 0$, $c > 0$, and $b > a + c$. The claim for non-square discriminants then follows by a lengthy computation using Theorem 12.10. If δ is a square we use explicit representatives for the associated class of quadratic forms and represent the cycle integral as an L-function. □

We discuss additional aspects of cycle integrals as related to mock modular forms in Chapter 16.

CHAPTER 13

Related Automorphic Forms

13.1. Introduction

In the previous chapters, we presented the theory of harmonic Maass forms and the analogies they share with classical modular and Jacobi forms. Along the way, we developed the key theoretical tools needed to make sense of Ramanujan's mock theta functions. In particular, we discussed how to complete them to obtain harmonic Maass forms. In the forthcoming chapters we shall take a close look at applications of these (and other) harmonic Maass forms to various other areas of number theory, combinatorics, Lie algebras, and physics.

There are further extensions of the classical theory of modular forms other than those considered above. Some of these other generalizations come equipped with their own natural applications. In this chapter, we survey some of these extensions which generalize the harmonic Maass forms considered in the previous chapters. Along the way, we highlight applications to combinatorics, divisors of classical modular forms, and geometry.

We begin by summarizing the contents of this chapter. In Section 13.2, we recall the important topic of mixed mock modular forms, which are essentially products of modular and mock modular forms. Although their completions no longer satisfy a simple differential equation, they are easily expressed in terms of functions which do, and they are natural in many combinatorial applications as they often arise as the generating series of natural partition functions and q-hypergeometric series. We shall also discuss closely related objects, the almost harmonic Maass forms, which include products of harmonic Maass forms and quasimodular forms. It turns out that these latter forms encode the structure of Kac-Wakimoto characters and coefficients of meromorphic Jacobi forms (cf. Subsection 20.2.2). Their completions are natural in the context of indefinite theta functions, as Zwegers' indefinite theta functions of Section 8.2 are generically mixed mock modular forms, and can only occasionally be written as simple quotients of mock modular forms with modular forms (cf. Theorem 8.30 and the second remark following Proposition 8.33).

In Section 13.3 we survey the theory of polar harmonic Maass forms, which are allowed to have poles at points in the upper half-plane. These have applications to the study of divisors of classical modular forms. That is, the natural method to investigate divisors of modular forms is to take their logarithmic derivatives, yielding weight 2 meromorphic modular forms with simple poles. Just as Maass-Poincaré series are used to give convenient bases for spaces of weakly holomorphic modular forms, in the case of meromorphic modular forms we require non-meromorphic Poincaré series to construct similar bases.

Finally, we conclude in Section 13.4 with a discussion of locally harmonic Maass forms, which can have jump discontinuities along hyperbolic geodesics. These new types of functions first arose in connection with functions analogous to the kernel functions of the Shimura and Shintani lifts, and are connected with the theory of modular L-functions. We note that in contrast to previous chapters in this book where z denotes an elliptic variable of a Jacobi form, throughout this chapter z and τ both lie in $\mathbb{H}$ and are considered as modular variables.

13.2. Mixed mock modular forms

In this section, we investigate automorphic objects known as *mixed mock modular forms.* In the theory of classical modular forms, it is often natural to study the graded algebra of modular forms; in particular a product of modular forms yields a new modular form. Although a product of harmonic Maass forms is typically not harmonic, there are special examples of such products which arise frequently enough in key examples to deserve their own consideration. These "mixed" objects are essentially (linear combinations of) products of harmonic Maass forms with ordinary modular forms. The following language is originally due to Zagier, although, as we shall discuss; we modify it slightly.

DEFINITION 13.1. A **mixed harmonic Maass form** of weight (k, ℓ) is a function h of the form

$$h(\tau) = \sum_{j=1}^{N} f_j(\tau) g_j(\tau),$$

where for each j, $f_j \in M_k^!$ and $g_j \in H_\ell^!$. We define the **holomorphic part** of h as $h^+ := \sum_{j=1}^{N} f_j g_j^+$, and refer to it as a **mixed mock modular form** of weight (k, ℓ). If only one weight k is specified, a mixed mock modular form of **total weight** k refers to a linear combination of mixed mock modular forms of weights (k_j, ℓ_j), where $k_j + \ell_j = k$. We further call h a **strong mixed harmonic Maass form** if h has at most polynomial growth at the cusps. The corresponding h^+ is a **strong mixed mock modular form**. We also denote the space of mixed harmonic Maass forms of weight (k, ℓ) by $\mathbb{H}_{k,\ell}^!$ and call the image of a form $h \in \mathbb{H}_{k,\ell}^!$ under ξ_ℓ the **shadow** of h^+.

REMARK. The definition of mixed mock modular forms is due to Zagier (we caution the reader that we label the indices of mock modular forms differently). Note that the original definition of mixed mock modular forms only includes those which we call strong mixed mock modular forms (cf. Section 7.3 of **[158]**). However, many natural examples arising in combinatorics lack the polynomial growth condition, and the language in the above definition has since become more standard in the literature. It is also common to abuse terminology and refer to a mixed mock modular form times a rational power of q and possibly plus a constant as a mixed mock modular form.

We have already seen several examples of mixed mock modular forms in this book, some of which we now review. In Theorem 8.30, we saw that mixed mock modular forms naturally arise as specializations of indefinite theta functions of Zwegers. Indeed, Ramanujan's examples of mock theta functions from this perspective are all first written as mixed mock modular forms, but satisfy the special condition that the modular forms arising in their decompositions as in Definition 13.1 happen to be equal and can thus be divided out to obtain (pure) mock modular forms. Hence, we see that these are somewhat special examples and that the modular-type objects obtained by the general indefinite theta function construction are usually legitimately mixed mock modular forms. There is in fact another explanation for the structure behind the definition of mixed mock modular forms, due to Zagier and Zwegers **[522]** (see also **[494]**).

DEFINITION 13.2 (Zagier and Zwegers). Let $M_k^0 := M_k$, the space of modular forms of weight k. For $d \in \mathbb{N}$, define the space $M_{k,k_1,\ldots,k_j}^d$ of **depth d harmonic Maass forms** to be the space of functions which transform like modular forms and whose images under the lowering operator L_k are linear combinations of elements of spaces of the form

$$\overline{M_{2-k}} \otimes M_{k-k_s,k_1,\ldots,k_{j-1}}^{d-1}.$$

For indefinite theta series of more general signatures than the type $(r-1,1)$ series of Zwegers, considered in Section 8.2, it is now known that the general modularity is explained by the structure of such higher depth forms. This gives a hint as to why mixed harmonic Maass forms are often more natural in a general sense than harmonic Maass forms in many contexts. Westerholt-Raum's work in Section 5 of **[494]** was the first instance where special cases of higher depth forms were characterized, in the context of a theoretical explanation of why Zwegers' $\widehat{\mu}$ function "splits" into a two-variable Jacobi form and a one-variable Maass Jacobi form (we described another explanation of this fact following Zagier in Theorem 8.18). Recently, Alexandrov, Banerjee, Manschot, and Pioline **[10]** explained the case of modular completions for indefinite theta series of type $(r-2,2)$, and Nazaroglu **[396]** followed their roadmap to complete the following new key result. Related results along these lines will also appear in forthcoming work of Zagier and Zwegers.

THEOREM 13.3. *For integral indefinite quadratic forms of type $(r-s,s)$, there is a general theory of modular completions which fits theta series associated to quadratic forms with such a signature within the framework of depth s harmonic Maass forms.*

REMARK. The construction of these more general modular completions, like those of Zwegers, can be given using Vignéras' useful Theorem 8.27. This perspective was taken in **[10]**, where instead of the special function E defined in (8.2) which Zwegers used in the type $(r-1,1)$ case, they required "double error functions", which for them arose from string-theoretic considerations as certain Penrose-type integrals. Alternative perspectives on these functions have also arisen. In particular, Kudla **[334]** realized the functions of **[10]** geometrically as theta integrals (see also related discussions of Westerholt-Raum **[493]**).

We conclude this section with a brief discussion of several examples of mixed mock modular forms. We first consider the $s=0$ case of Theorem 13.3. We obtain classical theta functions associated to positive definite quadratic forms which are modular (and hence of depth zero). For $s=1$, we obtain Zwegers' indefinite theta series which produce mixed harmonic modular forms (i.e., forms of depth one). Thus, these new results provide us with a new class of modular-type objects which have a general "mixed mock" shape, and given that there are many known examples of such indefinite theta series in combinatorics, Lie theory, and knot theory, it is natural to study the structure of these newly uncovered spaces.

Another natural use of products of modular forms with mock modular forms is in the context of holomorphic projection and recurrence formulas for coefficients of mock modular forms. The most important, first instance of such a recurrence is the famous Hurwitz-Kronecker class number relation, which we saw in Chapter 10, that states

$$\sum_{s\in\mathbb{Z}} H\left(4n-s^2\right) - 2\lambda_1(n) = 2\sigma_1(n),$$

where $H(n)$ is the Hurwitz-Kronecker class number, $\sigma_1(n)$ is the usual sum of divisors function, and λ_k is defined in (10.4). The beautiful mock explanation of this fact, due to Zagier **[512]**, is that Zagier's weight 3/2 Eisenstein series is a harmonic Maass form which grows moderately at the cusps, so the product of the generating function of $H(n)$ with the ordinary Jacobi theta series is a strong mixed mock modular form. This gives exactly the right conditions, as we see in Chapter 10 to take a holomorphic projection, from which the Hurwitz-Kronecker class number relations fall out from standard integral evaluations. A similar construction yields many interesting recurrence formulae for coefficients of mock modular forms. As we shall see in Section 20.3, these recurrences play a key role in the recent proof of the Umbral Moonshine Conjectures and make a connection to the classical moonshine idea of replicable functions. As with the Hurwitz-Kronecker class number relation proof, strong mixed modularity is very useful for convergence of the holomorphic projection. In fact, the most natural, and indeed one of the most common, examples of strong mixed mock modular forms one finds in this manner are products of mock modular forms with their shadows.

As mentioned in Chapter 10, a careful study of holomorphic projections of mixed mock modular forms by Mertens in **[385]** parameterized all such similar relations, including many classically known ones and several new ones. Mertens proceeded to show in Section 7 of **[385]** how his results recover the Eichler-Selberg trace formula on $\mathrm{SL}_2(\mathbb{Z})$ and allows for a new proof of a number of previously known class number relations, as well as infinitely many directly computable class number relations.

Other natural examples of mixed mock modular forms show up in combinatorics and representation theory. For example, products of mock modular forms with modular forms often arise from q-hypergeometric series manipulations. An early example was described by Andrews in **[27]**. He considered the number of partitions which do not contain any consecutive integers as parts, which have important applications in probability models (and in particular, their asymptotics are very interesting as they can be used to study threshold growth in cellular automata **[273]**). Calling the number of such partitions $p_2(n)$, Andrews showed that

$$\sum_{n=0}^{\infty} p_2(n)q^n = \frac{\left(-q^3;q^3\right)_\infty}{\left(q^2;q^2\right)_\infty}\chi(q),$$

where

$$\chi(q) := \sum_{n=0}^{\infty} \frac{q^{n^2}(-q)_n}{\left(-q^3;q^3\right)_n}$$

is one of Ramanujan's original third order mock theta functions. As we shall see in Chapter 15, the realization of such generating series as mixed mock modular forms allows one to use powerful modifications of the Circle Method, the first example of such an analysis being performed by Parberry in his thesis **[413]**.

Mixed mock modular forms arise in many further combinatorial examples. For example, we see in Theorem 14.21 below that they show up when describing the generating function of strongly unimodal sequences, and they are closely tied to the study of concave compositions **[38]**. It is also not surprising that they arise in so many q-hypergeometric examples given their prominent role in the theory of Bailey pairs. Namely, the theory of Bailey chains allows one to compute families of q-hypergeometric identities and tends to preserve modularity of inputs. Such theory is applied to the structure of mock theta functions in two seminal papers **[365, 364]** by Lovejoy and Osburn. Essentially, one begins with an indefinite theta function and iterates the "Bailey machinery". As remarked above, it is only in very special instances that indefinite theta functions of type $(n, 1)$ turn out to be mock modular forms times a fixed theta function, and they are more generically mixed mock modular forms. For similar reasons, the Bailey chain in many standard instances preserves mixed mock modularity, but not ordinary mock modularity.

13.3. Polar harmonic Maass forms

In this section, we consider polar harmonic Maass forms, which differ from ordinary harmonic Maass forms in that they are allowed to have poles in the upper half-plane. Although this may seem like a moderate relaxation of conditions, many aspects of the theory are quite different, and care must be taken to properly write down bases using Poincaré series. Their importance is highlighted by the many applications to classical modular forms and other areas of number theory. We offer three applications: the study of divisors of modular forms, computation of Green's functions (in relation to a famous conjecture of Gross-Zagier), and coefficients of meromorphic modular forms (expanding upon conjectures of Ramanujan and work of Hardy-Ramanujan). We begin this section by reviewing the first two of these applications; the third is outlined as a natural generalization of asymptotic methods for modular and mock modular forms in Section 15.5.

13.3.1. Divisors of modular forms. Our first motivation comes from the study of *divisors* (i.e., the sets of all zeros and poles together with their orders) of modular forms. As one learns early on in any modular forms course, a modular form's divisor is one of its fundamental characteristics. In level one, divisors of modular forms can be studied using only ordinary modular forms (i.e., without extending to nonholomorphic functions as will be made apparent shortly), but this study already reveals some very important classical objects. As we shall see below, the obstacle in higher levels is that the required meromorphic modular forms no longer exist, but suitable harmonic analogues do. This is analogous to the use of the Maass-Poincaré series throughout this book to write a full basis for weakly holomorphic modular forms of negative weight via principal parts, which is not possible using weakly holomorphic forms alone.

We now outline the $\mathrm{SL}_2(\mathbb{Z})$ case. As usual, let j be the modular j-invariant and E_k the normalized, weight k Eisenstein series (defined in (1.1)). The j-function plays a fundamental role in Monstrous Moonshine (cf. Section 20.1 for a brief survey) and is an important McKay-Thompson series. In the theory of moonshine, one also encounters the Koike-Norton-Zagier product expansion ($z, \mathfrak{z} \in \mathbb{H}$), where $J(z) := j(z) - 744 = \sum_{n=-1}^{\infty} b(n)e^{2\pi i n z}$,

$$J(z) - J(\mathfrak{z}) = e^{-2\pi i z} \prod_{m>0,\, n\in\mathbb{Z}} \left(1 - e^{2\pi i m z} e^{2\pi i n \mathfrak{z}}\right)^{b(mn)}. \tag{13.1}$$

This turns out to be the *denominator formula* for the Monster Lie algebra. This formula is equivalent (cf. Theorem 3 of [**41**]) to the following identity of Asai, Kaneko, and Ninomiya ($z, \mathfrak{z} \in \mathbb{H}$):

$$(13.2)\qquad H_z(\mathfrak{z}) := \sum_{n=0}^{\infty} j_n(z)e^{2\pi i n\mathfrak{z}} = \frac{E_4(\mathfrak{z})^2E_6(\mathfrak{z})}{\Delta(\mathfrak{z})}\frac{1}{j(\mathfrak{z})-j(z)} = -\frac{1}{2\pi i}\frac{j'(\mathfrak{z})}{j(\mathfrak{z})-j(z)}.$$

The functions j_n form a Hecke system. That is, if we let $j_0 := 1$ and $j_1 := j - 744$, then the others are obtained by applying the Hecke operators T_n (defined in (7.2))

$$(13.3)\qquad j_n(z) := nj_1(z) \mid T_n.$$

REMARK. The functions H_z and j_n played central roles in Zagier's theory of singular moduli [**516**], which is discussed in Chapter 16.

If $z \in \mathbb{H}$, then H_z is a weight 2 meromorphic modular form on $\mathrm{SL}_2(\mathbb{Z})$ with a single pole (modulo $\mathrm{SL}_2(\mathbb{Z})$) at the point z. In [**125**], it was shown how this observation can be utilized to study divisors of modular forms. The extension to $\Gamma_0(N)$ was recently offered in [**100**], the main results of which we now recall. For each N, we can also define a Hecke system of $\Gamma_0(N)$ harmonic Maass functions $j_{N,n}$ generalizing the j_n Hecke system. These are constructed as analytic continuations of *Niebur's Poincaré series* [**397**]. Namely, for $n \in \mathbb{N}$ and $\mathrm{Re}(s) > 1$ we set

$$(13.4)\qquad \mathfrak{F}_{N,-n,s}(z) := \sum_{M\in\Gamma_\infty\backslash\Gamma_0(N)} e\left(-n\mathrm{Re}(Mz)\right)\mathrm{Im}(Mz)^{\frac{1}{2}}I_{s-\frac{1}{2}}\left(2\pi n\mathrm{Im}(Mz)\right),$$

where $\Gamma_\infty = \{\pm\left(\begin{smallmatrix}1 & n\\ 0 & 1\end{smallmatrix}\right) : n \in \mathbb{Z}\}$ and I_s is the usual I-Bessel function of order n. These functions are weak Maass forms of weight 0 and eigenvalue $s(1-s)$. The analytic continuation to $s = 1$ may be obtained by computing a Fourier expansion and taking limits in the resulting expansions. This analytic continuation provides a definition of the analogous $j_{N,n}$ functions for $\Gamma_0(N)$. Specifically, we set $j_{N,n} := 2\pi\sqrt{n}\mathfrak{F}_{N,-n,1}$.

The analogous weight 2 forms $H^*_{N,z}$ are defined below in (13.10). These are *polar harmonic Maass forms* (cf. Definition 13.11), and they are distinguished by having only a simple pole at the one point z in the modular curve $X_0(N)$. The main result is then the following, where throughout we write $z = x+iy$, $\mathfrak{z} = \mathfrak{z}_1+i\mathfrak{z}_2$.

THEOREM 13.4. *If $z \in \mathbb{H}$, then the following are true:*

i) *If $z \in \mathbb{H}$ and $\mathfrak{z}_2 > \max\{y, \frac{1}{y}\}$, then we have that*

$$H^*_{N,z}(\mathfrak{z}) = \frac{3}{\pi\left[\mathrm{SL}_2(\mathbb{Z}) : \Gamma_0(N)\right]\mathfrak{z}_2} + \sum_{n=1}^{\infty} j_{N,n}(z)e^{2\pi i n\mathfrak{z}}.$$

ii) *For $\gcd(N,n) = 1$, we have $j_{N,n} = nj_{N,1} \mid T_n$.*

iii) *For $n|N$, we have $j_{N,n}(\mathfrak{z}) = j_{\frac{N}{n},1}(n\mathfrak{z})$.*

iv) *As $n \to \infty$, we have*

$$j_{N,n}(\mathfrak{z}) = \sum_{\substack{\lambda\in\Lambda_{\mathfrak{z}}\\ \lambda\le n}}\ \sum_{(c,d)\in S_\lambda} e\left(-\frac{n}{\lambda}r_{\mathfrak{z}}(c,d)\right)e^{\frac{2\pi n\mathfrak{z}_2}{\lambda}} + O_{\mathfrak{z}}(n).$$

Here,

$$\Lambda_{\mathfrak{z}} := \left\{\alpha^2|\mathfrak{z}|^2 + \beta\mathfrak{z}_1 + \gamma^2 > 0 \ : \ \alpha,\beta,\gamma \in \mathbb{Z}\right\},$$

S_λ is a certain set of solutions of the equation $Q_{\mathfrak{z}}(c,d) = \lambda$ for some quadratic form $Q_{\mathfrak{z}}$, and

$$r_{\mathfrak{z}}(c,d) := ac|\mathfrak{z}|^2 + (ad+bc)\mathfrak{z}_1 + bd$$

for any solutions $a, d \in \mathbb{Z}$ to $ad - bc = 1$.

REMARK. Prototypes for the sums in part iv) were introduced by Hardy and Ramanujan **[253]** (see also **[53, 62]**) in their study of $1/E_6$. We discuss this further in Section 15.5.

We next describe how these special functions can be used to provide a framework for describing divisors of modular forms. Suppose that f is a weight k meromorphic modular form on $\Gamma_0(N)$. We define the *divisor polar harmonic Maass form* to be

$$f^{\mathrm{div}}(\mathfrak{z}) := \sum_{z \in X_0(N)} e_{N,z} \mathrm{ord}_z(f) H^*_{N,z}(\mathfrak{z}), \tag{13.5}$$

where

$$e_{N,z} := \frac{2}{\#\mathrm{Stab}_z(\Gamma_0(N))} \tag{13.6}$$

(we let $e_{N,\rho} := 1$ for a cusp ρ). Recalling that $D = \frac{1}{2\pi i}\frac{\partial}{\partial \tau}$, we find that f^{div} is almost the logarithmic derivative of f.

THEOREM 13.5. *If f is a meromorphic modular form of weight k on $\Gamma_0(N)$, then*

$$f^{\mathrm{div}}(\mathfrak{z}) \equiv \frac{k}{4\pi \mathfrak{z}_2} - \frac{D(f(\mathfrak{z}))}{f(\mathfrak{z})} \pmod{S_2(\Gamma_0(N))}.$$

Theorem 13.5, whose proof we discuss in Subsection 13.3.2, can be used to numerically compute divisors of meromorphic modular forms f. This is well known to be a difficult task to perform directly. This numerical procedure is facilitated by the following asymptotic expansion, whose proof we omit.

LEMMA 13.6. *If $y \geq \mathrm{Im}(Mz)$ for all $M = \left(\begin{smallmatrix} a & b \\ c & d \end{smallmatrix}\right) \in \Gamma_0(N)$, then we have the approximation*

$$j_{N,n}(z) \approx e^{-2\pi i n z} + e^{2\pi i n \overline{z}} \sum_{\substack{c \geq 1 \\ N | c}} \sum_{\substack{d \in \mathbb{Z} \\ \gcd(c,d)=1 \\ |cz+d|^2=1}} e\left(n \frac{d-a}{c}\right).$$

The idea for applying these results to a numerical investigation of divisors is simple. The logarithmic derivative of f converts the points $z \in \mathbb{H}$ in the divisor of f into simple poles. These can be identified using the asymptotic properties of the coefficients of $H^*_{N,z}$ given in Theorem 13.4. This follows from Theorem 13.5 and the fact that coefficients of cusp forms only grow polynomially (cf. Deligne's bound in Theorem 15.3). This chain of thought directly yields as a first example the following.

COROLLARY 13.7. *Suppose that f is a meromorphic modular form of weight k on $\Gamma_0(N)$ whose divisor is not supported at cusps. Let y_1 be the largest imaginary part of any points in the divisor of f lying in $\mathbb{H}$. Then if*

$$-\frac{D(f(\mathfrak{z}))}{f(\mathfrak{z})} =: \sum_{n=-1}^{\infty} c(n) q^n,$$

we have that

$$y_1 = \limsup_{n\to\infty} \frac{\log |c(n)|}{2\pi n}.$$

EXAMPLE 13.8. We consider

$$f(\mathfrak{z}) := E_4(\mathfrak{z}) + \frac{\eta^{16}(2\mathfrak{z})}{\eta^8(\mathfrak{z})},$$

where η is Dedekind's eta-function. By the valence formula for $\Gamma_0(2)$, it has a single zero, say $\mathfrak{z}_0$, in $X_0(2)$. We find that

$$-\frac{D(f(\mathfrak{z}))}{f(\mathfrak{z})} = -e^{2\pi i\mathfrak{z}} - 495e^{4\pi i\mathfrak{z}} + 659e^{6\pi i\mathfrak{z}} + 113233e^{8\pi i\mathfrak{z}} - 261211e^{10\pi i\mathfrak{z}} + \cdots.$$

After the first 3000 terms the sequence $\log|c(n)|/(2\pi n)$ stabilizes and offers $\mathrm{Im}(\mathfrak{z}_0) \approx 0.4357$. As f has real coefficients and there is only one zero, $-\overline{\mathfrak{z}}_0$ must be $\Gamma_0(2)$-equivalent to $\mathfrak{z}_0$. By writing down an explicit fundamental domain, we find that either $\mathrm{Re}(\mathfrak{z}) = 0$, $1/2$ or z lies on the arc $|2\mathfrak{z} - 1| = 1$. The first two cases are easily excluded by the sign patterns of $c(n)$, and the zero on the arc is easily approximated as $\mathfrak{z}_0 \approx 0.2547 + 0.4357i$.

13.3.2. Definitions of the functions in Theorem 13.4 and the proof of Theorem 13.5. We begin with the construction of the functions $H^*_{N,z}$. This is similar in spirit to the general construction of the Poincaré series of Subsection 13.3.4, but the situation here is more technical since we require weight 2 (necessitating analytic continuations) and a higher level. For $z, \mathfrak{z} \in \mathbb{H}$ and $s \in \mathbb{C}$ with $\mathrm{Re}(s) > 0$, we set

$$P_{N,s}(\mathfrak{z}, z) := \sum_{M\in\Gamma_0(N)} \frac{\varphi_s(M\mathfrak{z}, z)}{j(M,\mathfrak{z})^2 \, |j(M,\mathfrak{z})|^{2s}} \tag{13.7}$$

with $j(\left(\begin{smallmatrix} a & b \\ c & d\end{smallmatrix}\right), \mathfrak{z}) = (c\mathfrak{z} + d)$ and

$$\varphi_s(\mathfrak{z}, z) := y^{1+s}(\mathfrak{z} - z)^{-1}(\mathfrak{z} - \overline{z})^{-1}|\mathfrak{z} - \overline{z}|^{-2s}.$$

These functions were studied in the z-aspect in **[98]**, where it was shown that they are polar harmonic Maass forms (cf. Definition 13.11). Here, we require their properties as functions of $\mathfrak{z}$. A long computation in **[100]** showed that $P_{N,s}(\mathfrak{z}, z)$ has an analytic continuation to $s = 0$, which we denote by $y\Psi_{2,N}(\mathfrak{z}, z)$. Lemma 4.4 of **[98]** then states that $z \mapsto y\Psi_{2,N}(\mathfrak{z}, z)$ is a polar harmonic Maass form of weight 0 on $\Gamma_0(N)$, and one can show that in $\mathfrak{z}$ they are weight 2 polar harmonic Maass forms with only a simple pole at the one point $z \in X_0(N)$. For $z \in \mathbb{H}$, we then set

$$H^*_{N,z}(\mathfrak{z}) := -\frac{y}{2\pi}\Psi_{2,N}(\mathfrak{z}, z).$$

In particular, in a sufficiently small neighborhood of z, we have

$$H^*_{N,z}(\mathfrak{z}) = \frac{1}{2\pi i e_{N,z}} \frac{1}{\mathfrak{z} - z} + O(1). \tag{13.8}$$

We also define $H^*_{N,z}$ when z is a cusp. For any cusp ρ of $\Gamma_0(N)$, let M_ρ be a matrix in $\mathrm{SL}_2(\mathbb{Z})$ with $\rho = M_\rho i\infty$. For $\mathrm{Re}(s) > 0$, define

$$E^*_{2,N,\rho,s}(\mathfrak{z}) := \sum_{M\in\Gamma_\rho\backslash\Gamma_0(N)} j(M_\rho M, \mathfrak{z})^{-2} |j(M_\rho M, \mathfrak{z})|^{-2s}, \tag{13.9}$$

where Γ_ρ is the stabilizer of ρ in $\Gamma_0(N)$. Using the Hecke trick (see also Lemma 6.2 and its sketch of proof), it is well known (cf. Satz 6 of **[259]**) that $E^*_{2,N,\rho,s}(\mathfrak{z})$ has an analytic continuation to $s = 0$, denoted by $E^*_{2,N,\rho}(\mathfrak{z})$. We are then able to define $H^*_{N,Z}$ in general by

$$H^*_{N,z}(\mathfrak{z}) := \begin{cases} -\frac{y}{2\pi}\Psi_{2,N}(\mathfrak{z}, z) & \text{if } z \in \mathbb{H}, \\ -E^*_{2,N,\rho}(\mathfrak{z}) & \text{if } z = \rho \in \mathbb{P}^1(\mathbb{Q}). \end{cases} \tag{13.10}$$

The proof of Theorem 13.4 then follows from the use of these functions after a long, careful series of calculations which we omit. Finally, we sketch the proof of the main application to divisors of modular forms.

SKETCH OF PROOF OF THEOREM 13.5. We show that the difference of both sides has no poles in $\mathbb{H}$ and decays towards the cusps. We start by considering the points in $\mathbb{H}$. One easily computes that the residue of $-\frac{D(f(\mathfrak{z}))}{f(\mathfrak{z})}$ at $\mathfrak{z} = z$ equals $\frac{1}{2\pi i}\mathrm{ord}_z(f)$. Using (13.8) gives that the principal parts at z agree. In a cusp ρ one similarly sees that $\frac{D(f(\mathfrak{z}))}{f(\mathfrak{z})}$ has no pole and its constant term equals $\mathrm{ord}_\rho(f)$. Using that the constant term of $H^*_{N,z}(\mathfrak{z})$ at ρ is -1 then gives the claim. □

13.3.3. Green's functions. Our second motivation for the theory of polar harmonic Maass forms comes from another distinguished class of meromorphic modular forms. These are analogous to Zagier's $f_{k,D}$ functions, which are discussed in detail in Section 13.4 (and in particular their important role in Shimura and Shintani lifts is described there). In the case when $D < 0$ (Zagier's discriminants were all positive) Bengoechea showed in her thesis **[51]** that one obtains interesting meromorphic modular forms (and in particular showed convergence and determined their Fourier expansions). The importance of these functions comes to the forefront when one integrates them, which can be used to give evaluations of Green's functions at CM points **[101]**. We briefly recall the definition of these functions. In what follows, $d(z, \mathfrak{z})$ denotes the hyperbolic distance between z and $\mathfrak{z}$ and

$$\omega_\mathfrak{z} := \#\Gamma_\mathfrak{z}. \tag{13.11}$$

DEFINITION 13.9. For $k \in \mathbb{N}_{>1}$ and $N \in \mathbb{N}$, the **higher Green's function** $\mathcal{G}_k\colon \mathbb{H} \times \mathbb{H} \to \mathbb{C}$ is uniquely characterized by the following properties, where $\Gamma_\mathfrak{z}$ is the stabilizer of $\mathfrak{z}$ under the action of $\mathrm{PSL}_2(\mathbb{Z})$ on $\mathbb{H}$.

i) The function $\mathcal{G}_k$ is smooth and real-valued on $\mathbb{H} \times \mathbb{H} \setminus \{(z, \gamma z) : \gamma \in \Gamma_0(N), z \in \mathbb{H}\}$.
ii) For $\gamma_1, \gamma_2 \in \Gamma$, we have $\mathcal{G}_k(\gamma_1 z, \gamma_2 \mathfrak{z}) = \mathcal{G}_k(z, \mathfrak{z})$.
iii) We have

$$\Delta_{0,z}\left(\mathcal{G}_k\left(z, \mathfrak{z}\right)\right) = k(1-k)\mathcal{G}_k\left(z, \mathfrak{z}\right) = \Delta_{0,\mathfrak{z}}\left(\mathcal{G}_k\left(z, \mathfrak{z}\right)\right).$$

iv) As $z \to \mathfrak{z}$, we have

$$\mathcal{G}_k(z, \mathfrak{z}) = \omega_\mathfrak{z} \log\left(r_\mathfrak{z}\left(z\right)\right) + O(1),$$

where

$$r_\mathfrak{z}(z) := \tanh\left(\frac{d(z, \mathfrak{z})}{2}\right).$$

v) As z approaches a cusp, we have $\mathcal{G}_k(z, \mathfrak{z}) \to 0$.

These functions have a long history, appearing as special cases of the resolvent kernel studied by Fay **[192]** and described in more detail by Hejhal **[261]**. Gross and Zagier conjectured **[242]** that their evaluations at CM-points are essentially logarithms of algebraic numbers. Specifically, in the special case when the space of weight $2k$ cusp forms on $\Gamma_0(N)$ is trivial, their conjecture states that

$$\mathcal{G}_k(z,\mathfrak{z}) = (D_1D_2)^{\frac{1-k}{2}}\log(\alpha)$$

for CM-points $z,\mathfrak{z}$ of discriminants D_1 and D_2, respectively and α is an algebraic number. Various cases of this conjecture have been solved. For example Mellit, in his Ph.D. thesis **[382]**, proved the case $k=2, \mathfrak{z}=i$ and also gave an interpretation of α as a certain intersection number of certain higher Chow cycles. Viazovska then proved the conjecture when the two CM points lie in the same imaginary quadratic field **[482]**.

We now consider the functions

$$f_{\mathcal{A}}(z) := D^{\frac{k}{2}}\sum_{Q\in\mathcal{A}} Q(z,1)^{-k},$$

where $\mathcal{A}$ is an $\mathrm{SL}_2(\mathbb{Z})$-equivalence class of integral binary quadratic forms of negative discriminant $-D$. The Petersson inner product of two $f_{\mathcal{A}}$ functions, suitably regularized and studied using the theory of polar harmonic Maass forms, gives evaluations of the Green's functions. In what follows, we denote by τ_Q the unique root of $Q(X,1)$ in the upper half-plane.

THEOREM 13.10. *Suppose that $-D_1,-D_2<0$ are two discriminants. For any two Q_1,Q_2 in distinct quadratic form classes $\mathcal{A}_1,\mathcal{A}_2$ of discriminants D_1 and D_2, we have*

$$\langle f_{\mathcal{A}_1}, f_{\mathcal{A}_2}\rangle = -\frac{\pi(-4)^{1-k}}{(2k-1)\beta(k,k)\omega_{\tau_{Q_1}}\omega_{\tau_{Q_2}}}\mathcal{G}_k\left(\tau_{Q_1},\tau_{Q_2}\right),$$

where for $\mathrm{Re}(s),\mathrm{Re}(w)>0$, *we have the beta-function*

$$\beta(s,w) := \int_0^1 u^{s-1}(1-u)^{w-1}du = \frac{\Gamma(s)\Gamma(w)}{\Gamma(s+w)}.$$

REMARK. It is possible that this result will shed new light on and offer an alternative approach to solving Gross and Zagier's conjecture.

13.3.4. Definition and construction of polar harmonic Maass forms. With the aforementioned applications in mind, we now describe the theory of polar harmonic Maass forms. We begin with their definition.

DEFINITION 13.11. For $\kappa\in\frac{1}{2}\mathbb{Z}$, a **polar harmonic Maass form of weight κ on a congruence subgroup** Γ is a function $F\colon\mathbb{H}\to\mathbb{C}$ which is real-analytic outside a discrete set of points and satisfies the following conditions:

i) For every $M\in\Gamma$, we have $F|_\kappa M = F$.
ii) We have $\Delta_\kappa(F)=0$.
iii) For every $\mathfrak{z}\in\mathbb{H}$, there exists $n\in\mathbb{N}_0$ such that $(z-\mathfrak{z})^nF(z)$ is bounded in some neighborhood of $\mathfrak{z}$.
iv) The function F has at most linear exponential growth at $i\infty$. Analogous conditions are required at the other cusps of Γ.

Denote by $\mathcal{H}_\kappa(\Gamma)$ the space of polar harmonic Maass forms of weight κ with respect to Γ, and write $\mathscr{H}_\kappa\subset\mathcal{H}_\kappa$ for the subspace of forms bounded at all cusps.

REMARK. If ii) is replaced by $\Delta_\kappa(F) = \lambda F$, then F is called a *polar Maass form with eigenvalue* λ.

REMARK. Analogous definitions may be made for forms with multiplier systems.

We now turn to the problem of constructing polar harmonic Maass forms, and in particular finding a convenient basis for them. For this purpose, it is natural to consider *elliptic expansions* of meromorphic modular forms of weight κ of the shape

$$f(z) = (z - \overline{\mathfrak{z}})^{-\kappa} \sum_{n \gg -\infty} c_{f,\mathfrak{z}}(n) X_{\mathfrak{z}}^n(z),$$

where

$$X_{\mathfrak{z}}(z) := \frac{z - \mathfrak{z}}{z - \overline{\mathfrak{z}}},$$

as opposed to the more common parabolic expansions, or Fourier series in q. Just as harmonic Maass forms have two parts to their Fourier expansions at the cusp $i\infty$ (cf. Lemma 4.3), polar harmonic Maass forms have expansions around points in the upper half-plane which break into two pieces. Specifically, polar harmonic Maass forms F of weight $2k$ have expansions around each point $\mathfrak{z} \in \mathbb{H}$ of the form $F = F_{\mathfrak{z}}^+ + F_{\mathfrak{z}}^-$, where the *meromorphic part* $F_{\mathfrak{z}}^+$ has a similar expansion to ordinary meromorphic modular forms,

$$F_{\mathfrak{z}}^+(z) := (z - \overline{\mathfrak{z}})^{-2k} \sum_{n \gg -\infty} c_{F,\mathfrak{z}}^+(n) X_{\mathfrak{z}}^n(z),$$

and where the *non-mermomorphic part* $F_{\mathfrak{z}}^-$ has an expansion of the form

$$F_{\mathfrak{z}}^-(z) := (z - \overline{\mathfrak{z}})^{-2k} \sum_{n \ll \infty} c_{F,\mathfrak{z}}^-(n) \beta_0\left(1 - r_{\mathfrak{z}}^2(z); 1 - 2k, -n\right) X_{\mathfrak{z}}^n(z).$$

REMARK. Throughout this section, such expansions for F^+ and F^- only exist for $y \gg 1$, due to the presence of poles.

Here we have

$$\beta_0(y; a, b) := \beta(y; a, b) - \mathcal{C}_{a,b},$$

where the *incomplete β-function* is defined by $\beta(w; a, b) := \int_0^w t^{a-1}(1-t)^{b-1} dt$ and where

$$\mathcal{C}_{a,b} := \sum_{\substack{0 \le j \le a-1 \\ j \ne -b}} \binom{a-1}{j} \frac{(-1)^j}{j+b}.$$

The *meromorphic elliptic Poincaré series* of Petersson, for $\mathfrak{z} \in \mathbb{H}$, $n \in \mathbb{Z}$, and $k \in \mathbb{N}_{>1}$ are given by

$$\Psi_{2k,n}^{\mathfrak{z}}(z) = \Psi_{2k,n}(z, \mathfrak{z}) := \sum_{M \in \mathrm{SL}_2(\mathbb{Z})} \psi_{2k,n}^{\mathfrak{z}}(z) \Big|_{2k} M,$$

where

$$\psi_{2k,n}^{\mathfrak{z}}(z) := (z - \overline{\mathfrak{z}})^{-2k} X_{\mathfrak{z}}^n(z).$$

For $n \ge 0$, this yields a cusp form, and for $n < 0$, $\Psi_{2k,n}^{\mathfrak{z}}$ is orthogonal to cusp forms (with respect to a suitable regularized inner product) and has a principal part around $z = \mathfrak{z}$ equal to

$$2\omega_{\mathfrak{z}} \psi_{2k,n}^{\mathfrak{z}}(z).$$

Similarly, for $k \in -\mathbb{N}$, the corresponding elliptic Maass-Poincaré series are defined by

$$\mathbb{P}_{2k,n}(z,\mathfrak{z}) := \sum_{M\in \mathrm{SL}_2(\mathbb{Z})} \varphi^{\mathfrak{z}}_{2k,n}(z)\Big|_{2k} M,$$

where

$$\varphi^{\mathfrak{z}}_{2k,n}(z) := (z-\overline{\mathfrak{z}})^{-2k}\beta\left(1-r_{\mathfrak{z}}^2(z);1-2k,-n\right)X_{\mathfrak{z}}^n(z).$$

The main structural result making many key applications such as those highlighted above basically follows from work of Fay **[192]** (see also **[95]**).

THEOREM 13.12. *Suppose that $k \in -2\mathbb{N}$. Then the following hold:*

i) *If $n \in -\mathbb{N}$, then the Poincaré series $z \mapsto \mathbb{P}_{2k,n}(z,\mathfrak{z})$ is the unique polar harmonic Maass form of weight $2k$ with principal part*

$$2\omega_{\mathfrak{z}}\mathcal{C}_{2k-1,-n}(z-\overline{\mathfrak{z}})^{2k-2}X_{\mathfrak{z}}^n(z)$$

at $\mathfrak{z}$. Moreover, its shadow is given by

$$\xi_{2k}\left(\mathbb{P}^{\mathfrak{z}}_{2k,n}\right) = (4\mathfrak{z}_2)^{1-2k}\Psi^{\mathfrak{z}}_{2-2k,-n-1}.$$

ii) *Every $F \in \mathscr{H}_{2k}$ can be written as a linear combination of the functions from the set $\{\mathbb{P}_{2k,n}(z,\mathfrak{z}) : \mathfrak{z} \in \mathbb{H}, n \in \mathbb{Z}\}$. Moreover, if the only poles of F in $\mathbb{H}$ occur at points equivalent to $\mathfrak{z}$ under the action of $\mathrm{SL}_2(\mathbb{Z})$, then F is a linear combination of functions from $\{\mathbb{P}_{2k,n}(z,\mathfrak{z}) : n \in \mathbb{Z}\}$.*

Moreover, the images of these Poincaré series under the differential operators ξ_{2-2k} and D^{2k-1} may be explicitly determined in terms of dual weight meromorphic Poincaré series, and results for the image of these maps on these spaces may be determined in analogy with Theorem 5.9.

13.4. Locally harmonic Maass forms

In this section, we briefly discuss locally harmonic Maass forms, which are obtained by relaxing the harmonicity condition in Definition 4.2 to allow discontinuities on the upper half-plane. Roughly speaking, we permit jump discontinuities to occur along certain Heegner cycles (see Example 13.14 below for a specific instance of these special geodesics), corresponding to quadratic forms of a given discriminant. Such functions were recently independently studied by several people: Hövel, in his thesis written under Bruinier **[275]**, constructed locally harmonic Maass forms as outputs of theta lifts in the case of weight 0, and Pioline **[420]** studied them in the setting of string theory.

Locally harmonic Maass forms were originally defined in **[99]** in the context of the functions

$$f_{k,D}(z) := \sum_{Q\in\mathcal{Q}_D} Q(z,1)^{-k}, \qquad (13.12)$$

where $\mathcal{Q}_D$ is the set of integral binary quadratic forms of discriminant $D > 0$. It can be shown directly that the $f_{k,D}$'s are cusp forms of weight $2k$. These functions were introduced by Zagier **[507]** in his study of the Doi-Naganuma lift, and were shown in important work of Katok **[301]** to be simple sums of hyperbolic Poincaré series. The interested reader is referred to the excellent article of Imamoḡlu and O'Sullivan **[277]** for a discussion of various types of Poincaré series, including the "parabolic series" studied in Section 6.3 and the analogous hyperbolic series related to the functions discussed here.

Zagier's functions also arise naturally as coefficients of the kernel functions for the Shimura and Shintani lifts. To describe this, we follow Kohnen **[324]**. For even k, Kohnen considered the function (here we renormalize his function slightly)

$$\Omega_k(z,\tau) := \sum_{\substack{m\geq 1\\ m\equiv 0,1 \pmod 4}} m^{k-\frac{1}{2}} f_{k,m}(z) e^{2\pi i m\tau}.$$

After one shows suitable convergence properties, it is clear that Ω_k is modular of weight $2k$ in z. It turns out to be modular in the τ variable as well. To show this, Kohnen used explicit Fourier expansions to prove that $\Omega_k(z,\tau)$ can be expanded as a Fourier expansion in z, up to a constant multiple, as the sum

$$\sum_{n=1}^{\infty} n^{k-1} \left(\sum_{d|n} \left(\frac{n}{d}\right)^k P_{k+\frac{1}{2},\frac{n^2}{d^2},4}(\tau) \right) e^{2\pi i n z}.$$

Here $P_{k+\frac{1}{2},m,4}$ is the Poincaré series of weight k and level 4 projected to the plus space (defined in (6.11)). Kohnen's main result is then that this function interpolates the Shimura and Shintani lifts, which also map between weights $k+1/2$ and $2k$.

THEOREM 13.13 (Kohnen). *Suppose that* $f \in S_{2k}$ *and* $g \in S^+_{k+\frac{1}{2}}(\Gamma_0(4))$. *Then the Shimura lift of* g *(cf.* (12.16) *with* $D=1$*) is given by*

$$\langle g, \Omega_k(-\overline{z}, \cdot)\rangle,$$

and the Shintani lift of f *is essentially*

$$\langle f, \Omega_k(\cdot, -\overline{\tau})\rangle.$$

Given the importance of the $f_{k,D}$ functions of Zagier in the construction of the kernel of the Shintani lift and given the usefulness of finding harmonic "lifts" of classical cusp forms, it is natural to ask for a preimage to $f_{k,D}$ under ξ_{2-2k}. Of course, the surjectivity of $\xi_{2-2k}\colon H_{2-2k} \to S_{2k}$ guarantees that such lifts exist, but they appear to be unnatural if the question is naively answered using the standard methods. For, as we have seen, the surjectivity of this map is understood directly via the theory of (parabolic) Poincaré series, and decomposing the (hyperbolic) $f_{k,D}$'s in terms of these Poincaré cusp forms hopelessly confounds the arithmetic. The locally harmonic Maass forms were thus first constructed to give such lifts under ξ_{2-2k}, and the singularities on the upper half-plane are artifacts of "regularization" procedures on a certain theta lift (which follow a similar philosophy as the regularization in Definition 5.6). One potential benefit of this construction as a whole is that it opens the door for possible geometric interpretations (and Hövel discussed this in the analogous case of $k=1$ in **[275]**).

In addition to the full theory of locally harmonic Maass forms, a simpler object of this flavor may be found in the recent subject of locally modular polynomials, which Zagier also previously considered in relation to applications of physics. We begin by recalling that polynomials can almost never be modular forms. One can easily construct polynomials for non-congruence subgroups (for example, τ^2+1 is a modular form of weight -2 for the subgroup $\{1,S\} \subseteq \mathrm{SL}_2(\mathbb{Z})$, where $S := \left(\begin{smallmatrix} 0 & -1 \\ 1 & 0\end{smallmatrix}\right)$), but Knopp showed **[314]** that this phenomenon is essentially relegated to non-discontinuous subgroups (with the trivial exception of constant functions).

In fact, the definition for $f_{k,D}$ can be viewed as an attempt to start with such basic polynomials and form a modular object, although one necessarily obtains infinite sums in this way. Another way to construct modular objects out of polynomials, without the use of infinite summation, is to allow discontinuities, so that we consider functions which are locally polynomial and which have jump discontinuities along the $\mathrm{SL}_2(\mathbb{Z})$-orbits of finitely many geodesics. The spaces of such functions, along with algorithms to compute them, were described in [**97**]. For the interested reader, we provide one detailed example.

EXAMPLE 13.14. We consider the case $D = 5$. Then the set of allowable discontinuities is formed by (the $\mathrm{SL}_2(\mathbb{Z})$-orbits of) two semicircles S_{Q_1} and S_{Q_2}, which are the geodesics connecting the real roots of the quadratic forms $Q_1 := [1, 1, -1]$ and $Q_2 := [1, -1, -1]$. These two semicircles intersect at i and partition the fundamental domain of $\mathrm{SL}_2(\mathbb{Z})$ into three components, which we call $\mathcal{C}_\infty$, $\mathcal{C}_\rho$, and $\mathcal{C}_{\rho-1}$, respectively, as in the following figure:

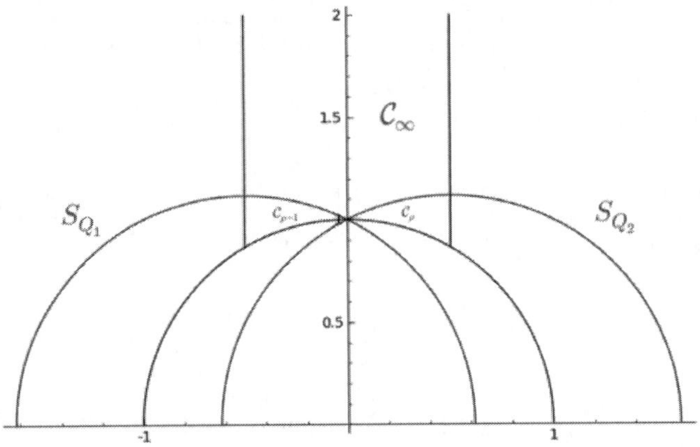

FIGURE 1. Geodesics of discontinuities for $D = 5$

The main results of [**97**] then show that the space of modular forms of weight -2 and which are locally polynomials on each of these three components is two dimensional. As we run through pairs of complex numbers α, β, these local polynomials are of the shape

$$\begin{cases} \alpha & \text{if } \tau \in \mathcal{C}_\infty, \\ \beta\left(\tau^2 - \tau + 1\right) & \text{if } \tau \in \mathcal{C}_\rho, \\ \beta\left(\tau^2 + \tau + 1\right) & \text{if } \tau \in \mathcal{C}_{\rho-1}. \end{cases}$$

An elementary calculation then shows that each of these local polynomials, away from the geodesics in Figure 1, is modular of weight -2.

Part 3

Applications

Here we discuss a number of recent applications of the theory of harmonic Maass forms. The subjects discussed here are not meant to be a complete treatment of the literature. Instead, Part 3 is a sampling of topics which reflect the interests of the authors. For the sake of brevity we only give expository introductions to these applications. Interested readers should consult the cited references for further details.

CHAPTER 14

Partitions and Unimodal Sequences

The partition function $p(n)$ plays a central role in the theory of modular forms. It has provided a fertile test case for many of the standard techniques in the theory of modular forms, owing to the fact that its generating function is closely related to the weight $-1/2$ weakly holomorphic modular form

$$\frac{1}{\eta(24\tau)} = \sum_{n=0}^{\infty} p(n) q^{24n-1}.$$

Indeed, the so-called Circle Method was developed to study asymptotics of $p(n)$, and the Deligne-Serre theory of modular ℓ-adic Galois representations was combined with the theory of the Shimura correspondence to obtain its general congruences.

In spite of this success in applying modular forms to the study of partitions, many deeper problems are not accessible using the classical theory of modular forms alone. For example, the elucidation of important properties of so-called ranks and cranks requires a deep understanding of Ramanujan's mock theta functions. Thanks to such problems, the work of Zwegers on Ramanujan's mock theta functions has reignited interest in some old questions about partitions, and explicit calculations with the relevant harmonic Maass forms have provided many new results.

In this chapter, we discuss some of the theorems where harmonic Maass forms play a fundamental role. The main results correspond to the theory of ranks and cranks, which were invented to study congruences for $p(n)$, and the more recent investigation of unimodal sequences and the Andrews spt-function. In particular, this chapter describes congruences, asymptotics, and a canonical example of Zagier's theory of quantum modular forms.

14.1. Asymptotic formulas for partitions

We begin by recalling a few basics about partitions.

DEFINITION 14.1. A **partition** of $n \in \mathbb{N}_0$ is a non-increasing sequence of positive integers, called its **parts**, which sum to n. We let $p(n)$ denote the number of partitions of n and set $p(0) := 1$.

EXAMPLE 14.2. The partitions of 5 are given by

$$5, \qquad 4+1, \qquad 3+2, \qquad 3+1+1, \qquad 2+2+1,$$
$$2+1+1+1, \qquad 1+1+1+1+1,$$

and so we see that $p(5) = 7$.

REMARK. We note that it is common to call the sum corresponding to a sequence in Definition 14.1 a partition as well, and we frequently do so throughout the remainder of this chapter. There are several excellent surveys on partitions. For example we refer readers to **[25]**.

Euler proved that the partition generating function has a nice representation as an infinite product. Namely, we have that

$$P(q) := \sum_{n=0}^{\infty} p(n)q^n = \prod_{n=1}^{\infty} \frac{1}{1-q^n} = \frac{q^{\frac{1}{24}}}{\eta(\tau)} \tag{14.1}$$

which we can then recognize (up to the factor $q^{\frac{1}{24}}$) as a modular form. This product representation is key to most deeper investigations of the partition function. For example, as Euler famously showed, this product representation allows one to deduce a recurrence formula for $p(n)$ which enables one to compute much higher values of $p(n)$ than by more naive methods. Even better, Rademacher used a careful study of the modularity of this product formula to give **[423]** an exact, infinite summation formula for $p(n)$, as in the following theorem. Before stating the result, we recall that $A_k(n)$ is the Kloosterman sum which is defined in (9.19), and $I_{\frac{3}{2}}$ is the usual modified Bessel function (of the first kind).

THEOREM 14.3 (Rademacher). *If n is a positive integer, then*

$$p(n) = \frac{2\pi}{(24n-1)^{\frac{3}{4}}} \sum_{k=1}^{\infty} \frac{A_k(n)}{k} I_{\frac{3}{2}}\left(\frac{\pi\sqrt{24n-1}}{6k}\right). \tag{14.2}$$

In particular, as $n \to \infty$ we have that

$$p(n) \sim \frac{1}{4\sqrt{3}n} e^{\pi\sqrt{\frac{2n}{3}}}. \tag{14.3}$$

REMARK. A more conceptual proof writes $P(q)$ as a Maass-Poincaré series using the η-multiplier (compare (9.20)). We also study more general asymptotic techniques which are more modern in detail in Chapter 15.

REMARK. It is natural to ask about the size of the error of (14.2) truncated after N terms. This difficult question has been studied by many authors, and much recent progress has been made; the interested reader is referred to **[5, 16, 200, 345, 346, 423, 425]**.

SKETCH OF PROOF OF THEOREM 14.3. Hardy and Ramanujan **[252]** introduced the so-called Circle Method, now called the *Hardy-Ramanujan Circle Method*, to prove (14.3). To summarize the idea, suppose that one is interested in the asymptotic behavior of some sequence $\{a(n)\}$ as $n \to \infty$. One builds a generating function of out this sequence

$$A(q) := \sum_{n=0}^{\infty} a(n)q^n$$

which is supposed to be scaled so that A has radius of convergence equal to 1. Cauchy's Theorem then gives, for $n \in \mathbb{N}$, the formula

$$a(n) = \frac{1}{2\pi i} \int_C \frac{A(q)}{q^{n+1}} \, dq, \tag{14.4}$$

where C is an arbitrary path inside the unit disk that loops around 0 in the counterclockwise direction exactly once.

When one considers $\{a(n)\} = \{p(n)\}$ (and thus $A(q) = P(q)$), note that due to the singularities at roots of unity of the product formula (14.1), one cannot take the integral in (14.4) over the whole unit circle. For many interesting sequences $\{a(n)\}$, including the case of the sequence $\{p(n)\}$ of partition numbers, the singularities of the generating function A on the unit circle are well understood and occur at roots of unity q. One can often find nice approximations of A near these points. For example, in the case of the sequence of partition numbers $\{p(n)\}$, one sees that all roots of unity lead to poles and that $q = 1$ is the "dominant" pole, i.e., leads to the biggest growth. The asymptotic formula (14.6) gives us the appropriate approximation, and the leftover term only contributes to a much smaller error term.

Hardy and Ramanujan actually proved a much stronger result than (14.3), namely a full asymptotic expansion for $p(n)$ which is however divergent. Rademacher [**424**] perfected the Circle Method to obtain (14.2). In particular, he used stronger bounds for Kloosterman sums and let a certain parameter that was a fixed number depending on n in the Hardy-Ramanujan Circle Method tend to infinity.

The claim that (14.3) is a consequence of (14.2) follows from the asymptotic behavior of the I-Bessel function

$$I_\alpha(x) \sim \frac{e^x}{\sqrt{2\pi x}} \qquad (\text{as } x \to \infty), \tag{14.5}$$

applied to the $k = 1$ term in (14.2). □

An alternative proof of the main asymptotic term of $p(n)$, (14.3), utilizes the following Tauberian Theorem (cf. [**279**]), which may be applied to derive the asymptotic properties of q-series with monotonically increasing coefficients. Such monotonicity properties are frequently known for combinatorial sequences, and the flexibility of this method is that one need only know the asymptotic expansion near $q = 1$, and estimates near poles at other roots of unity are not needed.

THEOREM 14.4 (Ingham). *Assume that $f(q) = \sum_{n=0}^{\infty} C_f(n)q^n$ is a power series with weakly increasing non-negative coefficients and radius of convergence equal to 1. If there exist constants $A > 0$, $\lambda, \alpha \in \mathbb{R}$ such that, as $t \to 0^+$, we have*

$$f\left(e^{-t}\right) \sim \lambda t^\alpha \exp\left(\frac{A}{t}\right),$$

then we have the asymptotic

$$C_f(n) \sim \frac{\lambda}{2\sqrt{\pi}} \frac{A^{\frac{\alpha}{2}+\frac{1}{4}}}{n^{\frac{\alpha}{2}+\frac{3}{4}}} \exp\left(2\sqrt{An}\right) \qquad (\text{as } n \to \infty).$$

In particular, using the modularity of $P(q)$ one sees, as $t \to 0^+$, that

$$P\left(e^{-t}\right) \sim \sqrt{\frac{t}{2\pi}} e^{\frac{\pi^2}{6t}}. \tag{14.6}$$

Since it is not hard to see that $p(n)$ is non-negative and increasing (since if λ is a partition of $n \in \mathbb{N}$, then $\lambda + 1$ is a partition of $n + 1$), we obtain, by Theorem 14.4, the asymptotic formula (14.3).

REMARK. A further application of the Circle Method is Waring's problem. Waring asked whether any positive integer can be represented by a finite sum of k-th powers. Hilbert answered this question affirmatively in 1909. The question here is whether for every $k \in \mathbb{N}$ there exists $\ell \in \mathbb{N}$ such that every natural number is the sum of at most ℓ many k-th powers of natural numbers.

REMARK. There is also a modified version of the Hardy-Ramanujan Circle Method going back to Wright [**498**]. This usually gives less precise results but requires less precise asymptotic approximations of the generating functions. In particular his method is powerful enough to provide an asymptotic expansion for the coefficients, but flexible enough that it applies to non-modular generating functions. We describe this method in Section 15.7.

In recent work [**128**], Bruinier and one of the authors obtained a formula of a different flavor, which expresses $p(n)$ as a finite sum of algebraic numbers. To describe this formula, we consider the weight -2 weakly holomorphic modular form

$$G(\tau) := \frac{1}{2}\frac{E_2(\tau) - 2E_2(2\tau) - 3E_2(3\tau) + 6E_2(6\tau)}{\eta(\tau)^2\eta(2\tau)^2\eta(3\tau)^2\eta(6\tau)^2}$$

defined on $\Gamma_0(6)$ and define the weak Maass form

$$\mathcal{P} := \frac{1}{4\pi}R_{-2}(G),$$

where R_{-2} is the weight -2 raising operator from Definition 5.1. From Lemma 5.2, it follows that $\mathcal{P}$ has weight 0 and eigenvalue -2 under Δ_0. Consider the set of binary quadratic forms given by

$$\mathcal{Q}_{D,6,1} := \left\{Q = [a,b,c] : a,b,c \in \mathbb{Z}, b^2 - 4ac = -24D + 1, 6|a, a > 0,\ b \equiv 1 \pmod{12}\right\}$$

and the corresponding CM points $\tau_Q \in \mathbb{H}$ for which $a\tau^2 + b\tau + c = 0$. We then define the *trace*

$$\operatorname{Tr}(n) := \sum_{Q \in \mathcal{Q}_{n,6,1}} \mathcal{P}\left(\tau_Q\right).$$

The following theorem gives the "finite algebraic" formula for the partition function.

THEOREM 14.5. *If $n \in \mathbb{N}$, then*

$$p(n) = \frac{1}{24n - 1}\operatorname{Tr}(n).$$

Moreover, $(24n-1)\mathcal{P}(\tau_Q)$ is always an algebraic integer.

REMARK. The full strength of the bound on the denominator of the $\mathcal{P}(\tau_Q)$'s implied by the algebraic integrality claim in Theorem 14.5 was proven in [**343**].

REMARK. The "algebraic formula" of Theorem 14.5 was also used to deduce the Hardy-Ramanujan asymptotic (14.3) by Dewar and Murty in [**161**].

REMARK. In Chapter 16, we will discuss the context and theory of such "traces of singular moduli" in detail.

14.2. Ramanujan's partition congruences

Ramanujan [**431, 428**] famously proved the following elegant partition congruences, valid for all $n \in \mathbb{N}_0$:

$$\begin{aligned} p(5n+4) &\equiv 0 \pmod{5}, \\ p(7n+5) &\equiv 0 \pmod{7}, \\ p(11n+6) &\equiv 0 \pmod{11}. \end{aligned} \tag{14.7}$$

Atkin and Watson [**42, 490**] extended these congruences, proving generalizations of (14.7) modulo powers of $5, 7$, and 11 (cf. [**43**] for congruences modulo other small primes). Using Shimura's theory of half-integral weight modular forms and Serre's theory of p-adic modular forms, one of the authors [**404**] proved that there are infinitely many congruences for all primes at least 5, albeit of a more complicated shape than Ramanujan's congruences.

THEOREM 14.6. *Suppose $m \geq 5$ is prime and $k \in \mathbb{N}$. Then a positive proportion of primes ℓ satisfy the congruence*

$$p\left(\frac{m^k\ell^3 n+1}{24}\right) \equiv 0 \pmod{m}$$

for all $n \in \mathbb{N}_0$ which are coprime to ℓ.

This theorem implies the existence of infinitely many linear congruences for every prime $m \geq 5$. For example, we have the congruence

$$p\left(59^4 \cdot 13n + 111247\right) \equiv 0 \pmod{13}.$$

Theorem 14.6 was generalized by Ahlgren and one of the authors [**4, 9**], and it is now known that if $\gcd(m, 6) = 1$, then there are infinitely many non-nested arithmetic progressions $An + B$ for which

$$p(An + B) \equiv 0 \pmod{m}.$$

By *non-nested* progressions, we mean arithmetic progressions $An + B$ with $0 \leq B < A$, none of which is contained in another progression.

Finally, we note that the condition that $\gcd(m, 6) = 1$ in the theorems of Ahlgren and one of the authors cannot be relaxed, as it is known that there are no such congruences for $p(n)$ modulo 2 or 3. This was established by Radu in [**427**] using clever and technical methods combined with a deep theorem of Deligne and Rapoport. It is expected that partitions are uniformly distributed modulo 2 and 3. This question remains far out of reach, and even showing the existence of a positive proportion of numbers n for which $p(n)$ is even (resp. odd) remains an open problem.

14.3. Ranks and cranks

14.3.1. Definition and generating functions. Enamored with Ramanujan's congruences (14.7) for $p(n)$, Dyson [**180**] introduced the so-called rank of a partition. His goal was to define a statistic on partitions which explains the congruences. For example, he wanted this statistic to divide the partitions of $5n + 4$ into five groups of equal size. His careful investigations revealed an important new object in number theory, which he called the rank function.

Partition	Rank	Rank (mod 5)
4	$4-1=3$	3
$3+1$	$3-2=1$	1
$2+2$	$2-2=0$	0
$2+1+1$	$2-3=-1$	4
$1+1+1+1$	$1-4=-3$	2

DEFINITION 14.7. The **rank** of a partition is its largest part minus the number of its parts.

EXAMPLE 14.8. Here we list the partitions of 4 and their associated ranks. The ranks of these partitions modulo 5 cover each residue class exactly once.

After discovering this function, Dyson made many further computations of the ranks of other partitions of numbers of the form $5n + 4$ (resp. $7n + 5$), and such calculations led him to conjecture that the corresponding ranks reduced modulo 5 (resp. modulo 7) divide the partitions into 5 (resp. 7) groups of equal size.

REMARK. We note that Dyson's definition of the partition rank has various generalizations in additive number theory. For example, Lovejoy defined a generalized rank function for the so-called overpartition function [**361**]. Other works such as those by Berkovich, Garvan, Lovejoy, and Osburn [**52, 362, 363**] include further examples of partition ranks.

Dyson's conjecture gives a proposed solution to his own problem of finding a combinatorial explanation of Ramanujan's congruences modulo 5 and 7, namely one which avoids the use of generating function identities. This conjecture was proven in 1954 by Atkin and Swinnerton-Dyer [**45**]. These works have inspired several further research papers on Dyson's partition ranks. For example, rank identities for the moduli 8, 9, and 12 have been found by Lewis and Santa-Gadea [**354, 355, 443**].

A short computation reveals that Dyson's rank does not explain Ramanujan's modulo 11 partition congruence. In other words, the partitions of $11n + 6$ are not generally divided into 11 groups of equal size when sorted by ranks modulo 11. As a natural follow-up to his conjecture, Dyson postulated the existence of yet another statistic, the so-called crank of a partition, and he predicted that the crank explains all three of Ramanujan's congruences in (14.7). Garvan found a suitable crank statistic for vector partitions [**218**], and, together with Andrews [**33**], formulated the definition of the crank which they proved confirms Dyson's speculation. In other words, Ramanujan's congruences (14.7) are indeed encoded by their crank statistic modulo 5, 7, and 11.

DEFINITION 14.9. For a partition λ, let $o(\lambda)$ denote the number of ones in λ, and define $\mu(\lambda)$ as the number of parts strictly larger than $o(\lambda)$. Then the **crank** of λ is defined as

$$\operatorname{crank}(\lambda) := \begin{cases} \text{largest part of } \lambda & \text{if } o(\lambda) = 0, \\ \mu(\lambda) - o(\lambda) & \text{if } o(\lambda) > 0. \end{cases}$$

EXAMPLE 14.10. Here we illustrate the cranks of the partitions of 6.

We note that the cranks of these partitions modulo 11 divide the partitions of 6 into 11 groups of equal size (namely one).

REMARK. We also note that work of Garvan, Kim, and Stanton [**223**] gives a different crank for several other partition congruences, and in particular for Ramanujan's congruence modulo 25.

In view of the role of the rank and the crank for partition congruences, it is natural to study their general properties. To this end, we make use of two important generating functions in the variables ζ and q. It turns out that these functions are intimately connected to modular forms and mock modular forms. Let $M(m, n)$

Partition	$o(\lambda)$	$\mu(\lambda)$	Crank	Crank (mod 11)
6	0	1	6	6
$5+1$	1	1	$1-1=0$	0
$4+2$	0	2	4	4
$4+1+1$	2	1	$1-2=-1$	10
$3+3$	0	2	3	3
$3+2+1$	1	2	$2-1=1$	1
$3+1+1+1$	3	0	$0-3=-3$	8
$2+2+2$	0	3	2	2
$2+2+1+1$	2	0	$0-2=-2$	9
$2+1+1+1+1$	4	0	$0-4=-4$	7
$1+1+1+1+1+1$	6	0	$0-6=-6$	5

(resp. $N(m,n)$) be the number of partitions of n with crank (resp. rank) m. Then, aside from the anomalous case of $M(m,1)$ (where the correct values are $M(0,1):=1$ and $M(m,1):=0$, for all $m\neq 0$), the two-parameter generating functions may be written as **[33, 45]**

$$C(\zeta;q) := \sum_{\substack{m\in\mathbb{Z}\\ n\geq 0}} M(m,n)\zeta^m q^n = \prod_{n=1}^{\infty}\frac{1-q^n}{(1-\zeta q^n)(1-\zeta^{-1}q^n)} = \frac{1-\zeta}{(q)_\infty}\sum_{n\in\mathbb{Z}}\frac{(-1)^n q^{\frac{n(n+1)}{2}}}{1-\zeta q^n}, \tag{14.8}$$

(14.9)

$$R(\zeta;q) := \sum_{\substack{m\in\mathbb{Z}\\ n\geq 0}} N(m,n)\zeta^m q^n = \sum_{n=0}^{\infty}\frac{q^{n^2}}{(\zeta q;q)_n(\zeta^{-1}q;q)_n} = \frac{1-\zeta}{(q;q)_\infty}\sum_{n\in\mathbb{Z}}\frac{(-1)^n q^{\frac{n(3n+1)}{2}}}{1-\zeta q^n}.$$

REMARK. Two specializations of the rank generating function should be highlighted. Thanks to a well-known identity arising from counting partitions according to the sizes of their so-called Durfee squares (which is also important in the proof of (14.9)), one easily finds our first example, which states that

$$P(q) = R(1;q) = \sum_{n=0}^{\infty}\frac{q^{n^2}}{(q;q)_n^2}, \tag{14.10}$$

where P is the partition generating function defined in (14.1). A second example is given by Ramanujan's third order mock theta function f (cf. (9.1)), whose definition may be restated as

$$f(q) = R(-1;q).$$

This gives a combinatorial interpretation of the coefficients of f as the number of partitions with even rank minus the number of those with odd rank. As we see below, the modularity properties of these two specializations are not coincidences, but manifestations of the rank generating function's nature as a "mock Jacobi form".

We also note that we are abusing notation slightly and in the remainder of this chapter $R(\zeta;q)$ refers only to the rank generating function, as opposed to Zwegers' completion R of (8.1). Although the generating functions in (14.8) and (14.9) are nearly identical, their modularity properties are very different. Indeed, these q-series, written as Lambert-type series, are identical apart from differing powers of q in their numerators. The crank generating function (14.8) is closely related to a Jacobi form, while the rank generating function (14.9) is essentially a mock Jacobi form. More specifically, in terms of Dedekind's η-function, the ϑ-function (cf. Definition 2.1), and the Zwegers μ-function (defined in Section 8.1), we have the following lemma (recall that $q = e(\tau)$ and $\zeta = e(z)$).

LEMMA 14.11. *The crank and rank generating functions can be expressed in terms of Jacobi and mock Jacobi forms as follows:*

i) *For $z \notin \mathbb{Z} + \mathbb{Z}\tau$, we have*

$$C(\zeta; q) = -\frac{2\sin(\pi z) q^{\frac{1}{24}} \eta^2(\tau)}{\vartheta(z;\tau)}.$$

ii) *If $z \notin \mathbb{Z} + \mathbb{Z}\tau$, we have*

$$R(\zeta; q) = -2\sin(\pi z)\left(\frac{q^{\frac{1}{24}}\eta(3\tau)^3}{\eta(\tau)\vartheta(3z;3\tau)} - q^{-\frac{1}{8}}\zeta^{-1}\mu(3z,-\tau;3\tau) + q^{-\frac{1}{8}}\zeta\mu(3z,\tau;3\tau)\right).$$

REMARK. We note that the statement of Lemma 14.11 ii) corrects a minor typographical error in [**520**]. Moreover, we make use of Zwegers' theory of Appell functions and offer a proof of Lemma 14.11 ii) different from the one appearing in [**520**].

SKETCH OF PROOF OF LEMMA 14.11. First we consider i). As explained in [**33**], an interpretation of the combinatorial definition of the crank functions leads to (14.8), which yields the claim using the definitions of η and ϑ. (The condition given in the statement of i) just avoids the poles arising when $\vartheta(z;\tau)$ has a zero.)

Now we turn to the proof of ii). Using Lemma 8.20, we can write

$$\begin{aligned} R(\zeta; q) &= \frac{1-\zeta}{\zeta^{\frac{3}{2}}(q)_\infty} A_3(z, -\tau; \tau) \\ &= \frac{1-\zeta}{\zeta^{\frac{3}{2}}(q)_\infty}\left(A_1(3z,-\tau;3\tau) + \zeta A_3(3z,0;3\tau) + \zeta^2 A_1(3z,\tau;3\tau)\right). \end{aligned}$$

A short computation then shows that

$$\begin{aligned} A_3(3z,0;3\tau) &= \frac{-i\eta^3(3\tau)}{\vartheta(3z;3\tau)}, \\ A_1(3z,\pm\tau;3\tau) &= \vartheta(\pm\tau;3\tau)\mu(3z,\pm\tau;3\tau), \\ \vartheta(\pm\tau;3\tau) &= \mp i q^{-\frac{1}{6}}\eta(\tau), \end{aligned}$$

which yields the claim when combined with the expression above for R. □

Using the properties of η and ϑ, it follows that C is essentially (that is, up to rational powers of ζ and q) a Jacobi form of weight $-1/2$ and index $-1/2$. Similarly, using these properties along with those of the Zwegers μ-function (given in Definition 8.1), it follows that R is essentially a mock Jacobi form. Using these (mock) modularity properties, one can prove many theorems about congruences and asymptotic properties of crank and rank partition functions, as we shall see in the next few subsections.

14.3.2. Properties of the crank partition function. Since the crank generating function is essentially a Jacobi form, we begin our discussion with this simpler case. We first consider certain crank statistics which sift partitions according to the residues of their cranks modulo t. More precisely, for any integer r and positive integer t, let $M(r,t;n)$ be the number of partitions of n whose crank is congruent to $r \pmod{t}$:

$$M(r,t;n) := \sum_{m \equiv r \pmod{t}} M(n,n).$$

For fixed r and odd t, Mahlburg **[373]** proved that these partition functions also satisfy Ramanujan-type congruences. In fact, he showed that such congruences hold for all arithmetic progressions $r \pmod{t}$ for any fixed odd t, thereby giving a Dyson-style explanation for partition function congruences of the form $(\mathcal{Q} > 3)$

$$p(An+B) \equiv 0 \pmod{\mathcal{Q}^j}.$$

Namely, these congruences follow directly from congruences for $M(r,t;n)$ thanks to the tautological identity

$$p(n) = \sum_{0 \le r \le t-1} M(r,t;n).$$

THEOREM 14.12 (Mahlburg). *Suppose that $t \geq 1$ is odd and that $\mathcal{Q} \nmid 6t$ is prime. If j is a positive integer, then there are infinitely many non-nested arithmetic progressions $An+B$ such that for every $0 \le r < t$ we have*

$$M(r,t;An+B) \equiv 0 \pmod{\mathcal{Q}^j}.$$

In particular, we have that

$$p(An+B) \equiv 0 \pmod{\mathcal{Q}^j}.$$

SKETCH OF PROOF. We first reduce the problem to a study of congruence properties of classical modular forms. By Lemma 14.11 i) we obtain, from Theorem 2.15, that specializing the generating function C at any torsion point gives a modular form on some congruence subgroup. The theorem then follows the proof of Theorem 14.6 *mutatis mutandis*. More precisely, the claimed congruences follow from the fact that a positive proportion of Hecke operators annihilate modular forms obtained by sieving the coefficients of the above specializations of C modulo $\mathcal{Q}^j$. This method of filtering out certain residue classes of coefficients has been used in many papers; see for example **[404]**. The existence of such primes follows from the Chebotarev density theorem, and the connection to the Hecke operators is made via the theory of ℓ-adic Galois representations and the Hecke equivariance of the Shimura correspondence. $\square$

Reformulating the conjecture of Dyson described above in our new notation, note that Andrews' and Garvan's combinatorial explanation for Ramanujan's congruences may be summarized in the following relations, valid for all non-negative integers n:

$$M(0,5;5n+4) = M(1,5;5n+4) = \cdots = M(3,5;5n+4) = M(4,5;5n+4),$$
$$M(0,7,7n+5) = M(1,7;7n+5) = \cdots = M(5,7;7n+5) = M(6,7;7n+5),$$
$$M(0,11;11n+6) = M(1,11;11n+6) = \cdots = M(10,11;11n+6).$$

The previous theorem demonstrates that the crank also plays a role in infinitely many partition congruences, but in a very different way. There we saw that the crank partition functions satisfy congruences themselves, which in turn imply congruences for $p(n)$. In light of these results, it is very natural to ask about the extent to which there are further identities for the sifted crank statistics, and what is the general theory underlying such identities. Much is now known in this direction, and there is indeed a rich structure of crank identities. For example, for n odd, we have

$$M(0,8;n)+M(1,8;n)=M(3,8;n)+M(4,8;n),$$

and related identities are known for the moduli $5,7,8,9,10$, and 11 (cf. [**217, 218, 219**]).

Such identities are rare. It turns out that cranks are not generally uniformly distributed among residue classes modulo t. In this direction, we have the following conjecture [**37**] in the special case of cranks modulo 3.

CONJECTURE 14.13 (Andrews and Lewis). *If $n \in \mathbb{N}$, then we have*

$$\begin{aligned} M\left(0,3;3n\right) &> M\left(1,3;3n\right), \\ M\left(0,3;3n+1\right) &< M\left(1,3;3n+1\right), \\ M\left(0,3;3n+2\right) &< M\left(0,3;3n+2\right) \quad \textit{unless } n \in \{1,4,5\}. \end{aligned}$$

This conjecture was proven by Kane [**295**], using the classical Circle Method. In particular, he determined the asymptotic behavior of $M(0,3;n)$ and $M(1,3;n)$ and explicitly bounded the resulting error terms. More general asymptotics for cranks were determined by Zapata [**523**].

Given the plethora of asymptotic results on cranks and related functions, the natural problem is to explain these as special cases of a general phenomenon. Motivated by heuristic integral calculations inspired by standard techniques in physics, Dyson envisioned [**181**] the general asymptotic underlying many of these results.

CONJECTURE 14.14 (Dyson). *As $n \to \infty$, we have*

$$M\left(m,n\right) \sim \frac{1}{4}\beta \operatorname{sech}^2\left(\frac{1}{2}\beta m\right) p(n),$$

where $\beta := \frac{\pi}{\sqrt{6n}}$.

We prove this conjecture in Chapter 15 below.

14.3.3. Properties of the rank partition function. In this subsection, we establish congruences for and inequalities between Dyson's rank functions. To do so, we make use of mock modular forms. Analogous to the sifted crank statistics, if r and t are integers, we let $N(r,t;n)$ be the number of partitions of n whose rank is $r \pmod{t}$. In analogy with Mahlburg's Theorem 14.12, we have the following result for rank congruences [**110**].

THEOREM 14.15. *Let t be a positive odd integer, and let $Q \nmid 6t$ be prime. If j is a positive integer, then there are infinitely many non-nested arithmetic progressions $An+B$ such that for every $0 \leq r < t$ we have*

$$N(r,t;An+B) \equiv 0 \pmod{Q^j}.$$

SKETCH OF PROOF. We begin as in the proof of Theorem 14.12. Namely, by Lemma 14.11 ii) and by a modified version of Theorem 2.15, we find that specializations of the rank generating function to torsion points are mock modular forms (using Corollary 8.15). In particular, using the usual orthogonality relations of roots of unity gives that for $t \in \mathbb{N}$ and $r \in \mathbb{Z}$

$$\sum_{n=0}^{\infty} N(r,t;n)q^n = \frac{1}{t}\sum_{n=0}^{\infty} p(n)q^n + \frac{1}{t}\sum_{j=1}^{t-1} \zeta_t^{-rj} R\left(\zeta_t^j;q\right), \tag{14.11}$$

where for $N \in \mathbb{N}$, $\zeta_N = e^{2\pi i/N}$. Thus a classical dissection argument then implies that the function

$$\sum_{n=0}^{\infty} \left(N(r,t;n) - \frac{1}{t}p(n)\right) q^{n-\frac{1}{24}}$$

is a mock theta function whose shadow is, up to a multiplicative constant,

$$\sum_{j=0}^{1} \sum_{n\equiv 2r+(-1)^j \pmod{2t}} (-1)^j \left(\frac{12}{n}\right) nq^{\frac{n^2}{24}}.$$

We are now able to reduce the problem to one involving classical modular forms. The key observation is that the shadow is always supported on finitely many square classes. Thus, we may sieve out the coefficients to restrict the Fourier expansion to run over appropriate arithmetic progressions away from the support of the shadow. This yields the holomorphic part of a harmonic Maass form (on a higher level congruence subgroup) with vanishing shadow, thus giving a classical weakly holomorphic modular form. To be more precise, the function defined by

$$\sum_{\substack{n\geq 0 \\ \left(\frac{1-24n}{Q}\right)=-1}} \left(N(r,t;n) - \frac{1}{t}p(n)\right) q^{n-\frac{1}{24}},$$

where the sum runs over those n that satisfy $\left(\frac{1-24n}{Q}\right) = -1$, is weakly holomorphic. Now the claim follows as in the proof of Theorem 14.6. □

REMARK. Similar congruences can be proven for general mock theta functions. This is because twisting operators always annihilate the shadow on certain arithmetic progressions and since the general theory of weakly holomorphic modular forms guarantees the existence of many congruences for classical modular forms (cf. [**478**]). It is expected that essentially all congruences of mock modular forms arise in this manner; for results in this direction see [**8, 17, 402**].

We now turn to inequalities for ranks. Early work in this direction was given in [**80**], where a modification of the method from [**20, 164**] determined the asymptotic expansion for $M(m,n)$ with m fixed. Since the formulas are quite lengthy, we do not state them here, and instead give an application. Using (14.11), one may directly obtain the asymptotic behavior of $N(r,s;n)$. This may then be used to prove a conjecture of Andrews and Lewis [**37**]. Analogous to the above inequalities for the crank function, using clever combinatorial methods, they showed that

$$\begin{aligned} N(0,2;2n) &< N(1,2;2n) && \text{if } n \geq 1,\\ N(0,4;n) &> N(2,4;n) && \text{if } 26 < n \equiv 0,1 \pmod 4,\\ N(0,4;n) &< N(2,4;n) && \text{if } 26 < n \equiv 2,3 \pmod 4. \end{aligned}$$

Moreover, they conjectured the following inequalities.

CONJECTURE 14.16 (Andrews and Lewis [**37**]). *For all $n \in \mathbb{N}$, we have*

$$\begin{aligned} N(0,3;n) &< N(1,3;n) \quad \text{if } n \equiv 0 \text{ or } 2 \pmod 3, \\ N(0,3;n) &> N(1,3;n) \quad \text{if } n \equiv 1 \pmod 3. \end{aligned} \tag{14.12}$$

A careful analysis of the asymptotic behavior of $N(j,3;n)$ for $j \in \{1,2\}$ gives the following theorem [**80**].

THEOREM 14.17. *The Andrews-Lewis Conjecture is true for all $n \notin \{3,9,21\}$, in which case we have equality in (*14.12*).*

REMARK. Note that refinements of Conjectures 14.13 and 14.16 were given by Chan and Mao [**140**].

The limiting shape of the rank generating function strongly resembles that of the crank and is given in Theorem 15.21 below. We remark in passing that this close asymptotic relationship is an early foreshadowing of deeper connections between modular forms, mock modular forms, and quantum modular forms (cf. Theorem 14.25).

14.4. Unimodal sequences

The previous sections describe many famous theorems concerning the rank and crank partition functions. Recent research has revealed that these two partition functions are intimately related to a third combinatorial function, which enumerates unimodal sequences. Here we recall the basic properties of the unimodal sequence generating function, and we show how its properties illuminate the relationship between modular forms, mock modular forms, and quantum modular forms.

DEFINITION 14.18. A finite sequence of positive integers $\{a_j\}_{j=1}^{s}$ is called a **unimodal sequence of size** n if there exists a positive integer k such that

$$1 \leq a_1 \leq a_2 \leq \cdots \leq a_k \geq a_{k+1} \geq \cdots \geq a_s \geq 1 \tag{14.13}$$

and $a_1 + \cdots + a_s = n$. Furthermore, such a sequence is called **strongly unimodal** if there exists a positive integer k such that

$$0 < a_1 < a_2 < \cdots < a_k > a_{k+1} > \cdots > a_s > 0. \tag{14.14}$$

We denote by $u^*(n)$ the number of unimodal sequences of size n and by $u(n)$ the number of strongly unimodal sequences of n.

REMARK. Examples of unimodal sequences are provided by the Stirling numbers of the first and second kind. We refer the reader to Stanley's articles [**469, 470**] for a broad overview of the various combinatorial appearances and applications of unimodal sequences.

The generating function of $u^*(n)$ is essentially a product of a modular form and a partial theta function (cf. [**471**]).

THEOREM 14.19. *We have that*

$$U^*(q) := \sum_{n=0}^{\infty} u^*(n)q^n = \frac{q^{\frac{1}{12}}}{\eta(\tau)^2} \sum_{n=1}^{\infty} (-1)^{n+1} q^{\frac{n(n+1)}{2}}.$$

The following theorem by Auluck and Wright, whose proof is given in Chapter 15 below, gives the asymptotic behavior of $u^*(n)$ (cf. [**46, 499, 500**]).

THEOREM 14.20 (Auluck and Wright). *As $n \to \infty$, we have that*

$$u^*(n) = 8^{-1}\left(3^3 n^5\right)^{-\frac{1}{4}} \exp\left(2\pi\sqrt{\frac{n}{3}}\right)\left(1 + O\left(n^{\frac{1}{2}}\right)\right).$$

The generating function for strongly unimodal sequences has some interesting mock behavior. Denote the relevant generating function by

$$U(q) := \sum_{n=0}^{\infty} u(n) q^n. \tag{14.15}$$

Rhoades proved [**436**] the following important identity (note, however, that Rhoades' notation for $u^*(n)$ and $u(n)$ is flipped with respect to the notation here).

THEOREM 14.21 (Rhoades). *We have*

$$4U(q) = -f(q) + (-q)_\infty^2 g(q),$$

where f is Ramanujan's third order mock theta function defined in (9.1) and

$$g(q) := \frac{2}{\left(q^2;q^2\right)_\infty (q)_\infty} \sum_{n\in\mathbb{Z}} \frac{q^{\frac{n(n+1)}{2}}}{1+q^n}$$

is another mock theta function.

REMARK. Andrews [**29**] found a related identity, namely

$$U(q) = \psi(-q) + 2(-q)_\infty^2 \alpha(-q),$$

where

$$\psi(q) := \sum_{n=1}^{\infty} \frac{q^{n^2}}{\left(q;q^2\right)_n}$$

is a third order mock theta function of Ramanujan and

$$\alpha(q) := \sum_{n=0}^{\infty} \frac{\left(-q^2;q^2\right)_n q^{n+1}}{\left(q;q^2\right)_{n+1}}$$

is a second order mock theta function [**380**] which also appears in Ramanujan's lost notebook.

SKETCH OF PROOF OF THEOREM 14.21. An elementary combinatorial argument shows that

$$U(q) = \sum_{n=0}^{\infty} (-q)_{n-1}^2 q^n. \tag{14.16}$$

A key step in the proof makes use of the following bilateral series identity from Entry 3.4.7 of Ramanujan's "lost" notebook (cf. p. 67 of [**30**]):

$$\sum_{n=0}^{\infty} \frac{(\xi_1\xi_2)^n q^{n^2}}{(\xi_1 q)_n (\xi_2 q)_n} + \sum_{n=1}^{\infty} \left(\xi_1^{-1}\right)_n \left(\xi_2^{-1}\right)_n q^n = -iq^{\frac{1}{8}} \xi_1^{\frac{1}{2}} \xi_2^{\frac{1}{2}} \left(\xi_1^{-1}\right)_\infty \left(\xi_2^{-1}\right)_\infty \mu\left(z_1, z_2; \tau\right), \tag{14.17}$$

where $\xi_j := e^{2\pi i z_j}$. The proof is completed by setting $\xi_1 = \xi_2 = -1$ and performing a short calculation. □

Theorem 14.21 is the key device which makes it possible to determine the asymptotic behavior of $u(n)$, as in Corollary 15.16.

REMARK. The strongly unimodal generating function is the $U(-1;q)$ case of the two-variable function defined in (14.18) below. To motivate the following discussion on ranks of unimodal sequences, we note that the specialization $U(1;q)$ is an elegant indefinite Hecke sum. Namely, we have (cf. Theorem 1.5 of **[133]**)

$$U(1;q) = \sum_{n=0}^{\infty} \sum_{6n \geq |6j+1|} (-1)^{j+1} q^{2n^2 - \frac{j(3j+1)}{2}}$$
$$+ 2 \sum_{n,m=0}^{\infty} \sum_{6n \geq |6j+1|} (-1)^{j+1} q^{2n^2+mn-\frac{j(3j+1)}{2}}.$$

Such Hecke sums are useful for obtaining congruences for $u(n)$. For example, for $3 < \ell \not\equiv 23 \pmod{24}$ prime and $\ell \nmid k$, the following congruences hold for all $n \in \mathbb{N}$:

$$u\left(\ell^2 n + k\ell - \frac{\ell^2 - 1}{24}\right) \equiv 0 \pmod{2}.$$

In analogy with Dyson's partition ranks, there is a notion of rank for strongly unimodal sequences. It turns out that these ranks are central in explaining the aforementioned role which unimodal sequences play in connecting certain modular forms, mock theta functions, and quantum modular forms. Here we describe how these objects come together in terms of q-series identities and asymptotic expansions. We begin with the definition of the rank of a unimodal sequence.

DEFINITION 14.22. In the notation above, the **rank** of a strongly unimodal sequence is equal to $s - 2k + 1$, i.e., it is the number of terms after the maximal term minus the number of terms that precede it.

REMARK. Further unimodal type statistics were recently introduced by Kim and Lovejoy **[306]**.

EXAMPLE 14.23. We consider the 6 strongly unimodal sequences of size 5:

Sequence	Rank
$\{5\}$	$0-0=0$
$\{1,4\}$	$0-1=-1$
$\{4,1\}$	$1-0=1$
$\{1,3,1\}$	$1-1=0$
$\{2,3\}$	$0-1=-1$
$\{3,2\}$	$1-0=1$

The generating function for $u(m,n)$, the number of size n and rank m strongly unimodal sequences, is given by

$$U(\zeta;q) := \sum_{n=0}^{\infty} \sum_{m \in \mathbb{Z}} u(m,n)(-\zeta)^m q^n = \sum_{n=0}^{\infty} (\zeta q;q)_n \left(\zeta^{-1}q;q\right)_n q^{n+1}. \tag{14.18}$$

This strongly unimodal rank generating function satisfies modular-like properties. For example, specializing $\zeta = -1$ yields the generating function for strongly unimodal sequences,

$$U(-1;q) = U(q)$$

defined in (14.16). Theorem 14.21 reveals that it is a mixed mock modular form. If $\zeta = \pm i$, we obtain a third order mock theta function of Ramanujan,

$$U(\pm i;q) = \Psi(q) := \sum_{n=1}^{\infty} \frac{q^{n^2}}{(q;q^2)_n}.$$

More generally, the generating function $U(\zeta;q)$ has (mock) Jacobi transformation properties. To state it, we let

$$H(z;\tau) := \frac{i}{2}\frac{\vartheta(z;\tau)}{\eta(\tau)}h(2z;\tau) - g(z;\tau),$$

where

$$g(z;\tau) := \frac{i}{\sqrt{3}}\int_{\mathbb{R}} e^{\frac{\pi i\tau t^2}{3} - 2\pi zt}\frac{\sinh\left(\frac{2\pi t}{3}\right)}{\cosh(\pi t)}dt$$

and h denotes the Mordell integral of Definition 8.4. Moreover, let

$$Y^{+}(z,\tau) := -2i\sin(\pi z)q^{-\frac{1}{24}}U\left(e^{2\pi iz}; e^{2\pi i\tau}\right).$$

REMARK. We warn the reader that the symbol Y^{+} is also used for the Jacobi raising operator.

THEOREM 14.24. *Assuming the notation above, we have the following transformation properties:*

$$\begin{aligned}
&Y^{+}(z;\tau) - e^{\frac{\pi i}{12}}Y^{+}(z;\tau+1) = 0,\\
&Y^{+}(z;\tau) + ie^{\frac{3\pi iz^2}{\tau}}(-i\tau)^{-\frac{1}{2}}Y^{+}\left(\frac{z}{\tau};-\frac{1}{\tau}\right) = -H(z;\tau),\\
&Y^{+}(z;\tau) + Y^{+}(z+1;\tau) = 0,\\
&Y^{+}(z;\tau) + e^{-6\pi iz-3\pi i\tau}Y^{+}(z+\tau;\tau)\\
&= e^{-5\pi iz-\frac{25\pi i\tau}{12}}\left(1-e^{4\pi iz+2\pi i\tau}\right) - \frac{i\vartheta(z;\tau)}{\eta(\tau)}\left(e^{-2\pi iz-\frac{\pi i\tau}{4}} - e^{-6\pi iz-\frac{9\pi i\tau}{4}}\right).
\end{aligned}$$

SKETCH OF PROOF. One may decompose

$$\begin{aligned}
Y^{+}(z,\tau) = i\zeta^{-1}q^{-\frac{1}{6}}\mu(3z,-\tau;3\tau) - i\zeta q^{-\frac{1}{6}}\mu(3z,\tau;3\tau)\\
- i\frac{\eta^3(3\tau)}{\eta(\tau)\vartheta(3z;3\tau)} + \frac{\vartheta(z;\tau)}{\eta(\tau)}\mu(z,-z;\tau).
\end{aligned}$$

The claimed properties then follow from the properties of μ, namely those listed in Propositions 8.2 and 8.7. □

REMARK. The function

$$\widehat{X}(z;\tau) := \left(\widehat{A}_3(z,0;\tau) - \widehat{A}_1(z,-z;\tau)\right)e^{-\frac{3\pi z^2}{2v}}$$

(where $\widehat{A}_3$ is the completed level 3 Appell function as in Definition 8.21) transforms as a weight 1 index 0 Jacobi form, and $Y^{+}(z;\tau)$ is the holomorphic part of

$$\widehat{X}(z;\tau)e^{\frac{3\pi z^2}{2v}}\frac{1}{\eta(\tau)}.$$

This combinatorial q-series enjoys some unexpected properties in terms of its behavior at roots of unity. To explain this, we first consider the special case where $\zeta = -1$, which was considered by Bryson, Pitman, Rhoades and one of the authors in **[133]**. This generating function has the following simple description:[1]

$$U(1;q) = \sum_{n=1}^{\infty} \sum_{m\in\mathbb{Z}} (-1)^m u(m,n) q^n = \sum_{n=1}^{\infty} \left(u_e(n) - u_o(n)\right) q^n,$$

where $u_e(n)$ (resp. $u_o(n)$) denotes the number of unimodal sequences of size n with even (resp. odd) rank. Therefore, it is the generating function for the difference between the number of size n strongly unimodal sequences with even rank minus the number with odd rank. In **[133]**, the authors showed that for every root of unity ζ,

$$U(1;\zeta) = F\left(\zeta^{-1}\right), \tag{14.19}$$

where Kontsevich's function $F(q)$ is defined by

$$F(q) := \sum_{n=0}^{\infty} (q;q)_n. \tag{14.20}$$

By **[266, 270]**, equation (14.19) can also be interpreted in terms of quantum topology. In particular, using colored Jones polynomials for certain torus knots at roots of unity (see also **[269]** for related work), the authors in **[266, 270]** introduced infinite families of "quantum modular forms" including $F(q)$ and $U(1;q)$. These quantum modular forms are instances of a newly defined theory of Zagier, which describes various almost modular-type objects and which are discussed in detail in Chapter 21. As with the generating function U^*, there is a connection between F and partial theta functions, but one which is only visible in the limit towards roots of unity. Namely, Zagier **[515]** proved that this function satisfies the "strange identity"

$$F(q) \text{``}=\text{''} -\frac{1}{2} \sum_{n=1}^{\infty} n\chi_{12}(n) q^{\frac{n^2-1}{24}}. \tag{14.21}$$

This is a first hint of a mock modular picture. More precisely, note that the right-hand side of (14.21) is an Eichler integral of the Dedekind η-function. That is, since the weight of η is $1/2$, the expected Eichler integral should be the analogue of an inverse image under the "Bol operator" of Lemma 5.3 which formally should become $D^{\frac{1}{2}}$. Thus, one might not be surprised to see a "half-derivative" of η arise. The connection with a half-derivative is made via the Pentagonal Number Theorem, which implies that

$$\eta(\tau) = \sum_{n=1}^{\infty} \chi_{12}(n) q^{\frac{n^2}{24}}.$$

Thus, one should expect a half-derivative to act on q expansions by multiplying each coefficient of q^n by $n^{\frac{1}{2}}$, and given the quadratic exponents in the Pentagonal

[1]The reader should note that our normalization of $U(\zeta;q)$ differs in the first argument by a minus sign from the notation of **[133]**.

Number Theorem, a reasonable "half-derivative" of η can be given by (a multiple of)

$$\sum_{n=1}^{\infty} n\chi_{12}(n)q^{\frac{n^2}{24}}.$$

The identity (14.21) is termed "strange" since the two sides never converge simultaneously. Indeed, the right-hand side of (14.21) converges in the unit disk $|q| < 1$, but nowhere on the unit circle, whereas the left-hand side converges only at roots of unity (where it is clearly a finite sum). The identity (14.21) is properly interpreted as a shorthand notation which means that at roots of unity ζ, $F(\zeta)$ agrees with the limit as q approaches ζ radially within the unit disk of the function on the right-hand side of (14.21). Moreover, Zagier proved that for $x \in \mathbb{Q} \setminus \{0\}$

$$\phi(x) + (ix)^{-\frac{3}{2}}\phi\left(-\frac{1}{x}\right) = \frac{\sqrt{3i}}{2\pi}\int_0^{i\infty} \frac{\eta(w)}{(w-x)^{\frac{3}{2}}}dw, \tag{14.22}$$

where $\phi(x) := e^{\frac{\pi i x}{12}}F(e^{2\pi i x})$. In particular this yields that F, and thus $U(1;q)$, is a quantum modular form. We offer more details on (14.22) and its proof in Chapter 21.

Zagier used (14.21) to determine the asymptotic behavior of $F(e^{-t})$ as $t \to 0^+$, thereby greatly improving previous bounds in knot theory towards the open problem of determining the number of linearly independent Vassiliev invariants of degree D. To be more precise, Zagier proved that

$$e^{-\frac{t}{24}}F\left(e^{-t}\right) = \sum_{n=0}^{\infty} \frac{T_n}{n!}\left(\frac{t}{24}\right)^n, \tag{14.23}$$

where T_n denotes the n-th Glaisher number (which is essentially a special value of the Dirichlet L-function for χ_{12}). This allows us to understand certain unimodal sequence generating functions, as (14.19) then implies a similar asymptotic for $U(-1;q)$. We discuss results related to this problem in Chapter 21.

REMARK. Note that general Eichler integrals of half-integral weight modular forms were considered by two of the authors [**116**]. A similar connection to L-values of twisted modular L-functions holds in general.

We conclude this section by giving our first result bridging the modular, mock modular, and quantum modular worlds. The next theorem establishes that the rank, the crank, and the unimodal rank generating functions are nicely linked by their radial limits at suitable roots of unity. This result was obtained by Rhoades and two of the authors [**203**], and it represents a prototypical example of the role of so-called quantum modular forms in the theory of modular forms. The proof is given in Chapter 21 below.

THEOREM 14.25. *Let* $1 \leq a < b$, $1 \leq h < k$ *with* $\gcd(a,b) = \gcd(h,k) = 1$, $b|h$ *and* $h' \in \mathbb{Z}$ *with* $hh' \equiv -1 \pmod{k}$. *Then, as* $q \to \zeta_k^h$ *radially within the unit disc, we have that*

$$\lim_{q \to \zeta_k^h} \left(R\left(\zeta_b^a;q\right) - \zeta_{b^2}^{-a^2h'k}C\left(\zeta_b^a;q\right)\right) = -\left(1-\zeta_b^a\right)\left(1-\zeta_b^{-a}\right)U\left(\zeta_b^a;\zeta_k^h\right).$$

REMARK. The conditions on a, b, h, k in the previous theorem are quite natural, as they determine exactly when the Appell-Lerch-type sum representations of the rank and crank generating functions have a pole in the denominator.

14.5. Andrews' spt-function

We conclude this chapter with one last application to a relatively new combinatorial object. Recently, Andrews **[28]** introduced an interesting partition statistic which is closely related to the story of ranks and cranks. To describe this, let $\mathrm{spt}(n)$ denote the number of smallest parts in the partitions of n. For example, there are 7 partitions of 5,

$$5,\ 4+1,\ 3+2,\ 3+1+1,\ 2+2+1,\ 2+1+1+1,\ 1+1+1+1+1,$$

and one counts $\mathrm{spt}(5) = 14$.

Our first examples of the arithmetic significance of this new partition function are the following Ramanujan-type congruences due to Andrews **[28]**.

THEOREM 14.26. *For all $n \in \mathbb{N}_0$, the following congruences hold:*

$$\begin{aligned} \mathrm{spt}(5\mathrm{n}+4) &\equiv 0 \pmod 5, \\ \mathrm{spt}(7\mathrm{n}+5) &\equiv 0 \pmod 7, \\ \mathrm{spt}(13\mathrm{n}+6) &\equiv 0 \pmod{13}. \end{aligned}$$

REMARK. Recently, a combinatorial interpretation of the first two congruences in Theorem 14.26 was found by Andrews, Garvan, and Liang **[35]**.

The main idea in Andrews' proof of Theorem 14.26, was to relate the spt-function to rank and crank moments, which we now recall **[44]**.

DEFINITION 14.27. For a non-negative integer k, define the k-th **crank moment** as

$$M_k(n) := \sum_{m\in\mathbb{Z}} m^k\, M(m,n), \tag{14.24}$$

and the k-th **rank moment** as

$$N_k(n) := \sum_{m\in\mathbb{Z}} m^k\, N(m,n). \tag{14.25}$$

REMARK. Due to the symmetries of the statistics, the crank and rank moments vanish if k is odd. However, in this case it makes sense to consider asymmetricized versions of odd moments, where the sums in (14.24) and (14.25) run only through $\mathbb{N}$. In this case, one loses modularity properties, and so the problems of determining their congruences and asymptotics become much more subtle. Such moments are considered in, for example, **[31, 103, 140]**.

We have already seen that the connection between ranks and cranks underlies important ties between the theories of modular, mock modular, and quantum modular forms. It turns out that the connection between ranks and cranks also runs very deep. Besides asymptotic relationships, there are also beautiful identities relating the two generating functions, which were first discovered by Atkin and Garvan **[44]**. In the context of moments, this allowed them to show that

$$N_{2k}(n) = P_k(n)N_2(n) + \sum_{j=1}^{k} Q_{k,j}(n)M_{2j}(n),$$

where P_k is a polynomial of degree $k-1$ and $Q_{k,j}$ is a polynomial of degree $k-j$. For example, we have the simple relation

$$N_4(n) = -\left(2n+\frac{2}{3}\right)M_2(n) + \frac{8}{3}M_4(n) + (1-12n)N_2(n).$$

A key to proving such identities, as well as congruences, is a partial differential equation relating the rank and the crank, which we now recall. To state this, it is most convenient to work with the normalized versions of the rank and the crank generating functions, which are given by

$$R^*(\zeta;q) := \frac{\zeta^{\frac{1}{2}}q^{-\frac{1}{24}}R(\zeta;q)}{1-\zeta}, \qquad C^*(\zeta;q) := \frac{\zeta^{\frac{1}{2}}q^{-\frac{1}{24}}C(\zeta;q)}{1-\zeta}. \tag{14.26}$$

Atkin and Garvan [**44**] then proved the following differential equation called the *rank-crank PDE*, where for a general variable x, we set $\mathcal{D}_x := x\frac{\partial}{\partial x}$. Before stating the result, we note that the differential operator in the PDE is a certain heat operator, which also plays a role in the classical theory of the Jacobi theta function and the general structure of Jacobi forms.

THEOREM 14.28. *Assuming the notation above, we have*

$$2\eta(\tau)^2C^*(\zeta;q)^3 = \left(6\mathcal{D}_q + \mathcal{D}_\zeta^2\right)R^*(\zeta;q).$$

REMARK. Such identities can be generalized in the framework of nonholomorphic Jacobi forms. The crux of the argument lies in the fact that the heat operator annihilates the nonholomorphic part of certain nonholomorphic Jacobi forms.

REMARK. In view of the results of Section 11.3, we now know that such differential equations have natural interpretations in the context of meromorphic Jacobi forms. In fact, they are direct consequences of structural theorems which state that an arbitrary meromorphic Jacobi form of negative index may be decomposed as a sum over its poles in a fundamental parallelogram, multiplied with derivatives of certain canonical Appell-Lerch-type sums.

Zwegers [**530**] nicely generalized Theorem 14.28 for arbitrary odd powers of the crank generating function using the theory of elliptic forms. For similar results giving a direct and clever proof of families of rank-crank-type PDEs using the theory of elliptic functions, see Chan–Dixit–Garvan [**139**].

Returning to Andrews' function spt(n), Andrews beautifully related it to ranks in Theorem 3 of [**28**], which then directly implies Theorem 14.26.

THEOREM 14.29. *For any $n \in \mathbb{N}$, we have*

$$\operatorname{spt}(n) = np(n) - \frac{1}{2}N_2(n).$$

REMARK. Note in particular that

$$np(n) = \frac{1}{2}M_2(n).$$

REMARK. There is also a generalization of Theorem 14.29 to higher order spt-functions [**220**]. These are related to the differences

$$M_{2k}(n) - N_{2k}(n).$$

Further beautiful congruences for the spt-function were found by Garvan [**221**]. Later, one of the authors discovered using the theory of harmonic Maass forms that such congruences exist for any prime at least 5 [**409**] (which can be generalized to all prime powers). Concerning its proof, we merely remark that the philosophy is very similar to the proof of Theorem 14.15 above. Namely, once the connection to mock modular forms is known (cf. Theorem 14.31 below), one applies sieving operators to reduce the relevant mock modular forms to weakly holomorphic modular forms, and in this case there are congruences for all primes thanks to the fact that the corresponding spaces of cusp forms always vanish. Such systematic congruences of the spt-function have also been derived using the structure of Hecke operators action on so-called mock-modular grids in [**6**].

THEOREM 14.30. *For any prime $\ell \geq 5$ and any positive integers m, n, the following are true:*

i) *If $\left(\frac{-n}{\ell}\right) = 1$, then we have*

$$\operatorname{spt}\left(\frac{\ell^{2m}n+1}{24}\right) \equiv 0 \pmod{\ell^m}.$$

ii) *We have*

$$\operatorname{spt}\left(\frac{\ell^{2m+1}n+1}{24}\right) \equiv \left(\frac{3}{\ell}\right)\operatorname{spt}\left(\frac{\ell^{2m-1}n+1}{24}\right) \pmod{\ell^m}.$$

We conclude our discussion of the congruences for the spt function by noting that in contrast to the partition function, a great deal is known about the residues of $\operatorname{spt}(n)$ modulo 2 and 3. In particular, two of the authors determined simple conditions to decide when $\operatorname{spt}(n)$ is odd and gave elegant congruences modulo 3 in [**202**]. The criterion of [**202**] was proven using q-series techniques in [**36**] (note that they correct a slight misstatement of Theorem 1.2 of [**202**]).

Following our usual mantra, the above congruences fall into place once we have elucidated the automorphic properties of the spt-generating function. To describe the relation to the main objects of this book, we first denote

$$S(q) := \sum_{n=0}^{\infty} \operatorname{spt}(n)q^n.$$

The relation we are searching for is given by the following theorem [**79**].

THEOREM 14.31. *The completed generating function*

$$\mathcal{M}(\tau) := q^{-1}S\left(q^{24}\right) - \frac{E_2(24\tau)}{12\eta(24\tau)} + N(\tau)$$

with

$$N(\tau) := -\frac{i}{4\pi\sqrt{2}}\int_{-\overline{\tau}}^{i\infty} \frac{\eta(24w)}{(-i(\tau+w))^{\frac{3}{2}}}dw$$

is a harmonic Maass form of weight $3/2$ on $\Gamma_0(576)$ with Nebentypus $\left(\frac{12}{\cdot}\right)$.

Theorem 14.31 also allows us to determine the asymptotic behavior of $\operatorname{spt}(n)$. Namely, standard calculations reveal that [**79**]

$$\mathrm{spt(n)} \sim \frac{1}{\pi\sqrt{2\mathrm{n}}}\mathrm{e}^{\pi\sqrt{\frac{2\mathrm{n}}{3}}} \quad (\mathrm{n} \to \infty). \tag{14.27}$$

REMARK. We note that (14.27) is rather surprising as one can show that

$$N_2(n), M_2(n) \sim \frac{1}{2\sqrt{3}} e^{\pi\sqrt{\frac{2n}{3}}}.$$

This implies in particular that there is a lot of cancellation in the asymptotics of rank and crank moments.

CHAPTER 15

Asymptotics for Coefficients of Modular-type Functions

15.1. Prologue

The determination of asymptotic behavior for functions and sequences is an old subject of broad importance in mathematics. Problems pertaining to asymptotic expansions have been a staple of analysis, probability, combinatorics, and statistical mechanics for hundreds of years and are related to many of the central problems of analytic number theory. A classical example, for the sequence of integer partitions, is given in Theorem 14.3. There we state Rademacher's exact formula and Hardy-Ramanujan's main asymptotic expression for $p(n)$. The proof of the Hardy-Ramanujan asymptotic for $p(n)$, given in Section 14.1, uses the Circle Method and Ingham's Tauberian Theorem. In both cases, the key idea is that growth of a sequence can be deduced from asymptotic properties of its generating function

$$P(q) = \sum_{n=0}^{\infty} p(n)q^n$$

near the boundary of its circle of convergence. In the case of $p(n)$, the asymptotic expansion of P is determined by an elementary computation using its modular transformation properties. In particular, the rapid growth of the numbers $p(n)$ is a direct consequence of the fact that P has poles on the unit circle. In fact, such estimates of modular forms near the real line (or the unit circle, depending on whether one uses the variable τ or $q = e^{2\pi i\tau}$) are deeply tied to Ramanujan's original interest in and discovery of the mock theta functions.

To describe this, we recall that it is straightforward to determine the asymptotic expansions of modular forms of positive weight, as Ramanujan knew. For example, for a holomorphic modular form f of positive weight k, the function f grows at most like v^{-k} as τ approaches zero. Moreover, in the full asymptotic expansion of f, this is the only term, meaning that upon subtracting this term and approaching 0, f decays rapidly. Similar statements hold for the other cusps. Thus, finding the asymptotic expansions of holomorphic positive weight modular forms near the cusps is relatively elementary.

Given the prominent place of many q-hypergeometric modular form examples within number theory, it is natural to try to detect modularity of such series in a routine fashion. This is well known to be a difficult question. For example, a famous open problem of Andrews (cf. Section 6 of [**26**]) asks whether one can directly prove the modularity of the partition generating function using its Durfee square representation given in (14.10). As Ramanujan noted in his last letter to Hardy, generic q-hypergeometric series have very different asymptotic behavior than modular forms do; instead of having just a single term in their asymptotic expansions, they have complicated infinite expansions. It turns out that q-hypergeometric sums which asymptotically resemble modular forms are exceedingly rare; for example, Zagier used the Euler-Maclaurin summation method (a classical tool of asymptotic analysis which we encounter in Section 15.7) in **[518]** to establish a special case of Nahm's conjecture (which Nahm gave in **[393, 394, 395]**) connecting modular q-series to torsion elements of the Bloch group. In particular, Zagier showed that there are just seven rational choices of A, B, C for which

$$\sum_{n=0}^{\infty} \frac{q^{\frac{A}{2}n^2+Bn+C}}{(q)_n}$$

is a modular form (in particular, they are all modular functions). The mock theta functions stand out among q-hypergeometric series due to their peculiar intermediary asymptotic properties lying between the single term expansions of modular forms and the complicated expansions of generic q-series. Specifically, there is more than one term in their asymptotic expansions near certain cusps, and so they cannot be modular forms. However, they have a finite asymptotic expansion whose main order of growth is determined by one term in the expansion (which matches the expansion of a weakly holomorphic modular form). That is, for any mock theta function f, and for any root of unity ζ, there exists a weakly holomorphic modular form g such that as $q \to \zeta$ radially, we have

$$f(q) - g(q) = O(1).$$

Ramanujan's attempt to codify this analytic structure is Definition 9.1. We remark that the proof that Ramanujan's examples do indeed satisfy his own definition is given in Section 9.8 using the modern theory of harmonic Maass forms, and further extensions of this theory are connected to "quantum modular forms" in Section 21.4. Given their history, it is therefore not surprising that mock modular forms play a deep role in the asymptotic analysis of many sequences arising in combinatorics and physics. In this chapter, we study the asymptotics of coefficients of harmonic Maass forms (in fact, as we see below, this is also useful even in the case of weakly holomorphic modular forms), as well as various analytic tools which are useful for studying a variety of mock modular-type objects which frequently arise in combinatorial applications. We continue in the next section with an outline of these techniques and applications.

15.2. Asymptotic methods

In this chapter we consider various types of modular objects, and we discuss some of the main asymptotic techniques which can be applied to them. We begin in Section 15.3 with a discussion of the upper bounds and distributions of coefficients of holomorphic modular forms. Since the coefficients of Eisenstein series are not difficult to understand, the main interest lies in the study of coefficients of cusp forms; this is a deep and difficult problem. In contrast to the case of the partition function, the coefficients of these forms oscillate, and their growth is at most polynomial in magnitude. In Section 15.3, we review some of the main results, which rely on tools from arithmetic geometry and analytic number theory.

Modulo the difficult problem of optimally bounding coefficients of cusp forms, the remainder of the modular-type objects we consider here have rapidly growing Fourier coefficients, which are studied by completely different methods. This book highlights three such techniques, each of which has advantages in different situations. Firstly, one can use results such as Ingham's Tauberian Theorem, stated in Theorem 14.4. Tauberian Theorems (partial converses to Abel's theorem on limits of power series towards points on the boundary of their radii of convergence) are useful for studying combinatorial sequences. In particular, Theorem 14.4 is applicable to sequences, such as the integer partition function, which are monotonically increasing. Many natural sequences enumerating quantities in combinatorics or physics have such monotonic behavior. We also note that Tauberian Theorems are extremely powerful tools, allowing one to give elegant proofs of the prime number theorem, for example.

Secondly, we see in the sketch of proof of Theorem 14.3 that Hardy and Ramanujan invented the Circle Method for estimating the size of $p(n)$. This simple idea requires a delicate analysis and choice of integrals "near" poles on the unit circle, and can be modified to yield quite flexible tools. As further developed by Hardy and Littlewood, this method is also useful in the study of many problems not directly related to modular forms, such as Diophantine problems (much work has been done using the Circle Method to study Waring's problem, as Hardy and Littewood did themselves; see also the excellent exposition in Vaughan's monograph [**481**]), and played an integral role in Helfgott's recent proof of the weak Goldbach conjecture [**262**]. In Section 15.6, we review the application of a version of the Circle Method to the mixed mock modular forms of Section 13.2. We discuss another modification of the Circle Method, due to Wright, in Section 15.7. There we highlight its application to the resolution of a problem of Dyson on the asymptotics of cranks of partitions (which are studied in Section 14.3). Our third and last main technique is to employ the theory of Poincaré series, such as the Maass-Poincaré series of Section 6.3. These are especially convenient as the analysis required to estimate their coefficients can be "pre-computed" for all Maass-Poincaré series in a uniform fashion, although it can require significant effort, especially in "boundary cases" when one requires the spectral theory of automorphic forms (see for example the fundamental work of Goldfeld and Sarnak [**225**]). Such estimates, and in fact exact Rademacher-type formulas analogous to the full asymptotic series for $p(n)$ in Theorem 14.3, are given in the general case of Maass-Poincaré series in Theorem 6.11. The idea is that once this calculation is performed, every harmonic Maass form can be directly expressed as a linear combination of these Poincaré series, with the coefficients in this linear combination only coming from the principal part of the original form. Analogous Poincaré series for modular and harmonic forms with poles in the upper half-plane are discussed in Section 15.5, where we study their application to quotients of modular forms investigated by Hardy-Ramanujan and Berndt-Bialek-Yee. These polar Maass-Poincaré series are examples of the polar harmonic Maass forms of Section 13.3; we also discussed how these Poincaré series are analogous to the Asai-Kaneko-Ninomiya functions arising in the denominator formula of the Monster Lie algebra (cf. Section 20.1 for a discussion of Monstrous Moonshine).

15.3. Classical holomorphic modular forms

Here we recall what is known about asymptotics and estimates for coefficients of holomorphic modular forms. More details and related results can also be found in Iwaniec's book [**282**], in particular Chapter 5 therein. The simplest modular forms are the Eisenstein series. Since their coefficients are essentially divisor sums, they are relatively easy to understand. For example, one has the elementary estimate ($r \in \mathbb{N}_0$)

$$n^r \leq \sigma_r(n) \leq n^r \zeta(r), \tag{15.1}$$

where, as before, $\sigma_r(n)$ is the sum of r-th powers of divisors of n, and ζ denotes the Riemann ζ-function. Thus, by Corollary 1.7, the Fourier coefficients of the level 1 Eisenstein series E_{2k} grow at the same rate as n^{2k-1}. Estimates for Eisenstein series on congruence subgroups and of half-integral weight can be proven similarly.

Since all holomorphic modular forms are sums of Eisenstein series and cusp forms, in order to fully describe the coefficients of modular forms, we are reduced to studying the coefficients of cusp forms. This is a more subtle problem, and the best known results require many powerful techniques from analytic number theory, algebraic geometry, and Galois representations. Cusp form coefficients are also, of course, extremely important to understand. For example, the Modularity Theorem of Wiles and Breuil-Conrad-Diamond-Taylor, previously conjectured by Shimura, Taniyama, and Weil, shows that the Hasse-Weil L-functions of elliptic curves are encoded by weight 2 newforms (cf. Chapter 19 for a more detailed discussion of the Modularity Theorem). The coefficients of cusp forms grow much slower than the coefficients of Eisenstein series, and they also oscillate strongly. Although they do not have simple asymptotic expansions like Eisenstein series coefficients, it is natural to ask what upper bounds for cusp form coefficients can be established. It turns out that coefficients of cusp forms grow relatively slowly. Roughly speaking, this is due to the fact that they decay rapidly at the cusps. Specifically, one can show the well-known *Hecke bound* in an elementary manner. In the following, we let $c_f(n)$ denote the n-th Fourier coefficient of a cusp form f.

THEOREM 15.1. *For $f \in S_k(\Gamma_0(N))$ with $k \in \frac{1}{2}\mathbb{N}$, we have*

$$c_f(n) \ll_f n^{\frac{k}{2}}.$$

PROOF. We follow the proof of Proposition 8 of **[519]**. First, consider the product $v^{\frac{k}{2}}|f(\tau)|$, which transforms as a modular function. Since f is a cusp form, this function is bounded. Thus, we have that $|f(\tau)| \leq Cv^{-\frac{k}{2}}$ for all $\tau \in \mathbb{H}$ for some constant C. Expressing the coefficients of f as (for any $t > 0$)

$$c_f(n) = e^{2\pi nt} \int_0^1 f(u+it)e^{-2\pi inu}du, \tag{15.2}$$

we find that $|c_f(n)| \leq Ct^{-\frac{k}{2}}e^{2\pi nt}$. Choosing $t = n^{-1}$ produces the claim. □

Although the proof of Theorem 15.1 is elementary, it does not capture the true order of growth of cusp form coefficients. Refining this estimate to obtain optimal bounds is a hard problem. We begin by describing the integral weight case, for which much more is known. By the Atkin-Lehner-Li theory of newforms, every cusp form can be written as a sum of Hecke eigenforms, which have the advantageous property that their L-functions have Euler products. As we see below, this is necessary in order to use connections to geometry, and in particular Deligne's proof of the Weil conjectures. Thus, we can and often do restrict our study to newforms.

The first case to consider in integral weight is weight 2. In this case, the Eichler-Shimura Theorem (given as Theorem 1.17, which is a simplified version of

Theorem 11.74 of [**311**]) gives a converse of the Modularity Theorem, stating that every weight two newform f with integer Fourier coefficients encodes the L-function of an elliptic curve over $\mathbb{Q}$. Here, we can apply the *Hasse bound* which says (cf. [**462**], Chapter V, Theorem 1.1) that for any elliptic curve $E/\mathbb{Q}$ and for any prime p of good reduction for E, we have

$$|c_f(p)| = |\#E(\mathbb{F}_p) - (p+1)| \leq 2\sqrt{p}, \tag{15.3}$$

where $\#E(\mathbb{F}_p)$ is the number of points of E over $\mathbb{F}_p$. Using the multiplicativity relations between Hecke operators, this directly implies that the coefficients of any weight two cusp form $f \in S_2^{\text{new}}(\Gamma_0(N))$ satisfy

$$c_f(n) \ll_f n^{\frac{1}{2}+\varepsilon},$$

which gives a power saving of $1/2 - \varepsilon$ over the Hecke bound.

Analogous results in higher weights are more difficult to prove, as one can no longer apply the theory of elliptic curves. A classical question concerns the case of the first level 1 cusp form, $\Delta \in S_{12}$, given explicitly by

$$\Delta(\tau) = q \prod_{n=1}^{\infty} (1-q^n) =: \sum_{n=1}^{\infty} \tau(n) q^n.$$

In this case, the *Ramanujan-Petersson Conjecture* (posed on page 176 of [**430**] by Ramanujan for Δ and more generally by Petersson in [**418**]) states the following.

CONJECTURE 15.2 (Ramanujan and Petersson). *For any prime p, we have*

$$|\tau(p)| \leq 2p^{\frac{11}{2}}.$$

This follows as a consequence of Deligne's monumental proof in [**159**] of the Weil conjectures (although this conclusion also took considerable effort and followed from a combination of work and ideas of Deligne, Ihara, Kuga, Sato, and Shimura). These conjectures concern the specific structure of local zeta functions of algebraic varieties, which are built out of point counts of the reductions of a variety over all finite fields of a fixed characteristic, like the point count in (15.3). We see several instances of the Hasse-Weil L-function for elliptic curves in this book; in this case and in general, these factors may be used to define Euler products for "Hasse-Weil zeta functions" of varieties. In the case of fields of a fixed prime characteristic, the Weil conjectures postulate that the local zeta functions have a simple structure; for example, they are are rational functions satisfying a functional equation and a Riemann hypothesis (for elliptic curves, the Riemann hypothesis is equivalent to the Hasse bound). After connecting coefficients of cusp forms to algebraic varieties, Deligne's proof implies the following optimal result for integral weight cusp forms (for example, instead of being related to an elliptic curve, the coefficients of Δ are related to a Kuga-Sato variety).

THEOREM 15.3 (Deligne). *If $f \in S_k^{\text{new}}(\Gamma_0(N))$ with $k \in \mathbb{N}$, then*

$$|c_f(n)| \leq \sigma_0(n) n^{\frac{k-1}{2}}.$$

Although Theorem 15.3 describes an optimal upper bound for coefficients of cusp forms, it is natural to ask how the coefficients $c_f(n)$ distribute in the interval $[-\sigma_0(n)n^{\frac{k-1}{2}}, \sigma_0(n)n^{\frac{k-1}{2}}]$. As mentioned above, unlike the Eisenstein series, there is huge oscillation in the size of the coefficients and explicitly specifying this variation is quite difficult; for example Lehmer's open conjecture on the coefficients of Δ asks if the coefficients ever vanish.

In view of Theorem 15.3, we can phrase this question in terms of "angles" for primes p. That is, the theorem implies that for any prime p there is an angle $\theta_p = \theta_{p,f} \in [0, \pi)$ such that $c_f(p) = 2p^{\frac{k-1}{2}} \cos(\theta_p)$. There are two very different cases, in which completely different distributions arise. First consider the case when the newform f has CM, say by an order in an imaginary quadratic field K. Then $c_f(p) = 0$ exactly if p is inert or ramified in the ring of integers $\mathcal{O}_K$, and so only the case of split primes is interesting. In this case, Hecke's equidistribution result of **[258]** implies the following.

THEOREM 15.4 (Hecke). *Let $k \geq 2$ and suppose $f \in S_k^{\mathrm{new}}(\Gamma_0(N))$ has CM by an order in K. Then, for any $-1 \leq \alpha \leq \beta \leq 1$, we have that*

$$\lim_{n\to\infty} \frac{\#\{p \leq n : p \text{ splits in } \mathcal{O}_K,\ \alpha \leq \cos(\theta_p) \leq \beta\}}{\#\{p \leq n : p \text{ splits in } \mathcal{O}_K\}} = \frac{1}{\pi}\int_\alpha^\beta \frac{d\theta}{\sqrt{1-\theta^2}}.$$

The distribution of coefficients of newforms in the non-CM case is very different, and much more difficult to prove. According to the conjecture of Sato and Tate, limiting distributions of these angles are given by the following measure.

CONJECTURE 15.5 (Sato and Tate). *For every two numbers α, β with $0 \leq \alpha \leq \beta \leq \pi$, and for any $f \in S_k^{\mathrm{new}}(\Gamma_0(N))$ non-CM newform of integral weight k, we have*

$$\lim_{n\to\infty} \frac{\#\{p \leq n : \alpha \leq \theta_p \leq \beta\}}{\#\{p \leq n\}} = \frac{2}{\pi}\int_\alpha^\beta \sin^2(\theta) d\theta.$$

This was recently proven in a landmark paper by Barnet-Lamb, Geraghty, Harris, and Taylor **[47]**, where the authors generalized the Modularity Theorem to higher rank unitary groups.

THEOREM 15.6 (Barnet-Lamb, Geraghty, Harris and Taylor). *The Sato-Tate Conjecture is true.*

We now turn to the case of half-integral weight cusp forms, where much less is proven. The best general bound for all cusp forms of half-integral weight is (cf. the introduction to **[281]**)

$$c_f(n) \ll_{f,\varepsilon} n^{\frac{k}{2}-\frac{1}{4}+\varepsilon}.$$

This estimate cannot be improved for all cusp forms, as theta functions do have growth of this magnitude, so for example by the Serre-Stark basis theorem in weight $1/2$, there is no improvement to give. However, it is expected that the stricter Deligne bound of Theorem 15.3 holds if n is square-free (cf. Section 1 of **[281]**). The first non-trivial bounds were proven (uniformly in the level aspect) by Iwaniec **[281]** for square-free n if $k \geq 5/2$ and by Duke **[165]**, who obtained bounds for Fourier coefficients of Maass waveforms (cf. Chapter 3) in the more general case $k \geq 3/2$.

THEOREM 15.7 (Iwaniec). *If $f \in S_k(\Gamma_0(N))$ with $k \in 1/2 + \mathbb{Z}$ and $k \geq 5/2$, normalized to have Petersson norm 1, then for square-free n we have*

$$c_f(n) \ll_k \frac{1}{[\mathrm{SL}_2(\mathbb{Z}) : \Gamma_0(N)]} n^{\frac{k}{2}-\frac{2}{7}} \sigma_0(n)(\log(2n))^2.$$

REMARK. Note that the index of the congruence subgroup $\Gamma_0(N)$ appears in the above estimate since we use a different normalization for Petersson norms than in [**281**].

Iwaniecs's proof of Theorem 15.7 makes use of Poincaré series. This begins with the Petersson coefficient formula, which we state in Theorem 6.8 ii). This allows one to express the coefficients $c_f(n)$ which we wish to estimate in terms of the Kloosterman sums (6.13). The fundamental difference from the case of integral weight is that in half-integral weight there is an extra twist by $(\frac{c}{d})\varepsilon_d^{2k}$ in (6.13). To state Duke's result (cf. Theorem 5 of [**165**]), we note that the *discriminant* D of a Maass form describes an extra multiplier system in the modular transformations of these Maass waveforms. The definition of a spectral form is given in the last line of Section 2 of [**165**].

THEOREM 15.8 (Duke). *Suppose that f is a spectral Maass waveform of weight $k \in 1/2 + \mathbb{Z}$ with even discriminant D for $\Gamma_0(N)$ and $N \equiv 0 \pmod{D}$, and with eigenvalue $\lambda = 1/4 + t^2$. Further suppose that n is either square-free or a fundamental discriminant and consider an arbitrary $\varepsilon > 0$. Then, with $c_f(n)$ denoting the n-th Fourier coefficient of f (cf. Section 3.2), as $|n| \to \infty$, we have*

$$c_f(n) \ll_{k,D,\varepsilon} |\lambda|^{\frac{5}{4} - \frac{k\operatorname{sgn}(n)}{4}} \cosh\left(\frac{\pi t}{2}\right) |n|^{-\frac{2}{7}+\varepsilon}.$$

Bykovskiĭ [**136**] improved Theorem 15.7 to show that for $k \geq 5/2$, we have

$$c_f(n) \ll_{f,\varepsilon} n^{\frac{k}{2}-\frac{1}{4}-\frac{1}{16}+\varepsilon},$$

where the dependence on level and weight is unspecified. Blomer and Harcos [**66**] proved a bound with the same power savings but uniformly in the level and the weight. At present, the best result is due to Conrey and Iwaniec [**153**], and follows from their work in [**153**].

THEOREM 15.9 (Conrey and Iwaniec). *Let f be a half-integral weight $k \geq 7/2$ cusp form on $\Gamma_0(4)$. Then for any $\varepsilon > 0$, we have*

$$c_f(n) \ll_{f,\varepsilon} n^{\frac{k}{2}-\frac{1}{4}-\frac{1}{12}+\varepsilon}.$$

15.4. Weakly holomorphic modular forms and mock modular forms

In this section, we study the growth of Fourier coefficients of weakly holomorphic modular forms which have a pole at at least one cusp. Due to the presence of poles, their coefficients grow much faster than those of holomorphic modular forms (for comparison, we see in (14.4) and (15.2) that estimates for coefficients of modular forms are often obtained by expressing the coefficients as integrals of the original modular form). Our first example of such an estimate is considered in Section 14.1 in the special case of the partition generating function. The situation for general weakly holomorphic modular forms is similar; either the Circle Method (discussed in the proof of Theorem 14.3) or the Ingham Tauberian theorem (given in Theorem 14.4) can be applied to give the asymptotic expansion of any weakly holomorphic modular form. This calculation was performed by Rademacher and Zuckerman in [**426, 524, 525**], and yields the following result.

THEOREM 15.10. *If $f \in M_k^!$ with $k \in -\mathbb{N}_0$, then, for $n \in \mathbb{N}$, we have*

$$c_f(n) = 2\pi i^k \sum_{m<0} c_f(m) \left(\frac{|m|}{n}\right)^{\frac{1-k}{2}} \sum_{c=1}^{\infty} \frac{K_k(m,n;c)}{c} I_{1-k}\left(\frac{4\pi\sqrt{|mn|}}{c}\right). \tag{15.4}$$

Here $K_k(m,n;c)$ are the Kloosterman sums defined in (6.13) and I_k denotes the usual I-Bessel function. In particular, as $n \to \infty$, we have

$$c_f(n) \sim \frac{|n_0|^{\frac{1}{4}-\frac{k}{2}}}{2\sqrt{2}\pi} n^{\frac{k}{2}-\frac{3}{4}} e^{4\pi\sqrt{|n_0|n}}, \tag{15.5}$$

where n_0 is the smallest negative integer with $c_f(n_0) \neq 0$.

REMARK. Analogous exact formulas and asymptotics in terms hold for weakly holomorphic modular forms on congruence subgroups; in this case, one must consider the order of poles at all cusps. Contrasting with the proof given in Section 14.1 for $p(n)$, the primary difference is that in the main term in the Circle Method for general weakly holomorphic modular forms one obtains contributions from the principal parts of the weakly holomorphic forms at all cusps.

The same method can also be applied to mock modular forms. The set-up is the same as for the classical Circle Method. However, when applying modular transformations, one has to bound Mordell integrals which arise due to the "mockness" of the forms. For example, in the case of Zwegers' μ-function, we see in Proposition 8.7 ii) that the Mordell integrals $h(z;\tau)$ arise as error to modularity of μ. This was carried out by Dragonette **[164]** and then Andrews **[20]** to obtain (9.7), which we recall states

$$\alpha_f(n) = (-1)^{n+1} \frac{\sqrt{6}\exp\left(\frac{\pi}{12}\sqrt{24n-1}\right)}{\sqrt{24n-1}} + O\left(\frac{\exp\left(\frac{\pi}{24}\sqrt{24n-1}\right)}{\sqrt{24n-1}}\right),$$

where f is Ramanujan's third order mock theta function

$$f(q) = \sum_{n=0}^{\infty} \frac{q^{n^2}}{(-q;q)_n^2} = \sum_{n=0}^{\infty} \alpha_f(n) q^n.$$

The point is then that one can show that these "errors of modularity" of mock modular forms are small enough to be ignored. This roughly comes from the fact that the corresponding harmonic Maass forms only have growth in their holomorphic parts (this is implicit in iii) of Definition 4.2). Typical bounds are as given in the following lemma **[164]**, which Dragonette proved via an easy application of the Residue Theorem, where for $\mathrm{Re}(\alpha) > 0$, we let

$$\mathcal{J}(\alpha) := \int_0^{\infty} e^{-\frac{3\alpha t^2}{2}} \frac{\sinh(\alpha t)}{\sinh\left(\frac{3\alpha t}{2}\right)} dt.$$

LEMMA 15.11 (Dragonette). *There exists a $K > 0$ such that $|\alpha\mathcal{J}(\alpha)| < K$ for all $\alpha \in \mathbb{C}$ with $\mathrm{Re}(\alpha) > 0$.*

We now describe a second method for approximating coefficients of weakly holomorphic and harmonic Maass forms, that of Poincaré series. We see above that this method is used for the cuspidal Poincaré series as a crucial step in Iwaniec's proof of Theorem 15.7; thus, the idea of employing Poincaré series to establish asymptotic estimates for coefficients of modular forms is a classical one. We also discuss an example of the method of Maass-Poincaré series applied to one of Ramanujan's mock theta functions in Section 9.7.

Returning to the general case, for simplicity, we restrict to mock modular forms $f(\tau) = \sum_{n \geq n_0} c_f(n)q^n$ ($c_f(n_0) \neq 0$) of non-positive integral weight k such that their associated completion $\widehat{f}$ is in H_k. We then have the following extension of Theorem 15.10, which follows using the structure of Maass-Poincaré series to construct forms with prescribed principal parts.

THEOREM 15.12. *Theorem* 15.10 *also holds for* $f \in H_k$*, where* k *is a non-positive integer. That is, for any* $f \in H_k$ *the exact formula* (15.4) *and the estimate* (15.5) *hold.*

PROOF. By definition $f = \widehat{f}^{+}$, and we set

$$g(\tau) := \sum_{m<0} c_f(m) F_{k,m}(\tau),$$

where the $F_{k,m}$ are the Maass-Poincaré series from Definition 6.10.

Now, recall that the principal part of f is (for some $n_0 \in -\mathbb{N}$)

$$P_f(\tau) = \sum_{n_0 \leq n < 0} c_f(n) q^n.$$

By Theorem 6.11 $P_g = P_f$. Since there are no harmonic Maass forms in H_k with trivial (holomorphic) principal part, we must have $\widehat{f} = g$ and therefore $f = g^{+}$. The exact formula for $c_f(n)$ then follows from Theorem 6.11. The asymptotic statement comes from (14.5). □

REMARK. Similar Poincaré series constructions also yield exact formulas for harmonic Maass forms on congruence subgroups, and for forms in $H_k^!$.

15.5. Coefficients of meromorphic modular forms

We can also generalize the above investigations to include meromorphic modular forms. In this case, the polar harmonic Maass forms of Section 13.3 can be used to solve the problems of computing and estimating the resulting Fourier coefficents. Until recently, very little was known about coefficients of meromorphic cusp forms. Since Section 15.4 already analyzes harmonic Maass forms which grow at the cusps, to understand the general case of meromorphic modular forms it suffices to consider the case of *meromorphic cusp forms*, which are those meromorphic modular forms which decay like cusp forms at the cusps. Since Section 15.3 already treats the case of holomorphic modular forms, we also restrict our attention to the case of meromorphic cusp forms which do have a pole in $\mathbb{H}$.

Hardy and Ramanujan **[253]** considered the special case where the form has a unique simple pole in $\mathrm{SL}_2(\mathbb{Z})\backslash\mathbb{H}$. In particular, they found a formula for the reciprocal of the Eisenstein series E_6. Ramanujan **[429]** then conjectured further formulas for other examples of meromorphic modular forms, such as ($z \in \mathbb{H}$ throughout)

$$\frac{1}{E_4(z)} = \sum_{n=0}^{\infty} \beta_n e^{2\pi i n z},$$

where

$$\beta_n := (-1)^n \frac{3}{E_6(\rho)} \sum_{(\lambda)} \sum_{(c,d)} \frac{h_{(c,d)}(n)}{\lambda^3} e^{\frac{\pi n \sqrt{3}}{\lambda}}. \tag{15.6}$$

Here, ρ is the root of unity $\rho := e^{\pi i/3}$, λ runs over the integers of the form $\lambda = 3^a \prod_{j=1}^r p_j^{a_j}$ where $a = 0$ or 1, p_j is a prime of the form $6m+1$, and $a_j \in \mathbb{N}_0$. Furthermore, (c, d) runs over distinct solutions (modulo a simple equivalence relation) to $\lambda = c^2 - cd + d^2$ and $(a, b) \in \mathbb{Z}^2$ is any solution to $ad - bc = 1$. Finally, we set $h_{(1,0)}(n) := 1$, $h_{(2,1)}(n) := (-1)^n$, and for $\lambda \geq 7$

$$h_{(c,d)}(n) := 2\cos\left((ad + bc - 2ac - 2bd + \lambda)\frac{\pi n}{\lambda} - 6\arctan\left(\frac{c\sqrt{3}}{2d - c}\right)\right).$$

Expansions like (15.6) are very convenient since they turn out to converge extremely rapidly. After noting that the six terms with $\lambda = 1$ are equivalent, it is easy to check that the main asymptotic growth in (15.6) comes from the $(c, d) = (1, 0)$ term, yielding

$$\beta_n \sim (-1)^n \frac{3}{E_6(\rho)} e^{\pi n\sqrt{3}}.$$

Such estimates confirm that the coefficients of meromorphic modular forms grow much faster than the coefficients of weakly holomorphic modular forms, as a lengthy calculation shows. To give an impression of the rate of convergence, β_{15} is a 36 digit number, while the main term gives the first 25 digits accurately (cf. **[96]** for further examples and tables with specific data). Ramanujan claimed explicit formulas for meromorphic modular forms with simple poles. These were later proven by Bialek in his Ph.D. thesis written under Berndt **[62]**. Berndt, Bialek, and Yee **[53]** then pushed the theory further to study examples with second order poles, and they managed to prove the remaining cases of Ramanujan's formulas. The proofs of **[53, 62, 253]** all utilized a modification of the Hardy-Ramanujan Circle Method (which is discussed in Section 14.1), but the calculations rapidly become more difficult with rising pole orders.

As we see above, a powerful alternative to the Circle Method for computing coefficients of modular forms is the use of the Maass-Poincaré series of Section 6.3. These can often be used to provide a convenient basis for the entire space of modular-type objects in which one is interested. In this setting, the original idea for creating modified Poincaré series with poles on the upper half-plane goes back to Petersson **[419]**, who wrote down formulas resembling those of Hardy and Ramanujan in the case when all poles are simple. More generally, the Fourier coefficients of all meromorphic cusp forms have a similar shape, as can be shown by expressing them in terms of this basis. For example, we give the formula explicitly in the simplest case in Theorem 15.14. More specifically, all of the meromorphic modular forms investigated by Ramanujan may be written as linear combinations of the series (recall that $\mathfrak{z} = \mathfrak{z}_1 + i\mathfrak{z}_2$)

$$(15.7) \qquad f_{k,j,r}(\mathfrak{z}, z) := \mathfrak{z}_2^{-j} \sum_{m=0}^{\infty} \sum_{\mathfrak{b}\subseteq\mathbb{Z}[\mathfrak{z}]}{}^{*} \frac{C_k(\mathfrak{b}, m)}{N(\mathfrak{b})^{\frac{k}{2}-j}} (4\pi m)^r e^{\frac{2\pi m \mathfrak{z}_2}{N(\mathfrak{b})}} e^{2\pi i m z},$$

where $\mathfrak{z} \in \{i, \rho\}$, $\mathfrak{b}$ runs over primitive ideals, $N(\mathfrak{b})$ is the norm, and C_k are certain functions on ideals which we next describe. Here, for $\mathfrak{b} = (c\rho + d) \subset \mathbb{Z}[\mathfrak{z}]$, we define

$$C_{6m}(\mathfrak{b}, n) := \cos\left(\frac{\pi n}{N(\mathfrak{b})}(ad + bc - 2ac - 2bd) - 6m\arctan\left(\frac{c\sqrt{3}}{2d - c}\right)\right),$$

and we set $C_m(\mathfrak{b}, n) := 0$ if $6 \nmid m$. Similarly, for $\mathfrak{b} = (ci + d) \subseteq \mathbb{Z}[i]$, we let

$$C_{4m}(\mathfrak{b}, n) := \cos\left(\frac{2\pi n}{N(\mathfrak{b})}(ac+bd) + 4m \arctan\left(\frac{c}{d}\right)\right)$$

and $C_m(\mathfrak{b}, n) := 0$ if $4 \nmid m$. More generally, it was shown in Theorem 1.2 of **[96]** that all negative weight meromorphic cusp forms with poles at i and ρ have expansions as linear combinations of the functions $f_{k,j,r}$ defined in (15.7).

To describe the coefficients in these linear combinations, we need a basis of meromorphic Poincaré series originally discovered by Petersson. On the subgroup $\Gamma_0(N)$, for a weight $k \in \mathbb{N}_{>1}$, we define a 2-variable Poincaré series (for $k = 1$, a similar construction holds but an analytic continuation, as performed in **[98]**, is required). Setting

$$H_{2k,N}(\mathfrak{z}, z) := \sum_{\gamma \in \Gamma_\infty \backslash \Gamma_0(N)} \frac{1}{1 - e^{2\pi i(z-\mathfrak{z})}}\bigg|_{2k,\mathfrak{z}} \gamma,$$

where $|_{k,\mathfrak{z}}$ denotes the usual slash operator with respect to the variable $\mathfrak{z}$, we then let

$$\mathcal{H}_{2k,N}(\mathfrak{z}, z) := H_{2k}(\mathfrak{z}, z) + \sum_{r=0}^{2k-2} \frac{(2iy)^r}{r!} \frac{\partial^r}{\partial \overline{z}^r} H_{2k}(\mathfrak{z}, \overline{z}).$$

This function is then a weight $2k$ meromorphic cusp form on $\Gamma_0(N)$ with poles supported on $\mathbb{H}$ in the variable $\mathfrak{z}$, and in z it is a polar harmonic Maass form of weight $2 - 2k$ with a cuspidal shadow. The following result, which follows by a careful computation and comparison of "principal parts" at poles on $\mathbb{H}$, shows that the functions $\mathcal{H}_{2k,N}$ provide bases for polar harmonic Maass forms. In what follows, $R_{k,\mathfrak{z}}$ denotes the usual Maass raising operator of Definition 5.1, but with respect to the variable $\mathfrak{z}$.

PROPOSITION 15.13. *If f is a polar harmonic Maass form of integral weight $2 - 2k < 0$ with cuspidal shadow, then we can decompose f as a sum of the form*

$$f(z) = \sum_{\ell=1}^{r} \sum_{n=0}^{n_\ell} a_{\ell,n} \left[R^n_{2k,\mathfrak{z}}\left(\mathcal{H}_{2k}(\mathfrak{z}, z)\right)\right]_{\mathfrak{z}=\tau_\ell},$$

where $a_{\ell,n} \in \mathbb{C}$ and $\tau_1, \ldots, \tau_r$ are the locations of the poles of f on $\mathbb{H}$.

REMARK. Alternatively one can work with the basis elements in Section 13.3, specifically those given by the Poincaré series in Theorem 13.12.

Proposition 15.13 then allows us to describe the Fourier expansions of meromorphic cusp forms.

THEOREM 15.14. *If f is a meromorphic cusp form of integral weight $2-2k < 0$ and $\mathfrak{z} \in \{i, \rho\}$ is the only pole of f in $\mathrm{SL}_2(\mathbb{Z}) \setminus \mathbb{H}$, then, with coefficients a_ℓ defined by the expansion in Proposition 15.13, for $y > \mathfrak{z}_2$, we have the following Fourier expansion:*

$$f(z) = 2e_{1,\mathfrak{z}} \sum_{n=0}^{n_0} a_n \sum_{j=0}^{n} \frac{(2k+n-1)!}{(2k+n-1-j)!} \binom{n}{j} f_{2k+2n,j,n-j}(\mathfrak{z}, z),$$

where $e_{N,z} = 2/\#\mathrm{Stab}_z(\Gamma_0(N))$.

REMARK. A more general formula for meromorphic cusp forms with arbitrary order poles at arbitrary points is given in Theorem 4.1 of **[96]**.

15.6. Mixed mock modular forms

We next consider mixed mock modular forms, which, as reviewed in Section 13.2, are important in many natural applications, for example combinatorics. Although these forms, which are built out of products of mock modular and modular forms, may seem similar to the cases considered above, the situation is in fact more complicated. The reason is that if we multiply a mock modular form by a modular form, then the nonholomorphic part of the mock modular form may inherit growth. Unlike the mock modular case where the nonholomorphic part is always of very small growth due to the fact that they are period integrals of holomorphic cusp forms, in this case terms arising from the nonholomorphic part may even become dominant. These problems can be overcome by using a refined version of the Hardy-Ramanujan Circle Method. The first example of such a modified version of the Circle Method was considered by Parberry in his thesis [**413**] (although Parberry did not use the same language that we use).

We illustrate this method with an example arising from supersymmetric gauge theory, which involves the class number generating function (6.3). In this case the method even yields an exact formula for certain Fourier coefficients. To be more precise, consider

$$F(\tau) := \frac{\mathcal{H}^+(\tau)|U_4}{\eta^6(\tau)} = \sum_{n=0}^{\infty} \beta(n) q^{n-\frac{1}{4}},$$

where $\mathcal{H}^+$ is the Hurwitz class number generating function, defined in (6.3), and U_4 the usual U-operator. This function is the generating function of certain weighted Euler numbers of moduli spaces of semistable vector bundles, and it arises in supersymmetric gauge theory (as conjectured in [**480**] and proven in [**104**]). Similarly, although we omit the details, one can write down an analogous function involving the coefficients of $\mathcal{H}^+$ in residue classes 3 (mod 4).

To describe the Fourier coefficients of F, we require some notation. For $k \in \mathbb{N}, g \in \mathbb{Z}$, and $t \in \mathbb{R}$, we let

$$f_{k,g}(t) := \begin{cases} \dfrac{\pi^2}{\sinh^2\left(\frac{\pi t}{k} - \frac{\pi i g}{2k}\right)} & \text{if } 2k \nmid g, \\[2ex] \dfrac{\pi^2}{\sinh^2\left(\frac{\pi t}{k}\right)} - \dfrac{k^2}{t^2} & \text{if } 2k | g. \end{cases}$$

Moreover the Kloosterman sums

$$\mathcal{K}_\ell(n, m; k) := \sum_{\substack{d \pmod{k}^* \\ dd' \equiv -1 \pmod{k}}} \psi_\ell(d, d', k) e_{4k}(-4dn + d'm)$$

occur, with ψ_ℓ a certain multiplier explicitly given in (3.4) of [**104**] and where $e_k(n) = e^{2\pi i n/k}$. Finally, we need the Bessel function integrals

$$\mathcal{I}_{k,g}(n) := \int_{-1}^{1} f_{k,g}\left(\frac{t}{2}\right) I_{\frac{7}{2}}\left(\frac{\pi}{k}\sqrt{(4n-1)(1-t^2)}\right) (1-t^2)^{\frac{7}{4}}\, dt.$$

We then have the following exact formula, which was proven in [**104**].

THEOREM 15.15. *The coefficients $\beta(n)$ of F are given by*

$$\beta(n) = -\frac{\pi}{6}(4n-1)^{-\frac{5}{4}} \sum_{k=1}^{\infty} \frac{\mathcal{K}_0(n,0;k)}{k} I_{\frac{5}{2}}\left(\frac{\pi}{k}\sqrt{4n-1}\right)$$
$$+ \frac{1}{\sqrt{2}}(4n-1)^{-\frac{3}{2}} \sum_{k=1}^{\infty} \frac{\mathcal{K}_0(n,0;k)}{\sqrt{k}} I_3\left(\frac{\pi}{k}\sqrt{4n-1}\right)$$
$$- \frac{1}{8\pi}(4n-1)^{-\frac{7}{4}} \sum_{k=1}^{\infty} \sum_{\substack{\ell\in\{0,1\} \\ -k<g\leq k \\ g\equiv \ell \pmod 2}} \frac{\mathcal{K}_\ell(n,g^2;k)}{k^2} \mathcal{I}_{k,g}(n).$$

We note that the first two summands in Theorem 15.15 resemble coefficients of weakly holomorphic forms (compare with (14.2) for one such example). The third term yields a very different contribution. Mahlburg and one of the authors (cf. Proposition 5.2 of [**103**]) determined the asymptotic behavior of such integrals, which yields the following result.

COROLLARY 15.16. *The leading asymptotic terms of $\beta(n)$, as $n \to \infty$, are*

$$\beta(n) = \left(\frac{1}{96}n^{-\frac{3}{2}} - \frac{1}{32\pi}n^{-\frac{7}{4}} + O\left(n^{-2}\right)\right) e^{2\pi\sqrt{n}}.$$

REMARK. Proceeding as in [**103**], one could determine further polynomial lower order main terms.

REMARK. It would be interesting to represent F as a Poincaré series in the flavor of those in Section 6.3. However, it is not clear what the seed of such a Poincaré series should be.

SKETCH OF PROOF OF THEOREM 15.15. The idea is to use the modularity properties of $\mathcal{H}$ investigated in Subsection 6.1.2, and of η, employing the Hardy-Ramanujan Circle Method. The contribution from the holomorphic part is treated as in the case of weakly holomorphic modular forms. The main difference is that certain "error integrals" occur. In particular, a key step is to estimate integrals of the shape ($w \in \mathbb{C}$ with $\mathrm{Re}(w) > 0$, $k \in \mathbb{N}$, $g, b \in \mathbb{Z}$)

$$\mathcal{I}_{k,g,b}(w) := e^{\frac{2\pi b}{kw}} w^{\frac{5}{2}} \int_{-\infty}^{\infty} e^{-\frac{2\pi t^2}{kw}} f_{k,g}(t)dt.$$

One may show that for $b \leq 0$ one has

$$|\mathcal{I}_{k,g,b}(w)| \ll |w|^{\frac{5}{2}} \times \begin{cases} \frac{k^2}{g^2} & \text{if } g \neq 0, \\ 1 & \text{if } g = 0, \end{cases} \tag{15.8}$$

as $\mathrm{Re}(w) \to \infty$, so these are negligible. If $b > 0$, these integrals grow, and similar to the case of Fourier expansions, where there is a principal part, we are interested in the "principal integral part" contribution. To be more precise, for $b > 0$ and $g \in \mathbb{Z}$, we let

$$\mathcal{J}_{k,g,b}(w) := e^{\frac{2\pi b}{kw}} w^{\frac{5}{2}} \int_{-\sqrt{b}}^{\sqrt{b}} e^{-\frac{2\pi t^2}{kw}} f_{k,g}(t)dt.$$

If $b > 0$, then a straightforward residue calculation shows that

$$\mathcal{I}_{k,g,b}(w) = \mathcal{J}_{k,g,b}(w) + \mathcal{E}_{k,g,b}(w),$$

where the error $\mathcal{E}_{k,g,b}$ satisfies the same estimate as $\mathcal{I}_{k,g,b}$ in (15.8). □

15.7. The Wright Circle Method

As alluded to in Chapter 14, there is a modified version of the Circle Method due to Wright, developed in [**499, 500**]. This is often useful for proving asymptotics of combinatorial series and is especially convenient for studying products of weakly holomorphic modular forms with other convergent series. Here, we outline this method and its applications. In particular, we see below how it can be used to prove a conjecture of Dyson on bivariate distributions of cranks of partitions (cf. Conjecture 15.19 and Theorem 15.20).

The precise statement of Wright's method is technical, but, roughly speaking, it is a general method to study the Fourier coefficients of functions with known asymptotic expansions towards 0. This is similar to the situation of Tauberian theorems like Theorem 14.4, but we no longer require monotonic growth of the coefficients. Note that the method can be modified to allow growth at other cusps. For ordinary modular forms, Wright's method gives much weaker bounds than the Hardy-Ramanujan Circle Method, but is more flexible in its applicability to non-modular generating functions. Essentially, Wright's idea begins with the same set-up as the classical Circle Method which is discussed in the proof of Theorem 14.3. Namely, we begin by expressing the Fourier coefficients in question as certain integrals using Cauchy's theorem, and then split these integrals into pieces which may be analyzed using asymptotic properties of the generating series near roots of unity. As opposed to the Hardy-Ramanujan method, Wright's modified approach instead has just one "major arc", around the exponential singularity at the cusp $q = 1$, and is especially convenient as it can be applied to any series that can be expressed as the product of a weakly holomorphic modular form with poles at the cusps and a series with a convergent asymptotic expansion.

A typical situation in which the Wright Circle Method may be applied is that of products of partial (or false) theta functions and weakly holomorphic modular forms. For this, the asymptotic behavior of the partial theta functions needs to be investigated. Here the Euler-Maclaurin summation formula is a useful tool. We use it in a form stated by Zagier [**517**]. In particular, suppose that a smooth function $f\colon (0,\infty) \to \mathbb{C}$ has an asymptotic power series expansion around 0, which means that for any $N \in \mathbb{N}_0$ we have

$$f(t) = \sum_{n=0}^{N} b_n t^n + O\left(t^{N+1}\right) \tag{15.9}$$

as $t \to 0^+$. For any $a > 0$, Zagier then considered the summation

$$g_a(t) := \sum_{n=0}^{\infty} f\left((n+a)t\right), \tag{15.10}$$

and showed that its asymptotic behavior can be simply described in terms of the coefficients of the expansion (15.9). Recall that a function f is said to be of *rapid decay* at infinity if $\int_\ell^\infty |f(w)|dw$ converges for some $\ell > 0$. The following result is stated as the first generalization of Proposition 3 in [**517**].

PROPOSITION 15.17. *Suppose that f has the asymptotic expansion* (15.9) *and that f and all of its derivatives are of rapid decay at infinity. Suppose further that $I_f := \int_0^\infty f(w)dw$ converges. The function g_a as defined in* (15.10) *then has the* *asymptotic expansion (for any $N \in \mathbb{N}_0$)*

$$g_a(t) = \frac{I_f}{t} - \sum_{n=0}^{N} b_n \frac{B_{n+1}(a)}{n+1} t^n + O\left(t^{N+1}\right),$$

where $B_n(x)$ is the n-th Bernoulli polynomial.

REMARK. This statement corrects a sign error in [**517**] where the sum is added rather than subtracted.

We now consider a few explicit examples. Our first one concerns $u^*(n)$, the number of strongly unimodal sequences. As mentioned in Chapter 14, we have the following asymptotic behavior due to Auluck and Wright [**46, 499, 500**].

THEOREM 15.18 (Auluck and Wright). *As $n \to \infty$, we have that*

$$u^*(n) = 8^{-1}\left(3^3 n^5\right)^{-\frac{1}{4}} \exp\left(2\pi\sqrt{\frac{n}{3}}\right)\left(1 + O\left(n^{\frac{1}{2}}\right)\right).$$

SKETCH OF PROOF. The crux of the proof relies on the representation of the strongly unimodal generating function as the product given in Theorem 14.19,

$$U^*(q) := \sum_{n=0}^{\infty} u^*(n) q^n = \frac{1}{(q)_\infty^2} \sum_{n=1}^{\infty} (-1)^{n+1} q^{\frac{n(n+1)}{2}}. \tag{15.11}$$

That is, U^* is essentially a product of a modular form and a partial theta function. Since we know the asymptotic behavior of the modular form in front of the last product, it suffices to determine the asymptotics of the partial theta function. This function is not modular, but its asymptotic behavior towards 0 can easily be determined using Proposition 15.17. To be more precise we write

$$\sum_{n=1}^{\infty} (-1)^{n+1} q^{\frac{n(n+1)}{2}} = q^{-\frac{1}{8}}\left(\sum_{n=1}^{\infty} q^{2\left(n+\frac{3}{4}\right)^2} - \sum_{n=1}^{\infty} q^{2\left(n+\frac{5}{4}\right)^2}\right),$$

and then use Proposition 15.17 with $f(x) = e^{-x^2}$. □

As another application, we recall Dyson's conjecture [**181**] on the asymptotic behavior of the crank, mentioned in Chapter 14.

CONJECTURE 15.19 (Dyson). *As $n \to \infty$, we have*

$$M(m,n) \sim \frac{\beta}{4} \operatorname{sech}^2\left(\frac{\beta m}{2}\right) p(n),$$

where $\beta := \frac{\pi}{\sqrt{6n}}$.

Although the range in this asymptotic was not prescribed exactly, Dyson asked whether there is a suitable formulation which makes the preceding estimate true, and how precise the error bound could be made. Recently, Dousse and one of the authors proved this conjecture by establishing estimates necessary to apply Wright's Circle Method [**84**].

THEOREM 15.20. *Conjecture 15.19 is true. More precisely, if*

$$|m| \leq \frac{1}{\pi\sqrt{6}} \sqrt{n} \log(n),$$

then, as $n \to \infty$, we have

$$M(m,n) = \frac{\beta}{4}\operatorname{sech}^2\left(\frac{\beta m}{2}\right) p(n)\left(1 + O\left(\beta^{\frac{1}{2}}|m|^{\frac{1}{3}}\right)\right).$$

REMARK. Mahlburg, in private communication with the authors of [**84**], has pointed out that almost all partitions satisfy Dyson's asymptotic. To be more precise, he noted that

$$\#\left\{\lambda \vdash n : |\mathrm{crank}(\lambda)| \leq \frac{\sqrt{n}}{\pi\sqrt{6}}\log n\right\} \sim p(n),$$

where $\lambda \vdash n$ means that λ is a partition of n. Furthermore, we note that the range on m in Theorem 15.20 has been extended by Parry and Rhoades using different methods [**414**].

SKETCH OF PROOF OF THEOREM 15.20. We set

$$\mathcal{C}_m(q) := \sum_{n=0}^{\infty} M(m,n)\, q^n.$$

Then, by Cauchy's theorem and Lemma 14.11 i), we have that

$$\mathcal{C}_m(q) = \frac{2}{(q)_\infty}\int_0^{\frac{1}{2}} g(z;\tau)\cos(2\pi m z)dz, \tag{15.12}$$

where

$$g(z;\tau) := \frac{i\left(\zeta^{\frac{1}{2}} - \zeta^{-\frac{1}{2}}\right)\eta^3(\tau)}{\vartheta(z;\tau)}.$$

The first step is to determine the asymptotic behavior of $\mathcal{C}_m$ for q near an essential singularity on the unit circle. It turns out that the dominant pole lies at $q = 1$. To be more precise, using the transformation law of the eta- and the theta-functions, we approximate g near the dominant pole 1 by

$$\frac{i\sin(\pi z)}{\tau\sinh\left(\frac{\pi i z}{\tau}\right)} e^{\frac{\pi i z^2}{\tau}}. \tag{15.13}$$

Inserting the Taylor expansion of $\sin(\pi z)e^{\frac{\pi i z^2}{\tau}}\cos(2\pi m z)$, one obtains integrals of the shape ($\ell \in \mathbb{N}_0$)

$$\mathcal{I}_\ell := \int_0^{\frac{1}{2}} \frac{z^{2\ell+1}}{\sinh\left(\frac{\pi i z}{\tau}\right)} dw,$$

which are closely related to the *Euler polynomials* $E_n(x)$ defined by the generating function

$$\frac{e^{xw}}{1+e^w} =: \sum_{n=0}^{\infty} E_n(x)\frac{w^n}{n!}.$$

To be more precise, we have

$$\int_0^\infty \frac{w^{2\ell+1}}{\sinh(\pi w)} dw = \frac{(-1)^{\ell+1}E_{2\ell+1}(0)}{2}.$$

Making all the error bounds explicit then gives

$$\mathcal{C}_m(q) = (-i\tau)^{\frac{3}{2}}\frac{\pi}{2}\operatorname{sech}^2\left(\frac{\beta m}{2}\right)e^{\frac{\pi^2}{6z}} + O\left(\beta^{\frac{5}{2}}m^{\frac{2}{3}}\operatorname{sech}^2\left(\frac{\beta m}{2}\right)e^{\pi\sqrt{\frac{n}{6}}}\right).$$

One continues by carefully bounding $\mathcal{C}_m$ away from the dominant pole, and the proof then follows by an application of Wright's Circle Method. $\square$

As described in Chapter 14, the rank has the same distribution as the crank. The following theorem proved by Douse, Mertens, Parry and Rhoades [**163, 414**] gives the bivariate distribution for the rank partition function.

THEOREM 15.21. *For* $|m| \leq \frac{1}{\pi\sqrt{6}}\sqrt{n}\log(n)$, *we have, as* $n \to \infty$, *that*

$$N(m,n) = \frac{\beta}{4}\operatorname{sech}^2\left(\frac{\beta m}{2}\right) p(n) \left(1 + O\left(\beta^{\frac{1}{2}}|m|^{\frac{1}{3}}\right)\right).$$

REMARK. The two approaches in [**163**] and [**414**] are very different. In [**163**] the authors used the modularity of the generating function, whereas [**414**] employed probabilistic methods. Here we give a sketch of the proof using modularity.

SKETCH OF PROOF OF THEOREM 15.21. The starting point is again the analogue of (15.12), where now g is the rank generating function R. This is most conveniently studied using the decomposition into η-quotients and Zwegers μ function is given in Lemma 14.11 ii). Finding the analogue of (15.13) is more involved. This mainly stems from the fact that the crank generating function is modular, while the rank generating function is instead mock modular. In this case, one may use the transformation law of μ (cf. Section 8.1), and then one has to bound certain Mordell integrals (which of course arise from the mock modularity of the rank generating function). These turn out, in this situation, to decay exponentially. The proof then follows in a similar way as the proof of Theorem 15.20. $\square$

CHAPTER 16

Harmonic Maass Forms as Arithmetic and Geometric Generating Functions

In this chapter, we discuss various arithmetic and geometric applications of harmonic Maass forms. We begin by reviewing a number of results, new and old, on traces of singular moduli. Broadly speaking, these are the traces (originally considered in the algebraic number theoretic sense of the word) of the values of weight zero modular-type functions at CM points, which are important values in the theory of complex multiplication.

We start by reviewing the breakthrough work of Zagier **[516]** in Section 16.1. This sets up much of the notation and background we shall need in the remaining sections, as well as connections to the Hilbert class polynomials and Borcherds products, which will be studied in detail in Chapter 18. As Zagier indicated in his original paper **[516]**, many possible generalizations of his original theory exist (such as for higher level congruence subgroups and nonholomorphic functions arising from derivatives of negative weight weakly holomorphic modular forms). In fact, we have already seen a special case of such generalizations in an application to partitions in Theorem 14.5.

In order to discuss such theories, two perspectives are useful. In particular, the analytic perspective of Poincaré series is reviewed in Section 16.2, and the representations of trace generating functions as theta lifts is described in Section 16.3. The various generalizations alluded to here are described at the end of Section 16.3. We continue by discussing the relation to the famous results of Gross, Kohnen, and Zagier **[239]** in Section 16.4.

We conclude with a summary of two cutting-edge extensions of this theory. Specifically, in Section 16.5, we discuss newly-defined analogues of traces of singular moduli associated to positive discriminants, due to Duke, Imamoḡlu, and Tóth **[172]**. This utilizes certain cycle integrals, generalizing the cycle integrals discussed in Section 12.3. Finally, we discuss recent theory independently developed by Duke-Li **[176]**, Ehlen **[187]**, and Viazovska **[483]** on harmonic Maass forms of weight one in Section 16.6.

16.1. Zagier's work on traces of singular moduli

In this section, we follow Zagier's exposition in **[516]**. Let j be the usual modular function for $\mathrm{SL}_2(\mathbb{Z})$ whose q-expansion begins

$$j(\tau) = q^{-1} + 744 + 196884q + 21493760q^2 + \cdots .$$

The values of j (or other modular functions) at imaginary quadratic arguments in $\mathbb{H}$ are known as *singular moduli*. For instance we have

$$j(i) = 1728, \qquad j\left(\frac{1+i\sqrt{3}}{2}\right) = 0, \qquad j\left(i\sqrt{2}\right) = 8000.$$

Singular moduli for j are algebraic integers which play many roles in number theory. For example, they generate class fields of imaginary quadratic fields, and can in fact be applied to solve Hilbert's 12th problem over imaginary quadratic fields. One important use of singular moduli occurs in Zagier's **[516]** proof of Borcherds' theorem on the infinite product expansions of integer weight modular forms on $\mathrm{SL}_2(\mathbb{Z})$ with Heegner divisor (cf. Chapter 18). More precisely, Zagier studied *traces of singular moduli*, where by trace of an algebraic integer, we mean the sum of the roots of its minimal polynomial. Zagier's work on traces complements the work of Gross-Zagier **[241]** on norms of singular moduli (and their differences). A key to Zagier's proof of Borcherds' theorem is the beautiful observation that the generating functions for traces of singular moduli are essentially weight 3/2 weakly holomorphic modular forms.

To consider traces of the *normalized Hauptmodul*

$$J(\tau) := j(\tau) - 744,$$

we require some notation, some of which we recall from previous chapters. In general, by a *Hauptmodul*, we mean the unique generator of a modular function field with leading coefficient 1, a simple pole at $i\infty$, and vanishing constant term, which exists if the corresponding modular curve has genus zero. Note that our choice of notation below differs slightly from some notation used in **[108]** and **[516]**, sources which we reference throughout this chapter.

We let $d \equiv 0, 1 \pmod 4$ be a negative integer and recall that $H(-d)$ denotes the Hurwitz class number of discriminant d. For $d < 0$, we let $\mathcal{Q}_d$ denote the set of positive definite integral binary quadratic forms $Q(X,Y) = aX^2 + bXY + cY^2$ of discriminant $d = b^2 - 4ac$ (note that positive definiteness is equivalent to $a > 0$). Moreover, for each Q let τ_Q be the unique root of $Q(X,1) = 0$. Furthermore, the numbers ω_Q are defined as $\omega_Q := \omega_{\tau_Q}$, where ω_τ is as in (13.11), and throughout this section $\Gamma := \mathrm{PSL}_2(\mathbb{Z})$.

DEFINITION 16.1. For $d \in -\mathbb{N}$, the **trace of the singular moduli** of discriminant d is the sum

$$\mathrm{Tr}_d = \mathrm{Tr}_d(J) := \sum_{Q \in \mathcal{Q}_d/\Gamma} \frac{J(\tau_Q)}{\omega_Q}.$$

Similarly, for an $\mathrm{SL}_2(\mathbb{Z})$-invariant function $f\colon \mathbb{H} \to \mathbb{C}$, we define the **trace**

$$\mathrm{Tr}_d(f) := \sum_{Q \in \mathcal{Q}_d/\Gamma} \frac{f(\tau_Q)}{\omega_Q}.$$

Note that the traces defined above are indeed traces in the algebraic sense of the definition. The value of j at τ_Q depends only on the equivalence class of Q. Moreover, there are $\omega_Q H(-d)$ equivalence classes of primitive quadratic forms in $\mathcal{Q}_d$ which together form a full set of conjugates.

Zagier **[516]** studied the generating function for the traces as d varies, and proved that it is a modular form. In fact, he explicitly identified this modular form as

$$g(\tau) := \theta_1(\tau)\frac{E_4(4\tau)}{\eta^6(4\tau)} \qquad \text{with} \qquad \theta_1(\tau) := \sum_{n\in\mathbb{Z}}(-1)^n q^{n^2}.$$

To precisely state Zagier's result, we let $B(n)$ denote the Fourier coefficients of g:

$$g(\tau) =: \sum_{n=-1}^{\infty} B(n)q^n. \tag{16.1}$$

Zagier's result may be summarized in the following theorem, where for the proof the important *Hilbert class polynomials* are defined for $d \in -\mathbb{N}$, $d \equiv 0, 1 \pmod 4$ by

$$\mathcal{H}_d(X) := \prod_{Q\in\mathcal{Q}_d/\mathrm{SL}_2(\mathbb{Z})} (X - j(\tau_Q))^{\frac{1}{\omega_Q}}. \tag{16.2}$$

THEOREM 16.2. *For $d \in -\mathbb{N}$, we have that*

$$\mathrm{Tr}_d = -B(-d).$$

SKETCH OF PROOF. The idea of proof is to show that Tr and $-B$ satisfy the same recursions. To be more precise, for $n \in \mathbb{N}$ they both obey

$$\begin{aligned} c(4n) &= -2 \sum_{1\le r\le\sqrt{4n+1}} c\left(4n - r^2\right), \\ c(4n-1) &= 240\sigma_3(n) - \sum_{2\le r\le\sqrt{4n+1}} r^2 c\left(4n - r^2\right). \end{aligned}$$

For the coefficients B these follow by using that $(g\theta)\big|U_4 = 0$, whereas $[g,\theta]_1$, the first Rankin-Cohen bracket defined in (2.37), is a multiple of E_4 which one can easily determine. For Tr the recursions follow by analyzing the Hilbert class polynomials of (16.2). □

We note that Poincaré series may also be used to study Zagier's Theorem 16.2, which we discuss in the next section. It is actually true that Tr_d is related to the first Fourier coefficient of $\mathcal{H}_d(j(\tau))$, where $\mathcal{H}_d(X)$ is defined in (16.2). Zagier's proof of this fact also gives a new proof of one of Borcherds' famous theorems on infinite product automorphic forms which we now recall.

To state it we need certain special modular forms. Recall that for every $d \in -\mathbb{N}_0$ with $d \equiv 0, 1 \pmod 4$ there exists a unique $f_d \in \mathbb{M}^!_{\frac{1}{2}} := \mathbb{M}^!_{\frac{1}{2}}(\Gamma_0(4))$, the space of weight $1/2$ weakly holomorphic modular forms in Kohnen's plus space, with Fourier expansion of the form

$$f_d(\tau) = q^d + \sum_{n=1}^{\infty} A(n,d)q^n. \tag{16.3}$$

The first element f_0 is just the classical Jacobi theta function $\sum_{n\in\mathbb{Z}} q^{n^2}$. The first few coefficients in the Fourier expansions of the first few functions f_d are as follows:
(16.4)

$$\begin{aligned}
f_0(\tau) &= 1 + 2q + 2q^4 + 2q^9 + 2q^{16} + O\left(q^{12}\right),\\
f_{-3}(\tau) &= q^{-3} - 248q + 26752q^4 - 85995q^5 + 1707264q^8 - 4096248q^9 + O\left(q^{12}\right),\\
f_{-4}(\tau) &= q^{-4} + 492q + 143376q^4 + 565760q^5 + 18473000q^8 + O\left(q^9\right),\\
f_{-7}(\tau) &= q^{-7} - 4119q + 8288256q^4 - 52756480q^5 + 5734772736q^8 + O\left(q^9\right),\\
f_{-8}(\tau) &= q^{-8} + 7256q + 26124256q^4 + 190356480q^5 + 29071392966q^8 + O\left(q^9\right),\\
f_{-11}(\tau) &= q^{-11} - 33512q + 561346944q^4 - 5874905295q^5 + O\left(q^8\right).
\end{aligned}$$

The functions $f_0, f_{-3}, f_{-4}, \ldots$ constitute a basis of $\mathbb{M}^!_{\frac{1}{2}}$.

REMARK. These forms are simple to compute. We note that f_{-3} satisfies

$$f_{-3}(\tau) = \frac{D(f_0(\tau))E_{10}(4\tau)}{2\Delta(4\tau)} - \frac{f_0(\tau)D(E_{10}(\tau))|V_4}{10\Delta(4\tau)} - \frac{152}{5}f_0(\tau),$$

where, as usual, $D = q\frac{d}{dq}$ and $f|V_d(\tau) = f(d\tau)$. One can then compute the other f_d forms iteratively by considering $f_{d-4}(\tau)j(4\tau)$ followed by the obvious diagonalization argument to obtain expansions satisfying (16.3).

A special example of Borcherds' theorem **[74]** is then as follows.

THEOREM 16.3 (Borcherds). *Let $d \in -\mathbb{N}$, $d \equiv 0, 1 \pmod 4$. Then*

$$\mathcal{H}_d\left(j(\tau)\right) = q^{-H(-d)} \prod_{n=1}^{\infty} (1-q^n)^{A\left(n^2,d\right)}.$$

We shall discuss the infinite products of Theorem 16.3 in greater detail in Chapter 18. Moreover, we will discuss the generalization of this theorem which makes use of weight 1/2 mock modular forms.

REMARK. Noting that

$$\mathcal{H}_d\left(j(\tau)\right) = q^{-H(-d)}\left(1 - \mathrm{Tr}_d\, q + O\left(q^2\right)\right),$$

for $d \in -\mathbb{N}$ we have

$$\mathrm{Tr}_d = A(1,d).$$

Thus, by Theorem 16.2, the traces Tr_d occur as Fourier coefficients of the modular form g defined in (16.1). On the other hand they also show up as coefficients of the modular forms f_d. The other coefficients $A(D,d)$ which appear as exponents in Theorem 16.3 are in a similar way related to modular forms. To state this, define $g_D \in \mathbb{M}^!_{\frac{3}{2}}$ for $D \in \mathbb{N}$ with $D \equiv 0, 1 \pmod 4$ to be the unique form which has a Fourier expansion of the form

$$g_D(\tau) = q^{-D} + \sum_{n=0}^{\infty} B(D,n)q^n. \tag{16.5}$$

In particular $g_1 = g$. The Fourier expansions of the first few g_D are given by

$$g_1(\tau) = q^{-1} - 2 + 248q^3 - 492q^4 + 4119q^7 - 7256q^8 + 33512q^{11} - 53008q^{12} + O\left(q^{15}\right),$$
$$g_4(\tau) = q^{-4} - 2 - 26752q^3 - 143376q^4 - 8288256q^7 - 26124256q^8 + O\left(q^{11}\right),$$
$$g_5(\tau) = q^{-5} + 0 + 85995q^3 - 565760q^4 + 52756480q^7 - 190356480q^8 + O\left(q^{11}\right),$$
$$g_8(\tau) = q^{-8} + 0 - 1707264q^3 - 18473000q^4 - 5734772736q^7 - 29071392966q^8 + O\left(q^{11}\right),$$
$$g_9(\tau) = q^{-9} - 2 + 4096248q^3 - 51180012q^4 + 22505066244q^7 - 125891591256q^8 + O\left(q^{11}\right).$$

Inspecting the Fourier expansions of the f_d and g_D one sees that they obey a striking duality. The following theorem generalizes Theorem 16.2. For a proof of this celebrated result, we refer the reader to [**516**].

THEOREM 16.4. *For $d \in -\mathbb{N}$ with $d \equiv 0, 1 \pmod 4$ and $D \in \mathbb{N}$ with $D \equiv 0, 1 \pmod 4$, we have*

$$A(D, d) = -B(D, d).$$

Theorem 16.4 can also be proved using Poincaré series and properties of Kloosterman sums, which we discuss in the next section.

REMARK. Kim [**307, 308**] has established the modularity of traces of singular moduli on certain genus zero congruence subgroups. Using theta lifts, Bruinier and Funke [**121**] proved a more general theorem which holds for modular functions on modular curves of arbitrary genus (see Section 16.3). In all of these cases, the corresponding generating functions are weight 3/2 weakly holomorphic modular forms.

16.2. Maass-Poincaré series

In [**516**], Zagier included several generalizations of the results discussed in the previous section. Further generalizations and results are given in Section 16.3 below. Here we highlight two of these. The first concerns twisted traces of singular moduli. If $\lambda \in \mathbb{Z}$ and D_2 is a non-zero integer for which $(-1)^\lambda D_2 \equiv 0, 1 \pmod 4$ and $(-1)^\lambda D_1 D_2 < 0$, then define the *twisted trace of a modular invariant* f by

$$\operatorname{Tr}_{D_1, D_2}(f) := \sum_{Q \in \mathcal{Q}_{D_1 D_2}/\Gamma} \frac{\chi_{D_1}(Q) f(\tau_Q)}{\omega_Q}, \tag{16.6}$$

where χ_{D_1} is defined in (12.15).

REMARK. Note that $D_1 = 1$ implies that $D_2 < 0$ satisfies $D_2 \equiv 0, 1 \pmod 4$ and we recover the ordinary trace of Definition 16.1. That is, $\operatorname{Tr}_{1,D_2}(f) = \operatorname{Tr}_{D_2}(f)$.

Zagier proved that traces of weakly holomorphic modular forms of weight 0 on $\mathrm{SL}_2(\mathbb{Z})$ also occur as coefficients of weight 3/2 modular forms (cf. Theorem 5 of [**516**]). The second generalization was discussed in the last section of [**516**]. There Zagier considered traces of special nonholomorphic modular functions f. In these cases, the generating functions have weights $k \in \{-11/2, -7/2, -5/2, -3/2, -1/2\}$ (cf. Theorems 10 and 11 of [**516**]). In [**108**], two of the authors explicitly represented the coefficients of half-integral weight Maass-Poincaré series as traces of singular moduli. The result (cf. Theorem 16.6 below) includes the results of Zagier described above and gives exact formulas for these traces. And, as in the case of weight 1/2 and 3/2, there are duality statements which do not depend on the existence of special bases.

To state these results, define for $\kappa \in 1/2+\mathbb{Z}$ and $m \in \mathbb{N}$ with $(-1)^{\kappa+\frac{1}{2}}m \equiv 0,1 \pmod 4$ the Poincaré series from Chapter 7:

$$G_{\kappa,m} := \begin{cases} P_{\kappa,-m} & \text{if } \kappa \geq \frac{5}{2}, \\ F_{\kappa,-m} & \text{if } \kappa \leq -\frac{1}{2}. \end{cases}$$

REMARK. For $\kappa = 1/2$ or $3/2$ one can similarly define Poincaré series by analytically continuing the Fourier expansions.

If $\kappa \geq -11/2$ with $\kappa \neq -9/2$, then the Poincaré series $G_{\kappa,m} = P_{\kappa,-m}$ are weakly holomorphic modular forms, which we write as

$$G_{\kappa,m}(\tau) = q^{-m} + \sum_{\substack{n\geq 0 \\ (-1)^{\kappa-\frac{1}{2}}n\equiv 0,1 \pmod 4}} b_\kappa(m,n)q^n.$$

Note that in particular

$$G_{\frac{3}{2},1}(\tau) = g_1(\tau),$$

the function defined in (16.6).

If $\kappa = -9/2$ or $\kappa \leq -13/2$, then $G_{\kappa,m} = F_{\kappa,-m}$ is a harmonic Maass form with Fourier expansion of the form

$$G_{\kappa,m}(\tau) = F^-_{\kappa,-m}(\tau) + q^{-m} + \sum_{\substack{n\geq 0 \\ (-1)^{\kappa-\frac{1}{2}}n\equiv 0,1 \pmod 4}} b_\kappa(m,n)q^n. \tag{16.7}$$

The duality illustrated in Theorem 16.4 is a special case of the following general result which also holds for the Fourier coefficients of the $F_{k,-m}$.

THEOREM 16.5. *Suppose that $k \in 1/2+\mathbb{N}$, and that $m \in \mathbb{N}$ satisfies $(-1)^{k+\frac{1}{2}}m\equiv 0,1 \pmod 4$. For $n \in \mathbb{N}$ satisfying $(-1)^{k-\frac{1}{2}}n\equiv 0,1 \pmod 4$, we have*

$$b_k(m,n) = -b_{2-k}(n,m).$$

SKETCH OF PROOF. The proof follows by a careful inspection of the Fourier expansion in Theorem 6.11 using that

$$K_k(m,n;c) = (-1)^{k-\frac{1}{2}} i K_{2-k}(-n,-m;c),$$

where the Kloosterman $K_k(m,n;c)$ sums are defined in (6.13). □

REMARK. Theorem 16.5 for $k = 3/2$ (resp. $k \in \{5/2, 7/2, 9/2, 11/2, 15/2\}$) is Theorem 4 of **[516]** (resp. Theorem 9 of **[516]**).

Motivated by Zagier's results, we now describe the coefficients of the $F_{k,-m}$ in terms of singular moduli. To state these results, we require the following Poincaré series defined by Niebur **[397]**. Throughout, let I_s denote the usual I-Bessel function. If $\mathrm{Re}(s) > 1$, then define the *Niebur Poincaré series*

$$\mathfrak{F}_s(\tau) := \pi \sum_{M\in\Gamma_\infty\backslash \mathrm{SL}_2(\mathbb{Z})} \mathrm{Im}(M\tau)^{\frac{1}{2}} I_{s-\frac{1}{2}}(2\pi \mathrm{Im}(M\tau)) e(-\mathrm{Re}(M\tau)). \tag{16.8}$$

Note that $\mathfrak{F}_s = \mathfrak{F}_{1,1s}$ with $\mathfrak{F}_{N,n,s}$ given in (13.4). (These Poincaré series should not be confused with the similarly named operator in Definition 5.14.)

REMARK. For $s = 1$, Niebur's **[397]** definition requires a careful argument involving analytic continuation. It turns out that $\mathfrak{F}_1(\tau) = \frac{1}{2}(j(\tau) - 744)$, and this is the continuation to $s = 1$ of the expansion

$$-12 + \pi \sum_{M \in \Gamma_\infty \backslash \mathrm{SL}_2(\mathbb{Z})} \mathrm{Im}(M\tau)^{\frac{1}{2}} I_{s-\frac{1}{2}}(2\pi \mathrm{Im}(M\tau)) e(-\mathrm{Re}(M\tau)).$$

The coefficients of $F_{k,-m}$ are traces and twisted traces of the singular moduli for $\mathfrak{F}_s$. In view of Theorem 16.5 on duality, to state this result we may without loss of generality assume that $k \geq 3/2$.

THEOREM 16.6. *Assume the notation above. For each positive integer n with $(-1)^{k-\frac{1}{2}} n \equiv 0, 1 \pmod 4$, we have*

$$b_k(m,n) = \frac{2(-1)^{\left[\frac{k}{2}+\frac{1}{4}\right]} n^{\frac{k}{2}-\frac{3}{4}}}{|m|^{\frac{k}{2}-\frac{1}{4}}} \mathrm{Tr}_{(-1)^{k+\frac{1}{2}} m, n}\left(\mathfrak{F}_{k-\frac{1}{2}}\right).$$

Theorem 16.6 is derived from the classical fact that half-integral weight Kloosterman sums are easily described in terms of simpler sums, the Salié sums. Suppose that $D_1 \equiv 0, 1 \pmod 4$ is a fundamental discriminant (including 1), $D_2 \neq 0$ an integer for which $(-1)^{k-\frac{1}{2}} D_2 \equiv 0, 1 \pmod 4$, and N is a positive multiple of 4. Then define the *generalized Salié sums* $S_k(D_1, D_2, N)$ by

$$S_k(D_1, D_2; N) := \sum_{\substack{d \pmod N \\ d^2 \equiv (-1)^{k-\frac{1}{2}} D_1 D_2 \pmod N}} \chi_{D_1}\left(\frac{N}{4}, d, \frac{d^2 - (-1)^{k-\frac{1}{2}} D_1 D_2}{N}\right) e\left(\frac{2d}{N}\right). \tag{16.9}$$

SKETCH OF PROOF OF THEOREM 16.6. One of the key steps of the proof is the fact that half-integral weight Kloosterman sums are essentially equal to Salié sums. Namely, we have (cf. Proposition 5 of **[324]**):

$$N^{-\frac{1}{2}}\left(1 - (-1)^{k-\frac{1}{2}} i\right)\left(1 + \delta_{\mathrm{odd}}\left(\frac{N}{4}\right)\right) K_k\left((-1)^{k-\frac{1}{2}} D_1, D_2; N\right) = S_k(D_1, D_2; N).$$

Here, for $j \in \mathbb{Z}$, we have that

$$\delta_{\mathrm{odd}}(j) := \begin{cases} 1 & \text{if } j \text{ is odd,} \\ 0 & \text{if } j \text{ is even.} \end{cases}$$

We then write the Salié sums as Poincaré-type series

$$S_k(D_1, D_2; 4a) = 2 \sum_{Q \in \mathcal{Q}_{D_1 D_2}/\Gamma} \frac{\chi_{D_1}(Q)}{\omega_Q} \sum_{\substack{M \in \Gamma_\infty \backslash \mathrm{SL}_2(\mathbb{Z}) \\ \mathrm{Im}(M\tau_Q) = \frac{\sqrt{|D_1 D_2|}}{2a}}} e\left(-\mathrm{Re}\left(M\tau_Q\right)\right).$$

The claim follows after inserting these Poincaré-type expressions for the Salié sums, as well as the expansions for the Niebur Poincaré series from (16.8), into the Fourier expansions in Theorem 6.11. □

16.3. Relation to (theta) lifts

Another way to understand the modularity of the trace generating functions of Sections 16.1 and 16.2 is to realize them as outputs of theta lifts. A good starting point for this perspective is to consider the constant function 1 and its trace generating function

$$\sum_{\substack{n\geq 1\\ n\equiv 0,3 \pmod 4}} \mathrm{Tr}_{-n}(1)q^n.$$

Note that $\mathrm{Tr}_{-n}(1)$ is exactly the Hurwitz class number $H(n)$. By Subsection 6.1.2 its generating function can be completed to a weight 3/2 harmonic Maass form, namely

$$\mathcal{H}(\tau) = \sum_{n=0}^{\infty} H(n)q^n + \frac{1}{4\sqrt{\pi}} \sum_{n=1}^{\infty} n\Gamma\left(-\frac{1}{2}, 4\pi n^2 v\right) q^{-n^2} + \frac{1}{8\pi\sqrt{v}}.$$

It was shown by Funke **[210]** that this Eisenstein series can in fact be obtained as a theta lift, and this may be understood as an extension of the Siegel-Weil formula **[491]**.

To explain Funke's result, we require a certain theta function (a special case of), the *Kudla-Millson theta function.* To this end, we define the *Kudla-Millson Schwartz function* for a quadratic form $[a,b,c]$ with discriminant $b^2 - 4ac = d$, $\tau = u + iv \in \mathbb{H}$ and $z = x + iy \in \mathbb{H}$ via

$$\varphi_{\mathrm{KM}}([a,b,c],\tau,z) := \left(\frac{4v}{y^2}\left(a|z|^2 + bx + c\right)^2 - \frac{1}{2\pi}\right) e^{-\frac{4\pi v\left(a|z|^2+bx+c\right)^2}{y^2} - 4\pi v d}.$$

Then the *Kudla-Millson theta function*

$$\theta_{\mathrm{KM}}(\tau,z) := \sum_{d\in\mathbb{Z}} \sum_{\substack{a,b,c\in\mathbb{Z}\\ b^2-4ac=d}} \varphi_{\mathrm{KM}}([a,b,c],\tau,z)q^{-d} \tag{16.10}$$

transforms as a modular form of weight 3/2 for $\Gamma_0(4)$ in τ, satisfying Kohnen's plus space condition, and is $\mathrm{SL}_2(\mathbb{Z})$-invariant in z. It follows from the very general and deep work of Kudla and Millson **[336]** that if $d = b^2 - 4ac < 0$, then the differential form

$$\omega_{[a,b,c]} := \varphi_{\mathrm{KM}}([a,b,c],\tau,z)\,\frac{dxdy}{y^2}$$

is a so-called *Poincaré dual form* of the point

$$\tau_{[a,b,c]} = \frac{-b+\sqrt{d}}{2a} \in \mathbb{H}.$$

Since $\omega_{[a,b,c]}$ also decays rapidly as y grows, one can show for a weakly holomorphic modular function f that

$$\int_{\mathrm{SL}_2(\mathbb{Z})\backslash\mathbb{H}} f(z)\,\omega_{[a,b,c]} = f(\tau_{[a,b,c]}). \tag{16.11}$$

Using these facts, one can prove **[122, 210]** that

$$\frac{1}{2}\int_{\mathrm{SL}_2(\mathbb{Z})\backslash\mathbb{H}} 1\cdot\theta_{\mathrm{KM}}(\tau,z)\frac{dxdy}{y^2} = \mathcal{H}(\tau).$$

Bruinier and Funke **[122]** demonstrated that the same method works for all harmonic Maass forms of weight zero, giving a unified proof for the modularity of $\mathcal{H}$ and of Theorem 16.2.

Using the theta lift framework, we now sketch how to obtain generalizations of Zagier's results to:

i) groups other than $\mathrm{SL}_2(\mathbb{Z})$,
ii) other weights,
iii) the twisted traces (as in (16.6)), and
iv) indefinite forms yielding cycle integrals.

i): Bruinier and Funke in **[122]** considered modular functions for arbitrary arithmetic subgroups of $\mathrm{SL}_2(\mathbb{Q})$.

ii): One can further study a weakly holomorphic modular form or harmonic Maass form f of negative weight $2k$ and then apply the Maass raising operator (defined in Section 5.1) $-k$ times to obtain a weak Maass form $R_{2k}^{-k}(f)$ of weight 0 and eigenvalue $k(1-k) < 0$ and then consider the traces of $R_{2k}^{-k}(f)$. Zagier gave examples for $k = -1, -2, -3, -4, -6$ which clearly indicate that similar modularity statements must hold. This was first systematically studied by Duke and Jenkins **[175]** for weakly holomorphic f using Poincaré series and extended to harmonic Maass forms by Guerzhoy, Kane, and one of the authors **[92]**.

Independently, Alfes **[11, 12]** studied these generalizations using the theta lift approach described above. Moreover, as we have seen in Theorem 14.5, such traces can be applied to give a formula for the number $p(n)$ of integer partitions when we consider a special weakly holomorphic modular form of weight -2.

The basic idea of Alfes is to consider the Kudla-Millson theta lift of $R_{2k}^{-k}(f)$ and then to apply lowering or raising operators to return to a function with eigenvalue 0 under Δ_0. In the following, we restrict to the case that $k \in -2\mathbb{N}$. For $f \in H_{2k}$ define

$$I_{\mathrm{KM}}(\tau, f) := R_{\frac{3}{2}}^{-\frac{k}{2}} \left(\int_{\mathrm{SL}_2(\mathbb{Z})\backslash\mathbb{H}} R_{2k}^{-k}\left(f(z)\right) \theta_{\mathrm{KM}}(\tau, z) \frac{dxdy}{y^2} \right).$$

The following theorem is contained in Theorem 1 of **[11]** and in **[92]**. The proof works essentially in the same way as for weight 0 in **[122]** because (16.11) still holds for $R_{2k}^{-k}(f)$.

THEOREM 16.7. *The function $\tau \mapsto I_{\mathrm{KM}}(\tau, f)$ is a weakly holomorphic modular form of weight $3/2-k$ for $\Gamma_0(4)$ in Kohnen's plus space. Moreover, its n-th ($n \in \mathbb{N}$) Fourier coefficient is given by*

$$(-4\pi n)^{-\frac{k}{2}} \operatorname{Tr}_{-n}\left(R_{2k}^{-k}(f)\right).$$

iii): In order to obtain a twisted version of the above results, the technique introduced in **[13]** can be used. The authors of that paper showed that the genus character can be employed to define an intertwining operator that gives rise to a twisted version of the theta kernel. This has been used in **[13]** to generalize Zagier's result on the twisted traces of singular moduli to $\Gamma_0(N)$ for arbitrary N and in **[11]** to obtain a twisted version of Theorem 16.7 (also for $\Gamma_0(N)$).

iv): We cover the traces corresponding to indefinite quadratic forms in Section 16.5. For the theta lift approach to these, we refer the reader to **[11, 12]**.

16.4. Gross-Kohnen-Zagier and generalized Jacobians

In addition to Zagier's weight 3/2 Eisenstein series there is another famous generating series of weight 3/2 which also arises in this context. Let p be a prime and denote by $X_0^+(p)$ the compactified modular curve corresponding to $\Gamma_0^+(p)$, the extension of $\Gamma_0(p)$ by the *Fricke involution* $\tau \mapsto -\frac{1}{p\tau}$. Note that $\Gamma_0^+(p)$ has only one cusp. We denote by $\mathcal{Q}_{d,p}$ the set of integral binary quadratic forms $[pa, b, c]$ of discriminant d, which are positive definite if $d < 0$. The group $\Gamma_0^+(p)$ acts on $\mathcal{Q}_{d,p}$ in the usual way. Specifically, for $[a, b, c]$ a quadratic form and $\left(\begin{smallmatrix}\alpha & \beta \\ \gamma & \delta\end{smallmatrix}\right) \in \mathrm{SL}_2(\mathbb{Z})$, we set

$$[a,b,c] \circ \left(\begin{smallmatrix}\alpha & \beta \\ \gamma & \delta\end{smallmatrix}\right)(X,Y) := [a,b,c](\alpha X + \beta Y, \gamma X + \delta Y), \tag{16.12}$$

and if the leading coefficient of Q is divisible by p, then the Fricke involution W_p acts as

$$[pa,b,c] \circ W_p(X,Y) := \frac{1}{p}[pa,b,c](-Y,pX) = [cp,-b,a](X,Y).$$

This action decomposes $\mathcal{Q}_{d,p}$ into finitely many orbits if $d \neq 0$. Recall that $X_0^+(p)$ can be viewed as a non-singular projective algebraic curve defined over $\mathbb{Q}$. The Jacobian J of $X_0^+(p)$ is a commutative algebraic group defined over $\mathbb{Q}$ whose $\mathbb{Q}$-rational points $J(\mathbb{Q})$ are isomorphic to the group of degree 0 divisors defined over $\mathbb{Q}$ modulo the group of divisors of rational functions in $\mathbb{Q}(X_0^+(p))^\times$. By the Mordell-Weil Theorem $J(\mathbb{Q})$ is finitely generated.

In their influential paper **[239]**, Gross, Kohnen, and Zagier considered the degree 0 divisors $Z(d) := Y(d) - \deg(Y(d))(\infty)$ with

$$Y(d) := \sum_{Q \in \mathcal{Q}_{d,p}/\Gamma_0^+(p)} \frac{1}{|\Gamma_0^+(p)_Q|}(\tau_Q),$$

where $\Gamma_0^+(p)_Q$ denotes the stabilizer of Q in $\Gamma_0^+(p)$. By the theory of complex multiplication, $Z(d)$ is defined over $\mathbb{Q}$. We denote by $[Z(d)]$ its class in $J(\mathbb{Q})_\mathbb{C} = J(\mathbb{Q}) \otimes_\mathbb{Z} \mathbb{C}$. The corresponding generating function has modular properties.

THEOREM 16.8. *The generating function*

$$A(\tau) := \sum_{\substack{n>0 \\ n \equiv 0,3 \pmod 4}} [Z(-n)]q^n$$

is the Fourier expansion of a cusp form of weight $3/2$ *for* $\Gamma_0(4p)$ *in Kohnen's plus space, i.e., it lies in the space* $S_{\frac{3}{2}}^+(\Gamma_0(4p)) \otimes_\mathbb{C} J(\mathbb{Q})_\mathbb{C}$. *In other words, for any* $\mathbb{C}$*-linear map* $\lambda : J(\mathbb{Q})_\mathbb{C} \to \mathbb{C}$, *the* q*-series*

$$\sum_{\substack{n \geq 0 \\ n \equiv 0,3 \pmod 4}} \lambda([Z(-n)])q^n$$

is the Fourier expansion of a modular form contained in $S_{\frac{3}{2}}^+(\Gamma_0(4p))$.

Recently, Gross **[240]** explained how this generating series and the results on traces of singular moduli are connected. He showed that the traces of singular moduli on $X(1)$ can be interpreted in terms of Heegner divisors in a *generalized Jacobian* of the curve $X(1)$. Building on this work of Gross, Bruinier and Li **[126]** defined classes of Heegner divisors for all integral discriminants in the generalized Jacobian of a modular curve of arbitrary level with cuspidal modulus. They proved a modularity result for the generating series of these classes by generalizing Borcherds' proof of the Gross-Kohnen-Zagier Theorem. This result implies modularity of singular moduli and the Gross-Kohnen-Zagier Theorem as consequences.

To explain this, we let $m \in \mathbb{N}_0$ and consider the effective divisor $\mathfrak{m} = m \cdot (\infty)$ on $X_0^+(p)$. The generalized Jacobian **[454]** (or rather its rational points as an algebraic group) $J_{\mathfrak{m}}(\mathbb{Q})$ associated with the modulus $\mathfrak{m}$ is the abelian group given by classes of degree 0 divisors defined over $\mathbb{Q}$ on $X_0^+(p)$ modulo the subgroup

$$P_{\mathfrak{m}}(\mathbb{Q}) := \left\{ \operatorname{div}(f) : f \in \mathbb{Q}(X_0^+(p))^\times \text{ with } q^{-\operatorname{ord}_\infty(f)} f(q) = 1 + O(q^m) \right\},$$

where $\mathbb{Q}(X_0^+(p))$ denotes the field of rational functions defined over $\mathbb{Q}$ on $X_0^+(p)$, $\operatorname{div}(f)$ the divisor of f, and $\operatorname{ord}_\infty(f)$ is the order of f at the cusp $i\infty$. We denote the class in $J_{\mathfrak{m}}(\mathbb{Q})$ of a degree zero divisor Z defined over $\mathbb{Q}$ by $[Z]_{\mathfrak{m}}$. Since $P_{\mathfrak{m}}(\mathbb{Q})$ is a subgroup of the group $P(\mathbb{Q})$ of principal divisors defined over $\mathbb{Q}$, we obtain a natural map $J_{\mathfrak{m}}(\mathbb{Q}) \to J(\mathbb{Q})$.

We need to define some classes for positive discriminants in this case. If $d = b^2$ for $b \neq 0$, then any quadratic form of discriminant d is $\Gamma_0^+(p)$-equivalent to $[0, b, c]$ with $c \in \mathbb{Z}/b\mathbb{Z}$. We let $h_Q \in \mathbb{Q}(X_0^+(p))^\times$ be a rational function on $X_0^+(p)$, satisfying

$$h_Q(q) = 1 - q^b + O(q^m),$$

and define

$$[Z(d)]_{\mathfrak{m}} := \sum_{Q \in \mathcal{Q}_{d,p}/\Gamma_0^+(p)} \frac{1}{b} [\operatorname{div}(h_Q)]_{\mathfrak{m}} = [\operatorname{div}(h_{[0,b,0]})]_{\mathfrak{m}}.$$

Note that this vanishes if $d > m^2$. If $d > 0$ is not a square, then we set $[Z(d)]_{\mathfrak{m}} = 0$. When $d = 0$, we define $[Z(0)]_{\mathfrak{m}}$ as the class of the line bundle of modular forms of weight -1 on $X_0^+(p)$; see **[126]**.

THEOREM 16.9. *The generating series*

$$A_{\mathfrak{m}}(\tau) := \sum_{n=-m^2}^{\infty} [Z(-n)]_{\mathfrak{m}} q^n$$

is the Fourier expansion of a weakly holomorphic modular form of weight 3/2 for $\Gamma_0(4p)$, i.e., it lies in the space $\mathbb{M}^!_{\frac{3}{2}}(\Gamma_0(4p)) \otimes_{\mathbb{Z}} J_{\mathfrak{m}}(\mathbb{Q})$.

To recover Theorem 16.8, we only need to apply the natural map $J_{\mathfrak{m}}(\mathbb{Q}) \to J(\mathbb{Q})$ and note that $[Z(-d)] = 0$ for $d \leq 0$.

Moreover, we can also conclude Zagier's result on traces of singular moduli. The point is that harmonic Maass functions with vanishing constant terms define functionals on the generalized Jacobian. More precisely, suppose that $F \in H_0(\Gamma_0^+(p))$ is a harmonic Maass function and assume that $c_F^+(-n) = 0$ for $n > m$ and $c_F^+(0) = 0$. Then the map $\operatorname{tr}_F : J_{\mathfrak{m}}(\mathbb{Q}) \to \mathbb{C}$ defined by

$$[D]_{\mathfrak{m}} \mapsto \sum_{z \in \operatorname{supp}(D) \setminus \{\infty\}} n_a F(z)$$

for $D = \sum_{z \in X(1)} n_a(z)$ is well-defined and linear. Here, $\operatorname{supp}(D)$ is the support of D, that is, the set of all $z \in X_0^+(p)$, such that $n_z \neq 0$. Applying this map, we recover g_1 as $g_1 = \operatorname{tr}_J(A_{\mathfrak{m}})$ for $p = 1$ and $\mathfrak{m} = 2 \cdot (\infty)$.

16.5. Cycle integrals and mock modular forms

In this section, we primarily discuss beautiful work of Duke, Imamoğlu, and Tóth from **[172]**, which extends much of the work described above in this chapter to positive discriminants, and incorporates mock modular forms into the theory. We note also that related observations on "real quadratic values" of $j(\tau)$ were independently made by Kaneko in **[297]**.

Recall the functions $j_m, m \in \mathbb{N}_0$, defined in (13.3). Each j_m is the unique basis element for $\mathbb{C}[j]$ satisfying

$$j_m(\tau) = q^{-m} + O(q).$$

Then we have that $j_0 = 1$, $j_1 = j - 744$, $j_2 = j^2 - 1488j + 159768$, and so on. These functions are a Hecke system; that is, $j_m = mj|T_m$.

For negative discriminants D_1 and positive fundamental discriminants D_2, we define the *twisted traces* of these functions by

$$\mathcal{T}_{D_1,D_2}(j_m) := \sum_{Q \in \mathcal{Q}_{D_1 D_2}/\Gamma} \frac{\chi_{D_2}(Q) j_m(\tau_Q)}{\omega_Q}. \tag{16.13}$$

These traces are similar to those given above in (16.6); we use (16.13) in order to keep consistent with definitions and notation in **[172]**, whose results we discuss in this section. In addition to the results described in Section 16.2, Zagier **[516]** also showed that the twisted traces in (16.13) may be expressed in terms of Fourier coefficients of modular forms. That is,

$$\mathcal{T}_{D_1,D_2}(j_m) = D_2^{\frac{1}{2}} \sum_{n|m} \left(\frac{D_2}{m/n}\right) nA\left(n^2 D_2, D_1\right), \tag{16.14}$$

where the coefficients $A(D,d)$ $(d<0)$ arise from the basis $\{f_d\}$ of weakly holomorphic modular forms of weight $1/2$ from Section 16.1. Equivalently, by Theorem 16.4, these coefficients may be viewed as coming from the basis $\{g_D\}$ of weakly holomorphic forms of weight $3/2$ (also discussed in Section 16.1).

It is natural to ask if there are analogues to Zagier's result (16.14) for positive discriminants D_1, D_2. One first needs an appropriate analogue of (16.13) in this setting, and then a modular family which encodes the traces in their Fourier coefficients. Duke, Imamoğlu, and Tóth answered this question in **[172]**. To do so, they first defined an extension of the function in (16.13) to positive discriminants D_1, using cycle integrals. Note that cycle integrals were described in Section 12.3. As usual, $\mathcal{Q}_\Delta$ denotes the set of integral binary quadratic forms of discriminant Δ (positive definite if $\Delta < 0$). We recall that $D_2 > 0$ and set

(16.15)

$$\widetilde{\mathcal{T}}_{D_1,D_2}(j_m) := \begin{cases} D_2^{-\frac{1}{2}} \mathcal{T}_{D_1,D_2}(j_m) & \text{if } D_1 < 0, \\ \frac{1}{2\pi} \sum\limits_{Q \in \mathcal{Q}_{D_1 D_2}/\Gamma} \chi_{D_2}(Q) \int_{C_Q} j_m(\tau) Q(\tau,1)^{-1} d\tau & \text{if } D_1 > 0. \end{cases}$$

REMARK. In forthcoming work by Duke, Imamoğlu, and Tóth **[170]**, the analogous case when $D_1, D_2 < 0$ will be developed, which requires an additional differential procedure. It would also be interesting to study the problem from a vector-valued perspective to see what such a repackaging of language might yield; Reihill has announced forthcoming work on this subject **[434]**.

The cycle integral of j_m above is taken over any smooth curve C_Q from a point $z \in \mathbb{H}$ to the point $\gamma_Q z$, where γ_Q is a certain generator of Γ_Q. Precisely, if Δ is not a square, then the (infinite cyclic) group of automorphs of an indefinite binary quadratic form $Q = [a, b, c] \in \mathcal{Q}_\delta$ is generated by

$$\gamma_Q := \begin{pmatrix} \frac{t+bn}{2} & cn \\ -an & \frac{t-bn}{2} \end{pmatrix},$$

where (t, n) is the smallest positive solution to the Pell equation $t^2 - \Delta n^2 = 4$. The automorphs are trivial if Δ is a square, and we let γ_Q be the identity in this case. Some of the main results from **[172]** show how these traces defined in (16.15) for positive discriminants are related to Fourier coefficients arising from a basis of mock modular forms. We summarize some results of Duke, Imamoğlu, and Tóth from **[172]** in the following theorem. Note that we have re-written the error function $\mathcal{E}$ from **[172]** in terms of the incomplete Γ-function in part ii) of Theorem 16.10 below.

THEOREM 16.10. *Assuming the notation above, the following are true:*

i) *For each positive discriminant d there exists a unique mock modular form f_d with shadow g_d (defined in (16.5)), with Fourier expansion of the form*

$$f_d(\tau) = \sum_{n=1}^{\infty} A(n, d) q^n. \tag{16.16}$$

Moreover, the set $\{f_d\}_{d \in \mathbb{N}}$ forms a basis for the space of mock modular forms of weight $1/2$ and level 4 which lie in the Kohnen plus space (i.e., their coefficients are supported on the same arithmetic progression defining the plus space for weakly holomorphic modular forms).

ii) *In particular, the function*

$$\sum_{d=1}^{\infty} \widetilde{\mathcal{T}}_{d,1}(j_1) q^d + \frac{i}{\sqrt{\pi}} \Gamma\left(\tfrac{1}{2}, -4\pi v\right) q - iq - 8\sqrt{v}$$
$$+ \pi^{-\frac{1}{2}} \sum_{d=1}^{\infty} \widetilde{\mathcal{T}}_{-d,1}(j_1) d^{-\frac{1}{2}} \Gamma\left(\tfrac{1}{2}, 4\pi d v\right) q^{-d}$$

is a harmonic Maass form of weight $1/2$ on $\Gamma_0(4)$ in the plus space with shadow

$$-q^{-1} + 2 + \sum_{d=0}^{\infty} \mathrm{Tr}_{-d}(j_1) q^d = -g_1(\tau),$$

where g_1 is as in (16.5).

iii) *For any integer $D_1 \equiv 0, 1 \pmod 4$ and positive fundamental discriminant D_2 with $D_1 D_2$ not a square, we have*

$$\widetilde{\mathcal{T}}_{D_1,D_2}(j_m) = \sum_{n \mid m} \left(\frac{D_2}{m/n}\right) n A\left(n^2 D_2, D_1\right),$$

where for $D_1 < 0$, the coefficients $A(n, D_1)$ are as defined in (16.3), and for $D_1 > 0$, the coefficients $A(n, D_1)$ are given in (16.16).

REMARK. The case of square discriminants requires more care, due to the fact that the corresponding geodesics are infinite, and hence that the corresponding cycle integrals diverge. Bruinier, Funke, and Imamoğlu showed in **[123]** that convergent expressions for traces at square discriminants could be obtained by utilizing suitably regularized integrals. Later, Andersen in **[18]** gave another definition of the traces at square discriminants, giving a uniform description of both the square and non-square cases.

REMARK. Our notation for g_D differs by a sign from the functions in (1.10) of **[172]**).

Rephrasing this another way, Theorem 16.10 gives a certain basis of mock modular forms f_d of weight 1/2 whose shadows are Zagier's weakly holomorphic modular forms g_d of weight 3/2, and whose Fourier coefficients are given in terms of cycle integrals of the modular j-function. In the case of j_1, we see in part ii) of Theorem 16.10 an explicit Maass form linking the traces defined in (16.15): the nonholomorphic part of this function is defined in terms of Zagier's traces $\widetilde{\mathcal{T}}_{d,1}(d < 0)$, and the holomorphic part is defined in terms of the cycle integrals $\widetilde{\mathcal{T}}_{d,1}(d > 0)$. Moreover, these results generalize Zagier's results for negative discriminants, including (16.14), to also include positive discriminants.

SKETCH OF PROOF OF THEOREM 16.10. Theorem 16.10 i) was proven in **[172]** by first explicitly constructing a basis $\{h_d\}_{d\equiv 0,1 \pmod 4}$ for a certain subspace $\mathbb{H}^!_{\frac{1}{2}}(\Gamma_0(4)) \subseteq H^!_{\frac{1}{2}}(\Gamma_0(4))$ defined by

$$\mathbb{H}^!_{\frac{1}{2}}(\Gamma_0(4)) := \left\{ f \in H^!_{\frac{1}{2}}(\Gamma_0(4)) \colon c_f^{\pm}(n) = 0 \text{ unless } n \equiv 0, 1 \pmod 4 \right\}.$$

Note that in **[172]**, forms in this space were called *weakly harmonic modular forms.* This basis is constructed using Poincaré series which are closely related to the Maass-Poincaré series of (6.10). There are delicate convergence issues to be dealt with, however, so the Poincaré series used here are given through analytic continuation. For $d \leq 0$, the Poincaré series $h_d = f_d$, where f_d is as defined in (16.4). For $d > 0$, the function f_d is the holomorphic part of the Poincaré series h_d.

Part iii) of Theorem 16.10 is proved by evaluating the traces and cycle integrals $\widetilde{\mathcal{T}}_{D_1,D_2}(j_m)$ introduced in (16.15) in terms of the Fourier coefficients of the basis $\{h_d\}$. To do so, the authors made use of Fay's Poincaré series from **[192]** to define functions $j_m(\tau, s)$ for $m \in \mathbb{N}$ and $\mathrm{Re}(s) > 1$. These functions are analytically continued to $\mathrm{Re}(s) > \frac{3}{4}$, from which it can be deduced that $j_m(\tau, 1) = j_m(\tau)$. The traces and cycle integrals of the functions $j_m(\tau, s)$ for $\mathrm{Re}(s) > 1$ were first computed in terms of Kloosterman sums, using methods similar to those first applied by Zagier **[507]**. The result then follows after a continuation argument.

Parts i) and iii) imply part ii) of Theorem 16.10. The proof uses Poincaré series and properties of Kloosterman sums. □

It is also natural to ask if there is an appropriate analogue to Theorem 16.10 ii) when $m = 0$, i.e., for $\mathrm{Tr}_d(1)$. This trace is given in terms of the Hurwitz class number $H(-d)$ when $d \leq 0$, and the narrow class number $h(d)$ of $Q(\sqrt{d})$ when $d > 0$, by

$$\widetilde{\mathrm{Tr}}_d(1) = \begin{cases} \mathrm{Tr}_d(1) = H(-d) & \text{if } d \leq 0, \\ \pi^{-1} i d^{-\frac{1}{2}} h(d) \log(\eta_d) & \text{if } d > 0 \text{ is a fundamental discriminant.} \end{cases}$$

Here, η_d denotes the smallest unit greater than 1 with norm 1. By work of Zagier, as explained in Theorem 6.3 (more generally, see Section 16.3), the generating function for $\mathrm{Tr}_d(1), d \leq 0$, is a mock modular form. As a companion to Theorem 16.10 ii), we have the following theorem from **[172]**, which was proved using the weight 1/2 Eisenstein series and limit formulas.

THEOREM 16.11. *The function*

$$\sum_{d=1}^{\infty} \widetilde{\mathrm{Tr}}_d(1)q^d + \frac{\sqrt{v}}{3} + \pi^{-\frac{1}{2}} \sum_{d=1}^{\infty} \widetilde{\mathrm{Tr}}_{-d}(1) d^{-\frac{1}{2}} \Gamma\left(\tfrac{1}{2}, 4\pi d v\right) q^{-d}$$
$$+ \frac{\sqrt{v}}{2\pi} \sum_{n \in \mathbb{Z}\setminus\{0\}} |n| q^{n^2} \int_0^{\infty} t^{-\frac{1}{2}} \log(1+t) e^{4\pi n^2 v t} dt - \frac{1}{4\pi}\log(v)$$

has weight 1/2 *for* $\Gamma_0(4)$.

In fact, the function from Theorem 16.11 gives a motivating example of a sesquiharmonic Maass form (cf. Definition 6.17). For more on sesquiharmonic Maass forms, see Section 6.3 and Section 12.3.

COROLLARY 16.12. *The function from Theorem* 16.11 *is a sesquiharmonic Maass form of weight* 1/2. *Moreover, its image under the* $\xi_{\frac{1}{2}}$*-operator is* $-2\mathcal{H}$, *where* $\mathcal{H}$ *is the Zagier Eisenstein series from Theorem* 6.3.

Another paper by Duke, Imamoḡlu, and Tóth, namely **[173]**, gives some interesting applications of some results in **[172]**, which again nicely link real quadratic and imaginary quadratic information. Their results surround the regularized inner products of the weakly holomorphic weight 3/2 forms g_D defined in (16.5),

$$\langle g_{D_1}, g_{D_2}\rangle = \lim_{T\to\infty} \int_{\mathcal{F}_4(T)} g_{D_1}(\tau)\overline{g_{D_2}(\tau)} v^{\frac{3}{2}} \frac{du dv}{v^2},$$

where $\mathcal{F}_4(T)$ is the standard truncated fundamental domain for $\Gamma_0(4)$ obtained by removing T-neighborhoods of the cusps. These "imaginary quadratic" functions g_d in fact contain real quadratic information, as exhibited by the next theorem from **[173]**. In what follows, $\rho := 2\tan^{-1}(\frac{n\sqrt{D_1D_2}}{t})$, where (t, n) is the integral solution to the Pell equation described earlier in this section. Theorem 16.13 shows how cycle integrals of j_1 may be expressed as regularized inner products of the weight 3/2 weakly holomorphic forms g_d.

THEOREM 16.13. *Let* D_1, D_2 *be distinct positive fundamental discriminants. Then we have that*

$$\langle g_{D_1}, g_{D_2}\rangle$$
$$= \frac{3}{16\pi}(D_1D_2)^{-\frac{1}{2}} \sum_{Q=[a,b,c]\in\mathcal{Q}_{D_1D_2}/\Gamma} \chi_{D_2}(Q) \int_{-\rho}^{\rho} j_1\left(-\frac{b}{2a} + \frac{i\sqrt{D_1D_2}}{2|a|}e^{i\theta}\right) \frac{d\theta}{\cos(\theta)}.$$

In **[173]**, an analogue to Theorem 16.13 was proved for $\langle g_{D_1}, g_0\rangle$; there, this inner product was expressed in terms of the class number $h(D_1)$. In **[82]**, Diamantis, Ehlen, and one of the authors extended work in **[173]** and computed $\langle g_1, g_1\rangle$ as an integral involving the j-function, and the digamma function.

The work in **[172]** by Duke, Imamoḡlu, and Tóth has inspired and is related to a number of additional interesting recent works including their two further works **[171, 174]**, and also **[18, 93, 123, 377]**. Moreover, the results from **[172]** and **[173]** are not limited to the results stated in this section.

The interesting work in **[174]**, among other things, examined the distribution properties of surfaces associated to ideal classes of real quadratic fields. In doing so, the authors of **[174]** provided some related nonholomorphic results, and in particular extended earlier work of Hecke **[260]**, Maass **[370]**, and Katok-Sarnak **[302]**. For example, they gave explicit expressions for products $c(D_1)\overline{c(D_2)}$ of Fourier coefficients of non-zero weight 1/2 forms in terms of Hecke-Maass cusp forms. The expressions given vary, depending on how the coprime fundamental discriminants are related to one another: $D_1, D_2 < 0, D_1, D_2 > 0$, or $D_1 D_2 < 0$. Note that in this analysis, there is a third case (namely $D_1, D_2 < 0$) which emerges, and in this case, one requires a different analysis.

16.6. Weight one harmonic Maass forms

Note that modular forms of weight one are somewhat special. To see this, recall that the Riemann-Roch Theorem yields a formula for the difference

$$\dim(S_k(\Gamma, \chi)) - \dim(M_{2-k}(\Gamma, \chi))$$

(for some congruence subgroup Γ and a character χ). Thus, if $k \geq 2$, one can determine $\dim(S_k(\Gamma, \chi))$ explicitly because $\dim(M_{2-k}(\Gamma, \chi))$ is zero for $k > 2$ and for $k = 2$ the space $M_{2-k}(\Gamma, \chi)$ is either zero or one dimensional and equal to $\mathbb{C}$ (if χ is trivial). For $k = 1$, however, we do not obtain an explicit dimension formula by using the Riemann-Roch Theorem.

In what follows we restrict to the case that $\Gamma = \Gamma_0(p)$ for a prime $p \equiv 3 \pmod 4$ with $p > 3$ and $\chi = \chi_{-p} = (\frac{-p}{\cdot})$. In this case (or in general for $\Gamma_0(-D)$ with $D < 0$ a discriminant and character $\chi = \chi_D = (\frac{D}{\cdot})$), we can at least construct some weight one modular forms rather explicitly using theta functions attached to binary quadratic forms of discriminant $D = -p$. Indeed, if $Q = [a, b, c] \in \mathcal{Q}_{-p}$ is a positive definite integral binary quadratic form of discriminant $-p$, then the theta function

$$\theta_Q(\tau) := \sum_{m,n \in \mathbb{Z}} q^{Q(m,n)} = \sum_{n=0}^{\infty} r_Q(n) q^n$$

is in $M_1(\Gamma_0(p), \chi_{-p})$, where $r_Q(n) := |\{(X, Y) \in \mathbb{Z}^2 : Q(X, Y) = n\}|$. We denote by $\Theta(-p) \subset M_1(\Gamma_0(p), \chi_{-p})$ the subspace of all weight one modular forms spanned by these theta functions. We note that the remaining modular forms of weight one, also called "exotic", are extremely mysterious. However, by the Deligne-Serre theorem **[160]** all newforms of weight one correspond to two-dimensional Galois representations. The newforms in $\Theta(-p)$ correspond to (imaginary) dihedral Galois representations. In this section, we concentrate on the space $\Theta(-p)$ because there has been recent progress in understanding harmonic Maass forms of weight one whose image under ξ_1 is contained in $\Theta(-p)$ which we now explain.

It is convenient to use the relation of positive definite integral binary quadratic forms to ideals in imaginary quadratic fields (cf. **[510]**). Recall that there is a bijection between $\mathcal{Q}_{-p}/\mathrm{SL}_2(\mathbb{Z})$. We write $\alpha \mapsto \alpha'$ for the non-trivial Galois automorphism of K, $\mathrm{N}(\alpha) = \alpha\alpha'$, and $\mathrm{Tr}(\alpha) = \alpha + \alpha'$. Then the correspondence from the class group of K to $\mathcal{Q}_{-p}/\mathrm{SL}_2(\mathbb{Z})$ is given by mapping a fractional ideal $\mathfrak{a} \subset K$ generated by α and β to the quadratic form

$$Q_{\mathfrak{a}} = \left[\frac{\mathrm{N}(\alpha)}{\mathrm{N}(\mathfrak{a})}, \frac{\mathrm{Tr}(\alpha\beta')}{\mathrm{N}(\mathfrak{a})}, \frac{\mathrm{N}(\beta)}{\mathrm{N}(\mathfrak{a})}\right].$$

We also write $\theta_{\mathfrak{a}}$ for the theta function corresponding to $Q_{\mathfrak{a}}$. Using this correspondence, it is not hard to show (cf. [**300**]) that the dimension of $\Theta(-p)$ is given by

$$\dim(\Theta(-p)) = \frac{h_{-p}+1}{2},$$

where h_{-p} is the class number of K, giving a lower bound for the dimension of $M_1(\Gamma_0(p), \chi_{-p})$. Moreover, the difference of any two theta functions in this space gives a cusp form, i.e., the intersection of $\Theta(-p)$ with the subspace of $M_1(\Gamma_0(p), \chi_{-p})$ spanned by Eisenstein series is one dimensional. It is spanned by an Eisenstein series that can be obtained in two ways:

i) The special value at $s = 0$ of the analytic continuation to $\mathbb{C}$ of

$$E_{K,s}(\tau) := v^{\frac{s}{2}} \sum_{\gamma \in \Gamma_\infty \backslash \mathrm{SL}_2(\mathbb{Z})} \frac{\Phi_p^+(\gamma)}{(c\tau+d)|c\tau+d|^s},$$

where, for $\gamma = \left(\begin{smallmatrix} a & b \\ c & d \end{smallmatrix}\right) \in \mathrm{SL}_2(\mathbb{Z})$, we let

$$\Phi_p^+(\gamma) := \begin{cases} \chi_{-p}(d) & \text{if } \gamma \in \Gamma_0(p), \\ ip^{-\frac{1}{2}}\chi_{-p}(c) & \text{if } \gamma \notin \Gamma_0(p). \end{cases}$$

ii) The Siegel-Weil formula states that

$$E_K(\tau) := E_{K,0}(\tau) = \frac{1}{h_{-p}} \sum_{[\mathfrak{a}] \subset \mathrm{Cl}(K)} \theta_{\mathfrak{a}}(\tau) = 1 + \frac{2}{h_{-p}} \sum_{n=1}^{\infty} \rho(n) q^n,$$

where $\mathrm{Cl}(K)$ denotes the class group of K and $\rho(n)$ is the number of integral ideals of norm n in K.

There is another Eisenstein series $\mathcal{E}_K(\tau)$, of weight one attached to K. It is a harmonic Maass form contained in $H_1^!(p, \chi_{-p})$ which has the property that

$$\xi_1(\mathcal{E}_K) = E_K.$$

Explicitly, $\mathcal{E}_K(\tau)$ is the value at $s = 0$ of the analytic continuation to $\mathbb{C}$ of the derivative of an *incoherent* Eisenstein series:

$$\mathcal{E}_K(\tau) := 2\frac{\partial}{\partial s} \left[v^{\frac{s}{2}} \sum_{\gamma \in \Gamma_\infty \backslash \mathrm{SL}_2(\mathbb{Z})} \frac{\Phi_p^-(\gamma)}{(c\tau+d)|c\tau+d|^s} \right]_{s=0},$$

where, for $\gamma = \left(\begin{smallmatrix} a & b \\ c & d \end{smallmatrix}\right) \in \mathrm{SL}_2(\mathbb{Z})$, we let

$$\Phi_p^-(\gamma) = \begin{cases} \chi_{-p}(d) & \text{if } \gamma \in \Gamma_0(p), \\ -ip^{-\frac{1}{2}}\chi_{-p}(c) & \text{if } \gamma \notin \Gamma_0(p). \end{cases}$$

Note that we consider the derivative of this Eisenstein series because its value at $s = 0$ vanishes because of the functional equation it satisfies. The Eisenstein series $\mathcal{E}_K(\tau)$ appeared in the work of Kudla, Rapoport, and Yang [**337**], and it turns out that the Fourier coefficients of its holomorphic part are related to certain special cycles on the stack of CM elliptic curves (with complex multiplication by the ring of integers in K). While the results of [**337**] are beyond the scope of this book, the Fourier expansion of the holomorphic part $\mathcal{E}_K^+$ has the simple shape

$$\mathcal{E}_K^+(\tau) = \log(p) + 2\frac{\Lambda'(1,\chi_{-p})}{\Lambda(1,\chi_{-p})}$$
$$-\frac{2}{h_{-p}}\sum_{n=1}^{\infty}\left(\log(p)(\mathrm{ord}_p(n)+1)\rho(n) + \sum_{\ell\neq p}\log(\ell)(\mathrm{ord}_\ell(n)+1)\rho\left(\tfrac{n}{\ell}\right)\right)q^n,$$

where the sum runs over all primes ℓ that are inert in K. Note that the sum over $\ell \neq p$ actually only contains at most one non-zero term, and the n-th Fourier coefficient ($n > 0$) of $\mathcal{E}_K^+$ is either equal to

$$-\tfrac{2}{h_{-p}}\log(p)(\mathrm{ord}_p(n)+1)\rho(n)$$

or there exists a unique prime ℓ that is inert in K such that it equals

$$-\tfrac{2}{h_{-p}}\log(\ell)(\mathrm{ord}_\ell(n)+1)\rho\left(\frac{n}{\ell}\right).$$

Moreover, note that while the coefficients are transcendental, they are essentially logarithms of integers whose factorization is rather simple and reflects the arithmetic of K.

Until recently, this was as much as was known about harmonic Maass forms of weight one. Thanks to independent results by Duke-Li [**176**], Ehlen [**187**], and Viazovska [**483**], we now know much more about the arithmetic nature of weight one harmonic Maass forms. Namely, we now have a lot of information about the Fourier coefficients of certain mock modular forms with shadows $\theta_{\mathfrak{a}}$. The Fourier coefficients of the holomorphic part of a certain normalized preimage under ξ_1 are all logarithms of algebraic integers which are contained in the Hilbert class field H of K. And, similar to the incoherent Eisenstein series $\mathcal{E}_K$, their prime factorizations can be made quite explicit.

The following result is contained in Theorems 1.2, 1.3, and Proposition 1.4 in [**187**].

THEOREM 16.14. *For every $[\mathfrak{a}] \in \mathrm{Cl}(K)$, there is a harmonic Maass form $F_{\mathfrak{a}}$ of weight one, such that the following are true:*

i) *We have $\xi_1(F_{\mathfrak{a}}) = \theta_{\mathfrak{a}}$.*
ii) *The following analogue of the Siegel-Weil formula holds:*

$$\frac{1}{h_{-p}}\sum_{[\mathfrak{a}]\in\mathrm{Cl}(K)} F_{\mathfrak{a}}(\tau) = \mathcal{E}_K(\tau).$$

iii) *There exists $r \in \mathbb{Z}$ such that for all $n \in \mathbb{N}$, there exists an algebraic integer $\alpha(\mathfrak{a}, n)$ contained in the Hilbert class field H of K, such that the Fourier coefficient $c_{\mathfrak{a}}^+(n)$ of $F_{\mathfrak{a}}^+$ equals*

$$c_{\mathfrak{a}}^+(n) = -\frac{2}{r}\log|\alpha(\mathfrak{a}, n)|.$$

For $n \in -\mathbb{N}$, we have $c_{\mathfrak{a}}^+(n) = \frac{2}{r}\log|\alpha(\mathfrak{a}, n)|$ for some unit $\alpha(\mathfrak{a}, n)$ in H.
iv) *Fix the embedding of K into $\mathbb{C}$ such that $\mathrm{Im}(\sqrt{-p}) > 0$ and identify $H = K(j)$ as a subfield of $\mathbb{C}$ where $j := j((1+\sqrt{-p})/2)$. There is a unique prime ℓ that is not split in K, such that $\alpha(\mathfrak{a}, n)$ is a unit at all places not above ℓ, and for the primes above ℓ, the following holds: If $\mathfrak{p}_0$ is the*

unique prime of H above ℓ, such that $\mathfrak{p}$ is fixed by complex conjugation and $\mathfrak{p} = \mathfrak{p}_0^\sigma$ for σ corresponding to $\mathfrak{a}$ under the Artin map of class field theory (cf. [**463**, Chapter II]*), then*

$$\mathrm{ord}_{\mathfrak{p}}(\alpha(\mathfrak{a}, n)) = 2r \cdot \nu_\ell(n)\rho\left(\frac{np}{\ell}, [\mathfrak{a}]^{-2}\right),$$

where $\rho(m, [\mathfrak{a}])$ is the number of integral ideals of K contained in the class $[\mathfrak{a}]$ of $\mathfrak{a}$ and $\nu_p(n) = \mathrm{ord}_p(n) + 1$ and $\nu_\ell(n) = \frac{1}{2}(\mathrm{ord}_\ell(n) + 1)$ for $\ell \neq p$.

We note that in the special case of a prime discriminant, i)–iii) are also contained in the work of Duke and Li [**176**] where the authors stated iv) (in a slightly different form) conjecturally.

The proof of Theorem 16.14 has essentially three parts: First, Ehlen proved a formula that relates the coefficients $c_{\mathfrak{a}}^+(n)$ to values at complex multiplication points of regularized theta lifts for the orthogonal group $O(2, n)$. In the special case that $n = 0$, this gives in particular a formula for $\langle f_m, \theta_{\mathfrak{a}} \rangle$, where $f_m \in H_1(\Gamma_0(p), \chi_{-p})$, such that the principal part of f_m is given by q^{-m} and f_m satisfies the plus space condition. Using Stokes' Theorem this is shown to be essentially equal to $c_{\mathfrak{a}}^+(m)$, plus some "error terms" that can be controlled.

The second step is to use a certain seesaw identity to relate the two cases for $n = 0$ and $n = 1$. For $n = 1$, the regularized theta lift is exactly the theta lift used for the Borcherds lift as in Chapter 18. The idea is roughly as follows. It is shown that if $f \in M_1^!(\Gamma_0(p), \chi_{-p})$ with vanishing constant term $c_f(0)$, then there exists $g \in M_{\frac{1}{2}}^!(\Gamma_0(p), \chi_{-p})$, with vanishing constant term, such that

$$\langle f, \theta_{\mathfrak{a}} \rangle^{\mathrm{reg}} = \log|\Psi(z_{-p}, g)|, \tag{16.17}$$

where $z_{-p} \in \mathbb{H}$ is a certain CM-point of discriminant D and $\Psi(z, g)$ is a modular function of level p, given by the Borcherds lift of g. Therefore (linear combinations of) the coefficients $c_{\mathfrak{a}}^+(m)$ can be related to logarithms of CM values of modular functions on $\Gamma_0(p)$.

These CM values are, much like the special values of j (cf. Section 16.1), algebraic integers, contained in the Hilbert class field H of K, and their prime factorization can be determined explicitly, using a generalization of the results of Gross and Zagier [**241**] to higher level. To obtain all the statements in Theorem 16.14, in particular iii) and iv), the form $F_{\mathfrak{a}}$ has to be normalized appropriately. The denominator r in iii) appears because we have to use weakly holomorphic modular forms with integral Fourier coefficients to obtain modular functions whose special values are algebraic integers in the seesaw argument. The maximal denominators that are required can be related to the orders of certain Borcherds products at the cusps, and it can be shown that the associated divisor on $X_0(p)$ is a rational torsion point in the Jacobian J of $X_0(p)$. Using the fact that $J(\mathbb{Q})$ is a finitely generated abelian group, we can obtain a uniform bound on these denominators.

We should also remark that Ehlen's proof of Theorem 16.14 shows that the harmonic Maass forms $F_{\mathfrak{a}}$ are related to arithmetic geometry. It is in fact shown in [**187**] and [**188**] that the factorization formula given in iv) is determined by certain special cycles on an arithmetic curve and the modularity of the degree generating series studied in [**337**] is recovered using a completely different proof.

CHAPTER 17

Shifted Convolution L-functions

Recent work by Hoffstein and Hulse **[272]** investigated generalizations of the important Rankin-Selberg convolution L-functions for modular forms. Symmetrized forms of their convolution L-functions are intimately related to mock modular forms. Here we describe these close connections. We begin by recalling the context of the work of Rankin and Selberg.

17.1. Rankin-Selberg convolutions

Rankin and Selberg are well known, among other things, for their introduction of the so-called Rankin-Selberg method. This method has its origins in Rankin's 1939 paper **[432]** on Ramanujan's tau-function, and in Selberg's papers **[448, 449]**. Roughly speaking, the Rankin-Selberg method is a powerful tool which allows one to obtain, under weak hypotheses, the meromorphic continuation and functional equation for the Mellin transform of the constant term of a Γ-invariant function, where Γ is an appropriate subgroup of the modular group.

Here we briefly describe the idea in the case of $\mathrm{SL}_2(\mathbb{Z})$. The idea is based on the analytic properties of the *nonholomorphic Eisenstein series* ($s \in \mathbb{C}$ with $\mathrm{Re}(s) > 1$)

$$E(\tau; s) := \frac{1}{2} \sum_{(m,n)\in\mathbb{Z}^2\setminus\{(0,0)\}} \frac{v^s}{|m\tau + n|^{2s}}.$$

As usual, here we have that $\tau = u + iv \in \mathbb{H}$ with $u, v \in \mathbb{R}$. It is an $\mathrm{SL}_2(\mathbb{Z})$-invariant function, and the *completed Eisenstein series*

$$E^*(\tau; s) := \pi^{-s}\Gamma(s)E(\tau; s)$$

has a meromorphic continuation in s to the entire complex plane. In fact, it is holomorphic except for simple poles at $s = 0, 1$, with residue $1/2$ at $s = 1$, and it satisfies the functional equation

$$E^*(\tau; 1 - s) = E^*(\tau; s).$$

Suppose that $r\colon \mathbb{H} \to \mathbb{C}$ is a smooth $\mathrm{SL}_2(\mathbb{Z})$-invariant function with Fourier expansion

$$r(\tau) = \sum_{n\in\mathbb{Z}} a(v; n)e^{2\pi i n u}.$$

Moreover, suppose there is an $\alpha > 0$ for which $r(\tau) = O(v^{-\alpha})$ as $v \to \infty$. The idea is to consider the Mellin transform of the constant term $a(v; 0)$, namely the function

$$A_0(s) := \int_0^\infty a(v; 0)v^{s-1}dv,$$

which makes sense when $0 < \mathrm{Re}(s) < \alpha$. For $1 < \mathrm{Re}(s) < \alpha + 1$, we have the key identity

$$\int_{\mathcal{F}} r(\tau)E(\tau; s)d\mu = \zeta(2s)A_0(s-1), \tag{17.1}$$

where $\mathcal{F}$ is a fundamental domain for $\mathrm{SL}_2(\mathbb{Z})$ and $d\mu = \frac{dudv}{v^2}$ is the usual $\mathrm{SL}_2(\mathbb{R})$-invariant measure.

The goal is to use this identity to obtain a meromorphic continuation and functional equation for $A_0(s)$. These are obtained by making use of the properties of $E(\tau; s)$. One finds that $\zeta(2s)A_0(s-1)$ has a meromorphic continuation in s with at most a simple pole at $s = 1$ with residue $\frac{\pi}{2}\int_{\mathcal{F}} r(\tau)d\mu$. Moreover, by inserting the appropriate Γ-factor, one obtains the desired functional equation for $\zeta(2s)A_0(s-1)$.

Rankin invented this method with the idea of bounding coefficients of cusp forms. Before Deligne's proof of the Weil Conjectures, the best bounds were obtained using (17.1). To see how this works, suppose that

$$f(\tau) = \sum_{n=1}^{\infty} c_f(n)q^n$$

is a weight k cusp form on $\mathrm{SL}_2(\mathbb{Z})$. Rankin let $r(\tau) := v^k|f(\tau)|^2$. In this case we have that

$$A_0(s) = \frac{\Gamma(s+k)}{(4\pi)^{s+k}} L(f \otimes f, s+k). \tag{17.2}$$

In (17.2), for $\mathrm{Re}(s) \gg 0$, we have the Rankin-Selberg convolution of two L-functions ($j \in \{1, 2\}$)

$$L(f_j, s) = \sum_{n=1}^{\infty} \frac{c_{f_j}(n)}{n^s}$$

arising from even integer weight cusp forms $f_j \in S_k(\Gamma_0(N))$, defined as follows (see also (3.5)):

$$L(f_1 \otimes f_2, s) = \sum_{n=1}^{\infty} \frac{c_{f_1}(n)\overline{c_{f_2}(n)}}{n^s}.$$

The function $L(f \otimes f, s)$ is also referred to as the *Rankin-Dirichlet series for f*.

The coefficients of $L(f \otimes f, s)$ are non-negative, and so by Landau's Theorem, it must have a singularity at its abscissa of convergence. Since $\zeta(s) \neq 0$ for $\mathrm{Re}(s) = 1$, it follows that the Rankin-Dirichlet series for f extends to a holomorphic function apart from a simple pole at $s = k$, with a residue which is a non-zero multiple of the square of the Petersson norm $\langle f, f \rangle$. Moreover, with the appropriate Γ-factors attached, it turns out that there is a functional equation relating s to $2k - 1 - s$. Then using standard techniques from analytic number theory, Rankin found that

$$\sum_{n \leq X} |c_f(n)|^2 = C(f)X^k + O\left(X^{k-\frac{2}{5}}\right),$$

where $C(f) > 0$. This implies the bound

$$|c_f(n)| = O\left(n^{\frac{k}{2}-\frac{1}{5}}\right).$$

This discussion highlights the important role of the Rankin-Dirichlet series. More generally, Rankin's and Selberg's works on the aptly named convolution function are among the most important contributions to the modern theory of automorphic forms.

17.2. Hoffstein-Hulse shifted convolution L-functions

In a recent paper, Hoffstein and Hulse **[272]** introduced generalizations of the Rankin-Selberg convolution Dirichlet series. For positive integers h, they defined[1] the *shifted convolution L-series*

$$D(f_1, f_2, h; s) := \sum_{n=1}^{\infty} \frac{c_{f_1}(n+h)\overline{c_{f_2}(n)}}{n^s}, \tag{17.3}$$

where the f_j are even weight k modular forms with square-free level. Using these shifted convolution Dirichlet series, by summing in h-aspect they defined the *shifted double Dirichlet series*

$$Z(f_1, f_2; s, w) := \sum_{h=1}^{\infty} \frac{D(f_1, f_2, h; s)}{h^w}. \tag{17.4}$$

They obtained **[272]** the meromorphic continuation of these series.

THEOREM 17.1 (Hoffstein and Hulse). *Assume the notation and hypotheses above. Then $D(f_1, f_2, h; s)$ has a meromorphic continuation to $\mathbb{C}$ and $Z(f_1, f_2; s, w)$ has a meromorphic continuation to $\mathbb{C}^2$.*

REMARK. As an important application, Hoffstein and Hulse used these results to obtain a Burgess-type bound for L-series associated to modular forms. To be more precise, if $f \in S_k(\Gamma_0(N))$, where k is even and N is square-free, then they proved that if χ is a Dirichlet character modulo $Q \in \mathbb{N}$, then

$$L\left(f \otimes \chi; \frac{k}{2}\right) \ll_{f,\varepsilon} Q^{\frac{3}{8}+\varepsilon}.$$

For much more general results, the readers should consult the papers of Beckwith, Blomer, Harcos, Maga, and Michel **[49, 65, 66, 67, 68, 69, 249, 250, 372]**.

17.3. Special values of shifted convolution L-functions

Here we consider a generalization of the shifted convolution L-functions defined by Hoffstein and Hulse. We investigate the *symmetrized shifted convolution series*

$$\widehat{D}(f_1, f_2, h; s) := D(f_1, f_2, h; s) - D\left(f_2^c, f_1^c, -h; s\right).$$

Here f^c denotes the usual operation of conjugation of Fourier coefficients, as in Lemma 5.17. We require that the weight k is even, but don't require the levels to be square-free. The generating function for these series at $s = k-1$ in h-aspect, defined by

$$\mathbb{L}(f_1, f_2; \tau) := \sum_{h=1}^{\infty} \widehat{D}(f_1, f_2, h; k-1) q^h, \tag{17.5}$$

is a weight 2 modular object. To be more precise, it is the sum of a weight 2 mixed mock modular form and a weight 2 quasimodular form.

To make this precise, we let $\widetilde{M_k^!}(\Gamma_0(N))$ be the extension of $M_k^!(\Gamma_0(N))$ by E_2. The following theorem is a special case of Theorem 1.1 of **[386]**.

[1] They used a slightly different normalization.

THEOREM 17.2. *Assuming the notation and hypotheses above, we have that*

$$\mathbb{L}(f_2, f_1; \tau) = -\frac{1}{(k_1-2)!} M^+_{f_1}(\tau) f_2(\tau) + F(\tau),$$

where $F \in \widetilde{M}^!_2(\Gamma_0(N))$, *and* $M^+_{f_1}$ *is a mock modular form with shadow* f_1.

REMARK. The reader should be aware of the ordering of the forms f_1 and f_2 above. In particular, note that F depends on the ordering of these forms.

EXAMPLE 17.3. Here we consider the case where $f_1 = f_2 = \Delta$, the normalized cusp form of weight 12 on $\mathrm{SL}_2(\mathbb{Z})$. The following table gives numerical approximations for the first few shifted convolution values $\widehat{D}(\Delta, \Delta, h; 11)$:

h	1	2	3	4
$\widehat{D}(\Delta, \Delta, h; 11)$	$-33.383\ldots$	$266.439\ldots$	$-1519.218\ldots$	$4827.434\ldots$

In terms of the Poincaré series of Chapter 6, we have that $\Delta = \frac{1}{\beta} P_{12,1}$, where β can be described in terms of a Petersson norm, or as an infinite sum of Kloosterman sums weighted by J-Bessel functions as follows

$$\beta := \frac{(4\pi)^{11}}{10!} \|P_{12,1}\|^2 = 1 + 2\pi \sum_{c=1}^{\infty} \frac{K(1,1;c)}{c} J_{11}\left(\frac{4\pi}{c}\right) = 2.8402\cdots.$$

There are no weight 2 holomorphic modular forms on $\mathrm{SL}_2(\mathbb{Z})$, and so it follows that

$$\begin{aligned}\mathbb{L}(\Delta, \Delta; \tau) &= \frac{F^+_{-10,1}(\tau)\Delta(\tau)}{\beta} - \frac{E_2(\tau)}{\beta} \\ &= -33.383\ldots q + 266.439\ldots q^2 - 1519.218\ldots q^3 + 4827.434\ldots q^4 - \cdots.\end{aligned}$$

Since the expressions for the coefficients of $F^+_{-10,1}$ are rapidly convergent, this q-series identity provides an efficient method for computing the values $\widehat{D}(\Delta, \Delta, h; 11)$. Alternatively, if we let $\alpha = 106.10455\ldots$ and define

$$\sum_{n=-1}^{\infty} r(n) q^n := -\Delta\left(j(\tau)^2 - 1464 j(\tau) - \alpha^2 + 1464\alpha\right),$$

then it turns out that

$$\mathbb{L}(\Delta, \Delta; \tau) = -\frac{\Delta(\tau)}{\beta}\left(\frac{65520}{691} - \sum_{n \in \mathbb{Z}\setminus\{0\}} \frac{r(n)}{n^{11}} q^n\right) - \frac{E_2(\tau)}{\beta}.$$

SKETCH OF PROOF OF THEOREM 17.2. To establish the convergence of the series $\widehat{D}(f_1, f_2, h; s)$ at the relevant s, we make use of deep estimates by Blomer[2] (cf. Corollary 1.4 of [**65**]). His work implies, for $X \in \mathbb{R}^+$, that

$$\sum_{n=1}^{X} c_{f_1}(n+h)\overline{c_{f_2}(n)} \ll_\epsilon X^{k-\delta},$$

for some fixed $\delta > 0$ if $h \leq X^{\frac{4}{3}-\varepsilon}$. With these estimates, we have convergence.

[2]The reader should note that Blomer only stated his result if $f_1 = f_2$. However, an inspection of his proof shows that his argument carries over directly to the present setting.

To complete the proof, it suffices to produce the generating function $\mathbb{L}(f_2, f_1; \tau)$ from the cusp forms f_1 and f_2 and then establish the required modularity properties. To this end, suppose that $M_{f_1} \in H_{2-k}(\Gamma_0(N))$ is a harmonic Maass form whose mock modular forms has shadow f_1. Then we consider the regularized holomorphic projection of the nonholomorphic modular form $M_{f_1} f_2$ (cf. Section 10.2).

By Proposition 10.4, we have that $\pi_{\mathrm{hol}}^{\mathrm{reg}}(M_{f_1} f_2)$ is a weakly holomorphic quasimodular form in $\widetilde{M}_2^!(\Gamma_0(N))$. To complete the proof, one then explicitly computes this holomorphic projection. The nonholomorphic part of M_{f_1} is

$$M_{f_1}^-(\tau) = \sum_{n=1}^{\infty} n^{1-k_1} \overline{c_{f_1}(n)} e^{4\pi n v} \Gamma(k_1 - 1, 4\pi n v) \overline{q}^{-n}.$$

A calculation then reveals that

$$\pi_{\mathrm{hol}}^{\mathrm{reg}}(M_{f_1} f_2) = M_{f_1}^+ f_2 + (k_1 - 2)! \mathbb{L}(f_2, f_1; \tau).$$

This completes the proof. □

REMARK. It would be very interesting to obtain asymptotics for the values $\widehat{D}(f_1, f_2, h; k_1 - 1)$ in terms of f_1, f_2 and h. Along these lines, Beckwith has proved (see Theorem 1.2 of **[49]**) that

$$\widehat{D}(f_1, f_2, h; k-1) \ll_{f_1, f_2} h^{\frac{k}{2}},$$

where $h \to \infty$ and f_1, f_2 are weight k cusp forms on the full modular group.

17.3.1. p-adic properties of special values.

In view of the numerics in Example 17.3, one might naturally assume that the values $\widehat{D}(f_2, f_1, h; k_1 - 1)$ are always mysterious transcendental numbers. It turns out that there are circumstances where these values have nice algebraic structure and can be interpreted in terms of p-adic modular forms.

To make this precise, we use holomorphic Eichler integrals, p-adic modular forms in the sense of Serre **[452]**, and weight 2 weakly holomorphic quasimodular forms. If

$$f(\tau) = \sum_{n \in \mathbb{Z}} c_f(n) q^n$$

is a weight k weakly holomorphic modular form, then recall that its holomorphic Eichler integral is given by

$$\mathcal{E}_f(\tau) = \sum_{n \neq 0} \frac{c_f(n)}{n^{k-1}} q^n.$$

Following the seminal work[3] of Serre **[452]**, we say that a q-series $G(q)$ is a *weakly holomorphic p-adic modular form* of weight k if there exists a sequence of weakly holomorphic modular forms whose coefficients p-adically tend to those of $G(q)$, with the additional property that their weights p-adically tend to k.

For even weight newforms f whose level is divisible by the square of a prime p, we have the following theorem (cf. Theorem 1.1 of **[105]**) about the p-adic properties of $\mathbb{L}(f, f; \tau)$.

[3]The notion used here is slightly different from Serre's original definition. We do not require that p-adic modular forms be limits of holomorphic modular forms.

THEOREM 17.4. *Let $f \in S_k(\Gamma_0(N))$ be an even weight newform. If p is a prime with $p^2|N$, then there exist constants $\delta_1, \delta_2 \in \mathbb{C}$, a weight 2 weakly holomorphic quasimodular form $\mathcal{Q}_f \in \widetilde{M}_2^!(\Gamma_0(N))$, and a weight $2-k$ weakly holomorphic p-adic modular form $\mathcal{L}_f$ such that*

$$\mathbb{L}(f, f; \tau) = \delta_1 f(\tau)\mathcal{L}_f(\tau) + \delta_2 f(\tau)\mathcal{E}_f(\tau) + \mathcal{Q}_f(\tau).$$

Moreover, if f has complex multiplication, then there are choices with $\delta_2 = 0$.

REMARK. The proof of this theorem establishes that $\mathcal{E}_f$ is also a weight $2-k$ p-adic modular form. It is a weight $2-k$ cuspidal p-adic modular form, a p-adic limit of cusp forms whose weights have p-adic limit $2-k$. Therefore, the mixed mock modular form $\delta_1 f\mathcal{L}_f + \delta_2 f\mathcal{E}_f$ is the linear combination of a weight 2 weakly holomorphic p-adic modular form and a weight 2 cuspidal p-adic modular form. Moreover, if f has CM, then $\mathbb{L}(f, f; \tau)$ is a linear combination of the weight 2 weakly holomorphic p-adic modular form $f\mathcal{L}_f$ and a quasimodular form $\mathcal{Q}_f$.

EXAMPLE 17.5. Consider the newform $f(\tau) := \eta(3\tau)^8 \in S_4(\Gamma_0(9))$. It has CM by $\mathbb{Q}(\sqrt{-3})$, and it is a multiple of the Poincaré series $P_{4,1,9}$ defined in Section 6.3, since this space of cusp forms is one dimensional. Theorem 17.4 applies for $p = 3$. Here are numerical approximations for the first few shifted convolution values:

h	3	6	9	12	15
$\widehat{D}(f, f, h; 3)$	$-10.7466\ldots$	$12.7931\ldots$	$6.4671\ldots$	$-79.2777\ldots$	$64.2494\ldots$

We note that $\widehat{D}(f, f, h; 3) = 0$ if n is not a multiple of 3. We define real numbers β, γ, and δ which are approximately

$$\beta := \frac{(4\pi)^3}{2}\|P_{4,1,9}\|^2 = 1.0468\ldots, \qquad \gamma = -0.0796\ldots, \qquad \delta = -0.8756\ldots.$$

These numbers may be described in terms of Petersson inner products. By Theorem 17.2, we have that

$$\mathbb{L}(f, f; \tau) = \frac{24 f(\tau) F^+_{4,-1,9}(\tau)}{\beta} + \gamma\left(1 - 24\sum_{n=1}^{\infty}\sum_{d|3n} dq^{3n}\right) + \delta\left(1 + 12\sum_{n=1}^{\infty}\sum_{\substack{d|3n\\3\nmid d}} dq^{3n}\right).$$

Since f has CM, we have given an identity where $\delta_2 = 0$. In particular, we have that $\delta_1 := 1/\beta$,

$$\mathcal{L}_f(\tau) := 24F^+_{4,-1,9}(\tau) = q^{-1} - \frac{1}{4}q^2 + \frac{49}{125}q^5 - \frac{3}{32}q^8 - \cdots,$$

and

$$\mathcal{Q}_f(\tau) = \sum_{n=1}^{\infty} b_f(n)q^n := \gamma\left(1 - 24\sum_{n=1}^{\infty}\sigma_1(3n)q^{3n}\right) + \delta\left(1 + 12\sum_{n=1}^{\infty}\sum_{\substack{d|3n\\3\nmid d}} dq^{3n}\right).$$

One can easily compute the rational q-series $\mathcal{L}_f$, using

$$M(\tau) := \left(\frac{\eta(\tau)^3}{\eta(9\tau)^3} + 3\right)^2 \eta(3\tau)^8 = q^{-1} + 2q^2 - 49q^5 + 48q^8 + \cdots.$$

It turns out that $\mathcal{L}_f = -\mathcal{E}_M$.

Theorem 17.4 shows that $\mathcal{L}_f$ is a weight -2 weakly holomorphic 3-adic modular form. In fact, it turns out that $f\mathcal{L}_f$ is a weight 2 cuspidal 3-adic modular form. A straightforward argument proves that this product as a q-series is congruent to the constant 1 modulo 3. This in turns implies that if $h \neq 0$, then

$$\widehat{D}(f,f,h;3) - b_f(h) \in \frac{3}{\beta}\mathbb{Z}_{(3)},$$

where $\mathbb{Z}_{(3)}$ denotes the localization of $\mathbb{Z}$ at 3, i.e., those rational numbers (in lowest terms) whose denominators are coprime to 3. Moreover, we have the following higher congruences for every non-negative integer n:

$$\widehat{D}(f,f,9n+6;3) - b_f(9n+6) \in \frac{9}{\beta}\mathbb{Z}_{(3)},$$

$$\widehat{D}(f,f,36n+30;3) - b_f(36n+30) \in \frac{27}{\beta}\mathbb{Z}_{(3)}.$$

There are infinitely many such congruences modulo any power of 3.

One surprising consequence of the fact that $\mathcal{L}_f$ is a cuspidal 3-adic modular form is that modulo any fixed power of 3, say 3^t, almost all of the coefficients of this p-adic modular form are divisible by 3^t. Here we use "almost all" in the sense of arithmetic density. This means that the rational numbers $\beta(\widehat{D}(f,f,h;3)-b_f(h))$ are almost always multiples of any fixed power of 3.

To illustrate this, we let

$$\pi(3^t;X) := \frac{\#\left\{1 \leq h \leq X \ : \ \beta\left(\widehat{D}(f,f,h;3) - b_f(h)\right) \equiv 0 \pmod{3^t}\right\}}{X}.$$

The table below indicates the slow rate of convergence to arithmetic density 1:

X	$\pi(3;X)$	$\pi(9;X)$	$\pi(27;X)$	$\pi(81;X)$	$\pi(243;X)$
3000	1	$0.912\ldots$	$0.784\ldots$	$0.705\ldots$	$0.676\ldots$
6000	1	$0.917\ldots$	$0.792\ldots$	$0.711\ldots$	$0.679\ldots$
9000	1	$0.920\ldots$	$0.798\ldots$	$0.716\ldots$	$0.680\ldots$
12000	1	$0.922\ldots$	$0.800\ldots$	$0.718\ldots$	$0.681\ldots$
15000	1	$0.923\ldots$	$0.803\ldots$	$0.720\ldots$	$0.683\ldots$
$\vdots$	$\vdots$	$\vdots$	$\vdots$	$\vdots$	$\vdots$
∞	1	1	1	1	1

SKETCH OF PROOF OF THEOREM 17.4. Let $f \in S_k(\Gamma_0(N))$ be a normalized newform of non-square-free level N whose coefficients lie in a number field K and let p be a prime with $p^2|N$. Furthermore, let $\mathcal{M}_f$ be a harmonic Maass form with holomorphic part

$$\mathcal{M}_f^+(\tau) = \sum_{n \gg -\infty} c^+(n)q^n$$

which satisfies Theorem 7.6 for f. Then by Theorem 17.2, we have that

$$\mathbb{L}(f,f;\tau) = \frac{(4\pi)^{k-1}\|f\|^2}{(k-2)!}\mathcal{M}_f^+(\tau)f(\tau) + \mathcal{Q}_f(\tau), \tag{17.6}$$

for a suitable weakly holomorphic quasimodular form $\mathcal{Q}_f \in \widetilde{M}_2^!(\Gamma_0(N))$.

For every $\alpha \in \mathbb{C}$ with $\alpha - c^+(1) \in K$, Theorem 7.10 shows that the coefficients of the *normalized mock modular form*

$$\mathcal{F}_\alpha(\tau) := M_f^+(\tau) - \alpha\mathcal{E}_f(\tau)$$

are in K as well. Note that we can always choose $\alpha = c^+(1)$. In the case where f has CM by a field of discriminant d, Theorem 7.7 tells us that we can even choose $\alpha = 0$ if we replace K by $K(\zeta_{Nd})$, the Nd-th cyclotomic field over K.

Thanks to Bol's identity for integer weight harmonic Maass forms (cf. Theorem 5.5), we find that $D^{k-1}(\mathcal{F}_\alpha) \in M_k^!(\Gamma_0(N))$. Therefore, we can write $\mathcal{F}_\alpha$ as an Eichler integral $\mathcal{F}_\alpha = \mathcal{E}_{g_\alpha}$ for some

$$g_\alpha(\tau) = \sum_{n \gg -\infty} a_\alpha(n) q^n \in M_k^!(\Gamma_0(N)).$$

Since $M_k^!(\Gamma_0(N))$ has a basis consisting of forms with integral Fourier coefficients, we know that the coefficients of g_α must have bounded denominators, in particular the p-adic valuation of g_α, $v_p(g_\alpha) := \inf v_p(a_\alpha(n))$, is bounded from below.

The hypothesis that $p^2|N$ plays a central role. By the theory of newforms, this condition implies that

$$f|U_p = 0.$$

Combining this with the properties of Poincaré series and Kloosterman sums, this vanishing implies much more. By Theorem 3.6 of **[105]**, it follows that $a_\alpha(n) = 0$ if $p|n$. This in turn implies that there cannot be arbitrarily high negative powers of p dividing the Fourier coefficients of $\mathcal{F}_\alpha$ either, so that we can assume without loss of generality that the coefficients are p-integral.

For n coprime to p and $t \in \mathbb{N}$, we have the congruence

$$n^{1-k} \equiv n^{(p-1)p^{t-1}+1-k} \pmod{p^t},$$

which implies that for every $r, t \in \mathbb{N}$ with $r\varphi(p^t) \geq k-1$, φ denoting Euler's φ-function, we have that

$$\mathcal{F}_\alpha \equiv D^{r(p-1)p^{t-1}-k+1}(g_\alpha) \pmod{p^t}.$$

Now g_α is a weakly holomorphic p-adic modular form of weight k in the sense of Serre **[452]** and by Théorème 5 (p. 211) of **[452]**, we know that the operator D maps p-adic modular forms of weight k to p-adic modular forms of weight $k+2$. This follows immediately from the fact that the *Serre derivative*

$$\theta(g) := D(g) - \tfrac{k}{12}E_2 g$$

maps modular forms of weight k to modular forms of weight $k+2$ and that E_2 is a p-adic modular form of weight 2. But this means that we have found a p-adically convergent sequence of p-adic modular forms which converges to $\mathcal{F}_\alpha$, so $\mathcal{F}_\alpha$ is a p-adic modular form as well. Since the sequence of weights $2 - k + r(p-1)p^{t-1}$ p-adically converges to $2-k$, the weight of $\mathcal{F}_\alpha$ as a p-adic modular form is indeed $2-k$, so that

$$\mathcal{F}_\alpha f = \mathcal{M}_f^+ f - \alpha\mathcal{E}_f f$$

becomes a p-adic modular form of weight 2. If we now apply this in (17.6), the theorem follows. $\square$

CHAPTER 18

Generalized Borcherds Products

At the International Congress of Mathematicians in 1994 in Zürich, Borcherds delivered an invited address on the infinite product expansions of certain modular forms with Heegner divisors. These results play an important role in the recent theory of automorphic forms. Loosely speaking, Borcherds proved **[72, 73]** that some modular forms with zeros and poles lying only at CM points and the cusps (i.e., Heegner divisors) admit infinite product expansions whose exponents are coefficients of weight 1/2 weakly holomorphic modular forms. Theorem 16.3 is a special case of his work.[1]

A few years after Borcherds' ICM lecture, Zagier revisited **[516]** some of these results, and he obtained new proofs in certain special cases such as forms with Heegner divisor over $\mathrm{SL}_2(\mathbb{Z})$. As described in Chapter 16, Zagier's approach utilized his theory of traces of singular moduli.

Here we recall the work of Borcherds and Zagier on these infinite products, and we then explain how these results naturally extend to the setting of harmonic Maass forms and mock modular forms. It turns out that one can also obtain modular forms with Heegner divisors as infinite products whose exponents are coefficients of suitable weight 1/2 mock modular forms. This extension was obtained by Bruinier and one of the authors **[127]**, and these results play an important role in the study of central derivatives and values of L-functions of elliptic curves, as we shall see in Chapter 19.

18.1. The simplest Borcherds products

We begin by recalling the work of Borcherds and Zagier. The Delta-function

$$\Delta(\tau) = q\prod_{n=1}^{\infty}(1-q^n)^{24}$$

is the simplest "Borcherds product". The exponents in this product all equal 24. Further note that the coefficients corresponding to the positive square exponents in the Fourier expansion of the weight 1/2 theta function

$$12\theta(\tau) = 12\sum_{n\in\mathbb{Z}} q^{n^2} = 12 + 24q + 24q^4 + 24q^9 + 24q^{16} + \cdots$$

also all equal 24. Moreover, the constant term of this theta function is 12, which happens to be the weight of Δ. Borcherds proved that these observations are more than mere coincidences. They are glimpses of a general theory.

[1] In **[74]**, Borcherds proved much more by establishing infinite product expansions of modular forms on orthogonal groups.

Works by Borcherds [**72, 73, 74**] (cf. [**119**]) concern infinite product expansions of those modular forms with a "Heegner divisor", and his deep results show that these observations are the simplest examples of a general phenomenon.

As another example of his theory, define integers $c(n)$ using the formal infinite product expansion

$$\begin{aligned} E_4(\tau) &= 1 + 240\sum_{n=1}^{\infty}\sigma_3(n)q^n \\ &= (1-q)^{-240}\left(1-q^2\right)^{26760}\left(1-q^3\right)^{-4096240}\cdots =: \prod_{n=1}^{\infty}(1-q^n)^{c(n)}. \end{aligned} \tag{18.1}$$

In the fundamental domain $\mathcal{F}$ for the action of $\mathrm{SL}_2(\mathbb{Z})$ on $\mathbb{H}$, the Eisenstein series E_4 has a single zero, and it is located at the CM point $\tau = e^{\frac{2\pi i}{3}} = \frac{-1+\sqrt{-3}}{2}$. Borcherds' theorem implies that there exists a weight $1/2$ meromorphic modular form G

$$G(\tau) = \sum_{n=-3}^{\infty} c_G(n)q^n = q^{-3} + 4 - 240q + 26760q^4 + \cdots - 4096240q^9 + \cdots$$

on $\Gamma_0(4)$ with the property that $c(n) = c_G(n^2)$ for every positive integer n. Moreover, his work implies that the weight of E_4, the number 4, is the constant term of G.

We make some of Borcherds' work more precise, and we then fit it into a larger framework offered by the theory of harmonic Maass forms. To do so, we first recall the notion of a Heegner divisor. We refer to a complex number $\alpha \in \mathbb{H}$ of the form

$$\alpha = \frac{-b+\sqrt{b^2-4ac}}{2a}$$

with $a, b, c \in \mathbb{Z}$, $\gcd(a, b, c) = 1$, and $b^2 - 4ac < 0$ as a *CM point*, and we denote its discriminant by the integer $d_\alpha := b^2 - 4ac$.

DEFINITION 18.1. A meromorphic modular form f on $\mathrm{SL}_2(\mathbb{Z})$ is said to have a **Heegner divisor** if its zeros and poles are supported at the cusp $i\infty$ and at CM points.

Borcherds obtained a description of the infinite product expansion of those meromorphic modular forms on $\mathrm{SL}_2(\mathbb{Z})$ with a Heegner divisor. The examples above for Δ and E_4 are simple illustrations of his theory.

We follow the treatment of Borcherds products as presented by Zagier [**516**]. This is closely related to his theory of traces of singular moduli, which was discussed in Section 16.1. To state Borcherds' results, we require a special sequence of weight $1/2$ modular forms on $\Gamma_0(4)$, which was also given in Section 16.1. Recall that $\mathrm{M}^!_{\frac{1}{2}}$ is the Kohnen plus space of weakly holomorphic modular forms of weight $1/2$ on $\Gamma_0(4)$, which are those weakly holomorphic modular forms f with a Fourier expansion of the form

$$\sum_{n\equiv 0,1 \pmod 4} c_f(n)q^n.$$

Now recall that for each $d \in -\mathbb{N}_0$ with $d \equiv 0, 1 \pmod 4$, we defined in Section 16.1 the unique functions $f_d \in \mathbb{M}^!_{\frac{1}{2}}$ with Fourier expansions of the form

$$f_d(\tau) = q^d + \sum_{n=1}^{\infty} A(n,d)q^n.$$

As we saw in (16.3), these functions form a basis of $\mathbb{M}^!_{\frac{1}{2}}$.

Using these weakly holomorphic modular forms, it is simple to state Borcherds' infinite product expansions of modular forms with Heegner divisor. Let $\mathcal{M}_H$ denote the set of integer weight meromorphic modular forms on $\mathrm{SL}_2(\mathbb{Z})$ with a Heegner divisor, integer coefficients, and leading coefficient 1. Obviously, $\mathcal{M}_H$ is closed under multiplication.

If $H(n)$ denotes the usual Hurwitz class number of discriminant $-n$, then recall that the Eisenstein-Hurwitz mock modular form $\mathcal{H}^+(\tau)$ is defined by (cf. (6.3))

$$\begin{aligned}\mathcal{H}^+(\tau) &= -\frac{1}{12} + \sum_{1<n\equiv 0,3 \pmod 4} H(n)q^n \\ &= -\frac{1}{12} + \frac{1}{3}q^3 + \frac{1}{2}q^4 + q^7 + q^8 + q^{11} + \frac{4}{3}q^{12} + \cdots.\end{aligned}$$

If $f(\tau) = \sum_{n=n_0}^{\infty} c_f(n)q^n \in \mathbb{M}^!_{\frac{1}{2}}$ with $c_f(n) \in \mathbb{Z}$, then define $\Psi(f(\tau))$ by

$$\Psi(f(\tau)) := q^{-h} \prod_{n=1}^{\infty} (1-q^n)^{c_f(n^2)}, \tag{18.2}$$

where h is the constant term of $f\mathcal{H}^+$. The following result gives some of Borcherds' infinite product expansions (cf. [**73, 72**]).

THEOREM 18.2 (Borcherds). *For forms $f \in \mathbb{M}^!_{\frac{1}{2}}$ with integer Fourier coefficients, the function $\Psi(f)$ lies in $\mathcal{M}_H$. Furthermore, Ψ induces an isomorphism*

$$\Psi \colon \mathbb{M}^!_{\frac{1}{2}} \longrightarrow \mathcal{M}_H.$$

The weight of $\Psi(f)$ is $c_f(0)$, and the multiplicity of the zero of $\Psi(f)$ at a CM point of discriminant $D<0$ equals

$$\sum_{n=1}^{\infty} c_f\left(Dn^2\right).$$

REMARK. The isomorphism Ψ respects the additive structure of $\mathbb{M}^!_{\frac{1}{2}}$ and the multiplicative structure of $\mathcal{M}_H$.

EXAMPLE 18.3. If $f = 12f_0 = 12\theta$, then

$$f(\tau) = \sum_{n=0}^{\infty} c_f(n)q^n = 12 + 24q + 24q^4 + 24q^9 + 24q^{16} + \cdots.$$

Therefore, it follows that

$$\Psi(f(\tau)) = q\prod_{n=1}^{\infty}(1-q^n)^{24}.$$

Obviously, $\Psi(f)$ is the Δ-function. Its weight is $c_f(0) = 12$, and its divisor is supported at $i\infty$; this fact is confirmed by the observation that f has no non-zero Fourier coefficients with negative exponents.

EXAMPLE 18.4. If $f = 3f_{-3}$, then

$$\begin{aligned} f(\tau) &= \sum_{n=-3}^{\infty} c_f(n) q^n \\ &= 3q^{-3} - 744q + 80256q^4 - 257985q^5 + 5121792q^8 - 12288744q^9 + \cdots . \end{aligned}$$

By Theorem 18.2, it follows that $\Psi(f)$ is a weight $c_f(0) = 0$ meromorphic modular form whose divisor is supported at the cusp $i\infty$ and a triple zero at $\omega = e^{\frac{2\pi i}{3}}$. Therefore, it follows that $\Psi(f)$ must be j, and this is confirmed by

$$\begin{aligned} \Psi(f(\tau)) &= q^{-1}(1-q)^{-744}\left(1-q^2\right)^{80256}\left(1-q^3\right)^{-12288744}\cdots \\ &= q^{-1} + 744 + 196884q + 21493760q^3 + \cdots . \end{aligned}$$

EXAMPLE 18.5. If $f = 4f_0 + f_{-3}$, then

$$f(\tau) = \sum_{n=-3}^{\infty} c_f(n) q^n = q^{-3} + 4 - 240q + 26760q^4 - 85995q^5 + \cdots - 4096240q^9 + \cdots .$$

Theorem 18.2 implies that $\Psi(f)$ is a holomorphic modular form of weight 4 with leading coefficient 1. Therefore, we must have that $\Psi(f) = E_4$. This confirms our earlier example in (18.1).

REMARK. Theorem 18.2 leads to an effective and simple procedure for computing class polynomials. In particular, it implies Theorem 16.3, which we recall states that

$$\mathcal{H}_d(j(\tau)) = q^{-H(-d)} \prod_{n=1}^{\infty} (1-q^n)^{A\left(n^2, d\right)},$$

where $\mathcal{H}_d$ is the Hilbert class polynomial defined in (16.2). Moreover, we saw in Section 16.1 that by using the duality of the coefficients $A(n,d) = -B(n,d)$ with coefficients of a basis for $\mathbb{M}^!_{\frac{3}{2}}$, this shows that traces of singular moduli of the j-invariant are coefficients of special weight $3/2$ weakly holomorphic modular forms.

18.2. Twisted Borcherds products

As we discussed in Section 16.2, Zagier **[516]** defined twisted traces of singular moduli. Zagier also extended the theory of Borcherds products in Theorem 18.2 to the context of twisted products. To make this precise, we let D and d be positive and negative fundamental discriminants, respectively. For convenience, suppose that they are relatively prime. The genus character $\chi = \chi_{D,d}$ was defined in Section 16.2, and here we find it convenient to define this for a quadratic form Q with discriminant dD by $\chi(Q) := \left(\frac{D}{p}\right) = \left(\frac{d}{p}\right)$, where $p \nmid Dd$ is any prime represented by Q (this definition does not depend on the choice of p).

Zagier's twisted products are defined using the rational functions

$$P_D(t) := \exp\left(-\sqrt{D}\sum_{r=1}^{\infty}\left(\frac{D}{r}\right)\frac{t^r}{r}\right) \in \mathbb{Q}\left(\sqrt{D}\right)[[t]]. \tag{18.3}$$

These rational functions and the modular forms in Borcherds' basis of $\mathbb{M}^!_{\frac{1}{2}}$ give an infinite product expansion for the modular functions

$$\mathcal{H}_{D,d}(X) := \prod_{Q \in \mathcal{Q}_{Dd}/\mathrm{SL}_2(\mathbb{Z})} (X - j(\tau_Q))^{\chi(Q)}. \tag{18.4}$$

As before, here the quadratic forms Q vary over $\mathrm{SL}_2(\mathbb{Z})$-equivalence classes of discriminant Dd, and the τ_Q are the roots of $Q(X,1)=0$ in $\mathbb{H}$.

Using this notation, Zagier's twisted product theorem (cf. Theorem 7 of [**516**]), which generalizes Theorem 18.2, reads as follows.

THEOREM 18.6 (Zagier). *Let D, d and χ be as above. Then we have that*

$$\mathcal{H}_{D,d}(j(\tau)) = \prod_{n=1}^{\infty} P_D(q^n)^{A(n^2D,d)}.$$

EXAMPLE 18.7. Let $D = 5$ and $d = -3$. There are two classes of binary quadratic forms with discriminant -15, and we choose representatives

$$Q_1(X,Y) := X^2 + XY + 4Y^2 \quad \text{and} \quad Q_2(X,Y) = 2X^2 + XY + 2Y^2.$$

Their CM points are $\frac{-1+\sqrt{-15}}{2}$ and $\frac{-1+\sqrt{-15}}{4}$, and their corresponding singular values are

$$j\left(\frac{-1+\sqrt{-15}}{2}\right) = \frac{-191025 - 85995\sqrt{5}}{2}, \qquad j\left(\frac{-1+\sqrt{-15}}{4}\right) = \frac{-191025 + 85995\sqrt{5}}{2}.$$

Theorem 18.6 then implies that

$$\mathcal{H}_{5,-3}(j(\tau)) = \frac{j(\tau) + \frac{191025}{2} + \frac{85995}{2}\sqrt{5}}{j(\tau) + \frac{191025}{2} - \frac{85995}{2}\sqrt{5}} = \prod_{n=1}^{\infty} P_5(q^n)^{A(5n^2,-3)},$$

where

$$P_5(X) = \frac{1 + \frac{1-\sqrt{5}}{2}X + X^2}{1 + \frac{1+\sqrt{5}}{2}X + X^2}.$$

18.3. Generalization to the mock modular setting

It turns out that Theorems 18.2 and 18.6 are special cases of a much more general phenomenon, one which extends to the setting of harmonic Maass forms (cf. [**119, 127**]). In this case, the exponents of the infinite products arise as the coefficients of suitable weight 1/2 mock modular forms and weakly holomorphic modular forms.

The proper framework for describing these results makes use of the theory of vector-valued harmonic Maass forms and the formal structure of "twisted Heegner divisors". One particular advantage of writing modular forms in vector-valued notation, as opposed to scalar-valued notation, is that one can write higher level modular objects in terms of level one vector-valued objects.

For example, consider the Mock Theta Conjectures (discussed in Section 9.6). Using the theory of (scalar-valued) harmonic Maass forms with level structure, one of the authors reduced the proof of the Mock Theta Conjectures to a massive finite computation [**194**]. Such finite calculations are based on the principle that two modular forms are equal if and only if the "first few coefficients" of their respective Fourier expansions agree. In the literature, the number of such coefficients required for such proofs is often referred to as the "Sturm bound" because of the theorem proved in [**476**]. However, this bound for scalar-valued forms on $\Gamma_0(N)$ is often impractical in practice. By using vector-valued modular forms, one can dramatically reduce the number of coefficients one must check to prove identities between modular forms as well as mock modular forms. Andersen used this strategy in [**19**], and he reduced the massive calculation to the comparison of just one term of a vector-valued form. Theorems such as the main results of this section are similarly convenient to express in this language, and would be much more tedious to write down without it.

We begin by first recalling basic results and notation pertaining to vector-valued modular forms and Heegner divisors.

18.3.1. The Weil representation. Here we set up the preliminary definitions needed to state the generalization of Borcherds' results to the mock modular case. We closely follow the exposition of [**127**]. Further discussion and properties of many of the objects described here can also be found in [**119**] and [**239**].

We begin by describing vector-valued modular forms. To this end, suppose that (V, Q) is a non-degenerate rational quadratic space of signature (b^+, b^-), and let $L \subset V$ be an even lattice with dual L'. The *discriminant group*, L'/L, together with the $\mathbb{Q}/\mathbb{Z}$-valued quadratic form induced by Q, is called the *discriminant form* of L. A *vector-valued modular form* is then defined as a function

$$f\colon \mathbb{H} \to \mathbb{C}[L'/L]$$

whose modular transformations are dictated by special matrices. In fact, we have already seen that these vector-valued transformations arise naturally within the theory of harmonic Maass forms. Indeed, Zwegers' result on the transformations of Ramanujan's third order mock theta functions $f(q)$ and $\omega(q)$ (described in Theorem 9.5) is conveniently stated using such language.

The most important set of vector-valued modular forms which we require are those which transform with respect to the *Weil representation.* To describe this, we first denote by $\mathrm{Mp}_2(\mathbb{R})$ the metaplectic two-fold cover of $\mathrm{SL}_2(\mathbb{R})$. Explicitly, this is the set

$$\mathrm{Mp}_2(\mathbb{R}) := \left\{(\gamma, \varepsilon) : \gamma \in \mathrm{SL}_2(\mathbb{R}), \varepsilon(\tau)^2 = c\tau + d \right\},$$

that is, ε is a holomorphic choice of a square root of $c\tau + d$. Multiplication in $\mathrm{Mp}_2(\mathbb{R})$ works by setting

$$(\gamma_1, \varepsilon_1) \cdot (\gamma_2, \varepsilon_2) := (\gamma_1\gamma_2, \varepsilon),$$

where $\varepsilon(\tau) := \varepsilon_1(\gamma_2\tau)\varepsilon_2(\tau)$. It is easy to check that this product is well defined using the usual cocycle relation of $c\tau + d$ under multiplication of matrices in $\mathrm{SL}_2(\mathbb{R})$ and that this indeed gives a two-covering of $\mathrm{SL}_2(\mathbb{R})$.

The inverse image of $\mathrm{SL}_2(\mathbb{Z})$ under the associated covering map (which sends $(\gamma, \varepsilon) \mapsto \gamma$) is denoted by $\widetilde{\Gamma} := \mathrm{Mp}_2(\mathbb{Z})$. It is known that $\widetilde{\Gamma}$ is generated by $T := \left(\left(\begin{smallmatrix} 1 & 1 \\ 0 & 1 \end{smallmatrix}\right), 1\right)$ and $S := \left(\left(\begin{smallmatrix} 0 & -1 \\ 1 & 0 \end{smallmatrix}\right), \sqrt{\tau}\right)$. One has the relations $S^2 = (ST)^3 = Z$, where $Z := \left(\left(\begin{smallmatrix} -1 & 0 \\ 0 & -1 \end{smallmatrix}\right), i\right)$ is the standard generator of $\widetilde{\Gamma}$. Associated to the discriminant form L'/L, there is distinguished representation called the Weil representation. It is a representation of $\widetilde{\Gamma}$ on the group algebra $\mathbb{C}[L'/L]$.

18.3.2. The $\Gamma_0(N)$ set-up. Here we briefly describe the set-up that applies to the setting of scalar-valued modular forms on $\Gamma_0(N)$, which is of importance in this chapter. Consider the quadratic space

$$V := \{M \in \mathrm{Mat}_2(\mathbb{Q}) : \mathrm{tr}(M) = 0\}, \tag{18.5}$$

endowed with the quadratic form $Q(M) := -N\det(M)$. The signature of V is $(2, 1)$. The corresponding bilinear form is then $(M_1, M_2) = N\,\mathrm{tr}(M_1M_2)$. The even

Clifford algebra $C^0(V)$ of V can be identified with $\mathrm{Mat}_2(\mathbb{Q})$. The Clifford norm on $C^0(V)$ is identified with the determinant. The group $\mathrm{GL}_2(\mathbb{Q})$ acts on V by

$$\gamma \cdot X := \gamma X \gamma^{-1},$$

where $\gamma \in \mathrm{GL}_2(\mathbb{Q})$, leaving the quadratic form invariant. This induces the following isomorphisms of algebraic groups over $\mathbb{Q}$:

$$\mathrm{GL}_2 \cong \mathrm{GSpin}(V), \quad \mathrm{SL}_2 \cong \mathrm{Spin}(V).$$

There is an isometry from (V, Q) to the trace zero part of $(C^0(V), -N \det)$. Let L be the lattice

$$L := \left\{ \begin{pmatrix} b & -\frac{a}{N} \\ c & -b \end{pmatrix} : a, b, c \in \mathbb{Z} \right\}. \tag{18.6}$$

The dual lattice is then given by

$$L' := \left\{ \begin{pmatrix} \frac{b}{2N} & -\frac{a}{N} \\ c & -\frac{b}{2N} \end{pmatrix} : a, b, c \in \mathbb{Z} \right\}. \tag{18.7}$$

We identify L'/L with $\mathbb{Z}/2N\mathbb{Z}$. We also occasionally identify $\frac{h}{2N} \in \mathbb{Q}/\mathbb{Z}$ with $h \in \mathbb{Z}/2N\mathbb{Z}$. We identify the quadratic form on L'/L with the quadratic form $x \mapsto x^2$ on $\mathbb{Z}/2N\mathbb{Z}$. The level of L is $4N$.

If $D \in \mathbb{Z}$, then let L_D be the set

$$L_D := \left\{ X \in L' \ : \ Q(X) = \frac{D}{4N} \right\}. \tag{18.8}$$

Note that L_D is empty unless D is a square modulo $4N$. Now, for $r \in L'/L$ with $r^2 \equiv D \pmod{4N}$ we define

$$L_{D,r} := \left\{ X \in L' \ : \ Q(X) = \frac{D}{4N} \ \text{ and } \ X \equiv r \pmod{L} \right\}. \tag{18.9}$$

We write L_D^0 for the subset of *primitive* vectors in L_D, and $L_{D,r}^0$ for the primitive vectors in $L_{D,r}$ respectively.

If we have

$$M = \begin{pmatrix} \frac{b}{2N} & -\frac{a}{N} \\ c & -\frac{b}{2N} \end{pmatrix} \in L_{D,r},$$

then the matrix

$$\psi(M) := \begin{pmatrix} a & \frac{b}{2} \\ \frac{b}{2} & Nc \end{pmatrix} = M \begin{pmatrix} 0 & N \\ -N & 0 \end{pmatrix}$$

defines an integral binary quadratic form of discriminant $D = b^2 - 4Nac = 4NQ(M)$ with $b \equiv r \pmod{2N}$. It can be seen that the natural homomorphism

$$\mathrm{SO}(L) \to \mathrm{Aut}(L'/L)$$

is surjective. Its kernel is called the *discriminant kernel subgroup.* We write $\mathrm{SO}^+(L)$ for the intersection of $\mathrm{SO}(L)$ and the connected component of the identity of $\mathrm{SO}(V)(\mathbb{R})$. The group $\Gamma_0(N) \subset \mathrm{Spin}(V)$ takes L to itself and acts trivially on L'/L (we also refer the reader to Proposition 2.2 of **[127]**).

For a fundamental discriminant D and $r/2N \in L'/L$ with $r^2 \equiv D \pmod{4N}$, we let

$$\mathcal{Q}_{D,N,r} := \{Q = [Nc, b, a] : a, b, c \in \mathbb{Z},\ \mathrm{disc}(Q) = D,\ b \equiv r \pmod{2N}\}. \tag{18.10}$$

The group $\Gamma_0(N)$ acts on this set in the usual way (cf. (16.12)). Note that this action preserves discriminants. Assuming this discussion, we now can construct the Weil representation in this setting.

LEMMA 18.8. *Assuming the notation and hypotheses above, we have that the the Weil representation ρ_L associated with the discriminant form L'/L is explicitly given by*

$$\rho_L(T)(\mathfrak{e}_h) = e\left(\frac{h^2}{4N}\right)\mathfrak{e}_h$$

and

$$\rho_L(S)(\mathfrak{e}_h) = \frac{e\left(-\frac{1}{8}\right)}{\sqrt{2N}} \sum_{h' \in \mathbb{Z}/2N\mathbb{Z}} e\left(-\frac{hh'}{2N}\right)\mathfrak{e}_{h'}.$$

REMARK. In order to check that the Weil representation is indeed a representation of the metaplectic group, one must check that the values above on the generators $T = (\left(\begin{smallmatrix}1&1\\0&1\end{smallmatrix}\right), 1)$ and $S = (\left(\begin{smallmatrix}0&-1\\1&0\end{smallmatrix}\right), \sqrt{\tau})$ extend to a representation of the full group. This involves checking the resulting relations, an exercise which we leave to the reader.

We also need another lattice closely related to these lattices for $\Gamma_0(N)$. The cusps of $\Gamma_0(N)$ correspond to $\Gamma_0(N)$-classes of primitive isotropic vectors in L (recall that a vector λ is *isotropic* with respect to Q if $Q(\lambda) = 0$). Let $\ell \in L$ be a primitive isotropic vector, and take $\ell' \in L'$ with $(\ell, \ell') = 1$. The one-dimensional lattice

$$K := L \cap \ell'^{\perp} \cap \ell^{\perp}$$

is then positive definite. For simplicity we assume that $(\ell, L) = \mathbb{Z}$. In this case we may choose $\ell' \in L$ isotropic. Then L splits into

$$L = K \oplus \mathbb{Z}\ell' \oplus \mathbb{Z}\ell,$$

and $K'/K \cong L'/L$.

REMARK. If N is square-free, then any primitive isotropic vector $\ell \in L$ satisfies $(\ell, L) = \mathbb{Z}$, and the above assumption is not a restriction. For general N, the results still hold with appropriate modifications, but the formulas become somewhat longer.

18.3.3. Vector-valued harmonic Maass forms. Returning to the general situation of a Weil representation, if $f\colon \mathbb{H} \to \mathbb{C}[L'/L]$ is a function, then we write

$$f = \sum_{h \in L'/L} f_h \mathfrak{e}_h \tag{18.11}$$

for its decomposition into components. For $k \in \frac{1}{2}\mathbb{Z}$, let $M^!_{k,\rho_L}$ denote the space of $\mathbb{C}[L'/L]$-valued weakly holomorphic modular forms of weight k and type ρ_L for the group $\widetilde{\Gamma}$. The subspaces of holomorphic modular forms (resp. cusp forms) are denoted by M_{k,ρ_L} (resp. S_{k,ρ_L}).

DEFINITION 18.9. A smooth function $f\colon \mathbb{H} \to \mathbb{C}[L'/L]$ is called a **harmonic Maass form of weight k with respect to $\widetilde{\Gamma}$ and ρ_L** if it satisfies:

i) The function f satisfies the transformations $f(M\tau) = \phi(\tau)^{2k}\rho_L(M, \phi)f(\tau)$ for all $(M, \phi) \in \widetilde{\Gamma}$.
ii) The function f satisfies the differential equation $\Delta_k(f) = 0$.

iii) There is a vector-valued polynomial

$$P_f(\tau) = \sum_{h \in L'/L} \sum_{\substack{n \in \mathbb{Z}+Q(h) \\ -\infty \ll n \leq 0}} c_f^+(n,h) q^n \mathfrak{e}_h$$

for which there is a constant $C > 0$ with

$$f(\tau) = P_f(\tau) + O\left(e^{-Cv}\right)$$

as $v \to \infty$.

We denote the vector space of these harmonic Maass forms by H_{k,ρ_L}. As with scalar-valued harmonic Maass forms, the Fourier expansion of any $f \in H_{k,\rho_L}$ gives a unique decomposition $f = f^+ + f^-$ as a sum of a holomorphic and nonholomorphic part. Namely, we have the following lemma.

LEMMA 18.10. *Assuming the notation and hypotheses above, if $f \in H_{k,\rho_L}$, then $f = f^+ + f^-$, where*

$$f^+(\tau) = \sum_{h \in L'/L} \sum_{\substack{n \in \mathbb{Q}, \\ n \gg -\infty}} c_f^+(n,h) q^n \mathfrak{e}_h,$$

$$f^-(\tau) = \sum_{h \in L'/L} \sum_{\substack{n \in \mathbb{Q}, \\ n < 0}} c_f^-(n,h) \Gamma(1-k, -4\pi n v) \mathfrak{e}_h.$$

The polynomial P_f is uniquely determined by f and is called its *principal part.* In analogy with the setting of scalar-valued harmonic Maass forms, by Corollary 3.8 of **[121]** (see (5.13), where we stated the scalar-valued form of this equation), we have the following lemma.

LEMMA 18.11. *Assuming the notation and hypotheses above, we have the following exact sequence (where $\overline{\rho}_L$ is the complex conjugate of ρ_L):*

$$0 \to M^!_{k,\overline{\rho}_L} \to H_{k,\overline{\rho}_L} \to S_{2-k,\rho_L} \to 0.$$

18.3.4. Twisted Siegel theta functions. The theory of generalized Borcherds products follows as a corollary of the main result on a generalized theta lift, which itself is an adaptation of earlier work by Harvey and Moore **[254]** and Borcherds **[72, 73]**. To define these lifts, we must first consider twisted Siegel theta functions. We assume the notation and hypotheses from the previous subsection.

Let $\Delta \in \mathbb{Z}$ be a fundamental discriminant and let $r \in \mathbb{Z}$ such that $\Delta \equiv r^2 \pmod{4N}$. Following Gross, Kohnen, and Zagier **[239]** (as also discussed in (12.15)), we define a generalized genus character for $\lambda = \left(\begin{smallmatrix} b/2N & -a/N \\ c & -b/2N \end{smallmatrix}\right) \in L'$ by putting

$$\chi_\Delta(\lambda) = \chi_\Delta([a,b,Nc]) := \begin{cases} \left(\frac{\Delta}{n}\right) & \text{if } \Delta | b^2 - 4Nac \text{ and } (b^2-4Nac)/\Delta \text{ is a} \\ & \text{square modulo } 4N \text{ and } \gcd(a,b,c,\Delta) = 1, \\ 0 & \text{otherwise.} \end{cases}$$

Here $[a,b,Nc]$ is the integral binary quadratic form corresponding to λ, and n is any integer prime to Δ represented by one of the quadratic forms $[N_1 a, b, N_2 c]$ with $N_1 N_2 = N$ and $N_1, N_2 > 0$ (cf. Section 1.2 of **[239]**, and also Section 1 of **[466]**).

The function χ_Δ is invariant under the action of $\Gamma_0(N)$, as well as under the action of all the level N Atkin-Lehner involutions. Therefore, it is invariant under

$\mathrm{SO}^+(L)$. It can be computed by the following explicit formula (cf. Section I.2, Proposition 1 of **[239]**): If $\Delta = \Delta_1\Delta_2$ is a factorization of Δ into discriminants, and $N = N_1N_2$ is a factorization of N into positive factors such that $\gcd(\Delta_1, N_1a) = \gcd(\Delta_2, N_2c) = 1$, then

$$\chi_\Delta([a,b,Nc]) = \left(\frac{\Delta_1}{N_1a}\right)\left(\frac{\Delta_2}{N_2c}\right). \tag{18.12}$$

If no such factorizations of Δ and N exist, then we have $\chi_\Delta([a,b,Nc]) = 0$.

We define a *twisted Siegel theta function* for L as follows. For a coset $h \in L'/L$ and variables $\tau = u + iv \in \mathbb{H}$, $z \in \mathrm{Gr}(V)$ (for more details on working with $\mathrm{Gr}(V)$ in this situation, we refer the reader to the discussion immediately preceding Definition 18.14), we put

(18.13)

$$\begin{aligned}\theta_{\Delta,r,h}(\tau,z) &:= v^{\frac{1}{2}} \sum_{\substack{\lambda\in L+rh \\ Q(\lambda)\equiv\Delta Q(h) \pmod{\Delta}}} \chi_\Delta(\lambda)e\left(\frac{1}{|\Delta|}\frac{\lambda_z^2}{2}\tau + \frac{1}{|\Delta|}\frac{\lambda_{z^\perp}^2}{2}\bar{\tau}\right)\\ &= v^{\frac{1}{2}} \sum_{\substack{\lambda\in L+rh \\ Q(\lambda)\equiv\Delta Q(h) \pmod{\Delta}}} \chi_\Delta(\lambda)e\left(\frac{1}{|\Delta|}\frac{\lambda^2}{2}u + \frac{1}{|\Delta|}\left(\frac{\lambda_z^2}{2} - \frac{\lambda_{z^\perp}^2}{2}\right)iv\right),\end{aligned}$$

where λ_z (resp. $\lambda_{z^\perp}$) is the orthogonal projection of λ to z (resp. $z^\perp$).

Using these theta functions, we define a $\mathbb{C}[L'/L]$-valued Siegel theta function by the following formula:

$$\Theta_{\Delta,r}(\tau,z) := \sum_{h\in L'/L} \theta_{\Delta,r,h}(\tau,z)\mathfrak{e}_h. \tag{18.14}$$

In the variable z, the function $\Theta_{\Delta,r}(\tau,z)$ is invariant under $\Gamma_0(N)$. Bruinier and one of the authors proved the following theorem (cf. Theorem 4.1 of **[127]**) which describes the modularity of these functions with respect to the variable τ.

THEOREM 18.12. *The theta function $\Theta_{\Delta,r}(\tau,z)$ is a nonholomorphic $\mathbb{C}[L'/L]$-valued modular form for $\mathrm{Mp}_2(\mathbb{Z})$ of weight $1/2$. It transforms with the representation $\widetilde{\rho}_L$ given by ρ_L if $\Delta > 0$, and by $\bar{\rho}_L$ if $\Delta < 0$.*

18.3.5. Twisted Heegner divisors. In order to describe the general Borcherds products of the next subsection, we also require some preliminaries on Heegner divisors. Our setup follows that of Gross, Kohnen, and Zagier **[239]**.

We begin by first considering the special case of $X_0(p)$, where p is prime. Let $d < 0$ and $\Delta > 0$ be fundamental discriminants which are both squares modulo p. Recall that $\mathcal{Q}_{d,p}$ is the set of discriminant d positive definite integral binary quadratic forms $[a,b,c]$ with $p|a$.

DEFINITION 18.13. Assuming the notation above, we define the **twisted Heegner divisor** $Z_\Delta(d)$ by

$$Z_\Delta(d) := \sum_{Q\in\mathcal{Q}_{\Delta d,p}/\Gamma_0(p)} \chi_\Delta(Q)\frac{\tau_Q}{\omega_Q},$$

where we recall that τ_Q is the unique root of $Q(X,1)$ in the upper half-plane and ω_Q denotes the order of the stabilizer of Q in $\Gamma_0(p)$, and where χ_Δ is the generalized genus character defined in (12.15).

It turns out that $Z_\Delta(d)$ is a divisor on $X_0(p)$ defined over $\mathbb{Q}(\sqrt{\Delta})$ (cf. Lemma 5.1 of **[127]**). More generally, we require the notion of twisted Heegner divisors on the modular curve $X_0(N)$. We now assume the previous discussion where scalar-valued forms on $\Gamma_0(N)$ are packaged by means of the aforementioned corresponding Weil representation. For any vector $\lambda \in L'$ of negative norm, the orthogonal complement $\lambda^\perp \subset V(\mathbb{R}) := V \otimes \mathbb{R}$ defines a point $Z(\lambda)$ in the Grassmannian of b^+-dimensional positive definite subspaces of $V(\mathbb{R})$: $\mathrm{Gr}(V) \cong \mathbb{H}$ (recall that in our case $(b^+, b^-) = (2,1)$).

To further explain this identification, we take the isotropic vectors $\ell, \ell' \in L$:

$$\ell = \begin{pmatrix} 0 & \frac{1}{N} \\ 0 & 0 \end{pmatrix}, \qquad \ell' = \begin{pmatrix} 0 & 0 \\ 1 & 0 \end{pmatrix}.$$

Then we have that the lattice $K = L \cap \ell'^\perp \cap \ell^\perp$ above can be written as $K = \mathbb{Z}\left(\begin{smallmatrix} 1 & 0 \\ 0 & -1 \end{smallmatrix}\right)$. For $\lambda \in K \otimes \mathbb{R}$, it is thus natural to use the notation $\lambda > 0$ to describe those λ which are positive multiples of $\left(\begin{smallmatrix} 1 & 0 \\ 0 & -1 \end{smallmatrix}\right)$. As in Section 13 of **[74]** and Section 3.2 of **[119]**, we can identify $\mathbb{H}$ with an open subset of $K \otimes \mathbb{C}$ by mapping $t \in \mathbb{H}$ to $\left(\begin{smallmatrix} 1 & 0 \\ 0 & -1 \end{smallmatrix}\right) \otimes t$. Moreover, $\mathbb{H}$ is identified with $\mathrm{Gr}(V)$ by mapping $t \in \mathbb{H}$ to the positive definite subspace

$$z(t) = \mathbb{R}\mathrm{Re}\begin{pmatrix} t & -t^2 \\ 1 & -t \end{pmatrix} + \mathbb{R}\mathrm{Im}\begin{pmatrix} t & -t^2 \\ 1 & -t \end{pmatrix}$$

of $V(\mathbb{R})$. Furthermore, the action of $\mathrm{Spin}(V)$ on $\mathbb{H}$ via fractional linear transformations corresponds to the linear action on $\mathrm{Gr}(V)$ through $\mathrm{SO}(V)$.

Then we have the following definition.

DEFINITION 18.14. For $h \in L'/L$ and a negative rational number $m \in \mathbb{Z} + \mathrm{sgn}(\Delta)Q(h)$, we have the **twisted Heegner divisor**

$$Z_{\Delta,r}(m,h) := \sum_{\lambda \in L_{d\Delta,hr}/\Gamma_0(N)} \chi_\Delta(\lambda)\frac{Z(\lambda)}{\omega(\lambda)},$$

where $d := 4Nm\,\mathrm{sgn}(\Delta) \in \mathbb{Z}$ and $\omega(\lambda)$ is the order of the stabilizer of λ in $\Gamma_0(N)$. Moreover, for $m \in \mathbb{Q}$ and $h \in L'/L$, we set (compare with (18.8) and 18.9))

$$L_{m,h} := \{\lambda \in L + h : Q(\lambda) = m\},$$

and we let

$$L_m := \{\lambda \in L' : Q(\lambda) = m\}.$$

Then $Z_{\Delta,r}(m,h)$ defines a $\mathbb{Q}$-rational divisor on $X_0(N)$. In a related fashion, for any vector-valued harmonic Maass form f as in Lemma 18.10 with $\mathbb{Q}$-rational principal part, we define the *twisted Heegner divisor associated to* f by

$$Z_{\Delta,r}(f) := c_f^+(m,h)Z_{\Delta,r}(m,h). \tag{18.15}$$

We also require a divisor $C_{\Delta,r}(f)$, which is supported at the cusps (cf. 5.12 of **[127]**). The multiplicity of any cusp ℓ in the divisor $C_{\Delta,r}(f)$ is given by the *Weyl vector* $\rho_{f,\ell}$. This is a vector $\rho_{f,\ell} \in K' \otimes \mathbb{R}$ whose definition is given in (4.8) of **[127]**.

18.3.6. Generalized Borcherds products. Having set up the necessary theoretical groundwork, we are almost ready to state the theorem which generalizes Theorems 18.2 and 18.6. We closely follow the exposition in **[127]**.

We assume the notation above and hypotheses from the previous section. Let $f \in H_{\frac{1}{2},\widetilde{\rho}_L}$ be a harmonic Maass form with real coefficients $c_f^+(m,h)$ for all $m \in \mathbb{Q}$ and $h \in L'/L$. For this f we define the regularized theta integral

$$\Phi_{\Delta,r}(z,f) := \int_{\mathcal{F}}^{\mathrm{reg}} \langle f(\tau), \Theta_{\Delta,r}(\tau,z)\rangle v^{\frac{1}{2}} \frac{du dv}{v^2}. \tag{18.16}$$

Here $\mathcal{F}$ denotes the standard fundamental domain for the action of $\mathrm{SL}_2(\mathbb{Z})$ on $\mathbb{H}$, and the integral has to be regularized as in Definition 5.6. Furthermore, we note that $\langle \cdot,\cdot\rangle$ denotes the standard scalar product of vectors.

The following theorem offers the promised generalized Borcherds products obtained by Bruinier and one of the authors (cf. Theorem 6.1 of **[127]**).

THEOREM 18.15. *Assume the notation above and hypotheses in the previous subsection. Let $f \in H_{\frac{1}{2},\widetilde{\rho}_L}$ be a harmonic Maass form with real coefficients $c_f^+(m,h)$ for all $m \in \mathbb{Q}$ and $h \in L'/L$. Moreover, assume that $c_f^+(n,h) \in \mathbb{Z}$ for all $n \leq 0$. Then the infinite product in the variable $z = x+iy$ with $x, y \in \mathbb{R}$, defined by*

$$\Psi_{\Delta,r}(z,f) := e((\rho_{f,\ell},z)) \prod_{\substack{\lambda\in K' \\ \lambda>0}} \prod_{b \pmod{\Delta}} \left[1 - e\left((\lambda,z)+\frac{b}{\Delta}\right)\right]^{\left(\frac{\Delta}{b}\right)c_f^+\left(\frac{|\Delta|\lambda^2}{2}, r\lambda\right)},$$

converges for y sufficiently large and has a meromorphic continuation to all of $\mathbb{H}$ with the following properties:

i) *It is a meromorphic modular form for $\Gamma_0(N)$ with a unitary character σ which may have infinite order.*
ii) *The weight of $\Psi_{\Delta,r}(z,f)$ is $c_f^+(0,0)$ when $\Delta = 1$ and is 0 when $\Delta \neq 1$.*
iii) *The divisor of $\Psi_{\Delta,r}(z,f)$ on $X_0(N)$ is given by $Z_{\Delta,r}(f) + C_{\Delta,r}(f)$.*
iv) *The regularized theta integral $\Phi_{\Delta,r}(z,f)$ satisfies*

$$\Phi_{\Delta,r}(z,f) = \begin{cases} -c_f^+(0,0)(\log(4\pi N)+\Gamma'(1)) - 4\log\left|\Psi_{\Delta,r}(z,f)y^{\frac{c_f^+(0,0)}{2}}\right| & \text{if } \Delta = 1, \\ 2\sqrt{\Delta}c_f^+(0,0)L(1,\chi_\Delta) - 4\log|\Psi_{\Delta,r}(z,f)| & \text{if } \Delta \neq 1. \end{cases}$$

SKETCH OF PROOF. To prove this theorem, one makes use of the modularity of the twisted Siegel theta functions described in Theorem 18.12. This modularity can be used to prove that $\Phi_{\Delta,r}(z,f)$ is a $\Gamma_0(N)$-invariant function on $\mathbb{H}\backslash Z_{\Delta,r}(f)$ with a logarithmic singularity on the divisor $-4Z_{\Delta,r}(f)$.

A long involved calculation (see the proof of Theorem 5.3 of **[127]**) gives its Fourier expansion. For $z \in \mathbb{H}$ with $y \gg 0$, we find that

(18.17)

$$\Phi_{\Delta,r}(z,f) = -4 \sum_{\substack{\lambda\in K' \\ \lambda>0}} \sum_{b \pmod{\Delta}} \left(\frac{\Delta}{b}\right) c_f^+\left(\frac{|\Delta|\lambda^2}{2}, r\lambda\right) \log\left|1 - e\left((\lambda,z)+\frac{b}{\Delta}\right)\right|$$
$$+ \begin{cases} 2\sqrt{\Delta}c_f^+(0,0)L(1,\chi_\Delta) & \text{if } \Delta \neq 1, \\ 8\pi(\rho_{f,\ell},y) - c_f^+(0,0)\left(\log\left(4\pi N y^2\right)+\Gamma'(1)\right) & \text{if } \Delta = 1. \end{cases}$$

The claims about $\Psi_{\Delta,r}(z,f)$ then essentially follow by exponentiation. □

We can also use these Borcherds products to study rationality properties of coefficients of mock modular forms.

THEOREM 18.16. *Suppose that* $\Delta \neq 1$. *Let* $f \in H_{\frac{1}{2},\widetilde{\rho}_L}$ *be a harmonic Maass form with real coefficients* $c_f^+(m,h)$ *for all* $m \in \mathbb{Q}$ *and* $h \in L'/L$. *Moreover, assume that* $c_f^+(n,h) \in \mathbb{Z}$ *for all* $n \leq 0$. *The following are equivalent:*

i) *The character* σ *of the function* $\Psi_{\Delta,r}(z,f)$ *defined in Theorem* 18.15 *is of finite order.*
ii) *The coefficients* $c_f^+(|\Delta|\lambda^2/2, r\lambda)$ *are rational for all* $\lambda \in K'$.

SKETCH OF PROOF. If i) holds, then there is a positive integer M such that $\Psi_{\Delta,r}(z,f)^M$ is a rational function on $X_0(N)$ with divisor $MZ_{\Delta,r}(f)$. Theorem 5.5 of **[127]** then implies that $c_f^+(|\Delta|\lambda^2/2, r\lambda) \in \mathbb{Q}$ for all $\lambda \in K'$. Conversely, if ii) holds, then the same theorem implies that $MZ_{\Delta,r}(f)$ is the divisor of a rational function R on $X_0(N)$ for some positive integer M. This then implies that

$$\log|R| - M\log|\Psi_{\Delta,r}(z,f)|$$

is a harmonic function on $X_0(N)$ (without any singularities). By the maximum modulus principle, it is constant. Hence $R/\Psi_{\Delta,r}(z,f)^M$ is a holomorphic function on $\mathbb{H}$ with constant modulus, which must be constant. Consequently, σ^M must be the trivial character. □

18.4. Examples of generalized Borcherds products

We now offer a number of examples of generalized Borcherds products. In particular, we revisit Theorem 18.6, and we consider an example which arises from Ramanujan's third order mock theta functions.

18.4.1. Twisted Borcherds products revisited. Theorems 18.15 and 18.16 include Theorem 18.6. To see this, let $N = 1$. Then we have $L'/L \cong \mathbb{Z}/2\mathbb{Z}$. Moreover, $H_{\frac{1}{2},\rho_L} = M^!_{\frac{1}{2},\rho_L}$ and $H_{\frac{1}{2},\overline{\rho}_L} = 0$. Therefore we consider the case where $\Delta > 1$ is a positive fundamental discriminant. Let $r \in \mathbb{Z}$ such that $\Delta \equiv r^2 \pmod 4$. By Section 5 of **[191]**, the space $M^!_{\frac{1}{2},\rho_L}$ can be identified with the space $\mathbb{M}^!_{\frac{1}{2}}$.

Now let the $\{f_d\}$ be the weight 1/2 weakly holomorphic modular forms defined by (16.4). Moreover, recall that their Fourier expansions satisfy

$$f_d(\tau) = q^d + \sum_{n=1}^{\infty} A(n,d)q^n.$$

If d is a negative discriminant, then Theorems 18.15 and 18.16 give a meromorphic modular form

$$\Psi_\Delta(\tau, f_d) := \Psi_{\Delta,r}(\tau, f_d)$$

of weight 0 for the group $\Gamma = \mathrm{SL}_2(\mathbb{Z})$ whose divisor on $X(1)$ is given by

$$Z_\Delta(d) = Z_{\Delta,r}\left(\frac{d}{4}, \frac{d}{2}\right) = \sum_{\lambda \in L_{\Delta d}/\Gamma} \frac{\chi_\Delta(\lambda)}{\omega(\lambda)} Z(\lambda).$$

Following the discussion in Subsection 18.3.1, one can show that $L_{\Delta d}/\Gamma$ corresponds to the Γ-classes of integral binary quadratic forms of discriminant Δd. Moreover, we obtain the infinite product expansion

$$\Psi_\Delta(\tau, f_d) = \prod_{n=1}^{\infty} \prod_{b \pmod{\Delta}} \left[1 - e\left(n\tau + \frac{b}{\Delta}\right)\right]^{\left(\frac{\Delta}{b}\right)A(\Delta n^2, d)}. \tag{18.18}$$

We also remark that the products on the inside of the (18.18) can be expressed in terms of (18.3) using $P_\Delta(q^n)$. Note that here we have replaced z with τ in Theorem 18.15 for Ψ_Δ to be consistent with the bulk of this book. From these properties, combined with the fact that $j(\tau) - 744$ is the Hauptmodul for $\mathrm{SL}_2(\mathbb{Z})$, it follows that

$$\Psi_\Delta(\tau, f_d) = \prod_{\lambda \in L_{\Delta d}/\Gamma} (j(\tau) - j(Z(\lambda)))^{\chi_\Delta(\lambda)}. \tag{18.19}$$

This is Theorem 18.6.

18.4.2. Ramanujan's mock theta functions $f(q)$ and $\omega(q)$. In view of the theorems and examples above, it is natural to ask about general results for mock modular forms with algebraic Fourier coefficients. New, deep results of Bruinier and Schwagenscheidt **[130]** provide a satisfying, overarching picture. In particular, they constructed harmonic Maass forms which give preimages for arbitrary vector-valued unary theta functions of weights 1/2 or 3/2 (which, other than mock modular forms with CM shadows, are the only ones which are expected to have algebraic coefficients). For any number field K, this allowed them to show that the Fourier coefficients of mock modular forms of weight 1/2 with shadows lying in the space of unary theta functions and with K-rational principal parts are in fact all K-rational. Moreover, the explicit realization of these preimages allowed them to prove that mock modular forms of weight 1/2 with rational Fourier expansions have rational Weyl vectors.

Here we give an example of the theory of generalized Borcherds product arising from Ramanujan's third order mock theta functions. We first recall the modular transformation properties from Chapter 9 of f, and ω, defined in (9.1) and (9.2), respectively. We define F by

$$F(\tau) = (F_0(\tau), F_1(\tau), F_2(\tau))^T = \left(q^{-\frac{1}{24}} f(q), 2q^{\frac{1}{3}}\omega\left(q^{\frac{1}{2}}\right), 2q^{\frac{1}{3}}\omega\left(-q^{\frac{1}{2}}\right)\right)^T,$$

and we define the vector-valued nonholomorphic function G by

$$G(\tau) = (G_0(\tau), G_1(\tau), G_2(\tau))^T = 2i\sqrt{3}\int_{-\overline{\tau}}^{i\infty} \frac{(\mathcal{G}_1(w),\ \mathcal{G}_0(w),\ -\mathcal{G}_2(w))^T}{\sqrt{-i(\tau+w)}}dw,$$

where the $\mathcal{G}_j$ are the cuspidal weight 3/2 theta functions from (9.11). The function $H(\tau) = F(\tau) - G(\tau)$ is a vector-valued weight 1/2 harmonic Maass form (cf. Theorem 9.5). In particular, it satisfies

$$H(\tau+1) = \begin{pmatrix} \zeta_{24}^{-1} & 0 & 0 \\ 0 & 0 & \zeta_3 \\ 0 & \zeta_3 & 0 \end{pmatrix} H(\tau), \tag{18.20}$$

$$H\left(-\frac{1}{\tau}\right) = \sqrt{-i\tau}\begin{pmatrix} 0 & 1 & 0 \\ 1 & 0 & 0 \\ 0 & 0 & -1 \end{pmatrix} H(\tau). \tag{18.21}$$

Now let $N := 6$. One can check the following lemma which asserts that this representation of $\widetilde{\Gamma}$ is an irreducible piece of the Weil representation $\overline{\rho}_L$.

LEMMA 18.17. *Assume that* $H = (h_0, h_1, h_2)^T : \mathbb{H} \to \mathbb{C}^3$ *is a vector-valued modular form of weight* k *for* $\widetilde{\Gamma}$ *transforming with the representation defined by the transformation laws of Theorem* 9.5. *Then the function*

$$\widetilde{H} := (0, h_0, h_2 - h_1, 0, -h_1 - h_2, -h_0, 0, h_0, h_1 + h_2, 0, h_1 - h_2, -h_0)^T$$

is a vector-valued modular form of weight k *for* $\widetilde{\Gamma}$ *with representation* $\overline{\rho}_L$. *Here we identify* $\mathbb{C}[L'/L]$ *with* $\mathbb{C}^{12}$ *by mapping the standard basis vector of* $\mathbb{C}[L'/L]$ *corresponding to the coset* $j/12+\mathbb{Z} \in L'/L$ *to the standard basis vector* e_j *of* $\mathbb{C}^{12}$ *(where* $j = 0, \ldots, 11$*).*

This lemma shows that H gives rise to an element $\widetilde{H} \in H_{\frac{1}{2},\overline{\rho}_L}$. Let $c^{\pm}(m,h)$ be the coefficients of $\widetilde{H}$. For any fundamental discriminant $\Delta < 0$ and any integer r such that $\Delta \equiv r^2 \pmod{24}$, we obtain a twisted generalized Borcherds lift $\Psi_{\Delta,r}(z,\widetilde{H})$. By Theorems 18.15 and 18.16, it is a weight 0 meromorphic modular function on $X_0(6)$ with divisor

$$2Z_{\Delta,r}\left(-\tfrac{1}{24}, \tfrac{1}{12}\right) - 2Z_{\Delta,r}\left(-\tfrac{1}{24}, \tfrac{5}{12}\right).$$

Moreover, it has the infinite product expansion

$$\Psi_{\Delta,r}\left(z,\widetilde{H}\right) = \prod_{n=1}^{\infty} P_{\Delta}\left(q^n\right)^{c^+\left(\frac{|\Delta| n^2}{24}, \frac{rn}{12}\right)}. \tag{18.22}$$

For instance, let $\Delta := -8$ and $r := 4$. The set $L_{-8,4}/\Gamma_0(6)$ is represented by the binary quadratic forms $Q_1 = [6,4,1]$ and $Q_2 = [-6,4,-1]$, and $L_{-8,-4}/\Gamma_0(6)$ is represented by $-Q_1$ and $-Q_2$. The Heegner points in $\mathbb{H}$ corresponding to Q_1 and Q_2 respectively are

$$\alpha_1 = \frac{-2+\sqrt{-2}}{6}, \qquad \alpha_2 = \frac{2+\sqrt{-2}}{6}.$$

Consequently, the divisor of $\Psi_{-8,4}(z,\widetilde{H})$ on $X_0(6)$ is given by $2(\alpha_1) - 2(\alpha_2)$. In this case the infinite product expansion (18.22) only involves the coefficients of the components of $\widetilde{H}$ of the form $\pm(h_1 + h_2)$. To simplify the notation, we put

$$\begin{aligned} -2q^{\frac{1}{3}}\left(\omega\left(q^{\frac{1}{2}}\right) + \omega\left(-q^{\frac{1}{2}}\right)\right) &=: \sum_{n \in \frac{1}{3}+\mathbb{Z}} a(n) q^n \\ &= -4\,q^{\frac{1}{3}} - 12\,q^{\frac{4}{3}} - 24\,q^{\frac{7}{3}} - 40\,q^{\frac{10}{3}} - \cdots. \end{aligned}$$

We have

$$P_{-8}(X) = \frac{1+\sqrt{-2}X - X^2}{1-\sqrt{-2}X - X^2},$$

and the infinite product expansion (18.22) can be rewritten, after replacing z by τ to be consistent with the bulk of the book, as

$$\Psi_{-8,4}\left(\tau,\widetilde{H}\right) = \prod_{n=1}^{\infty} P_{-8}\left(q^n\right)^{\left(\frac{n}{3}\right) a\left(\frac{n^2}{3}\right)}. \tag{18.23}$$

One can easily work out a closed-form expression for $\Psi_{-8,4}(\tau, \widetilde{H})$. We use the Hauptmodul for $\Gamma_0^+(6)$, the extension of $\Gamma_0(6)$ by all Atkin-Lehner involutions, which is

$$\begin{aligned} j_6^+(\tau) &= \left(\frac{\eta(\tau)\eta(2\tau)}{\eta(3\tau)\eta(6\tau)}\right)^4 + 4 + 3^4\left(\frac{\eta(3\tau)\eta(6\tau)}{\eta(\tau)\eta(2\tau)}\right)^4 \\ &= q^{-1} + 79\,q + 352\,q^2 + 1431\,q^3 + \cdots . \end{aligned}$$

We have $j_6^+(\alpha_1) = j_6^+(\alpha_2) = -10$. Hence $j_6^+(\tau) + 10$ is a rational function on $X_0(6)$ whose divisor consists of the 4 cusps with multiplicity -1 and the points α_1, α_2 with multiplicity 2. The unique normalized cusp form of weight 4 for $\Gamma_0^+(6)$ is

$$\begin{aligned} \delta(\tau) :&= \eta(\tau)^2\eta(2\tau)^2\eta(3\tau)^2\eta(6\tau)^2 \\ &= q - 2\,q^2 - 3\,q^3 + 4\,q^4 + 6\,q^5 + 6\,q^6 - 16\,q^7 - 8\,q^8 + \cdots . \end{aligned}$$

Using these functions, we find that

$$\phi(\tau) := \Psi_{-8,4}\left(\tau, \widetilde{H}\right)(j_6^+(\tau) + 10)\delta(\tau)$$

is a holomorphic modular form of weight 4 for $\Gamma_0(6)$ with divisor $4(\alpha_1)$. Using the classical Eisenstein series, it turns out that

$$\begin{aligned} 450\phi(\tau) = {} & \left(3360 - 1920\sqrt{-2}\right)\delta(\tau) + \left(1 - 7\sqrt{-2}\right)E_4(\tau) + \left(4 - 28\sqrt{-2}\right)E_4(2\tau) \\ & + \left(89 + 7\sqrt{-2}\right)E_4(3\tau) + \left(356 + 28\sqrt{-2}\right)E_4(6\tau). \end{aligned}$$

Putting this all together, (18.23) becomes

$$\begin{aligned} \prod_{n=1}^{\infty}\left(\frac{1 + \sqrt{-2}q^n - q^{2n}}{1 - \sqrt{-2}q^n - q^{2n}}\right)^{\left(\frac{n}{3}\right)a\left(\frac{n^2}{3}\right)} &= \frac{\phi(\tau)}{(j_6^+(\tau) + 10)\delta(\tau)} \\ &= 1 - 8\sqrt{-2}q - \left(64 - 24\sqrt{-2}\right)q^2 + \left(384 + 168\sqrt{-2}\right)q^3 + \cdots . \end{aligned}$$

CHAPTER 19

Elliptic Curves over $\mathbb{Q}$

In their seminal works which resulted in the formulation of their celebrated conjecture on elliptic curves **[63, 64]**, Birch and Swinnerton-Dyer made important use of formulas which expressed the central values of Hasse-Weil L-functions of CM elliptic curves in terms of the corresponding Weierstrass zeta function.

In Section 6.2, we made use of the Weierstrass zeta-function in a different way in connection with modular elliptic curves. We showed how to construct canonical harmonic Maass functions for every elliptic curve $E/\mathbb{Q}$, not just those with complex multiplication. In view of these facts, it is natural to suspect the existence of a general theory in which harmonic Maass forms "encode" the vanishing and non-vanishing of the central values and derivatives of Hasse-Weil L-functions. Here we show that this is indeed the case. It turns out that the central values $L(E_D, 1)$ and central derivatives $L'(E_D, 1)$ for the quadratic twist elliptic curves E_D of all modular elliptic curves can be studied using the Fourier expansions of harmonic Maass forms. Thanks to Kolyvagin's Theorem on modular elliptic curves, these results can be applied to study the distribution of quadratic twists with Mordell-Weil rank ≤ 1. This theory can be thought of as an extension of important work of Kohnen **[322, 323, 324]** and Waldspurger **[487]** on the arithmetic of the Shimura correspondence **[456]** when specialized to the case of modular elliptic curves. The extension makes use of the work of Gross, Kohnen and Zagier (cf. Section 16.4).

19.1. The Birch and Swinnerton-Dyer Conjecture

We begin by briefly recalling essential features of the basic theory of elliptic curves. We refer the reader to standard texts such as **[311, 462, 463, 464]** for background on the arithmetic theory of elliptic curves.

19.1.1. Rational points on elliptic curves.

19.1.1. Rational points on elliptic curves. Here we collect standard facts about elliptic curves and the theory which motivates the Birch and Swinnerton-Dyer Conjecture and the content of this chapter.

DEFINITION 19.1. An **elliptic curve** E over a field K is a smooth, projective algebraic curve of genus one, on which there is a specified point O.

Rational points on elliptic curves form abelian groups with respect to the well-known "chord-tangent law". The determination of these groups in the case of elliptic curves over number fields is one of the fundamental problems in arithmetic geometry and number theory. The following classical theorem of Mordell and Weil reduces this problem to the calculation of "ranks" of Mordell-Weil groups and the classification of possible torsion subgroups.

THEOREM 19.2 (Mordell-Weil). *If E/K is an elliptic curve over a number field K, then the set of K-rational points $E(K)$ forms a finitely generated abelian group with respect to the chord-tangent law.*

If $K = \mathbb{Q}$, then the following famous theorem of Mazur **[379]** offers a classification of the possible torsion subgroups $E_{\mathrm{tor}}(\mathbb{Q})$ which arise.

THEOREM 19.3 (Mazur). *If $E/\mathbb{Q}$ is an elliptic curve, then*

$$E(\mathbb{Q}) \cong \mathbb{Z}^{\mathrm{rk}(E)} \times E_{\mathrm{tor}}(\mathbb{Q}),$$

where $\mathrm{rk}(E) \geq 0$ *is the rank of* $E/\mathbb{Q}$, *and*

$$E_{\mathrm{tor}}(\mathbb{Q}) \in \{\mathbb{Z}/n\mathbb{Z} \ : \ 1 \leq n \leq 10 \ \textit{or} \ n = 12\} \cup \{\mathbb{Z}/2\mathbb{Z} \times \mathbb{Z}/2n\mathbb{Z} \ : \ 1 \leq n \leq 4\}.$$

REMARK. Mazur's Theorem is a special case of the general fact that there are only finitely many possible candidates for the torsion subgroups $E_{\mathrm{tor}}(K)$ as E varies over all elliptic curves over a fixed number field K. In fact, a deep theorem of Merel **[383]** implies that there are finitely many possible such torsion subgroups as one varies over all elliptic curves over all number fields K with bounded degree over $\mathbb{Q}$. Merel's theorem is a confirmation of the "Uniform Boundedness Conjecture" in the case of elliptic curves.

In view of Mazur's Theorem, it is natural to ask about the distribution of the ranks $\mathrm{rk}(E)$ as one varies over all elliptic curves $E/\mathbb{Q}$. Recent work by Bhargava, Shankar, Skinner, and Zhang **[57, 58, 59, 60, 61]** has shed much light on this question.

To describe their work, we require a natural method for organizing elliptic curves. To this end, we begin by observing that every elliptic curve $E/\mathbb{Q}$ is isomorphic to a unique (minimal) curve of the form

$$E_{A,B}\colon Y^2 = X^3 + AX + B, \tag{19.1}$$

where A and B are integers with the property that for every prime p we have that $p^6 \nmid B$ whenever $p^4 | A$. The *naive height* $H(E_{A,B})$ of this curve is the integer

$$H(E_{A,B}) := \max\left\{4|A^3|, 27B^2\right\}. \tag{19.2}$$

It is widely believed that "half" of the elliptic curves $E/\mathbb{Q}$ have rank zero (resp. rank one). More precisely, we have the following "50-50 conjecture".

CONJECTURE 19.4. *For $r = 0, 1$, we have that*

$$\lim_{N\to+\infty} \frac{\#\{E_{A,B} :\ H(E_{A,B}) \leq N \ \textit{and} \ \mathrm{rk}(E_{A,B}) = r\}}{\#\{E_{A,B} :\ H(E_{A,B}) \leq N\}} = \frac{1}{2},$$

where the limit is taken over minimal elliptic curves $E_{A,B}$.

The works of Bhargava, Shankar, Skinner, and Zhang alluded to above combine to give the following theorem.

THEOREM 19.5 (Bhargava, Shankar, Skinner, and Zhang). *We have that*

$$\liminf_{N\to+\infty} \frac{\sum_{H(E_{A,B})\leq N} \mathrm{rk}(E_{A,B})}{\#\{H(E_{A,B}) \leq N\}} \geq 0.2068,$$

$$\limsup_{N\to+\infty} \frac{\sum_{H(E_{A,B})\leq N} \mathrm{rk}(E_{A,B})}{\#\{H(E_{A,B}) \leq N\}} \leq 0.885.$$

Furthermore, at least 16.5% *(resp.* 20.68%*) of elliptic curves have rank zero (resp. one).*

19.1.2. The Birch and Swinnerton-Dyer Conjecture. The reason that results such as Theorem 19.5 are fantastic follows from the unfortunate fact that there is no proven finite procedure which determines the Mordell-Weil group of an elliptic curve E. It is much simpler to prove that $E(K)$ is finitely generated. These difficulties provide the inspiration for the celebrated Birch and Swinnerton-Dyer Conjecture. Loosely speaking, for elliptic curves $E/\mathbb{Q}$, the conjecture asserts that the analytic behavior at $s = 1$ of the Hasse-Weil L-function $L(E, s)$ encodes $\mathrm{rk}(E)$ as well as other invariants including the order of $E_{\mathrm{tor}}(\mathbb{Q})$.

Suppose that $E/\mathbb{Q}$ is an elliptic curve with conductor N_E. For $\mathrm{Re}(s) \gg 0$, the *Hasse-Weil* L-function for E is the Dirichlet series

$$L(E,s) = \sum_{n=1}^{\infty} \frac{a_E(n)}{n^s}, \tag{19.3}$$

where for primes $p \nmid N_E$ we have $\#E(\mathbb{F}_p) =: p+1-a_E(p)$. For primes $p|N_E$, there is a well-known formula for $a_E(p)$ depending on the nature of the bad reduction at p. Here $\#E(\mathbb{F}_p)$ denotes the number of points on the reduction of E modulo p. The Dirichlet series is formally defined by the Euler product

$$L(E,s) := \prod_{p|N_E} \frac{1}{1-a_E(p)p^{-s}} \prod_{p\nmid N_E} \frac{1}{1-a_E(p)p^{-s}+p^{1-2s}}.$$

These L-functions are endowed with rich structure which is not apparent from their definition as Euler products. It turns out that they are L-functions for weight 2 newforms on $\Gamma_0(N_E)$. By the works of Breuil, Conrad, Diamond, Taylor, and Wiles [**78, 152, 477, 496**], combined with the standard Hecke theory of modular L-functions, we have the following theorem which can be thought of as the converse to the Eichler-Shimura theorem given as Theorem 1.17.

THEOREM 19.6 (Modularity Theorem). *Suppose that $E/\mathbb{Q}$ is an elliptic curve with conductor N_E. Then the following are true:*

i) *There is a non-constant map*

$$\phi_E : X_0(N) \twoheadrightarrow E.$$

Moreover, the pull-back of the unique (up to multiplication by scalar) holomorphic differential on E is the form $2\pi i F_E(\tau)d\tau$ for some weight 2 newform F_E on $\Gamma_0(N_E)$.

ii) *There is a newform $F_E \in S_2(\Gamma_0(N_E))$ for which*

$$L(E,s) = L(F_E,s).$$

iii) *The Hasse-Weil L-function $L(E,s)$ has an analytic continuation to $\mathbb{C}$. In particular, there is an $\varepsilon \in \{\pm 1\}$ for which*

$$\Lambda(E,s) = \varepsilon\Lambda(E,2-s),$$

where $\Lambda(E,s) := (2\pi)^{-s}\Gamma(s)N_E^{\frac{s}{2}}L(E,s)$.

REMARK. The number ε above is called the *sign of the functional equation* of $L(E,s)$. If $\varepsilon = 1$ (resp. $\varepsilon = -1$), then we have that $\mathrm{ord}_{s=1}(L(E,s))$ is even (resp. odd).

EXAMPLE 19.7. Consider the *congruent number elliptic curve* given by

$$E\colon Y^2 = X^3 - X.$$

One easily finds the three order 2 rational points $(0,0)$ and $(\pm 1, 0)$. It turns out that $E_{\text{tor}} \cong \mathbb{Z}/2\mathbb{Z} \times Z/2\mathbb{Z}$ and that $\text{rk}(E) = 0$. In terms of the Modularity Theorem, we have that $L(E,s) = L(F_E, s)$, where $F_E(\tau) := \eta(4\tau)^2\eta(8\tau)^2$ is the unique weight 2 newform with respect to $\Gamma_0(32)$.

The Birch and Swinnerton-Dyer Conjecture relates the analytic behavior of $L(E,s)$ at $s = 1$ to the arithmetic of E, which only makes sense thanks to Theorem 19.6.

CONJECTURE 19.8 (Birch and Swinnerton-Dyer). *Suppose that $E/\mathbb{Q}$ is an elliptic curve. Then the following are true:*

i) *We have that*
$$\text{ord}_{s=1}(L(E,s)) = \text{rk}(E).$$
ii) *If $Ш(E)$ is the Shafarevich-Tate group of $E/\mathbb{Q}$,* $\text{Tam}(E)$ *is the Tamagawa number for E, and $\Omega(E)$ is the real period of E, and $R(E)$ is the regulator of $E/\mathbb{Q}$, then*
$$\lim_{s\to 1}\left((s-1)^{-\text{rk}(E)}L(E,s)\right) = \frac{2^{\text{rk}(E)}\#Ш(E)\cdot \text{Tam}(E)\cdot \Omega(E)\cdot R(E)}{\#E_{\text{tor}}(\mathbb{Q})^2}.$$

REMARK. The recent work by Bhargava, Skinner, and Zhang **[61]** proves that Conjecture 19.8 i) is true for $> 66\%$ of the (minimal) elliptic curves $E_{A,B}$ ordered by height.

Although this conjecture remains open, a lot is known about its truth. In view of Conjecture 19.4, most of the time one expects $\text{ord}_{s=1}(L(E,s)) \leq 1$. Moreover, thanks to Theorem 19.5, there is ample theoretical evidence supporting its truth. Therefore, the following deep theorem of Kolyvagin **[329]** is particularly satisfying.

THEOREM 19.9 (Kolyvagin). *If $E/\mathbb{Q}$ satisfies $\text{ord}_{s=1}(L(E,s)) \leq 1$, then*
$$\text{rk}(E) = \text{ord}_{s=1}(L(E,s)).$$

REMARK. Strictly speaking, Kolyvagin proved Theorem 19.9 assuming various hypotheses which were later confirmed for all $E/\mathbb{Q}$ by Bump, Friedberg, and Hoffstein **[135]**, Murty and Murty **[391]**, and Waldspurger **[487]**.

To apply Theorem 19.9, it is important to have tools which can be used to determine the vanishing and non-vanishing of $L(E,1)$ and $L'(E,1)$. In the case of central derivatives, we have a deep theorem of Gross and Zagier **[242]**.

To state their theorem we first require some preliminaries. The modular curve $X_0(N)$ parameterizes isogeny classes of elliptic curves
$$\phi\colon\ E \to E'$$
for which $\ker(\phi) \cong \mathbb{Z}/N\mathbb{Z}$. In particular, we have the covering map
$$\pi_N\colon\ X_0(N) \to X_0(1)$$
which corresponds to the map
$$\mathbb{C}/(\mathbb{Z}+\mathbb{Z}\tau) \to \mathbb{C}/(\tfrac{1}{N}\mathbb{Z}+\mathbb{Z}\tau) \to \mathbb{C}/(\mathbb{Z}+\mathbb{Z}\tau).$$
Here we identify elliptic curves as usual as quotients of $\mathbb{C}$ by 2-dimensional lattices.

Suppose that $K = \mathbb{Q}(\sqrt{d})$ is an imaginary quadratic field for which every prime $p|N$ splits in K. Such a field K is said to satisfy the *Heegner hypotheses* for $X_0(N)$. Given these hypotheses, we have an ideal $\mathfrak{n} \subset \mathcal{O}_K$, the ring of integers of K, for which $\mathcal{O}_K/\mathfrak{n} \cong \mathbb{Z}/N\mathbb{Z}$. As a consequence, for any $\mathfrak{a} \subset \mathcal{O}_K$ we obtain a covering map

$$\mathbb{C}/\mathfrak{a} \to \mathbb{C}/\mathfrak{n}^{-1}\mathfrak{a} \in X_0(N)(\mathbb{C}).$$

Scaling an ideal $\mathfrak{a}$ by elements in $K^\times$ does not change an elliptic curve, and so this covering offers a well-defined map on ideal classes

$$\gamma_\mathfrak{n} \colon \mathrm{Cl}(K) \to X_0(N)(\mathbb{C}).$$

These points are the so-called *Heegner points* for K on $X_0(N)$. By the theory of complex multiplication, they are defined over the Hilbert class field of K.

Suppose now that $E/\mathbb{Q}$ is an elliptic curve with conductor $N_E = N$. Using the modular parameterization

$$\phi_E \colon X_0(N) \to E,$$

for each ideal $\mathfrak{a} \subset \mathcal{O}_K$, we obtain a point $\phi_E(\gamma_\mathfrak{n}([\mathfrak{a}])) \in E(H_K)$, where H_K is the Hilbert class field of K. Using this notation, we then define the point

$$P_{K,E} := \sum_{[\mathfrak{a}] \in \mathrm{Cl}(K)} \phi_E(\gamma_\mathfrak{n}([\mathfrak{a}])). \tag{19.4}$$

By Galois equivariance and the fact that $\mathrm{Gal}(H_K/K) \cong \mathrm{Cl}(K)$, it follows that $P_{K,E} \in E(K)$.

The celebrated *Gross-Zagier formula* [**242**] gives a formula for the derivative $L'(E/K, 1)$, where

$$L(E/K, s) := L(E, s)L(E_d, s). \tag{19.5}$$

The elliptic curve E_d is the d-th quadratic twist, curves which will be discussed in the next section. This formula involves the Petersson norm of the newform F_E, the height of the Heegner point $P_{K,E}$, and the degree of the modular parameterization ϕ_E. Namely, we have the following theorem.

THEOREM 19.10 (Gross and Zagier). *Assuming the notation and hypotheses above, we have that*

$$L'(E/K, 1) = \frac{32\pi^2 ||F_E||^2}{|\mathcal{O}_K^\times|^2 \sqrt{|d|} \deg(\phi(E))} h_E(P_{K,E}).$$

In particular, we have that $L'(E/K, 1) = 0$ if and only if $P_{K,E}$ is a torsion point.

REMARK. Thanks to (19.5), we note that the work of Gross and Zagier offers infinitely many formulas involving $L'(E, 1)$, one for each K which satisfies the Heegner hypotheses for $X_0(N)$. The case where the sign of the functional equation satisfies $\varepsilon = -1$ is of particular importance. In this case we have that $\mathrm{ord}_{s=1}(L(E, s))$ is odd, and so $L(E, 1) = 0$. In this case, thanks to a deep theorem of Waldspurger [**487**], there are infinitely many such $K = \mathbb{Q}(\sqrt{d})$ for which $L(E_d, 1) \neq 0$, thereby offering infinitely many genuine formulas for $L'(E, 1)$. The Gross-Zagier formula can be used to test the vanishing of $L'(E, 1)$ by determining whether or not $P_{K,E}$ is torsion.

19.2. Quadratic twists of elliptic curves

The arithmetic of half-integral weight modular forms and half-integral weight harmonic Maass forms encodes information about the central critical L-values and derivatives of families of quadratic twists of elliptic curves. Thanks to the results of the previous subsection, this analytic information can be used to study the distribution of rank zero and rank one elliptic curves in these families. To make this precise, we first recall basic facts about these families.

19.2.1. Quadratic twists. Throughout this section suppose that $E/\mathbb{Q}$ is an elliptic curve given by the Weierstrass model

$$E\colon Y^2 = X^3 + AX + B. \tag{19.6}$$

If $D \neq 1$ is a non-zero square-free integer, then E_D is the D-th *quadratic twist*

$$E_D\colon DY^2 = X^3 + AX + B. \tag{19.7}$$

We have that $E(\mathbb{Q}(\sqrt{D})) \cong E_D(\mathbb{Q}(\sqrt{D}))$. The following theorem is a recapitulation of Theorem 19.6 for quadratic twists.

THEOREM 19.11. *Suppose that $E/\mathbb{Q}$ is an elliptic curve with conductor N_E, and let $F_E(\tau) = \sum_{n=1}^{\infty} a_E(n)q^n \in S_2(\Gamma_0(N_E))$ be the corresponding weight 2 newform. If D is a fundamental discriminant for which* $\gcd(D, N_E) = 1$, *then the following are true:*

i) *We have that*
$$L(E_D, s) = L(f_{E_D}, s),$$
where $f_{E_D}(\tau) := \sum_{n=1}^{\infty} a_E(n)\left(\frac{D}{n}\right)q^n$.

ii) *If* $\varepsilon \in \{\pm 1\}$ *satisfies*
$$\Lambda(F_E, s) = \varepsilon\Lambda(F_E, 2-s),$$
then
$$\Lambda(f_{E_D}, s) = \varepsilon\left(\frac{D}{-N(E)}\right)\Lambda(f_{E_D}, 2-s).$$

The sign of the functional equation of E_D in the theorem above is given by $\varepsilon(\frac{D}{-N(E)})$. Therefore, the extended Law of Quadratic Reciprocity, implies that "half" of the fundamental discriminants D correspond to those Hasse-Weil L-functions $L(E_D, s)$ with sign of the functional equation $+1$ (resp. -1).

In analogy with Conjecture 19.4, the following conjecture by Goldfeld **[226]** predicts that the average rank of elliptic curves E_D in a family of quadratic twists is $1/2$. In view of the variation of the sign of the functional equation, this conjecture asserts that "half" of the E_D have rank zero (resp. one).

Goldfeld's conjecture is actually a claim about the vanishing of central values and derivatives of even weight newforms in families of quadratic twists. The application to elliptic curves is the case of weight 2 $\mathbb{Q}$-rational newforms. To make this precise, we require the following definition.

DEFINITION 19.12. If $f(\tau) = \sum_{n=1}^{\infty} c_f(n)q^n \in S_{2k}(\Gamma_0(M))$, then its D-th **quadratic twist** is a cusp form with Fourier expansion

$$f_D(\tau) := \sum_{n=1}^{\infty} \left(\frac{D}{n}\right)c_f(n)q^n.$$

REMARK. Theorem 19.11 is a special case of the obvious generalization which holds for all even weight newforms.

CONJECTURE 19.13 (Goldfeld). *The following are true:*

i) *If* $f \in S_{2k}(\Gamma_0(M))$ *is an even weight newform and* f_D *is its* D*-th quadratic twist, then*

$$\sum_{\substack{-N<D<N \\ \gcd(D,M)=1}} \operatorname{ord}_{s=k}(L(f_D, s))$$

$$\sim \frac{1}{2}\#\{D \;:\; -N < D < N \text{ and } \gcd(D, M) = 1\},$$

as $N \to +\infty$. *Here the* D *vary over fundamental discriminants.*

ii) *If* $E/\mathbb{Q}$ *is an elliptic curve with conductor* N_E, *then*

$$\sum_{\substack{-N<D<N \\ \gcd(D,N_E)=1}} \operatorname{rk}(E_D) \sim \frac{1}{2}\#\{-N < D < N \text{ and } \gcd(D, N_E) = 1\}.$$

Conjecture 19.13 remains open. In fact, if $r \in \{0, 1\}$, then for general f and $E/\mathbb{Q}$ it is not even known that

$$\#\{|D| \le N \;:\; \operatorname{ord}_{s=k}(L(f_D, s)) = r\} \gg_f N, \tag{19.8}$$

$$\#\{|D| \le N \;:\; \operatorname{rk}(E_D) = r\} \gg_E N. \tag{19.9}$$

Iwaniec, Katz, and Sarnak have conditional proofs of (19.8) in **[280, 303]**. There are some unconditional theorems for (19.8) and (19.9) for special f and E. In particular, for (19.8), Kohnen **[326]** proved the $r = 0$ case for certain level 1 newforms. Heath-Brown **[256, 257]**, using a careful analysis of 2-Selmer groups, has proven the $r = 0$ case of (19.9) for the congruent number elliptic curve

$$E\colon\; Y^2 = X^3 - X,$$

named for its connection to the ancient Diophantine problem about right triangles. Recall that a positive integer is *congruent* if it is the area of a right triangle with rational number sidelengths. It turns out that a positive square-free integer D is congruent if and only if $\operatorname{rk}(E_D)$ is positive (for example, see **[316]**). We shall discuss this problem further in Example 19.18.

REMARK. Kane **[296]** and Yu **[504]** established the $r = 0$ case of (19.9) more generally for elliptic curves with full 2-torsion over $\mathbb{Q}$.

In terms of general results, for $r = 0$ one of the authors and Skinner **[411]** showed that

$$\#\{|D| \le N \;:\; L(f_D, k) \neq 0\} \gg_f \frac{N}{\log(N)}. \tag{19.10}$$

For most elliptic curves $E/\mathbb{Q}$ without $\mathbb{Q}$-rational 2-torsion, this lower bound has been improved to $N/\log(N)^{\frac{5}{6}}$ by Kriz and Li **[331]**. For $r = 1$, Perelli and Pomykala **[416]** proved that

$$\#\{|D| \le N \;:\; \operatorname{ord}_{s=k}(L(f_D, s)) = 1\} \gg_f N^{1-\varepsilon}. \tag{19.11}$$

Kriz and Li also improved this lower bound to $N/\log(N)^{\frac{5}{6}}$ for most elliptic curves $E/\mathbb{Q}$ without $\mathbb{Q}$-rational 2-torsion in **[331]**. Estimate (19.10) was proved by making use of deep work of Waldspurger on Shimura's correspondence, the topic of the next subsection.

19.3. The Shimura correspondence

In 1973 Shimura **[456]** published his seminal paper on the theory of half-integral weight modular forms.[1] His theory offers a correspondence between half-integral weight cusp forms and even weight cusp forms which commutes with the corresponding Hecke algebras.

The "Shimura correspondence" is a family of maps which send modular forms of half-integral weight to those of integer weight. Although Shimura's original theorem was stated for half-integral weight eigenforms, the generalization we describe here follows in a fairly straightforward way.

Suppose that $g(\tau) = \sum_{n=1}^{\infty} c_g(n)q^n \in S_{k+\frac{1}{2}}(\Gamma_0(4N), \psi)$ is a cusp form where $k \geq 2$. If t is any positive square-free integer, then define $A_t(n)$ by

$$\sum_{n=1}^{\infty} \frac{A_t(n)}{n^s} := L\left(s-k+1, \psi\chi_{-1}^k\chi_t\right) \sum_{n=1}^{\infty} \frac{c_g\left(tn^2\right)}{n^s}. \tag{19.12}$$

Here χ_{-1} (resp. χ_t) is the Kronecker character for $\mathbb{Q}(i)$ (resp. $\mathbb{Q}(\sqrt{t})$).

THEOREM 19.14 (Shimura). *Assume the notation and hypotheses above. Then the following are true:*

i) *The numbers $A_t(n)$ defined in (19.12) are the Fourier coefficients of $\mathscr{S}_t(g)(\tau)$, a cusp form*

$$\mathscr{S}_t(g)(\tau) := \sum_{n=1}^{\infty} A_t(n)q^n$$

in $S_{2k}(\Gamma_0(4N), \psi^2)$.

ii) *The correspondence $\mathscr{S}_t$ commutes with the Hecke algebra. In particular, if $p \nmid 4N$ is prime, then*

$$\mathscr{S}_t\left(g|T_{p^2}\right) = \mathscr{S}_t(g)|T_p.$$

Here T_p (resp. T_{p^2}) denotes the usual Hecke operator acting on the space $S_{2k}(\Gamma_0(4N), \psi^2)$ (resp. $S_{k+\frac{1}{2}}(\Gamma_0(4N), \psi)$).

REMARK. Theorem 19.14 extends to include the case where $k = 1$. In particular, the theorem holds as stated for those weight $3/2$ cusp forms g which are orthogonal to the space of weight $3/2$ unary theta functions.

19.4. Central values of quadratic twist L-functions

In the 1980s, Kohnen and Zagier, and Waldspurger realized the strong interplay between the half-integral weight modular forms and central L-values of modular L-functions. Some of Kohnen's work in this area was previously discussed in Section 12.3. Here we recall the main features of their results with the application to elliptic curves in mind.

[1] Shimura was awarded the 1977 Cole Prize largely for this work.

19.4.1. A theorem of Kohnen and Zagier. Kohnen refined Shimura's theory with the idea of obtaining isomorphisms between spaces of half-integral weight cusp forms and even integral weight cusp forms. He successfully developed such a theory which we now describe.

Suppose that N is an odd and square-free integer and that $k \geq 1$. Kohnen's theory **[322, 323, 324]** offers an isomorphism between his half-integral weight "plus-spaces" of newforms $S^{\mathrm{new}}_{k+\frac{1}{2}}(\Gamma_0(4N))$ and classical even integer weight newform spaces $S^{\mathrm{new}}_{2k}(\Gamma_0(N))$. He proved that there is a finite linear combination of the maps $\mathscr{S}_t$ which establishes an isomorphism between these spaces.

Under this isomorphism, suppose that

$$g(\tau) = \sum_{(-1)^k n \equiv 0,1 \pmod 4} c_g(n) q^n \in S^{\mathrm{new}}_{k+\frac{1}{2}}(\Gamma_0(4N)) \tag{19.13}$$

is a Kohnen newform which corresponds to the even weight $2k$ newform $f \in S^{\mathrm{new}}_{2k}(\Gamma_0(N))$. Kohnen proved that the coefficients of $g(\tau)$ determine the central critical values of many of the quadratic twists $L(f_D, s)$. Recalling that $t(N)$ denotes the number of distinct prime divisors of N, let $\langle f, f\rangle$ (resp. $\langle g, g\rangle$) denote the Petersson inner product on $S_{2k}(\Gamma_0(N))$ (resp. $S_{k+\frac{1}{2}}(\Gamma_0(4N))$). Generalizing a theorem of his and Zagier's when $N = 1$ **[320]**, he proved the following important theorem (cf. Corollary 1 of **[324]**).

THEOREM 19.15 (Kohnen). *Assume the notation and hypotheses above. If $\ell | N$ is prime, then let $\lambda_\ell \in \{\pm 1\}$ be the eigenvalue of the Atkin-Lehner involution*

$$f|_{2k} W(Q_\ell) = \lambda_\ell f.$$

If $(-1)^k D > 0$ and D has the property that $\left(\frac{D}{\ell}\right) = \lambda_\ell$ for each prime $\ell | N$, then

$$L(f_D, k) = \frac{\langle f, f\rangle \pi^k}{2^{t(N)}(k-1)!|D|^{k-\frac{1}{2}}\langle g, g\rangle} |c_g(|D|)|^2.$$

For all other fundamental discriminants D with $(-1)^k D > 0$ we have $c_g(|D|) = 0$.

EXAMPLE 19.16. If $f = \Delta \in S_{12}$, then the associated Kohnen newform in $S^{\mathrm{new}}_{\frac{13}{2}}(\Gamma_0(4))$ is

$$g(\tau) = \sum_{n=1}^{\infty} c_g(n) q^n = \frac{E_4(4\tau)\frac{\partial}{\partial \tau}\theta(\tau)}{4\pi i} - \frac{\theta(\tau)\frac{\partial}{\partial \tau}E_4(4\tau)}{32\pi i} = q - 56q^4 + 120q^5 - \cdots.$$

If D is a positive fundamental discriminant, then Theorem 19.15 implies that

$$L(\Delta_D, 6) = \frac{\langle \Delta, \Delta\rangle \pi^6}{120 D^{\frac{11}{2}} \langle g, g\rangle} c_g(D)^2.$$

19.4.2. A theorem of Waldspurger. Waldspurger proved a theorem which is closely related to Theorem 19.15. He was primarily interested with the interpretation of the coefficients of all half-integral weight Hecke eigenforms in terms of central L-values. Here we describe his theorem which offers these formulas.

To describe his theorem, we first fix some notation. For every fundamental discriminant D, define D_0 by

$$D_0 := \begin{cases} |D| & \text{if } D \text{ is odd,} \\ \frac{|D|}{4} & \text{if } D \text{ is even.} \end{cases} \tag{19.14}$$

The following is a reformulation of one of the main results in **[487]**.

THEOREM 19.17 (Waldspurger). *If* $f(\tau) = \sum_{n=1}^{\infty} c_f(n)q^n \in S_{2k}^{\mathrm{new}}(\Gamma_0(M))$ *is an even weight newform and* $\varepsilon \in \{\pm 1\}$ *is the sign of the functional equation of* $L(f,s)$*, then there is a positive integer* N *with* $M|N$*, a Dirichlet character* χ *modulo* $4N$*, a non-zero complex number* Ω_f*, and a non-zero half-integral weight Hecke eigenform*

$$g_f(\tau) = \sum_{n=1}^{\infty} b_f(n)q^n \in S_{k+\frac{1}{2}}(\Gamma_0(4N), \chi)$$

with the property that there are arithmetic progressions of fundamental discriminants D *coprime to* $4N$ *for which* $\varepsilon D > 0$ *and*

$$b_f(D_0)^2 = \varepsilon_D \frac{L(f_D, k)D_0^{k-\frac{1}{2}}}{\Omega_f},$$

where ε_D *is algebraic. For all other* D *with* $\varepsilon D > 0$*, we have* $b_f(D_0) = 0$*. Moreover, the coefficients* $a_f(n), b_f(n)$ *and the values of* χ *are in* $\mathcal{O}_K$*, the ring of integers of some fixed number field* K*. In addition, if* $p \nmid 4N$ *is prime, then* $\lambda_p = \chi(p)c_f(p)$*, where* λ_p *is the eigenvalue of* $g_f(z)$ *for the Hecke operator* $T_{p^2,k,\chi}$*.*

EXAMPLE 19.18. We now revisit the congruent number problem using Theorems 19.9 and 19.17. As discussed earlier, this ancient Diophantine question asks for the classification of those positive integers N which are areas of right triangles with rational side lengths. A positive integer N is congruent if and only if the elliptic curve

$$E_N\colon\ Y^2 = X^3 - N^2X,$$

the N-th quadratic twist of E_1, has positive rank over $\mathbb{Q}$. Tunnell [**479**] used Theorem 19.17 and work of Coates and Wiles, which is superseded by Theorem 19.9, to provide a conditional solution to the congruent number problem. By Example 19.7, the newform $f_{E_1}(\tau) = \eta(4\tau)^2\eta(8\tau)^2 \in S_2^{\mathrm{new}}(\Gamma_0(32))$ has the property that $L(E,s) = L(f,s)$. Tunnell found that if $f_1(\tau) := \eta(8\tau)\eta(16\tau)$, then

$$f_1(\tau)\theta(2\tau) = \sum_{n=1}^{\infty} a(n)q^n \in S_{\frac{3}{2}}(\Gamma_0(128)),$$

$$f_1(\tau)\theta(4\tau) = \sum_{n=1}^{\infty} b(n)q^n \in S_{\frac{3}{2}}(\Gamma_0(128), (\tfrac{8}{\cdot}))$$

are Hecke eigenforms whose image under the Shimura correspondence is f_{E_1}. Using Theorem 19.17, he then proved that if d is an odd positive square-free integer, then

$$L(E_d, 1) = a(d)^2 \frac{\Omega}{4\sqrt{d}}, \qquad L(E_{2d}, 1) = b(d)^2 \frac{\Omega}{2\sqrt{2d}},$$

where

$$\Omega := \int_1^{\infty} \frac{dx}{\sqrt{x^3 - x}} \approx 2.622\ldots.$$

Suppose that d is a positive odd square-free integer. The Birch and Swinnerton-Dyer Conjecture implies that d (resp. $2d$) is congruent if and only if $a(d) \neq 0$ (resp. $b(d) \neq 0$). Unconditionally, Theorem 19.9 implies that d (resp. $2d$) is not congruent when $a(d) \neq 0$ (resp. $b(d) \neq 0$).

19.5. Harmonic Maass forms and quadratic twists of elliptic curves

As explained in the previous section, the non-vanishing of $L(E_D, 1)$ can be determined with the help of the Fourier expansions of special weight $3/2$ modular forms. More precisely, if the relevant coefficients which appear in Theorems 19.15 or 19.17 are nonzero, then Theorem 19.9 implies that $\mathrm{rk}(E_D) = 0$.

In view of Theorem 19.9, it is natural to ask about conditions which determine the non-vanishing of central derivatives $L'(E_D, 1)$ with the idea of quantifying the distribution of rank one quadratic twists. To this end we have the Gross-Zagier formula in Theorem 19.10, which formulates this problem in terms of heights of Heegner points $P_{K,E}$.

It is natural to ask whether there is an extension of Theorems 19.15 and 19.17 which also applies to the setting of central derivatives $L'(E_D, 1)$. It turns out that there is such an extension, and this theory relies on the theory of harmonic Maass forms. Bruinier and one of the authors **[127]** proved that the Fourier coefficients of certain harmonic Maass forms encode both central values and derivatives for L-functions for elliptic curves in a family of quadratic twists.

To simplify exposition, suppose that p is prime and that $E/\mathbb{Q}$ is an elliptic curve with conductor $N_E = p$. By Theorem 19.6, we let

$$F_E(\tau) = \sum_{n=1}^{\infty} a_E(n) q^n \in S_2^{\mathrm{new}}(\Gamma_0(p))$$

be the newform for which $L(F_E, s) = L(E, s)$. Finally, suppose that the sign of the functional equation of $L(E, s)$ is $\varepsilon = -1$, so that $L(E, 1) = 0$.

By Kohnen's theory of newforms **[322, 323, 324]**, there is a half-integral weight newform

$$g_E(\tau) = \sum_{n=1}^{\infty} b_E(n) q^n \in S_{\frac{3}{2}}^{\mathrm{new}}(\Gamma_0(4p)), \tag{19.15}$$

unique up to a multiplicative constant, which lifts to F_E under the Shimura correspondence. Choose g_E so that its coefficients are in $\mathbb{Q}$.

The following lemma is analogous to Theorem 7.6, which is a refinement of the surjectivity statement of Theorem 5.10 i).

LEMMA 19.19. *Assume the notation and hypotheses above. Then there is a weight 1/2 harmonic Maass form*

$$k_E(\tau) = \sum_{n \gg -\infty} c_E^+(n) q^n + \sum_{n<0} c_E^-(n) \Gamma\left(\frac{1}{2}, 4\pi |n| v\right) q^n \in H_{\frac{1}{2}}(\Gamma_0(4p))$$

with $\mathbb{Q}$-rational principal part, which also has the property that $\xi_{\frac{1}{2}}(k_E) = \frac{g_E}{\|g_E\|^2}$.

Using the definition of $\xi_{\frac{1}{2}}$, one checks that if $n > 0$, then

$$b_E(n) = -2\sqrt{\pi n}\|g\|^2 c_E^-(-n). \tag{19.16}$$

Therefore, Theorem 19.15 implies that the coefficients of the nonholomorphic part of k_E encode the central critical values $L(E_D, 1)$.

The coefficients $c_E^+(n)$ of the holomorphic part are much more mysterious at first glance. It turns out that these coefficients are also related to the quadratic twist L-functions $L(E_D, s)$. Theorem 1.1 of **[127]** offers the following relations.

THEOREM 19.20. *Assume the notation and hypotheses above. Then the following are true:*

i) *If $D<0$ is a fundamental discriminant for which $\left(\frac{D}{p}\right)=1$, then*

$$L(E_D,1)=8\pi^2\|F_E\|^2\|g_E\|^2\sqrt{\frac{|D|}{N}}c_E^-(D)^2.$$

ii) *If $D>0$ is a fundamental discriminant for which $\left(\frac{D}{p}\right)=1$, then $L'(E_D,1)=0$ if and only if $c_E^+(D)$ is algebraic.*

By Theorem 19.11 ii), the two cases in Theorem 19.20 correspond respectively to the two choices of signs of functional equations. Namely, if $\gcd(D,p)=1$, then the sign of the functional equation of $L(E_D,s)$ is $+1$ if and only if $D<0$. This follows because the theorem assumes that $E/\mathbb{Q}$ is an elliptic curve with conductor p with sign of the functional equation -1.

REMARKS. For simplicity, here we only offered a special case of these types of results with Theorem 19.20. The following remarks are intended for the interested reader who wishes to study the subject further.

(1) Section 7 of **[127]** gives the generalization of Theorem 19.20 which holds for arbitrary elliptic curves $E/\mathbb{Q}$, not just those with prime conductor and $\varepsilon=-1$. This generalization is phrased in terms of weight $1/2$ vector-valued harmonic Maass forms. In fact, this generalization includes central values and derivatives of L-functions for arbitrary weight 2 newforms, not just those associated to elliptic curves by Theorem 19.6.

(2) Although Theorem 19.20 ii) is not an identity, there is an identity which relates the Fourier coefficients $c_E^+(D)$ to periods. Work by Bruinier **[120]** establishes that

$$c_E^+(D)=\frac{\mathrm{Re}\left(\int_{C_{F_E}}\zeta_D(F_E)\right)}{\sqrt{D}\int_{C_{F_E}}\omega_{F_E}},$$

where $\zeta_D(F_E)$ is the normalized differential of the third kind for a certain divisor associated to F_E and $\omega_{F_E}:=2\pi iF_E(\tau)d\tau$. Here C_{F_E} is a generator of the F_E-isotypical component of the first homology of $X_0(N_E)$. The interested reader should consult **[120]** for further details.

(3) Theorem 19.20 and the work in **[127]** does not provide a canonical choice of the weight $1/2$ form $k_E(\tau)$. Alfes, Griffin, and two of the authors **[14]** have given a canonical choice of these forms for those elliptic curves $E/\mathbb{Q}$ with odd square-free conductor N_E. The construction makes use of the weight 0 harmonic Maass forms constructed in Section 6.2. Namely, a modified theta lift of the weight 0 harmonic Maass function Z_E-M_E in Theorem 6.4 serves as a suitable choice for k_E.

(4) In her Ph.D. thesis, Alfes **[11]** obtained a different proof of Theorem 19.20. Below, we sketch the original proof offered in **[127]**. This proof makes critical use of the generalized Borcherds products described in the previous chapter.

(5) There have been many important works extending and generalizing Theorem 19.20. The interested reader should consult papers by Bruinier, Howard, and Yang **[124, 132]**.

Theorem 19.20 is obtained by combining the Gross-Zagier formula, the generalized Borcherds products arising from harmonic Maass forms, and general transcendence theorems for canonical differentials of the third kind on modular curves.

SKETCH OF PROOF OF THEOREM 19.20. Theorem 19.20 i) follows from Theorem 19.15 and (19.16).

The proof of Theorem 19.20 ii) is more difficult, and it involves a study of Heegner divisors. We establish that the algebraicity of the coefficients $c_E^+(D)$ is dictated by the vanishing of certain twisted Heegner divisors in the Jacobian of $X_0(p)$. Combining this with the work of Gross and Zagier **[242]** and Scholl and Waldschmidt then implies Theorem 19.20 ii).

To make this precise, let $d < 0$ and $D > 0$ be fundamental discriminants which are both squares modulo p. Let $\mathcal{Q}_{d,p}$ be the set of discriminant $d = b^2 - 4ac$ integral binary quadratic forms $aX^2 + bXY + cY^2$ with the property that $p|a$. For these pairs of discriminants, we recall the twisted Heegner divisor, $Z_D(d)$, given by

$$Z_D(d) = \sum_{Q \in \mathcal{Q}_{Dd,p}/\Gamma_0(p)} \chi_D(Q) \frac{\tau_Q}{\omega_Q}, \tag{19.17}$$

where χ_D denotes the generalized genus character corresponding to the decomposition Dd as before, τ_Q is the unique root of $Q(X, 1)$ in $\mathbb{H}$ as before, and ω_Q denotes the order of the stabilizer of Q in $\Gamma_0(p)$. We have that $Z_D(d)$ is a divisor on $X_0(p)$ defined over $\mathbb{Q}(\sqrt{D})$. Using these twisted Heegner divisors, we define the degree 0 divisor

$$y_D(d) := Z_D(d) - \deg(Z_D(d)) \cdot \infty. \tag{19.18}$$

Finally, we assemble a divisor from k_E by letting

$$y_D(k_E) := \sum_{n<0} c_E^+(n) y_D(n) \in \mathrm{Div}^0(X_0(p)) \otimes F_E. \tag{19.19}$$

Let J be the Jacobian of $X_0(p)$, and let $J(F)$ denote the F-rational points of J. The Hecke algebra acts on $J(F) \otimes \mathbb{C}$, and the Mordell-Weil Theorem asserts that it is a finite-dimensional vector space.

It turns out that the point corresponding to $y_D(k_E)$ in $J(\mathbb{Q}(\sqrt{D})) \otimes \mathbb{C}$ is in its F_E-isotypical component. Moreover, the following can be shown to be equivalent:

i) The Heegner divisor $y_D(k_E)$ vanishes in $J(\mathbb{Q}(\sqrt{D})) \otimes \mathbb{C}$.
ii) The coefficient $c_E^+(D)$ is algebraic.
iii) The coefficient $c_E^+(D)$ is $\mathbb{Q}$-rational.

To obtain these results, one explicitly constructs modular functions with prescribed Heegner divisors using results related to Theorem 18.16. By work of Scholl **[446]** and Waldschmidt **[486]**, the vanishing of the points corresponding to these divisors is equivalent to the algebraicity of the Fourier expansions of these modular functions. Using the Hecke algebra and the explicit formulas provided by the generalized Borcherds product, one can show that this algebraicity is detectable from the single coefficient $c_E^+(D)$.

Generalizing a theorem of Gross, Kohnen, and Zagier **[239]**, it can be shown that

$$\sum_{n>0} y_D^{F_E}(-n) q^n = g_E(\tau) \otimes y_D(k_E) \in S_{\frac{3}{2}}^+(\Gamma_0(4p)) \otimes J\left(\mathbb{Q}\left(\sqrt{D}\right)\right),$$

where $y_D^{F_E}(-n)$ denotes the projection of $y_D(-n)$ onto its F_E-isotypical component (cf. Section 16.4). This result, when combined with Theorem 19.10, gives the conclusion that the Heegner divisor $y_D(k_E)$ vanishes in $J(\mathbb{Q}(\sqrt{D}))\otimes\mathbb{C}$ if and only if $L'(E_D,,1)=0$, thereby proving Theorem 19.20 ii) using the equivalence of the conditions i), ii), and iii) above. □

EXAMPLE 19.21. We consider the the relation for Heegner points of discriminant -139 on $X_0(37)$ found by Gross (cf. Section 4 of [**511**]). We let $N:=37, D:=-139$, and $r:=3$. In the notation from Chapter 18, we have that $L_{-139,3}/\Gamma_0(37)$ is represented by the quadratic forms

$$\begin{aligned} Q_1 &= [37,3,1], & Q_2 &= [185,151,31], & Q_3 &= [185,-71,7],\\ Q_1' &= [-37,3,-1], & Q_2' &= [-185,151,-31], & Q_3' &= [-185,-71,-7]. \end{aligned}$$

The corresponding points on $X_0(37)$ α_1, α_2, α_3 and α_1', α_2', α_3' satisfy

$$Z_{1,1}\left(-\tfrac{139}{4\cdot 37},\tfrac{3}{2\cdot 37}\right)=\alpha_1+\alpha_2+\alpha_3+\alpha_1'+\alpha_2'+\alpha_3',$$
$$Z_{-139,3}\left(-\tfrac{1}{4\cdot 37},\tfrac{1}{2\cdot 37}\right)=\alpha_1+\alpha_2+\alpha_3-\alpha_1'-\alpha_2'-\alpha_3'.$$

Gross proved that the function

$$r(\tau):=\frac{\eta(\tau)^2}{\eta(37\tau)^2}-\frac{3+\sqrt{-139}}{2}$$

on $X_0(37)$ has the divisor $(\alpha_1)+(\alpha_2)+(\alpha_3)-3(\infty)$. This implies that $\widetilde{r}(\tau)$, the image of $r(\tau)$ under complex conjugation, has the divisor

$$(\alpha_1')+(\alpha_2')+(\alpha_3')-3(\infty).$$

It turns out that $r(\tau)$ arises as a regularized theta lift. To illustrate this, let $f_{139}\in H_{\frac{1}{2},\rho_L}$ be the unique harmonic Maass form whose Fourier expansion is of the form

$$f_{139}=e\left(-\tfrac{139}{4\cdot 37}\tau\right)\mathfrak{e}_3+e\left(-\tfrac{139}{4\cdot 37}\tau\right)\mathfrak{e}_{-3}+O(e^{-\varepsilon v}),\qquad v\to\infty.$$

One can show that the dual space $S_{\frac{3}{2},\bar{\rho}_L}$ is one dimensional. Moreover, one can check that the coefficients with index $\frac{139}{4\cdot 37}$ of forms in this space must vanish. Therefore, we have that $\xi_{\frac{1}{2}}(f_{139})=0$, and so f_{139} is weakly holomorphic.

We now apply the theory of generalized Borcherds products. In this case, Theorem 18.15 implies that

$$\Psi_{1,1}(\tau,f_{139})=r(\tau)\widetilde{r}(\tau)\frac{\eta(37\tau)^2}{\eta(\tau)^2}. \tag{19.20}$$

On the other hand, one can also consider the unique harmonic Maass form $f_1\in H_{\frac{1}{2},\bar{\rho}_L}$ whose Fourier expansion satisfies

$$f_1=e\left(-\tfrac{1}{4\cdot 37}\tau\right)\mathfrak{e}_1-e\left(-\tfrac{1}{4\cdot 37}\tau\right)\mathfrak{e}_{-1}+O(e^{-\varepsilon v}),\qquad v\to\infty.$$

The dual space $S_{\frac{3}{2},\rho_L}$ is two dimensional. For a fixed λ_0 in the positive definite one dimensional sublattice $K\subset L$, the theta series

$$g_0(\tau)=\sum_{\lambda\in K'}(\lambda,\lambda_0)q^{Q(\lambda)}\mathfrak{e}_\lambda$$

is a non-zero element. Under the Shimura correspondence it is mapped to the Eisenstein series in $M_2^-(\Gamma_0(37))$. Here the superscript $-$ indicates the subspace of

$M_2(\Gamma_0(37))$ spanned by forms which are eigenfunctions of the Atkin-Lehner involution W_{37} with eigenvalue -1. Let g_1 be a generator of the orthogonal complement of g_0 in $S_{\frac{3}{2},\rho_L}$. Then g_1 is a Hecke eigenform which we choose to have rational coefficients. The Shimura lift of g_1 is the newform $F_E \in S_2^-(\Gamma_0(37))$, which corresponds to the conductor 37 elliptic curve

$$E\colon Y^2 = X^3 + 10X^2 - 20X + 8.$$

Its L-function has sign of functional equation $\varepsilon = 1$. One can easily check that $L(E,1) \neq 0$. The coefficients of g_1 with index $\frac{1}{4\cdot 37}$ do not all vanish, and so we find that

$$\xi_{\frac{1}{2}}(f_1) = c_0 g_0 + c_1 g_1,$$

with non-zero constants c_0 and c_1. So f_1 is not weakly holomorphic. Nevertheless, we may look at the twisted generalized Borcherds lift of f_1. We obtain that

$$\Psi_{-139,3}(\tau, f_1) = \frac{r(\tau)}{\widetilde{r}(\tau)} = \frac{\eta(\tau)^2 - \frac{3+\sqrt{-139}}{2}\eta(37\tau)^2}{\eta(\tau)^2 - \frac{3-\sqrt{-139}}{2}\eta(37\tau)^2}. \tag{19.21}$$

Strömberg computed a large number of coefficients of f_1 numerically. The first few coefficients of the holomorphic part of f_1 (indexed by the corresponding discriminants) are listed in Table 1. Details on the computations and some further results will be given in [**131**]. The rationality of the coefficients $c^+(139)$ and

TABLE 1. Coefficients of f_1

D	$c^+(-D)$
-3	$-0.32442836276932151751 8\ldots$
-4	$-0.25982119967765611249 0\ldots$
-7	$-0.43665675122666412619 5\ldots$
-11	$-0.13716672548383608172 0\ldots$
$\vdots$	$\vdots$
-136	$0.05357746688521873900 4\ldots$
-139	0
-151	$-0.27711896059755897348 8\ldots$
$\vdots$	$\vdots$
-815	$-0.35196562635635980371 4\ldots$
-823	-1
-824	$-0.66620220136583522452 5\ldots$

$c^+(823)$ corresponds to the vanishing of the Heegner divisors $Z_{-139,3}(-\frac{1}{4\cdot 37}, \frac{1}{2\cdot 37})$ and $Z_{-823,19}(-\frac{1}{4\cdot 37}, \frac{1}{2\cdot 37})$ in the Jacobian of $X_0(37)$.

To obtain an element of $H_{\frac{1}{2},\bar{\rho}_L}$ corresponding to g_1, we consider the unique harmonic Maass form $f_{12} \in H_{\frac{1}{2},\bar{\rho}_L}$ whose Fourier expansion is of the form

$$f_{12}(\tau) = e\left(-\tfrac{12}{4\cdot 37}\tau\right)\mathfrak{e}_{30} - e\left(-\tfrac{12}{4\cdot 37}\tau\right)\mathfrak{e}_{-30} + O\left(e^{-\varepsilon v}\right), \qquad v \to \infty.$$

Arguing as above we see that $\xi_{\frac{1}{2}}(f_{12})$ is a non-zero multiple of g_1. Table 2 includes some of the coefficients of f_{12} and the corresponding values of $L'(E_D,1)$ as numerically computed by Strömberg. We have that $L'(E_{-139},1) = L'(E_{-823},1) = 0$

as required by the Gross-Zagier Theorem 19.10. This is in agreement with the rationality of coefficients as dictated by the generalization of Theorem 19.20.

TABLE 2. Coefficients of f_{12}

D	$c^+(-D)$	$L'(E_D, 1)$
−3	1.026714911692035447445...	1.47929949207700...
−4	1.220536400967031662527...	1.81299789721820...
−7	1.690029746320007621414...	2.11071898017914...
−11	0.588499823548491754837...	3.65679089534028...
⋮	⋮	⋮
−136	−4.839267599344343782986...	5.73824076491330...
−139	−6	0
−151	−0.831356881792676920466...	6.69750855158616...
⋮	⋮	⋮
−815	121.9441031209309205888...	4.74925836934506...
−823	312	0
−824	−322.9986066040975056735...	17.5028741140542...

CHAPTER 20

Representation Theory and Mock Modular Forms

There is a rich history relating the areas of representation theory and modular forms, some of which we have already discussed at the end of Section 5.2. Other classical examples include the theory of integer partitions, which gave an early realization of the connection between modular forms and representations of the symmetric group.

This chapter offers an introduction to some of the recent applications of the theory of harmonic Maass forms to representation theory. These applications are born from a number of well-known works: the celebrated Macdonald identities **[371]**, the Weyl-Kac character formula **[288]**, the Lepowsky, Milne, and Wilson Lie-theoretic interpretation of the Rogers-Ramanujan identities **[348, 349, 350]**, and Monstrous Moonshine.

In Section 20.1, we begin by recalling the main results in the theory of Monstrous Moonshine. Loosely speaking, this theory shows that a natural representation of the Monster, the largest of the sporadic finite simple groups, is infinite dimensional and involves the arithmetic of modular functions. Similar roles for mock modular forms and representation theory largely remained unseen, until hints appeared recently in connection to the Mathieu group M_{24} and Umbral Moonshine, which we discuss in Section 20.3, and the affine Lie superalgebra $s\ell_2(m|n)^\wedge$, which we discuss in Section 20.2 – two very different contexts. In Section 20.2 we also see connections between representation theory and meromorphic Jacobi forms (cf. Chapter 11), the differential operators of which are related to universal enveloping algebras.

20.1. Monstrous Moonshine

One of the most beautiful and well-known results relating representations of groups and ordinary modular forms is given by "Monstrous Moonshine". We review the remarkable tale of Monstrous Moonshine in this section, in anticipation of the new "Umbral Moonshine" discussed in Section 20.3, which is related to mock modular forms. We recommend the excellent survey articles **[75]** and **[213]** by Borcherds and Gannon, respectively, for more detailed and more comprehensive treatments of Monstrous Moonshine than what follows.

The adjective "monstrous" arises from the *monster group* $\mathbb{M}$, the largest of the sporadic finite simple groups; its order is indeed monstrous, at a size of

$$\begin{aligned}&2^{46}\cdot 3^{20}\cdot 5^{9}\cdot 7^{6}\cdot 11^{2}\cdot 13^{3}\cdot 17\cdot 19\cdot 23\cdot 29\cdot 31\cdot 41\cdot 47\cdot 59\cdot 71\\&=808,017,424,794,512,875,886,459,904,961,710,757,005,754,368,000,000,000.\end{aligned}$$

In fact, Monstrous Moonshine began when Ogg **[401]** observed[1] that the set of prime factors of $|\mathbb{M}|$ is the same as the set of primes p for which the *Fricke group*

$$\Gamma_0^+(p) := \left\langle \Gamma_0(p), p^{-\frac{1}{2}} \begin{pmatrix} 0 & -1 \\ p & 0 \end{pmatrix} \right\rangle$$

has genus 0. Later, in **[154]**, Conway and Norton made a surprising conjecture, beginning with observations of McKay and Thompson, relating the dimensions of the irreducible representations of $\mathbb{M}$ to the Fourier coefficients of the modular j-function.

To describe this, let ρ_n denote the n-th smallest irreducible representation of $\mathbb{M}$, and let δ_n denote its dimension. Moreover, let $c(k)$ denote the coefficient of q^k in the Fourier expansion of the modular j-function

$$j(\tau) = q^{-1} + 744 + 196884q + \cdots .$$

McKay observed that $c(1) = \delta_1 + \delta_2$, that is,

$$196884 = 1 + 196883,$$

first relating these seemingly unrelated numbers. This relationship may seem striking, or perhaps it seems like "moonshine" (nonsense), as coined by Conway. The facts reveal, however, that further similar, intriguing, relationships between the dimensions δ_n and the coefficients $c(k)$ of the j-function exist: for example, McKay and Thompson also computed that

$$\begin{aligned} c(1) &= \delta_1 + \delta_2, & c(3) &= 2\delta_1 + 2\delta_2 + \delta_3 + \delta_4, \\ c(2) &= \delta_1 + \delta_2 + \delta_3, & c(4) &= 3\delta_1 + 3\delta_2 + \delta_3 + 2\delta_4 + \delta_5, \end{aligned}$$

where

$$\begin{aligned} \delta_1 &= 1, & c(-1) &= 1, \\ \delta_2 &= 196883, & c(1) &= 196884, \\ \delta_3 &= 21296876, & c(2) &= 21493760, \\ \delta_4 &= 842609326, & c(3) &= 864299970, \\ \delta_5 &= 18538750076, & c(4) &= 20245856256. \end{aligned}$$

It is interesting to note that at the time of these discoveries, the monster group $\mathbb{M}$ was not actually known to exist, although various (conjectural) statements about its structure, and character table, were known **[154, 231]**. Griess[2] ultimately proved its existence in 1980, seven years after independently conjecturing its existence and alleged character table with Fischer.

The observations above of McKay and Thompson can be interpreted as evidence of a grading, in which case dimension is the graded trace of the identity element. This led Thompson to surmise that studying the graded traces of other elements $g \in \mathbb{M}$ may also be of interest. To this end, after Thompson's suggestion, Conway and Norton **[154]** defined *head representations* H_n of the monster group $\mathbb{M}$ to be the representations suggested by the coefficients of j. That is, the sequence of

[1] Ogg also offered a bottle of Jack Daniels for an explanation.

[2] Griess was awarded the 2010 Leroy P. Steele Prize by the American Mathematical Society for this achievement.

representations which begins with

$$\begin{aligned} H_1 &= \rho_1 + \rho_2, \\ H_2 &= \rho_1 + \rho_2 + \rho_3, \\ H_3 &= 2\rho_1 + 2\rho_2 + \rho_3 + \rho_4, \\ \vdots \quad & \quad \vdots \end{aligned}$$

The Conway-Norton *McKay-Thompson series* for an element $g \in \mathbb{M}$ is defined by

$$T_g(q) := q^{-1} + \sum_{n=1}^{\infty} H_n(g) q^n,$$

so that $T_e(q) + 744$ should equal the modular j-function (where e is the identity in $\mathbb{M}$). It is interesting to note that there are 194 irreducible representations ρ_n of $\mathbb{M}$, while the coefficients $c(k)$ of the j-function grow exponentially. Rephrasing (and reindexing) it another way, there should be an infinite-dimensional graded module

$$V = \bigoplus_{n=0}^{\infty} V_n$$

for the Monster at work with characters given by the H_n as above. The head representations are associated to characters ch_{V_n}, as follows:

$$T_g(q) = q^{-1} \sum_{n=0}^{\infty} \mathrm{ch}_{V_n} q^n, \tag{20.1}$$

so that the module V should have graded dimension

$$\dim_V(q) = qT_e(q) = q(j(\tau) - 744).$$

With supporting computational evidence up to $O(q^{10})$ for every $g \in \mathbb{M}$, Conway and Norton **[154]** made the following famous conjecture.

CONJECTURE 20.1 (Monstrous Moonshine). *For each $g \in \mathbb{M}$, each McKay-Thompson series T_g is a Hauptmodul, that is, a generator of a genus zero modular function field for some subgroup $\Gamma_g \subseteq \mathrm{SL}_2(\mathbb{R})$ which contains a specific congruence subgroup as a normal subgroup.*

Atkin, Fong, and Smith **[468]** later verified that, indeed, a graded (virtual) infinite-dimensional module $V^\natural$ exists, in the sense of the prediction of Conway-Norton described above. Namely, they did not confirm the non-negativity of the multiplicities of each irreducible component of each V_n. By this time, Griess **[232, 233]** had constructed the monster group $\mathbb{M}$ by hand, by realizing it as a certain automorphism group. Thus, Thompson's conjecture was confirmed, though indirectly.

Moving forward in time, Frenkel, Lepowsky, and Meurman explicitly constructed a graded infinite-dimensional representation $V^\natural$ of $\mathbb{M}$ – a vertex operator algebra – in their deep work **[206, 207]**, and conjectured that it is the representation $V^\natural$ from the Conway-Norton conjectures above. In particular, the Frenkel-Lepowsky-Meurman $V^\natural$ should simultaneously have an automorphism group equal to $\mathbb{M}$, and have graded dimension equal to $q(j(\tau) - 744)$. Indeed, some of their key results are summarized in the following theorem **[206, 207]**.

THEOREM 20.2 (Frenkel, Lepowsky, and Meurman). *The Moonshine module $V^\natural$ (constructed explicitly) is a vertex operator algebra whose graded dimension is given by $q(j(\tau) - 744)$, which has automorphism group equal to $\mathbb{M}$.*

One might hope to prove the Monstrous Moonshine conjectures by somehow using the Frenkel-Lepowsky-Meurman $V^\natural$, combined with the fact that all Hauptmoduls satisfy *replication formulae*, meaning that their coefficients satisfy certain recursions. With respect to the j-function, the replication formulae can be simultaneously captured by the following beautiful infinite product identity observed by Koike, Norton, and Zagier (cf. **[41, 177, 328, 516]**). This was previously given in (13.1), but for completeness we recall that it states ($z, \mathfrak{z} \in \mathbb{H}$)

$$J(z) - J(\mathfrak{z}) = e^{-2\pi i z} \prod_{m>0,\, n\in\mathbb{Z}} \left(1 - e^{2\pi i m z} e^{2\pi i n \mathfrak{z}}\right)^{b(mn)},$$

where $b(n)$ denotes the n-th Fourier coefficient of $J = j - 744$. Conway and Norton conjectured analogous formulae for all McKay-Thompson series T_g, which would mean that each T_g is determined by only finitely many Fourier coefficients.

Finally, in 1992, Borcherds famously reconciled **[71]** these two objects, replication formulae and $V^\natural$, and approached the Conway-Norton conjectures by defining a Lie algebra $\mathfrak{m}$, the Monster Lie algebra, associated to the Frenkel-Lepowsky-Meurman $V^\natural$, and which inherits the action of $\mathbb{M}$ from $V^\natural$.

THEOREM 20.3 (Borcherds). *Let $V^\natural$ be the Frenkel-Lepowsky-Meurman Moonshine module, with automorphism group $\mathbb{M}$. Let T_g be given by (20.1) with $V = V^\natural$ for each $g \in \mathbb{M}$, and let Γ_g be as in Conjecture 20.1. Then T_g is the Hauptmodul for Γ_g.*

A remaining key to Borcherds' ultimate proof of Monstrous Moonshine came in the form of a denominator formula. One of the most well-known denominator formulas is Macdonald's identity for $A_1^{(1)}$,

$$\sum_{n\in\mathbb{Z}} (-1)^n \zeta^n q^{\frac{n^2}{2}} = \prod_{m=1}^{\infty} (1 - q^m)\left(1 - \zeta q^{m-\frac{1}{2}}\right)\left(1 - \zeta^{-1} q^{m-\frac{1}{2}}\right),$$

otherwise known as the Jacobi Triple Product Identity for the Jacobi theta function ϑ (cf. Lemma 2.3). It turns out that Borcherds' Monster Lie algebra $\mathfrak{m}$, associated to $V^\natural$, has denominator identity given by (13.1). By "twisting" this by each $g \in \mathbb{M}$, one obtains the other analogous conjectured replication identities for each T_g. Borcherds' celebrated proof of Monstrous Moonshine resulted from these denominator identities by comparing finitely many coefficients of finitely many T_g, one for each conjugacy class in $\mathbb{M}$.

To shed some light on the genus zero property of the McKay-Thompson series, one also has associated *modular equations*, which are satisfied by functions obeying replication formulae. In general, modular equations are two-variable polynomial relations satisfied by a function $f(\tau)$ and $f(m\tau)$ ($m \in \mathbb{N}$). It is well known that the Hauptmodul j satisfies a modular equation for each $m \in \mathbb{N}$ **[155]**, and in fact is the only "non-trivial" function to do so. More generally, we have the following theorem of Cummins and Gannon **[157]**.

THEOREM 20.4 (Cummins and Gannon). *A function $f(\tau) = q^{-1} + \sum_{n=1}^{\infty} b_n q^n$ obeys a modular equation for all $m \equiv 1 \pmod{N}$ if and only if it is either of the form $f(\tau) = q^{-1} + b_1 q$ or is a Hauptmodul for a modular group of Moonshine-type.*

Further generalizations and extensions of Monstrous Moonshine have since been explored (cf. [**177, 213**] for example), e.g. we have seen similar connections between modular-type objects and other sporadic groups, many of which are subgroups or quotients of the monster itself. With respect to mock modular forms, we discuss Umbral Moonshine at the end of this chapter in Section 20.3.

20.2. Kac-Wakimoto characters

An early appearance of mock modular forms in representation theory can be found in the 2001 work of Kac and Wakimoto [**290**]. Unlike Moonshine (discussed in Section 20.1), which is about the representation theory of finite groups, the work of Kac and Wakimoto which we discuss in this section is about affine Lie superalgebras (see also their earlier work [**289**]). In particular, Kac and Wakimoto studied characters $\operatorname{ch} L(\Lambda_{(s)}; m, n)$ and specialized characters $\operatorname{tr}_{L(\Lambda_{(s)};m,n)} q^{L_0}$ for irreducible highest weight $s\ell(m|n)^\wedge$-modules $L(\Lambda_{(s)})$ with highest weight $\Lambda_{(s)}$ for integers $m \geq 2$, $n \geq 1$, and $s \in \mathbb{Z}$. Interestingly, the modular-like properties of these characters, and their proofs, vary depending on the relationship between m and n. We treat the four cases ($n = 1$ and $m \geq 2$), ($m > n$), ($m < n$), and ($m = n$) in Subsections 20.2.1–20.2.4 below. In Subsection 20.2.5, we also discuss the more recent works of Kac-Wakimoto [**291, 292, 293**], which establish mock-modular-like properties of numerous other types of supercharacters.

20.2.1. The case with $n = 1, m \geq 2$. In the special case $(m, n) = (2, 1)$, Kac and Wakimoto's work provided a first connection to mock modular forms and the representation theory of $s\ell(2|1)^\wedge$ via the following character formula (cf. [**290**, (4.2)]):

$$\operatorname{ch} L(\Lambda_{(s)}; 2, 1) = e^{\Lambda_0 + s\varepsilon_3} q^{-\frac{s}{2}} \frac{(-\xi_1 q; q)_\infty (-\xi_1^{-1}; q)_\infty}{(q; q)_\infty^2} \sum_{k \in \mathbb{Z}} \frac{q^{\frac{k(k+1)}{2}} \xi_2^k}{1 + \xi_1^{-1} q^{k-s}}. \tag{20.2}$$

Above, Λ_0 is a fundamental weight, and ε_3 is a basis element of $\mathfrak{h}^*$, where $\mathfrak{h}$ is the Cartan subalgebra of $g\ell(2|1)$ consisting of all diagonal matrices (cf. [**290**, Section 2]). The sum on the right-hand side of (20.2) can be written in terms of a level one Appell sum A_1 (cf. Definition 8.19), and the factor before it is a Jacobi form (up to a simple multiplicative factor and after suitable changes of variables), which can be seen from Lemma 2.3 and (2.2). Hence, these $s\ell(2|1)^\wedge$ characters may be viewed as *mixed mock Jacobi forms,* using work of Zwegers (Theorem 8.13), which emerged at roughly the same time as the aforementioned work of Kac-Wakimoto. Mixed mock Jacobi forms are two-variable functions which exhibit appropriate Jacobi transformation properties, and they are defined in an analogous way to mixed mock modular forms (cf. Definition 13.1).

For $n = 1$ and integers $m \geq 2$, Kac and Wakimoto also obtained explicit character formulas for the specialized $s\ell(m|1)^\wedge$-characters $\operatorname{tr}_{L(\Lambda_{(s)};m,1)} q^{L_0}$ in terms of theta functions and the multivariable Appell sum ($N \in \mathbb{N}, \boldsymbol{k} = (k_1, k_2, \ldots, k_N)$, $|q| < 1$)

$$A_{B,\mathcal{L}}(w; \xi_1, \xi_2, \ldots, \xi_N; q) := \sum_{\boldsymbol{k} \in \mathbb{Z}^N} \frac{q^{\frac{1}{2}\boldsymbol{k}^T B \boldsymbol{k}} \xi_1^{k_1} \xi_2^{k_2} \cdots \xi_N^{k_N}}{1 + w q^{\mathcal{L}(\boldsymbol{k})}},$$

which is defined in terms of an $N \times N$ symmetric matrix B such that $\operatorname{Re}(B)$ is positive definite, and a linear function $\mathcal{L}$ on $\mathbb{C}^N$. If $N = 1$, then under suitable hypotheses, this function can be related to a level ℓ Appell sum A_ℓ from Definition 8.19. In particular, for integers $m \geq 2$, and $s \in \mathbb{Z}$, Kac-Wakimoto established the following specialized character formula (cf. [**290**, (4.7)])

$$\begin{aligned}(20.3)\qquad \operatorname{tr}_{L(\Lambda_{(s)};m,1)} q^{L_0} &= 2q^{-\frac{s}{2}} \frac{\left(q^2;q^2\right)_\infty^2}{(q;q)_\infty^{m+2}} A_{I_{m-1},|\cdot|}\left(q^{-s};q^{\frac{1}{2}},q^{\frac{1}{2}},\ldots,q^{\frac{1}{2}};q\right)\\ &= 2q^{-\frac{s}{2}} \frac{\left(q^2;q^2\right)_\infty^2}{(q;q)_\infty^{m+2}} \sum_{\boldsymbol{k}\in\mathbb{Z}^{m-1}} \frac{q^{\frac{1}{2}\sum_{j=1}^{m-1} k_j(k_j+1)}}{1+q^{|\boldsymbol{k}|-s}},\end{aligned}$$

where I_N denotes the $N \times N$ identity matrix, and for an N-dimensional vector $\boldsymbol{k}$, $|\boldsymbol{k}| := k_1 + k_2 + \cdots + k_N$. The operator L_0 is the *energy operator,* or *Hamiltonian* (cf. [**290**, Section 3]).

In view of Zwegers' Ph.D. thesis [**529**] and further developments in the overarching theory of harmonic Maass forms at that time [**520**], Kac (in a private communication to one of the authors) asked whether there may be a connection between the multivariable sum characters (20.3) and harmonic Maass forms. Two of the authors answered this question in [**109**]. Their results have since been generalized to characters $\operatorname{tr}_{L(\Lambda_{(s)};m,n)}$, where $m > n$, as we shall discuss in Subsection 20.2.2. However, the methods used in the general case $m > n$ to prove Theorem 20.8 are different from those originally used in [**109**] in the special case $n = 1$; for the sake of the interested reader, we discuss those methods briefly here.

To describe this, for $r \in \mathbb{N}$ and $s \in \mathbb{Z}$, let

$$K_{r,s}(\tau) := q^{-\frac{s}{2}} \frac{\eta^2(2\tau)}{\eta^{r+3}(\tau)} A_{I_r,|\cdot|}\left(q^{-s};q^{\frac{1}{2}},q^{\frac{1}{2}},\ldots,q^{\frac{1}{2}};q\right)$$

(where $q = e^{2\pi i\tau}, \tau \in \mathbb{H}$), so that by (20.3), we have

$$(20.4)\qquad K_{m-1,s}(\tau) = \frac{q^{\frac{2-m}{24}}}{2} \operatorname{tr}_{L(\Lambda_{(s)};m,1)} q^{L_0}.$$

The function $K_{r,s}$ is completed in [**109**] as follows:

$$(20.5)\qquad \widehat{K}_{r,s}(\tau) := q^{-\frac{s^2}{2r}} K_{r,s}(\tau) - 2^{r-1} q^{-\frac{s^2}{2r}} \frac{\eta^{2r+2}(2\tau)}{\eta^{2r+3}(\tau)} R(-s\tau; r\tau),$$

where the function $R(z;\tau)$ is the nonholomorphic function defined by Zwegers (cf. (8.1)). The main result from [**109**] is the following.

THEOREM 20.5. *If $r \in \mathbb{N}$ and $s \in \mathbb{Z}$, then $\widehat{K}_{r,s}$ is a nonholomorphic modular function.*

Using more recent terminology, Theorem 20.5 and (20.4) show that the Kac-Wakimoto characters $\operatorname{tr}_{L(\Lambda_{(s)};m,1)} q^{L_0}$ are mixed mock modular functions (i.e., weight 0), up to multiplication by a suitable power of q. In fact, $\eta^{2r+2}(2\tau)/\eta^{2r+3}(\tau)$ is non-vanishing on $\mathbb{H}$, so that

$$\frac{\eta^{2r+3}(\tau)}{\eta^{2r+2}(2\tau)} \widehat{K}_{r,s}(\tau)$$

is a weight $1/2$ harmonic Maass form.

SKETCH OF PROOF OF THEOREM 20.5. For full details, we refer the interested reader to [**109**]. In short, the result follows by first rewriting the multivariable Appell sum $A_{I_r,|\cdot|}$ from (20.3) as a sum over a finite cone of integers of Zwegers' one-variable level one Appell sums A_1 (cf. Definition 8.19), and then using their associated completions and transformation properties (cf. Theorem 8.22), as well as modular transformation properties of the factor $\eta^2(2\tau)/\eta^{r+3}(\tau)$. To be more precise, after a series of changes of variables in the indices of summation, it is shown in Theorem 2.1 of **[109]** that

(20.6)
$$A_{I_r,|\cdot|}\left(q^{-s};q^{\frac{1}{2}},q^{\frac{1}{2}},\ldots,q^{\frac{1}{2}};q\right)=-i\sum_{\boldsymbol{d}\in D_r}\widetilde{\vartheta}_{\boldsymbol{d}}(\tau)q^{\frac{d_r^2}{2r}+\frac{s}{2}}A_1\left(\tfrac{1}{2}-(d_r+s)\tau,\tfrac{1}{2}-d_r\tau;r\tau\right),$$

where the finite cone D_r is defined by

$$D_r:=\{\boldsymbol{d}\in\mathbb{N}^r:0\leq d_j\leq j-1\text{ for each }1\leq j\leq r\},$$

and the function $\widetilde{\vartheta}_{\boldsymbol{d}}$ is (up to multiplication by q-powers) a product of Jacobi theta functions. Precisely, $\widetilde{\vartheta}_{\boldsymbol{d}}$ is defined by

$$\widetilde{\vartheta}_{\boldsymbol{d}}(\tau):=\prod_{j=1}^{r-1}\vartheta(d_j,-d_{j+1},j,j+1;\tau),$$

where the function $\vartheta(a,b,c,d;\tau)$ is given by

$$\vartheta(a,b,c,d;\tau):=q^{\frac{cd}{2}\left(\frac{a}{c}+\frac{b}{d}-\frac{1}{2}\right)^2}\vartheta\left(\left(ad+bc-\tfrac{cd}{2}\right)\tau-\tfrac{1}{2};cd\tau\right),$$

in terms of the Jacobi theta function from Definition 2.1. Using Theorem 2.15, we have that $\vartheta(a,b,c,d;\tau)$ and $\widetilde{\vartheta}_{\boldsymbol{d}}$ are modular forms. Using Theorem 8.22, one can then complete $K_{r,s}$ using (20.6), a finite sum, to the nonholomorphic modular form $\widehat{K}_{r,s}$ in (20.5). □

Theorem 20.5 was refined in **[196**, Theorem 1.1], where it was shown that up to multiplication and addition by modular forms, the specialized Kac-Wakimoto characters are equal to specializations of universal mock theta functions g_2 and g_3 of Gordon-McIntosh (cf. Definition 9.9).

THEOREM 20.6. *For $s\in\mathbb{Z}$ and integers $m\geq 2$, we have that*

$$g_2\left(q^{\frac{m-1}{4}-\frac{s}{2}};q^{\frac{m-1}{2}}\right)=-2^{-m}q^{\frac{5}{24}-\frac{m}{6}}\frac{\eta^{2m+1}(\tau)}{\eta^{2m}(2\tau)}\operatorname{tr}_{L(\Lambda_{(s)};m,1)}q^{L_0}-f_{m,s}(\tau),$$

where the function $f_{m,s}$, explicitly given in **[196]**, *is a modular form of weight $1/2$ (up to multiplication by a power of q).*

REMARK. Further results in **[196]** also related the characters $\operatorname{tr}_{L(\Lambda_{(s)};m,1)}q^{L_0}$ to the universal mock theta function g_3 (cf. [**196**, Theorem 1.1 (ii), (iii)]). Moreover, under certain hypotheses on m and s, the characters reduce to ordinary modular forms (cf. [**196**, Corollary 1.2]).

Due to the fact **[228]** that each of Ramanujan's original mock theta functions can be expressed in terms of the universal mock theta functions g_2 or g_3 (see the Appendix), Theorem 20.6 and its counterparts in **[196]** pertaining to g_3 immediately lead to the following corollary: every original mock theta function of Ramanujan can be expressed in terms of Kac-Wakimoto characters $\operatorname{tr}_{L(\Lambda_{(s)};m,1)}q^{L_0}$. A complete dictionary giving a character formula for every original mock theta function of Ramanujan, as well as certain subsequent generalizations, can be found in the Appendix of **[196]**. Here, we illustrate this with the following example.

EXAMPLE 20.7. Recall the mock theta functions f and ϕ, defined by

$$f(q) = \sum_{n=0}^{\infty} \frac{q^{n^2}}{(-q;q)_n^2}, \qquad \phi(q) = \sum_{n=0}^{\infty} \frac{q^{\frac{n(n+1)}{2}}}{(q;q^2)_{n+1}}.$$

The third order mock theta function f originated in Ramanujan's last letter to Hardy, and the tenth order mock theta function ϕ appeared in Ramanujan's "lost" notebook. In terms of Kac-Wakimoto characters, we have that

$$f(-q) = \frac{q^{-\frac{23}{24}}}{2^{11}} \frac{\eta(\tau)^{27}}{\eta(2\tau)^{26}} \left(q^{\frac{1}{2}} \operatorname{tr}_{L(\Lambda_{(1)};13,1)} q^{L_0} + q^{-\frac{1}{2}} \operatorname{tr}_{L(\Lambda_{(5)};13,1)} q^{L_0} \right) + F_f(\tau),$$

$$\phi(q) = -\frac{q^{-\frac{5}{8}}}{2^{10}} \frac{\eta(\tau)^{23}}{\eta(2\tau)^{22}} \operatorname{tr}_{L(\Lambda_{(1)};11,1)} q^{L_0} + F_\phi(\tau),$$

where F_f and F_ϕ are weight 1/2 modular forms (up to multiplication by powers of q), explicitly given in **[196]**.

SKETCH OF PROOF OF THEOREM 20.6. We begin by applying (20.4), which relates the Kac-Wakimoto characters to the multivariable Appell sums $A_{I_{m-1},|\cdot|}$, followed by (20.6), which expresses these multivariable Appell sums in terms of level one Appell sums A_1. By definition, the level one Appell sums A_1 are, up to multiplication by the Jacobi theta function, equal to Zwegers' μ-functions (see the remark following Definition 8.19). Thus, the Kac-Wakimoto characters may be written in terms of Zwegers' μ-functions. To further re-write the characters in terms of the universal mock theta function g_2, we use Theorem 9.12 from Chapter 9 (due to Kang **[299]**), which under certain hypotheses expresses Zwegers' μ-function in terms of the universal mock theta function g_2. Theorem 20.6 eventually follows from these steps after lengthy but straightforward calculations, and simplifications of various mock and modular terms which emerge. □

20.2.2. The case with $m > n$. In this section we outline some methods and results from **[86]**. Unlike the case of $n = 1$ described in Subsection 20.2.1, one does not have the luxury of beginning with a multivariable Appell-Lerch sum expression for $\operatorname{tr}_{L(\Lambda_{(\ell)};m,n)} q^{L_0}$ for arbitrary n (and any $m > n$) as in (20.3). Instead, we consider the generating function $\mathrm{ch}F$ for the characters $\mathrm{ch}F_\ell$ given by Kac-Wakimoto:

(20.7)

$$\mathrm{ch}F := \sum_{\ell \in \mathbb{Z}} \mathrm{ch}F_\ell \zeta^\ell = e^{\Lambda_0} \prod_{k=1}^{\infty} \frac{\prod_{\ell=1}^{m} \left(1 + \zeta w_\ell q^{k-\frac{1}{2}}\right)\left(1 + \zeta^{-1} w_\ell^{-1} q^{k-\frac{1}{2}}\right)}{\prod_{j=1}^{n} \left(1 - \zeta w_{m+j} q^{k-\frac{1}{2}}\right)\left(1 - \zeta^{-1} w_{m+j}^{-1} q^{k-\frac{1}{2}}\right)},$$

where $\mathrm{ch}F = \mathrm{ch}F(z;\tau)$, $\mathrm{ch}F_\ell = \mathrm{ch}F_\ell(m,n;\tau)$ (with $q = e^{2\pi i \tau}, \zeta = e^{2\pi i z}$), and e^{Λ_0} and the w_s are certain operators, which are of significance in representation theory (cf. (3.15) of **[290]**). To simplify the situation, we set e^{Λ_0} and all w_s equal to 1, also guaranteeing convergence of (20.7). Something to note about the coefficient functions $\mathrm{ch}F_\ell$ is that they depend upon the range in which ζ is taken. A wall-crossing phenomenon occurs when this range is moved, a fact which is explained in work of Dabholkar, Murthy and Zagier **[158]** (cf. Section 11.2). The specialized characters $\operatorname{tr}_{L(\Lambda_{(\ell)};m,n)} q^{L_0}$ are related to the characters $\mathrm{ch}F_\ell$ under the aforementioned basic specialization by **[290]**

$$\mathrm{tr}_{L(\Lambda_{(\ell)};m,n)} q^{L_0} = \mathrm{ch}F_\ell \prod_{k=1}^{\infty}\left(1-q^k\right). \tag{20.8}$$

Using this relationship and the modularity of the η-function (defined in (2.2)), it is not difficult to deduce the automorphic properties of the specialized characters $\mathrm{tr}_{L(\Lambda_{(\ell)};m,n)} q^{L_0}$ from those of the $\mathrm{ch}F_\ell$. At first glance this appears promising because the generating function $\mathrm{ch}F$ for the $\mathrm{ch}F_\ell$ is essentially a Jacobi form. Precisely, with the same specializations mentioned above, we have that

$$\mathrm{ch}F = \Phi\left(z+\frac{\tau}{2};\tau\right)(-1)^m i^{-n}\zeta^{\frac{m-n}{2}} q^{\frac{m-n}{6}}\eta(\tau)^{n-m}, \tag{20.9}$$

with

$$\Phi(z;\tau) = \Phi_{m,n}(z;\tau) := \frac{\vartheta\left(z+\frac{1}{2}\right)^m}{\vartheta(z)^n}. \tag{20.10}$$

In light of (20.9), one may hope to appeal to methods from Chapter 2 to derive the modular properties of the $\mathrm{ch}F_\ell$, which are essentially the Fourier coefficients of a Jacobi form.

However, when viewed as a Jacobi form, the function in (20.9) is not holomorphic, so the theory developed in **[191]** (explained in Chapter 2) does not directly apply. It is meromorphic, yet its poles may be of arbitrarily high order n (where n is a fixed positive integer).

As outlined in Section 11.2, in **[86]**, two of the authors extended the work of Dabholkar-Murthy-Zagier **[158]** and Zwegers **[529]**, who developed some of the theory of meromorphic Jacobi forms with poles of order $n=1$ and $n=2$. In doing so, one encounters almost harmonic Maass forms (not appearing in **[158]** or **[529]**), as given by Definition 11.1. These functions generalize both harmonic Maass forms as defined in Chapter 4, and almost holomorphic modular forms defined by Kaneko and Zagier **[298]**. They are explicitly described in Definition 11.1, and are discussed in the context of Jacobi forms of positive index in Section 11.2. Loosely speaking, such functions may be viewed as sums of harmonic Maass forms under iterates of the raising operator (themselves therefore non-harmonic weak Maass forms), multiplied by almost holomorphic modular forms. In particular harmonic Maass forms may be viewed as almost harmonic Maass forms of depth $r=1$, and almost holomorphic modular forms may be viewed as almost harmonic Maass forms of depth $r=0$ or $r=1$. It turns out that the Kac-Wakimoto characters are holomorphic parts of almost harmonic Maass forms and are thus referred to as almost mock modular forms (cf. Section 11.2). For simplicity, the following theorem is restricted to the case of even integers m and n satisfying $m>n\geq 2$.

THEOREM 20.8. *Assume the restrictions above. The Kac-Wakimoto characters* $\mathrm{tr}_{L_{m,n}(\Lambda_{(\ell)})} q^{L_0}$ *are almost mock modular forms of weight* 0 *and depth* $n/2$.

The results and methods from **[86]** relating to Theorem 20.8 have since been generalized, and this is what is described in Section 11.2. In particular, Theorem 20.8 is proved in **[86]** exactly as in the proof of Theorem 11.4, for the specific Jacobi form arising from the product in (20.7).

REMARK. We point out that when $n = 2$, the proof of Theorem 20.8 shows that the Kac-Wakimoto characters have a simple description as mixed mock modular forms. Explicit examples in the cases $n = 2$ and $n = 4$ are given in [**86**, Section 4].

20.2.3. The case with $m < n$. If $m < n$, the Jacobi form $\Phi = \Phi_{m,n}$ in (20.10) which is essentially equal to the generating function for the Kac-Wakimoto characters is of negative index $\frac{m-n}{2}$. This seemingly subtle difference leads to a fairly significant change in the automorphic properties of the characters. This is described in [**81**] by Creutzig and two of the authors, and more generally in [**117**] by Zwegers and two of the authors.

Define the functions $\chi_{(m,n;r)}$ and $D_j = D_{j,m,n}$ by

$$\Phi_{m,n}(z;\tau) =: \sum_{r\in\frac{m-n}{2}+\mathbb{Z}} \chi_{(m,n;r)}(\tau)\zeta^r \tag{20.11}$$

$$=: \frac{D_n(\tau)}{(2\pi i z)^n} + \frac{D_{n-2}(\tau)}{(2\pi i z)^{n-2}} + \cdots + O(1). \tag{20.12}$$

In light of (20.9), in order to study the automorphic properties of the Kac-Wakimoto characters $\mathrm{ch}F_\ell$ when $m < n$ (or the $\mathrm{tr}_{L(\Lambda_{(\ell)};m,n)}q^{L_0}$ by (20.8)), it is enough to determine the automorphic properties of the functions $\chi_{(m,n;r)}$ from (20.11).

To describe this, we let δ_e be equal to 0 or 1, depending on whether n is odd or even, and let

$$\rho(r) = \rho_n(r) := \begin{cases} r & \text{if } r \geq \frac{n}{2}, \\ n - r & \text{if } r < \frac{n}{2}. \end{cases}$$

We also define the partial theta functions

$$\Theta_{\frac{1}{2}+\delta_e}(n,r;\tau) := \sum_{j=0}^{\infty}(-1)^{jn}\left(j+\frac{r}{n}\right)^{\delta_e} q^{\frac{n}{2}\left(j+\frac{r}{n}\right)^2}.$$

Using the automorphic properties of the functions $\Phi_{m,n}(z;\tau)$, the following theorem is established in [**81**] for τ satisfying $0 \leq y < v$, where $y = \mathrm{Im}(z)$, $v = \mathrm{Im}(\tau)$. Here, we let $\mathcal{D}_q := q\frac{\partial}{\partial q}$ (so as not to be confused with the functions D_j).

THEOREM 20.9. *For any $n \in \mathbb{N}, m \in \mathbb{N}_0, r \in \frac{N}{2} + \mathbb{Z}$, we have that*

$$\chi_{(2m,n+2m;r)}(\tau)$$
$$= (-1)^{1+\delta_e} q^{-\frac{r^2}{2n}} \sum_{j=0}^{\frac{n-1-\delta_e}{2}+m} \frac{D_{2j+\delta_e+1}(\tau)}{(2j+\delta_e)!} n^{j+\delta_e} 2^j \mathcal{D}_q^j \left(\Theta_{\frac{1}{2}+\delta_e}(n,\rho(r);\tau)\right).$$

Using Theorem 20.9, one can deduce the automorphic properties of the functions $\chi_{(m,n;r)}$, and hence, the automorphic properties of the Kac-Wakimoto characters $\mathrm{tr}_{L(\Lambda_{(\ell)};m,n)}q^{L_0}$. Firstly, we point out that it is not difficult to show using the Jacobi properties of $\Phi_{m,n}$ that the functions D_j defined by (20.12) are quasimodular forms. Explicitly, they can be computed in terms of the usual Eisenstein series G_k [**81, 86**]. Secondly, as discussed in Section 21.5, partial theta functions like those appearing in Theorem 20.9 are often quantum modular forms, a notion that we describe more precisely in Section 21.1. In particular, the quantum modularity of the partial theta functions $\Theta_{\frac{1}{2}}(n,r;\tau)$ ($\delta_e = 0$) in Theorem 20.9 is established in [**203**], and the quantum modularity of the $\Theta_{\frac{3}{2}}(n,r;\tau)$ ($\delta_e = 1$) is shown in [**81**].

Combining these facts with Theorem 20.9 shows how the Kac-Wakimoto characters in the case of $m < n$, i.e., the case in which their generating function is a negative index meromorphic Jacobi form, are expressed in terms of quasimodular forms and derivatives of quantum modular forms. As explained in [**117**] and mentioned in Section 11.3, it is not difficult to deduce a similar result for the Fourier coefficients of general negative index meromorphic Jacobi forms using Theorem 11.7.

We now discuss applications of some of these more general results from Section 11.3, relating to the Kac-Wakimoto characters with $m < n$, and some well-known combinatorial functions.

We emphasize the very special case when $m = 0$. In this case, we obtain equations relating important combinatorial objects, namely the *ranks* and *cranks* of partitions, which are discussed in Section 14.3, and whose generating functions are denoted by $R(\zeta; q)$ and $C(\zeta; q)$, defined in (14.9) and (14.8), respectively. We immediately see that

$$\Phi_{0,n}(z;\tau) = \frac{i^n \zeta^{\frac{n}{2}} q^{-\frac{n}{24}}}{\eta(\tau)^{2n}(1-\zeta)^n} C(\zeta;q)^n.$$

If $n = 1$, one could see Theorem 11.7 by direct calculation. In particular, by making use of a classical partial fraction identity for the crank generating function, which can be proven using a standard application of the Mittag-Leffler theorem, one can re-write C as in (14.8):

$$C(\zeta;q) = \frac{1-\zeta}{(q)_\infty} \sum_{n\in\mathbb{Z}} \frac{(-1)^n q^{\frac{n(n+1)}{2}}}{1-\zeta q^n}. \tag{20.13}$$

To see how the result of Theorem 11.7 is obtained in this case, we may perform a geometric series expansion on the denominator of (20.13) in an appropriate range.

We next consider the case of $n = 3$, which is of great historical interest. In this case, Theorem 11.5 is equivalent to the famous *rank-crank* PDE of Atkin and Garvan [**44**], which was given in Theorem 14.28. Recall that this states that

$$2\eta(\tau)^2 C^*(\zeta;q)^3 = \left(6\mathcal{D}_q + \mathcal{D}_\zeta^2\right) R^*(\zeta;q), \tag{20.14}$$

where for convenience we have used the normalized rank and crank generating functions R^* and C^* defined in (14.26). Hence, Theorem 11.5 can be viewed as providing the natural framework generalizing (20.14). Generalizations of (20.14) in a related context were also considered in [**81, 117, 139, 530**].

20.2.4. The case with $m = n$. If $m = n$, the automorphic properties of the Kac-Wakimoto characters are of a different, arguably simpler, type. To describe this, we define the functions $\chi_{(m,m;r)}$ as in (20.11) and the functions $D_j = D_{j,m,m}$ as in (20.12). In this case, we also make use of the implied $O(1)$ function from (20.12), which we denote by $D_0(\tau)$. The functions D_j in the case of $m = n$ are simply ordinary modular forms (which can be explicitly computed), as opposed to quasimodular forms as in Subsection 20.2.3, a fact which is explained in [**87**].

The following theorem of Mahlburg and two of the authors was established in [**87, 88**]. For simplicity, the result is stated for even $m \geq 2$; similar methods lead to a similar result for odd m. The functions E_{2j} appearing in this theorem are the usual Eisenstein series of weight $2j$.

THEOREM 20.10. *For $r \in \mathbb{Z}$, we have that*

$$\chi_{(m,m;r)}(\tau) = \frac{1}{1-q^r}\sum_{j=1}^{\frac{m}{2}} \frac{r^{2j-1}}{(2j-1)!}D_{2j}(\tau), \quad r \neq 0,$$

$$\chi_{(m,m;0)}(\tau) = D_0(\tau) + \sum_{j=1}^{\frac{m}{2}} \frac{B_{2j}}{(2j)!}D_{2j}(\tau)E_{2j}(\tau),$$

where for each $0 \leq j \leq \frac{m}{2}$, the function D_{2j} is a modular form of weight $-2j$.

Here, as usual, B_{2j} is used to denote the Bernoulli numbers. The proof of Theorem 20.10 is similar to the proofs of the results discussed in Subsections 20.2.2 and 20.2.3, and in Chapter 11, so omitted here for brevity.

20.2.5. Additional supercharacters. Recent results of Kac and Wakimoto [**291, 292, 293**] also showed how numerous other types of supercharacters can be modified to obtain modular invariant functions. Namely, the authors studied principal admissible modules over $\widehat{s\ell}(2|1)$ and $\widehat{ps\ell}(2|2)$ in [**291**], further integrable atypical modules over $\widehat{s\ell}(2|1)$ and $\widehat{\mathrm{osp}}(3|2)$ in [**292**], and a more general setting involving tame integrable modules over an affine Lie superalgebra associated to an arbitrary basic Lie superalgebra in [**293**]. Like some other results described in this chapter, results from [**291, 292, 293**] also used a Zwegers-type completion to correct the erroneous modular transformation properties of the associated characters. In particular, as in the aforementioned work [**86**], Kac and Wakimoto made use of the nonholomorphic function $R_{M,\ell}$ from (11.5). In most of the cases studied in [**291, 292, 293**], Kac and Wakimoto showed that their completed characters nicely exhibit weight 0 vector-valued Jacobi transformation properties on $\mathrm{SL}_2(\mathbb{Z})$, similar to the transformations induced by the Weil representation (see Proposition 2.22 for such a transformation in the weight 1/2 setting). The papers [**291, 292, 293**] are rich with interesting results of this type, and we refer the reader there for technical hypotheses and explicit details.

20.3. Umbral Moonshine

Since the time of Monstrous Moonshine, mock modular forms have emerged within a similar framework. The tale of mock modular Moonshine began with Eguchi, Ooguri, and Tachikawa [**184**], who made similar observations to those of McKay and Thompson, as described in Section 20.1. In this case, the largest sporadic simple Mathieu group M_{24} plays the role of the monster group $\mathbb{M}$. In [**184**], the authors observed that dimensions of representations of M_{24} are the multiplicities of superconformal algebra characters in the $K3$ elliptic genus.

To describe this, we first recall some background on complex $K3$ surfaces, which are certain compact connected dimension 2 complex manifolds M (see [**48**] for further background). In general, the elliptic genus of a complex manifold of dimension d is defined to be

$$Z_M = Z_M(z;\tau) := \int_M \mathrm{ch}(\mathbb{E}_{q,\zeta})\mathrm{td}(M),$$

where $\mathrm{td}(M)$ is the Todd class of M and $\mathrm{ch}(\mathbb{E}_{q,\zeta})$ is the Chern character of the power series

$$\mathbb{E}_{q,\zeta} := \zeta^{\frac{d}{2}} \bigotimes_{n=1}^{\infty} \left(\bigwedge\nolimits_{-\zeta^{-1}q^{n-1}} T_M \otimes \bigwedge\nolimits_{-\zeta q^n} T_M^* \otimes \bigvee\nolimits_{q^n} T_M \otimes \bigvee\nolimits_{q^n} T_M^* \right),$$

defined in terms of the holomorphic tangent bundle T_M and its dual T_M^*. The coefficients of this power series are obtained as sums of tensor products of the exterior and symmetric powers of T_M and T_M^*. It is well known **[238]** that the elliptic genus of a complex $K3$ surface transforms like a Jacobi form of weight 0 and index 1 in the usual variables $z \in \mathbb{C}$ and $\tau \in \mathbb{H}$, when q and ζ are specialized to $e^{2\pi i \tau}$ and $e^{2\pi i z}$, respectively. That is, for $\left(\begin{smallmatrix} a & b \\ c & d \end{smallmatrix}\right) \in \mathrm{SL}_2(\mathbb{Z})$ and $\lambda, \mu \in \mathbb{Z}$,

$$Z_M(z;\tau) = e^{-2\pi i \frac{cz^2}{c\tau+d}} Z_M\left(\frac{z}{c\tau+d}; \frac{a\tau+b}{c\tau+d}\right),$$
$$Z_M(z;\tau) = e^{2\pi i(\lambda^2\tau+2\lambda z)} Z_M\left(z+\lambda\tau+\mu;\tau\right).$$

The space of Jacobi forms to which Z_M belongs is one dimensional, so Z_M is independent of M and can be given explicitly in terms of Jacobi theta functions **[76, 185, 304]**.

The expansion of the $K3$ elliptic genus $Z_M(z;\tau)$ in terms of irreducible representations of $\mathcal{N} = 4$ superconformal algebras led to the mock modular form **[182, 184, 186]**

$$\begin{aligned} (20.15) \quad & -8i\Big(\mu\left(\tfrac{1}{2},\tfrac{1}{2};\tau\right) + \mu\left(\tfrac{1+\tau}{2},\tfrac{1+\tau}{2};\tau\right) + \mu\left(\tfrac{\tau}{2},\tfrac{\tau}{2};\tau\right)\Big) \\ & \qquad = 2q^{-\frac{1}{8}}\left(-1+45q+231q^2+770q^3+2277q^4+\cdots\right), \end{aligned}$$

expressed in terms of Zwegers' μ-functions (cf. Definition 8.1), which plays the role of the j-function in the original Moonshine. The first few coefficients $45, 231\ldots$ shown in (20.15) are dimensions of irreducible representations of the Mathieu group M_{24}. This led to the conjectured existence of an infinite-dimensional graded M_{24}-module, such that the graded dimensions are the coefficients of the mock modular form in (20.15); the analogous Mathieu McKay-Thompson series $H_g(\tau)$ formed using the graded traces for general $g \in M_{24}$ should be mock modular forms. Beautiful work by Gannon **[214]** proved the existence of this module, following work of Cheng **[141]**, Eguchi and Hikami **[183]**, and Gaberdiel, Hohenegger, and Volpato **[211, 212]**. We summarize one of Gannon's results in the following theorem.

THEOREM 20.11. *There is a graded M_{24}-module $K = \bigoplus_{n=1}^{\infty} K_{n-\frac{1}{8}}$ for which*

$$H_g(\tau) = -2q^{-\frac{1}{8}} + \sum_{n=1}^{\infty} \mathrm{tr}\left(g|K_{n-\frac{1}{8}}\right) q^{n-\frac{1}{8}}.$$

A concrete construction of K remains unknown.

Subsequently, Cheng, Duncan, and Harvey **[142, 143]** formulated the *Umbral Moonshine Conjectures*, which purport the existence of further graded infinite-dimensional modules relating finite groups and mock modular forms, extending the result just described. These new modules arise from *Niemeier lattices.*

Niemeier **[399]** showed that (up to isomorphism) there are 24 even unimodular positive-definite lattices of rank 24, one of which is the Leech lattice and the others of which have roots systems of full rank, called *Niemeier root systems*. Attached to such a root system X is an *umbral group* G^X, and vector-valued modular forms $(H^X_{g,r}(\tau))_r$, $g \in G^X$. The umbral group is the quotient of the automorphism group of the lattice by the Weyl group of the lattice.

In the ADE notation for root systems, when $X = A_1^{24}$, then $G^X \cong M_{24}$, and $(H^X_{g,r})_r$ is a four-dimensional vector indexed by $r \in \mathbb{Z}/4\mathbb{Z}$. For this X, $H^X_{g,r} = 0$ for $r \equiv 0 \pmod 2$, and $H^X_{g,r} = -H^X_{g,-r}$, thus, the entire vector is determined by $H^X_{g,1}$. For each $g \in M_{24}$, these are exactly the mock modular forms from [**141, 183, 211, 212**]. Generalizing work from Eguchi, Ooguri, and Tachikawa, we have the following "Umbral" conjecture of Cheng, Duncan, and Harvey [**142, 143, 177, 178**].

CONJECTURE 20.12 (Umbral Moonshine Conjecture). *Let X be a Niemeier root system, and $m = m^X$ the Coxeter number of any simple component of X. Then there is a bi-graded infinite-dimensional G^X-module*

$$\check{K}^X = \bigoplus_{r \in I^X \subseteq \mathbb{Z}/2m\mathbb{Z}} \bigoplus_{\substack{D \leq 0 \\ D \equiv r^2 \pmod{4m}}} \check{K}^X_{r,-\frac{D}{4m}}$$

such that the vector-valued mock modular form $(H^X_{g,r}(\tau))_r$ is a McKay-Thompson series for $\check{K}^X$ related to the graded trace of g on $\check{K}^X$ by

$$H^X_{g,r}(\tau) - 2q^{-\frac{1}{4m}}\delta_{r,1} + \sum_{\substack{D \leq 0 \\ D \equiv r^2 \pmod{4m}}} \operatorname{tr}\left(g|\check{K}^X_{r,-D/4m}\right) q^{-\frac{D}{4m}}.$$

As mentioned, Gannon [**214**] proved the conjecture in the case of $X = A_1^{24}$, and in [**178**], Duncan, Griffin, and one of the authors proved the remaining cases of the conjecture.

THEOREM 20.13. *Conjecture 20.12 is true.*

The proof of Theorem 20.13 in [**178**] adapted some of Gannon's methods from [**214**] which pertain to the case $X = A_1^{24}$; it also involves the explicit determination of each G^X-module $\check{K}^X$, done by computing the multiplicity of each irreducible component for each homogeneous subspace. The proof guarantees the existence and uniqueness of $\check{K}^X$ which are compatible with the representation theory of G^X, and also the Fourier expansions of the mock modular forms $(H^X_{g,r}(\tau))_r$. The proof from [**178**] additionally established the replicability of the mock modular forms $H^X_{g,r}$.

THEOREM 20.14. *For each $g \in G^X$ and for each $0 < r < m$, the mock modular form $H^X_{g,r}$ is replicable in the mock modular sense.*

By "replicability in the mock modular sense", we mean the recurrence relations which arise from the holomorphic projection of the product of these mock modular forms with the theta functions which define their corresponding shadows as in the discussion in Section 10.3. In fact, these relations are key to the proof of the theorem. To make this precise, we first recall that Borcherds' proof of the original Monstrous Moonshine Conjecture made use of replicability in Lie theoretic terms. The authors of [**178**] showed that the Umbral Moonshine mock modular replicability admits some weakly analogous properties.

To prove Theorem 20.14, the authors of [**178**] required recent work of Zwegers (in the form of unpublished notes), Imamoğlu-Raum-Richter [**278**] and Mertens

[385] on the principle of holomorphic projection, which stems from original work on the topic by Gross and Zagier **[242]**. We discussed this principle in Chapter 10. To give a taste of what holomorphic projection can lead to in the mock modular setting, we recall a recent result of Imamoḡlu, Raum, and Richter discussed in Chapter 10. For $n \in \mathbb{N}$, it was shown in **[278]** (cf. Theorem 10.5) that the coefficients $\alpha_f(n)$ of Ramanujan's mock theta function $f(q)$ from (9.1) satisfy, for all positive integers n,

(20.16)

$$\sum_{3m^2+m\leq 2n} \left(m+\tfrac{1}{6}\right) \alpha_f\left(n-\tfrac{3m^2}{2}-\tfrac{m}{2}\right) = \tfrac{4}{3}\sigma(n) - \tfrac{16}{3}\sigma\left(\tfrac{n}{2}\right) - 2\sum_{2n=ab} d\left(N_1, N_2, \tfrac{1}{6}, \tfrac{1}{6}\right).$$

(The numbers σ, d, N_1, N_2 are defined explicitly in Section 10.3.) Using the principle of holomorphic projection in a similar manner as it was used to prove (20.16) (i.e., Theorem 10.5), similar recursions are established in **[178]** for the coefficients of the mock modular forms in the Umbral Conjecture 20.12.

REMARK. Since the proof of Theorem 20.13, there have been further developments related to Umbral Moonshine. Moreover, there is still much to explore on the algebraic side of the story. In addition to the works already mentioned in this section, see, for example, recent work by Cheng and Harrison **[144]**, Griffin and Mertens **[236]**, Harvey and Rayhaun **[255]**, Trebat-Leder and two of the authors **[410]**, and Persson and Volpato **[417]**.

REMARK. In their seminal 1979 paper (cf. page 321 of **[154]**), Conway and Norton asked whether any of the sporadic finite simple groups which are not related to the Monster $\mathbb{M}$ possess moonshine properties. This problem refers to the *Pariah sporadic groups*: the Janko groups J_1, J_3, and J_4, the Lyons group Ly, the Rudvalis group Ru, and the O'Nan group $O'N$. In a recent paper **[179]**, Duncan, Mertens, and one of the authors establish moonshine for the O'Nan group, and the corresponding McKay-Thompson series are weight 3/2 mock modular forms.

CHAPTER 21

Quantum Modular Forms

21.1. Introduction to quantum modular forms

Many of the objects we have discussed in this book are modular to some extent. Namely, we have seen functions defined on the upper half-plane $\mathbb{H}$ which transform in various ways with respect to the action on $\mathbb{H}$ by an appropriate subgroup $\Gamma \subseteq \mathrm{SL}_2(\mathbb{Z})$ of the modular group. We have also found that these functions obey different degrees of holomorphicity, or lack thereof, on $\mathbb{H}$ and at the *cusps* of Γ, the equivalence classes of points in $\mathbb{P}^1(\mathbb{Q}) := \mathbb{Q} \cup \{i\infty\}$ under the action of Γ. The cusps are a boundary for the space $\Gamma\backslash\mathbb{H}$, and thus it is a natural question to study the modular properties of functions whose domains lie within $\mathbb{P}^1(\mathbb{Q})$. This question, although natural, is relatively new; it was posed by Zagier in 2010 **[521]**. The most naive definition one may consider along these lines is that of a $\mathbb{C}$-valued analytic function whose domain is some subset of $\mathbb{P}^1(\mathbb{Q})$, which exhibits a true modular transformation under the action of Γ (i.e., as in Definition 4.2 i), where $\tau \in \mathbb{P}^1(\mathbb{Q})$ as opposed to $\mathbb{H}$). However, requiring such functions to be analytic poses problems, as $\mathbb{P}^1(\mathbb{Q})$ admits the discrete topology. Requiring such functions to exhibit true modular transformations also poses problems, as Γ acts on $\mathbb{P}^1(\mathbb{Q})$ with only finitely many orbits.

To compensate, Zagier assembled a definition of functions he calls *quantum modular forms*, which fail to be modular in such a way that precisely offsets their failure to be analytic, and vice-versa.

DEFINITION 21.1. A **quantum modular form of weight** k ($k \in \frac{1}{2}\mathbb{Z}$) is a function $f : \mathbb{P}^1(\mathbb{Q})\backslash S \to \mathbb{C}$, for some discrete subset S, such that for all $\gamma = \left(\begin{smallmatrix} a & b \\ c & d \end{smallmatrix}\right) \in \Gamma$, a congruence subgroup of $\mathrm{SL}_2(\mathbb{Z})$, the functions

$$h_\gamma(x) = h_{f,\gamma}(x) := f(x) - \varepsilon^{-1}(\gamma)(cx+d)^{-k} f\left(\frac{ax+b}{cx+d}\right) \tag{21.1}$$

satisfy a suitable property of continuity or analyticity in $\mathbb{R}$ (as explained below).

REMARK. We have slightly modified Zagier's original definition to allow the set S to be discrete, $k \in \frac{1}{2}\mathbb{Z}$, subgroups $\Gamma \subseteq \mathrm{SL}_2(\mathbb{Z})$, and multiplier systems $\varepsilon(\gamma)$, in accordance with the theory of ordinary modular forms.

To elaborate on this definition, a typical example of a quantum modular form may be (but is not necessarily required to be) such that the corresponding "error functions" h_γ are defined on $\mathbb{Q}\backslash\{\gamma^{-1}(i\infty)\}$, and extend to functions which are C^∞, or perhaps analytic in $\mathbb{R} \setminus \{\gamma^{-1}(i\infty)\}$. Many of the examples we shall see in the remainder of this chapter in fact satisfy the stronger condition that (21.1) holds as an identity between formal power series, defined in complex disks of positive radii around the rational points x. If this is the case, Zagier refers to such functions f as *strong quantum modular forms*.

Zagier's definition, in particular the continuity or analyticity requirement of the functions h_γ, is intentionally vague so that it may encompass many diverse, interesting, examples. We make these notions more precise in the examples that follow. In the remainder of this chapter, we discuss some aspects of quantum modular forms which are related to topics in this book, including Maass forms, Eichler integrals, mock modular forms, radial limits, combinatorial generating functions, partial theta functions, meromorphic Jacobi forms, and vertex algebras. In addition to the references **[89, 106, 115, 116, 117, 133, 145, 199, 203, 268, 270, 287, 344, 353, 356, 515, 521]** given throughout this chapter, there are a number of other recent papers on quantum modular forms, including **[118, 246, 310, 439, 441]**.

21.2. Quantum modular forms and Maass forms

In Example 3.14, we discuss the q-hypergeometric functions

$$\sigma(q) = \sum_{n=0}^{\infty} \frac{q^{\frac{n(n+1)}{2}}}{(-q;q)_n} = \sum_{n=0}^{\infty} T(n) q^{\frac{n-1}{24}}, \tag{21.2}$$

$$\sigma^*(q) = 2\sum_{n=0}^{\infty} \frac{(-1)^n q^{n^2}}{(q;q^2)_n} = \sum_{n=-\infty}^{-1} T(n) q^{\frac{|n|-1}{24}}, \tag{21.3}$$

the first of which was studied in Ramanujan's lost notebook. In terms of their relationship to modular forms, these functions were related to indefinite theta functions in **[32]**, and their coefficients $T(n)$ were used by Cohen **[151]** to assemble a Maass cusp form.

Zagier revisited these functions in the context of quantum modular forms, beginning with some interesting observations of Cohen. First, due to Andrews and Cohen (respectively), one has the q-series identities

$$\sigma(q) = 1 + \sum_{n=0}^{\infty} (-1)^n q^{n+1} (q;q)_n, \quad \sigma^*(q) = -2\sum_{n=0}^{\infty} q^{n+1} \left(q^2;q^2\right)_n. \tag{21.4}$$

The right-hand sides of the two identities in (21.4) not only converge for $|q| < 1$, but also for roots of unity $q = \zeta_k^h$ (where $h \in \mathbb{Z}, k \in \mathbb{N}, \gcd(h,k) = 1$, and $\zeta_N = e^{2\pi i/N}, N \in \mathbb{N}$). Cohen observed, and subsequently proved using (21.4) and a delicate recursion, that for any root of unity ζ, the functions σ and σ^* are related by

$$\sigma(\zeta) = -\sigma^*\left(\zeta^{-1}\right).$$

This led Zagier to make the following definition, for $x \in \mathbb{Q}$:

$$f(x) := e^{\frac{\pi i x}{12}} \sigma(e^{2\pi i x}) = -e^{\frac{\pi i x}{12}} \sigma^*(e^{-2\pi i x}).$$

This function f is one of Zagier's first beautiful examples of a quantum modular form.

PROPOSITION 21.2. *The function $f\colon \mathbb{Q} \to \mathbb{C}$ satisfies*

$$f(x+1) = \zeta_{24} f(x), \quad f(x) - \zeta_{24}^{-1}(2x+1)^{-1} f\left(\frac{x}{2x+1}\right) = h(x),$$

where $h\colon \mathbb{R} \to \mathbb{C}$ is C^∞ on $\mathbb{R}$ and analytic on $\mathbb{R} \setminus \{-1/2\}$. In particular, f is a quantum modular form of weight 1 on $\mathbb{Q}$, with respect to $\Gamma_0(2)$.

REMARK. Proposition 21.2 in its original form may be found in **[521]**.

SKETCH OF PROOF OF PROPOSITION 21.2. The function f extends to a function $f\colon \mathbb{H}\cup\mathbb{H}^-\cup\mathbb{Q}\to\mathbb{C}$, where $\mathbb{H}^- := \{\tau\in\mathbb{C} : \mathrm{Im}(\tau)<0\}$, by defining

$$f(\tau) := \begin{cases} e^{\frac{\pi i\tau}{12}}\sigma\left(e^{2\pi i\tau}\right) & \text{if } \tau\in\mathbb{H}\cup\mathbb{Q},\\ -e^{\frac{\pi i\tau}{12}}\sigma^*\left(e^{-2\pi i\tau}\right) & \text{if } \tau\in\mathbb{H}^-\cup\mathbb{Q}.\end{cases}$$

Combining the Fourier expansions given on the right-hand sides of (21.2) and (21.3) with work of Lewis and Zagier **[353]** (appropriately extended to forms on congruence subgroups) implies that the function f, when restricted to $\mathbb{C}\setminus\mathbb{R}$, can be re-written in terms of the Maass cusp form c defined in (3.8) as

$$f(\tau) = \begin{cases} \int_\tau^{i\infty}\left[c(w), r_\tau(w)\right] & \text{if } \tau\in\mathbb{H},\\ -\int_{\overline{\tau}}^{i\infty}\left[r_\tau(w), c(w)\right] & \text{if } \tau\in\mathbb{H}^-,\end{cases}$$

where $r_\tau(w) := (\mathrm{Im}(w)/(w-\tau)(\overline{w}-\tau))^{\frac{1}{2}}$, and the Green's form $[\cdot,\cdot]$ is defined by

$$[f,g] := \frac{\partial f}{\partial w}g\ dw + f\frac{\partial g}{\partial\overline{w}}\ d\overline{w}.$$

Using the fact that the functions c and r_τ are both eigenfunctions of the Laplace operator, together with the modular transformation properties satisfied by the Maass form c, reveals that the error to modularity of the function f (in the sense of the second transformation in Proposition 21.2) when viewed as a function on $\mathbb{H}$ or $\mathbb{H}^-$ under the matrix $\left(\begin{smallmatrix}1&0\\2&1\end{smallmatrix}\right)$ is a constant multiple of the integral $\int_{-\frac{1}{2}}^{i\infty}[c(w), r_\tau(w)]$. The claimed analytic and continuity properties of this function are carefully discussed in **[356, 521]**. □

21.3. Quantum modular forms and Eichler integrals

In Subsection 21.3.1, we describe another of Zagier's examples of a quantum modular form, which arises from a function originally studied by Kontsevich. As we shall see below, Zagier's proof of quantum modularity for this function is rather different in nature than the proof of quantum modularity for the function f in the previous section, which ultimately relies upon its connection to a Maass form. In Subsection 21.3.2, we discuss how the method of proof in Subsection 21.3.1 can be generalized to establish the quantum modularity of Eichler integrals and partial theta functions associated to half-integral weight ordinary modular forms.

21.3.1. Kontsevich's function. Consider the function $\phi\colon\mathbb{Q}\to\mathbb{C}$ defined by

$$\phi(x) := e^{\frac{\pi ix}{12}}\sum_{n=0}^{\infty}\left(e^{2\pi ix}; e^{2\pi ix}\right)_n \tag{21.5}$$

(see also (14.22) and the discussion following). This function, originally studied by Kontsevich, has the curious property that it converges on no open subset in $\mathbb{C}$ and is only defined at points $x\in\mathbb{Q}$. Indeed, for rational x, it is not difficult to see that the infinite sum defining ϕ truncates, and becomes a polynomial in the root of unity $e^{2\pi ix}$. As it turns out, Kontsevich's function ϕ is another of Zagier's original examples of a quantum modular form in **[521]**. Together with the easily verified transformation under $x\mapsto x+1$, Zagier's result (14.22) can be incorporated into the following theorem.

THEOREM 21.3. *The function* $\phi\colon \mathbb{Q} \to \mathbb{C}$ *satisfies*

$$\phi(x+1) = \zeta_{24}\phi(x), \quad \phi(x) + (ix)^{-\frac{3}{2}}\, \phi\left(-\tfrac{1}{x}\right) = h(x),$$

where $h\colon \mathbb{R} \to \mathbb{C}$ *is analytic on* $\mathbb{R}\setminus\{0\}$. *In particular,* ϕ *is a quantum modular form of weight* $3/2$ *on* $\mathbb{Q}$ *with respect to* $\mathrm{SL}_2(\mathbb{Z})$.

SKETCH OF PROOF. Unless otherwise stated, we let $\tau \in \mathbb{H}$, $q = e^{2\pi i\tau}$. Here, we elaborate on the discussion following (14.22) in Chapter 14. Define the functions

$$\widetilde{\eta}(\tau) := \sum_{n=0}^{\infty} n\left(\frac{12}{n}\right) q^{\frac{n^2}{24}}, \quad \eta^*(\overline{\tau}) := \int_{\tau}^{i\infty} \frac{\eta(w)}{(w-\overline{\tau})^{\frac{3}{2}}} dw. \tag{21.6}$$

The function $\widetilde{\eta}$ is akin to a "half-derivative" of the Dedekind eta-function, noting that $\eta(\tau) = \sum_{n=0}^{\infty}\left(\frac{12}{n}\right) q^{\frac{n^2}{24}}$. It may also be viewed as a partial theta function, which, a priori, has no modular properties (cf. Section 21.5). Zagier first related the function ϕ to the function $\widetilde{\eta}$ in the following beautiful sums of tails identity:

$$\sum_{n=0}^{\infty}\left(q^{\frac{1}{24}}(q;q)_n - \eta(\tau)\right) = -\frac{1}{2}\widetilde{\eta}(\tau) + \eta(\tau)\left(\frac{1}{2} - \sum_{n=1}^{\infty}\frac{q^n}{1-q^n}\right). \tag{21.7}$$

The η-function vanishes to infinite order at any rational number. Moreover, the function $\sum_{n=1}^{\infty}\frac{q^n}{1-q^n}$ behaves at worst like $(1-|q|)^{-2}$; therefore, from (21.7), we see that

$$\phi(x) = -\frac{1}{2}\widetilde{\eta}(x) \tag{21.8}$$

for $x \in \mathbb{Q}$, where we interpret the right-hand side of (21.8) as an asymptotic statement, as $\tau \to x \in \mathbb{Q}$ from the upper half-plane.

On the other hand, as we shall make more precise in Subsection 21.3.2, one can show that asymptotically, towards rational numbers, the function η^* agrees with the function $\widetilde{\eta}$ (up to multiplication by a constant). The quantum modular transformation properties of ϕ may thus be deduced from η^*. In particular, it is not difficult to see using the definition of η^* and the modular transformation $\eta(-1/\tau) = \sqrt{-i\tau}\eta(\tau)$ that

$$\eta^*\left(-\frac{1}{\overline{\tau}}\right) - i^{-\frac{1}{2}}\overline{\tau}^{\frac{3}{2}}\eta^*(\overline{\tau}) = -i^{-\frac{1}{2}}\overline{\tau}^{\frac{3}{2}}\int_0^{i\infty}\frac{\eta(w)}{(w-\overline{\tau})^{\frac{3}{2}}}dw. \tag{21.9}$$

The error to modularity on the right-hand side of (21.9) above is a real-analytic function away from 0, a fact that we elaborate upon in Subsection 21.3.2. The result follows as in the previous example by analytic continuation. □

Some further results which relate to the above example may be found in **[85, 89, 116, 133, 145, 199, 270, 515]**. In particular, the work **[85]** defined the notion of a *quantum Jacobi form*, a natural hybrid of Definition 2.11 and Definition 21.1, and realizes the two-variable strongly unimodal rank function (14.18) as an example. In the next subsection, we elaborate upon the results in **[116]**.

21.3.2. Eichler integrals and partial theta functions. Not only does the proof of Theorem 21.3 show that Kontsevich's function ϕ is a quantum modular form, it also shows that the formal Eichler integral $\widetilde{\eta}$ of Dedekind's η-function (21.6) is a quantum modular form. Two of the authors generalized this phenomenon, and showed in Theorem 1.1 of **[116]** that if $f(\tau) = \sum_{n=1}^{\infty} c_f(n)q^n$ is a cusp form of level $N \in 4\mathbb{N}$ and weight $k \in 1/2 + \mathbb{N}_0$, then its Eichler integral

$$\widetilde{f}(\tau) := \sum_{n=1}^{\infty} c_f(n)n^{1-k}q^n$$

is a quantum modular form of weight $2-k$. In this section, the Dirichlet character $\chi_{-4}(\cdot)$ is equal to $\left(\frac{-4}{\cdot}\right)$.

THEOREM 21.4. *If $f \in S_k(\Gamma_0(N))$ with $k \in 1/2 + \mathbb{N}_0$ and $N \in 4\mathbb{N}$, then its Eichler integral $\widetilde{f}$ is a quantum modular form of weight $2-k$. In particular, for all $\gamma = \left(\begin{smallmatrix} a & b \\ c & d \end{smallmatrix}\right) \in \Gamma_0(N)$ and $x \in \mathbb{Q} \setminus \{\gamma^{-1}(i\infty)\}$, we have that*

$$\widetilde{f}(x) - \chi_{-4}(d)(cx+d)^{k-2}\widetilde{f}(\gamma x) = h_\gamma(x),$$

where h_γ is analytic on $\mathbb{R} \setminus \{\gamma^{-1}(i\infty)\}$.

REMARK. In the case of weights $1/2$ and $3/2$, the Eichler integrals $\widetilde{f}$ may be viewed as partial theta functions, a topic we discuss in Section 21.5 below.

REMARK. It turns out that the quantum modular forms from Theorem 21.4 may be interpreted as L-values. In particular, Corollary 1.2 of **[116]** shows that the twisted L-values

$$L_f\left(\zeta_m^\ell; k-1\right) := \sum_{n=1}^{\infty} \frac{c_f(n)\zeta_m^{\ell n}}{n^{k-1}}$$

are quantum modular forms if viewed as functions of $\ell/m \in \mathbb{Q}$.

SKETCH OF PROOF OF THEOREM 21.4. Similar to the proof of Theorem 21.3 described in the previous section (cf. **[268]**, **[344]**, and **[515]**), we introduce the nonholomorphic Eichler integral of the cusp form f, defined for $\tau \in \mathbb{H}$ by

$$f^*(\overline{\tau}) := \frac{(-2\pi i)^{k-1}}{\Gamma(k-1)} \int_{\tau}^{i\infty} \frac{f(w)}{(w-\overline{\tau})^{2-k}} dw.$$

The function $\widetilde{f}$ has a nice q-series expansion, and it is not difficult to see that the function f^* satisfies nice transformation properties (cf. (21.12)). These two seemingly different functions turn out to be asymptotically related at rational numbers, and so the modular transformation properties of f^* are inherited by $\widetilde{f}$ at rationals.

To make this more precise, by using Mellin transforms

$$\mathcal{M}(g)(s) := \int_0^{\infty} g(w)w^{s-1}dw,$$

after a technical argument (see also **[344**, p98 Proposition**]**), we arrive at the following asymptotic expansions at any rational number $\ell/m \in \mathbb{Q}$ for the functions $\widetilde{f}$ and f^* as $t \to 0^+$:

$$\widetilde{f}\left(\frac{\ell}{m} + \frac{it}{2\pi}\right) \sim \sum_{n=0}^{\infty} \frac{(-t)^n}{n!} L_f\left(\zeta_m^\ell; k-1-n\right), \tag{21.10}$$

$$f^*\left(\frac{\ell}{m} - \frac{it}{2\pi}\right) \sim \sum_{n=0}^{\infty} \frac{t^n}{n!} L_f\left(\zeta_m^\ell; k-1-n\right). \tag{21.11}$$

Moreover, a direct calculation, using a change of variable in the integral as well as the modular properties of f, reveals that

$$f^*(\overline{\tau}) - \chi_{-4}(d)(c\overline{\tau}+d)^{k-2}f^*(\gamma\overline{\tau}) = \frac{(-2\pi i)^{k-1}}{\Gamma(k-1)}\int_{-\frac{d}{c}}^{i\infty} \frac{f(w)}{(w-\overline{\tau})^{2-k}}dw. \tag{21.12}$$

The asymptotic expansion for f^* in (21.11) shows that

$$\lim_{t\to 0^+} f^*\left(\frac{\ell}{m} - \frac{it}{2\pi}\right) = L_f\left(\zeta_m^\ell; k-1\right),$$

hence, the left-hand side of (21.12) has a well-defined limit as $\overline{\tau} \to \ell/m \in \mathbb{Q}$ from the lower half-plane (where $\ell/m \neq -d/c = \gamma^{-1}(i\infty)$ if $k \in \{1/2, 3/2\}$).

We next use the fact that the function on the right hand side of (21.12) is analytic on $\mathbb{R}\setminus\{-d/c\}$ (as a function in the variable $\overline{\tau}$), which can be explained as follows (cf. Section 6 of **[515]**). Other than at the pole $\tau = -d/c$, the integrand is bounded in absolute value. One can then apply the Leibniz theorem for indefinite integrals to conclude that the integral is analytic as claimed.

Thus, we have that the nonholomorphic Eichler integral f^* is a quantum modular form of weight $2-k$. From (21.10), we see that the function $\widetilde{f}$ also extends to a function on $\mathbb{Q}$ which takes on the same values as f^*, and hence, is a quantum modular form which transforms as claimed in Theorem 21.4. □

21.4. Quantum modular forms and radial limits of mock modular forms

The example given in Section 21.2 shows a quantum modular form f whose modular-like properties on $\mathbb{Q}$ extend into $\mathbb{C}$. Loosely speaking, the example gives a function whose modular-like properties on $\mathbb{H}$ leak into $\mathbb{H}^-$ through $\mathbb{Q}$. The second example given in Subsection 21.3.1 pertaining to Kontsevich's function ϕ fits into a similar framework: the modular-like properties of ϕ exist in $\mathbb{Q}$ and extend into the lower half-plane $\mathbb{H}^-$ via a certain Eichler integral. In Subsection 21.4.1, we explain how these modular-like properties also extend into the upper half-plane via a combinatorial generating function studied in Chapter 14. In Subsection 21.4.2, we elaborate on some related results from Chapter 9 and Chapter 14, and discuss quantum modular forms and radial limits of mock modular forms.

21.4.1. A unimodal rank generating function. In Chapter 14, we discussed the combinatorial generating function

$$U(1;q) = \sum_{n=1}^{\infty}(u_e(n) - u_o(n))q^n = \sum_{n=0}^{\infty}(q;q)_n^2\, q^{n+1}, \tag{21.13}$$

where $u_e(n)$ (resp. $u_O(n)$) counts the number of strongly unimodal sequences of size n and even (resp. odd) rank. Bryson, Pitman, Rhoades, and one of the authors **[133]** related this function, perhaps surprisingly, to Kontsevich's function ϕ in the following identity:

$$e^{-\frac{\pi i x}{12}}U(1; e^{2\pi i x}) = \phi(-x), \tag{21.14}$$

which holds for $x \in \mathbb{Q}$. The following theorem, proved in **[133]**, exhibits modular transformation properties for $U(1;q)$ on both $\mathbb{H}$ and $\mathbb{Q}\setminus\{0\}$. In particular, Theorem 21.5 essentially shows that the function $e^{-\frac{\pi i x}{12}}U(1;e^{2\pi i x})$ is a quantum modular form. If combined with (21.14), these facts extend the transformation properties of Kontsevich's function ϕ into the upper half-plane.

THEOREM 21.5. *For $x \in \mathbb{H} \cup \mathbb{Q} \setminus \{0\}$, we have that*

$$e^{-\frac{\pi i x}{12}} U(1; e^{2\pi i x}) + (-ix)^{-\frac{3}{2}} e^{\frac{\pi i}{12x}} U\left(1; e^{-\frac{2\pi i}{x}}\right) = g(x),$$

where

$$g(x) := \frac{\sqrt{3}}{2\pi i} \int_0^{i\infty} \frac{\eta(w)}{(-i(x+w))^{\frac{3}{2}}} dw - \frac{i}{2} e^{\frac{\pi i x}{6}} (e^{2\pi i x}; e^{2\pi i x})_\infty^2 \int_0^{i\infty} \frac{\eta^3(w)}{(-i(x+w))^{\frac{1}{2}}} dw.$$

Moreover, the function g is analytic on $\mathbb{R} \setminus \{0\}$.

REMARK. More general results related to Theorem 21.5 may be found in [**85, 199**].

SKETCH OF PROOF OF THEOREM 21.5. A key ingredient towards the proof of Theorem 21.5 comes from the function ($q = e^{2\pi i \tau}$, $\tau \in \mathbb{H}$)

$$\widehat{f}(\tau) := U(1; q)(q; q)_\infty + \frac{\sqrt{3}}{2\pi i} \eta(\tau) \int_{-\overline{\tau}}^{i\infty} \frac{\eta(w) dw}{(-i(w+\tau))^{\frac{3}{2}}} - \frac{i}{2} \eta^3(\tau) \int_{-\overline{\tau}}^{i\infty} \frac{\eta^3(w) dw}{(-i(w+\tau))^{\frac{1}{2}}}.$$

In Theorem 1.1 of [**38**], the authors showed that $\widehat{f}$ transforms like a nonholomorphic modular form of weight 2 and level 1. Combining this fact with the modular transformation properties satisfied by the Dedekind η-function ultimately leads to the transformation given in the statement in Theorem 21.5 for $\tau \in \mathbb{H}$ after a short calculation.

The transformation extends to $\mathbb{Q} \setminus \{0\}$ as follows. The function $(e^{2\pi i x}; e^{2\pi i x})_\infty$ equals zero for any $x \in \mathbb{Q}$. Moreover, the function $U(1; e^{2\pi i x})$ is defined for any $x \in \mathbb{Q}$, which is not difficult to see using its definition. Finally, the remaining integral $\int_0^{i\infty} \eta(w)(-i(x+w))^{-\frac{3}{2}} dw$ is analytic as a function of x on $\mathbb{R} \setminus \{0\}$, as explained in the previous section (cf. Section 6 of [**515**]). □

21.4.2. Radial limits and quantum modular forms. In Section 14.4 we also discussed the more general two-variable unimodal rank generating function defined by

$$U(\zeta; q) = \sum_{n=0}^{\infty} \sum_{m \in \mathbb{Z}} u(m, n)(-\zeta)^m q^n = \sum_{n=0}^{\infty} (\zeta; q)_n (\zeta^{-1}; q)_n q^{n+1}, \tag{21.15}$$

where $u(m, n)$ counts the number of strongly unimodal sequences of size n and rank m. When viewed as a combinatorial generating function, the function U is similar to the two-variable partition rank and crank generating functions (previously discussed in detail in Section 14.3) given by

$$R(\zeta; q) = \sum_{n=0}^{\infty} \sum_{m \in \mathbb{Z}} N(m, n)\zeta^m q^n = \sum_{n=0}^{\infty} \frac{q^{n^2}}{(\zeta q; q)_n (\zeta^{-1} q; q)_n} \tag{21.16}$$

and

$$C(\zeta; q) = \sum_{n=0}^{\infty} \sum_{m \in \mathbb{Z}} M(m, n)\zeta^m q^n = \prod_{n=1}^{\infty} \frac{1 - q^n}{(1 - \zeta q^n)(1 - \zeta^{-1} q^n)}, \tag{21.17}$$

where $N(m, n)$ (resp. $M(m, n)$) counts the number of integer partitions of n with rank (resp. crank) equal to m.

Despite their similarities as combinatorial generating functions, the modular properties of the functions $U, R,$ and C are rather different. We have seen above and in Chapter 14 that, essentially, the function U is a quantum modular (and mixed mock modular) form, the function R is a mock Jacobi form, and the function C is an ordinary (meromorphic) Jacobi form. As mentioned in Section 14.4, Rhoades with two of the authors related these different types of modular combinatorial generating functions [**203**] by their radial limits, as shown in the following theorem.

THEOREM 21.6. *Let* $1 \leq a < b$, $1 \leq h < k$ *with* $\gcd(a,b) = \gcd(h,k) = 1$, $b|h$ *and* $h' \in \mathbb{Z}$ *with* $hh' \equiv -1 \pmod{k}$. *Then, as* $q \to \zeta_k^h$ *radially within the unit disc, we have that*

$$\lim_{q \to \zeta_k^h} \left(R\left(\zeta_b^a; q\right) - \zeta_{b^2}^{-a^2 h' k} C\left(\zeta_b^a; q\right) \right) = -\left(1 - \zeta_b^a\right)\left(1 - \zeta_b^{-a}\right) U\left(\zeta_b^a; \zeta_k^h\right).$$

It is not difficult to show that under the hypotheses given, the values $U(\zeta_b^a; \zeta_k^h)$ are non-singular, and in fact, can be explicitly evaluated as explicit (finite, Laurent) polynomials in ζ_b^a and ζ_k^h. Theorem 21.6 is a generalization of Theorem 21.7 below from [**203**], which is also discussed in Chapter 9. This result asymptotically relates Ramanujan's third order mock theta function f and the modular (up to multiplication by a suitable power of q) form b, given by

$$f(q) = \sum_{n=0}^{\infty} \frac{q^{n^2}}{(-q;q)_n^2}, \qquad b(q) := \prod_{n=0}^{\infty} \left(1 - q^{2n-1}\right) \sum_{m \in \mathbb{Z}} (-1)^m q^{m^2},$$

and the unimodal function $U(-1;q)$.

THEOREM 21.7. *If* ζ *is a primitive even-order* $2k$ *root of unity, then, as* $q \to \zeta$ *radially within the unit disc, we have that*

$$\lim_{q \to \zeta} \left(f(q) - (-1)^k b(q) \right) = -4U\left(-1; \zeta\right).$$

Theorem 21.7 was anticipated by Ramanujan, who, as described in Section 9.8, claimed in his last letter to Hardy that the mock theta function f and the modular form b satisfy

$$f(q) - (-1)^k b(q) = O(1)$$

as $q \to \zeta$, where ζ is a $2k$-th primitive root of unity. Ramanujan's claim concerns the asymptotic properties of the mock theta function f at its singularities. This example suggests that the modular form b is a near miss in terms of carving out the exponential singularities of the mock theta function b. While Watson's work [**489**] indicates how one could prove the boundedness of this radial limit, Theorem 21.7 goes on to show that the implied constants in Ramanujan's claim are in fact special values of the quantum modular form U. Theorem 21.6 generalizes this phenomenon (the special case $a = 1, b = 2$ is Theorem 21.7), asymptotically relating the mock modular rank generating function and the modular crank generating function.

For further generalizations and related results, see Section 9.8 and, for example, [**115, 145, 287**]. Here, we provide a sketch of the original proof of Theorem 21.6 which appeared in [**203**]. Alternative proofs and formulations of Theorem 21.6 and Theorem 21.7 may be found in [**204, 526**]. One may also proceed as in [**115**], which we discuss at the conclusion of this section.

SKETCH OF PROOF OF THEOREM 21.6. We begin with the beautiful basic hypergeometric identity (14.17), originally due to Ramanujan (cf. p. 67 in [**30**]) and later reinterpreted by Choi [**146**]. Observe that the two summands on the left-hand side of (14.17) are related to the rank generating function (21.16) and the unimodal rank moment generating function (21.15). This plays a fundamental role in our proof. With a, b as in the hypotheses of Theorem 21.6, we define a function m appearing as a factor of the right-hand side of (14.17) by

$$m(a, b; w) := ie^{\frac{\pi i w}{4}}(1 - \zeta_b^{-a})\zeta_b^a(\zeta_b^a e^{2\pi i w}; e^{2\pi i w})_\infty(\zeta_b^{-a}; e^{2\pi i w})_\infty,$$

which can be rewritten in terms of Jacobi's ϑ-function (cf. Definition 2.1) and the Dedekind η-function. We establish the following asymptotic results (cf. Theorem 3.2 and Proposition 3.3 of [**203**]) for the mixed mock modular form $m\mu$ appearing on the right-hand side of (14.17), and the crank moment generating function (21.17), in which we have set $q := e^{(2\pi i/k)(h+iz)}$, and $q_1 := e^{(2\pi i/k)(h'+i/z)}$. (We note that the variable z here should not be confused with the Jacobi variable z previously used.)

THEOREM 21.8. *Let $b|k, \gcd(a, b) = 1, \gcd(h, k) = 1$, where a, b, h, k are positive integers, and let b' and h' be positive integers such that $bb' = k$ and $hh' \equiv -1 \pmod{k}$. For $z \in \mathbb{R}^+$, as $z \to 0^+$, we have that*

(21.18)
$$m\left(a, b; \frac{1}{k}(h + iz)\right)\mu\left(-\frac{a}{b}, \frac{a}{b}; \frac{1}{k}(h + iz)\right)$$
$$= (-iz)^{-\frac{1}{2}}\psi(\gamma)^{-1}q^{\frac{1}{24}}q_1^{-\frac{1}{24}}(-1)^{ab'}\zeta_{2b}^{ah'-a}\zeta_{2b^2}^{-3a^2kh'}\frac{\zeta_b^a - 1}{1 - \zeta_b^{ah'}}\left(1 + O\left(q_1^\alpha\right)\right),$$

(21.19)
$$C\left(\zeta_b^a; \frac{1}{k}(h + iz)\right)$$
$$= (-iz)^{-\frac{1}{2}}\psi(\gamma)^{-1}q^{\frac{1}{24}}q_1^{-\frac{1}{24}}(-1)^{ab'}\zeta_{2b}^{ah'-a}\zeta_{2b^2}^{-a^2kh'}\frac{\zeta_b^a - 1}{1 - \zeta_b^{ah'}}\left(1 + O\left(q_1^\beta\right)\right),$$

for some constants $\alpha, \beta > 1/24$.

Overall, the strategy to confirm (21.18) and (21.19) is the same, however establishing one is much less straightforward than the other. Namely, to prove (21.19), we carefully exploit the (ordinary) Jacobi properties satisfied by the crank moment function; a technical but straightforward argument eventually leads to the equality.

Establishing (21.18) is more complicated; while the function m is essentially an ordinary Jacobi form (like the crank moment function), the function μ is a mock Jacobi form. In addition to exploiting the Jacobi transformation properties satisfied by the completed function $\widehat{\mu}$ (cf. Theorem 8.13), we must analyze terms involving the nonholomorphic function R in its definition (cf. Definition 8.8, which should not be confused with the function in (21.16)). We require some further technical lemmas related to these nonholomorphic terms, as well as certain Mordell integrals (cf. Definition 8.4) in order to establish (21.18). The interested reader can find these in [**203**].

From the asymptotic theorem above, we deduce that the mixed mock modular form $m\mu$ is asymptotic to a suitable multiple of the modular crank generating function C. That is, with notation as above, as $z \to 0^+$, we have from Theorem 21.8 that

$$m\left(a, b; \frac{1}{k}(h + iz)\right)\mu\left(-\frac{a}{b}, \frac{a}{b}; \frac{1}{k}(h + iz)\right) \sim \zeta_{b^2}^{-a^2h'k}C\left(\zeta_b^a; \frac{1}{k}(h + iz)\right). \tag{21.20}$$

Theorem 21.6 now follows from the fact that the function $U(\zeta_b^a;\zeta_k^h)$ is a finite convergent sum if $b|k$, combined with (14.17) and (21.20). □

As mentioned in Chapter 9, Theorem 21.6 and Theorem 21.7 are special cases of a more general theory. Proving a conjecture of Rhoades [**435**] on the universal mock theta function g_2, two of the authors [**115**] determined modular forms $f_{a,b,A,B,h,k}$ such that

$$Q_{a,b,A,B,h,k} := \lim_{q\to\zeta_k^h} \left(g_2\left(\zeta_b^a q^A; q^B\right) - f_{a,b,A,B,h,k}(q)\right)$$

is bounded as $q \to \zeta_k^h$ radially from within the unit disk. Moreover, they obtained explicit finite formulas for the constants $Q_{a,b,A,B,h,k}$, and, parallel to Theorem 21.6, showed that the functions $Q(h/k) := Q_{a,b,A,B,h,k}$ are quantum modular forms using results from [**145**]. Analogous statements in [**287**] (and also Theorem 21.6 via the Appell-Lerch-type sum representation of g_3 from Theorem 9.11 and (14.9)) pertain to the universal mock theta function g_3.

21.5. Quantum modular forms and partial theta functions

We have alluded to partial theta functions and their connections to quantum modular forms in various places in the narrative. To explain this relationship, we begin with the following definition.

DEFINITION 21.9. For a Dirichlet character χ, consider the modular theta functions ($\nu \in \{0,1\}$)

$$\sum_{n\in\mathbb{Z}} \chi(n) n^{\nu} q^{n^2}. \tag{21.21}$$

A **partial theta function** is any function resembling (21.21) with the summation over $\mathbb{Z}$ replaced by a partial lattice (e.g. $n \geq n_0$ for $n_0 \in \mathbb{Z}$).

A priori, partial theta functions (and their close relatives *false theta functions* [**21**]) possess no modular properties due to the incomplete lattices over which they are defined. Nevertheless these functions have a rich history and have been a topic of interest for decades. In particular, works by many authors including Alladi, Andrews, Berndt, Fine, Ramanujan, Rogers, and Yee (cf. [**15, 22, 55, 193**]) showed how partial or false theta functions may be related to q-hypergeometric series, a widely studied topic in analytic number theory.

In another direction, more recently, Hikami [**267**], Lawrence and Zagier [**344**], and Zagier [**515**] investigated partial theta functions in connection to topological invariants of 3-manifolds. In doing so, they were led to study the asymptotic properties of certain partial theta functions, including

$$\phi_m^{(a)}(\tau) := m \sum_{n=0}^{\infty} \chi_{2m}^{(a)}(n) q^{\frac{n^2}{4m}}$$

($m \in \mathbb{N}_{\geq 2}, a \in \{0,1,\ldots,m-2\}$), where $\chi_{2m}^{(a)}$ are certain characters. In [**267**], Hikami showed that these functions satisfy asymptotic properties towards rational numbers similar to those of $\widetilde{f}$ in (21.10), a result which in hindsight hints at quantum modular properties. For the remainder of this section we elaborate upon both topics, partial theta functions as related to q-hypergeometric series, and the asymptotics of partial theta functions, as we discuss the quantum modularity of partial theta functions; we begin with the latter.

In Subsection 21.3.1 and Subsection 21.3.2, indeed, following methods of Zagier, we established the quantum modularity of certain Eichler integrals, which, in the cases of weights $1/2$ and $3/2$, may also be viewed as partial theta functions. These results have already seen applications to diverse subjects. Below, we discuss applications to the subjects of vertex algebras and meromorphic Jacobi forms of negative index, and refer the reader to [**106**] and [**117**] for details.

As a first example, in [**106**], Milas with one of the authors showed that quantum modular forms naturally appear as module characters for many sufficiently nice irrational vertex algebras. To be more precise, the atypical irreducible modules $M_{r,s}$ of the $(1,p)$-singlet algebra [**156**] are studied, and by making use of their (non-obvious) connection to partial theta functions, it is shown that the numerators of the singlet algebra module characters $\mathrm{ch}_{M_{r,s}}$ are quantum modular forms of weight $1/2$ on $\mathbb{Q}$. The authors obtained similar results for the numerators of the (p_+,p_-)-singlet algebra characters [**2**], though in this setting, the forms are of mixed weights ($3/2$ and $1/2$).

As a second example, consider the meromorphic Jacobi forms of negative index considered in Section 11.3. It is natural to ask in this case whether the coefficient functions h_{ℓ,z_0} of the Jacobi forms exhibit modular properties, as they do in the case of positive index, as shown in Section 11.2. Zwegers and two of the authors addressed this question in [**117**]. Indeed, by combining Theorem 11.7 with a special case of Theorem 21.4 on the quantum modularity of a certain class of partial theta functions, as well as the fact that the Laurent coefficients $D_{n,w}$ are quasimodular forms (cf. Definition 2.35) in the suppressed variable τ, results in [**117**] show that the Fourier coefficients of a general negative index Jacobi form are expressible as derivatives of quantum modular forms multiplied by quasimodular forms

As a final example, we mention another result of Rhodes with two of the authors (cf. Theorem 1.3 in [**203**]), which also established the quantum modularity of certain families of partial theta functions, but by very different methods involving q-hypergeometric series. Consider Rogers' q-hypergeometric series [**437**]

$$F(\alpha,\beta,t;q) := \sum_{n=0}^{\infty} \frac{(\alpha q;q)_n}{(\beta q;q)_n} t^n.$$

These functions do not typically specialize to modular forms, but instead may sometimes be related to partial theta functions. For example, we have the following result of Rogers and Fine [**193**]:

$$\frac{1}{1+\xi} F(\xi q^{-1},-\xi,\xi;q) := \frac{1}{1+\xi}\sum_{n=0}^{\infty} \frac{(\xi;q)_n}{(-\xi q;q)_n}\xi^n = \sum_{n=0}^{\infty}(-1)^n \xi^{2n} q^{n^2}. \tag{21.22}$$

Typically, q-hypergeometric series converge inside the unit disk, that is, for $|q|<1$. However, it is sometimes the case that they possess the special property that they converge both inside ($|q|<1$) and outside ($|q|>1$) the unit disk. In fact, under suitable choices of parameters, the authors in [**203**] showed that these Rogers-Fine functions $F(\alpha,\beta,t;q)$ satisfy this special property. Inside the unit disk, the functions studied in [**203**] are related to partial theta functions via (21.22). Outside the unit disk, they are in fact modular objects. Namely, it was shown in [**203**] that they are related to mock modular forms. After establishing that the functions are well defined on certain subsets of $\mathbb{Q}$ as well – which in this case follows from their interpretation as q-hypergeometric series – the mock modular properties which occur outside the unit disk, or equivalently, for $\tau \in \mathbb{H}^-$ if $q = e^{2\pi i\tau}$, extend from the lower half-plane to $\mathbb{Q}$ via analytic continuation. Full details may be found in **[203]**.

21.5.1. Connections with the Habiro ring. We conclude by noting that many of the above examples of quantum modular forms tend to have similar q-series representations demonstrating their convergence at roots of unity. For example, in (21.4), (21.5), and (21.13), we find q-series which become finite sums at all roots of unity due to the presence of rising q-factorials in the summands. These are all instances of the important, but still mysterious, connection with the so-called *Habiro ring*.

This ring was introduced by Habiro **[248]** as a uniformizing structure in the study of Witten-Reshetikhin-Turaev (WRT) invariants of 3-manifolds, and were proposed as a promising setting for performing $\mathbb{F}_1$-geometry by Manin **[375]**. This ring is formally defined as the completion

$$\mathcal{H} := \varprojlim_{n \geq 0} \mathbb{Z}[q]/((q;q)_n),$$

which we can think of as the set of formal expansions of the form

$$\sum_{n=0}^{\infty} a_n(q)(q;q)_n,$$

where $a_n(q) \in \mathbb{Z}[q]$. This representation reveals an important structure about related quantum modular forms. To any element of the Habiro ring and to any root of unity ζ there is a power series expansion in $(\zeta - q)$ **[248]**. That is, there is a map

$$\phi_\zeta \colon \mathcal{H} \to \mathbb{Z}[\zeta][[\zeta - q]].$$

For example, the map ϕ_1 may be realized explicitly (and be computed efficiently) for any $F(q) = \sum_{n=0}^{\infty} a_n(q)(q;q)_n \in \mathcal{H}$ as the expansion $\sum_{n=0}^{\infty} c_n (1-q)^n$ by recursively solving for c_n in the expansion

$$\sum_{n=0}^{\infty} a_n(1-q)(1-q;1-q)_n = \sum_{n=0}^{\infty} c_n q^n.$$

The resulting expressions defining c_n always terminate, since

$$(1-q;1-q)_n = O(q^n).$$

It is straightforward to find examples of non-trivial representations of the function 1 in the Habiro ring, and so expressions for quantum modular forms like those in (21.4), (21.5), and (21.13) are not unique or canonical. However, a central result of Habiro (cf. Theorem 5.2 of **[248]**) states that the map ϕ_ζ is injective for any ζ, so that in fact, an element of the Habiro ring is uniquely determined by its power series expansion at any given root of unity. As Habiro pointed out, this is similar to the fact that a holomorphic function is determined by its power series expansion at a single point, so that $\mathcal{H}$ may be thought of as a ring of analytic functions on roots of unity. Thus, the power series coefficients around a fixed root of unity are unique, and it is natural to study these coefficients in hopes of finding additional structure in the theory of quantum modular forms. Note that Zagier's notion of strong quantum modular forms also considers these power series expansions.

In fact, it turns out to be very fruitful to study these. For example, if ϕ is Kontsevich's function, then the coefficients in Habiro's power series at 1 give rise to the combinatorially interesting Fishburn numbers, whose congruences were elegantly studied by Andrews and Sellers [**39**] and several others [**7, 222, 474**]. This phenomenon was studied in general in [**246**], where Guerzhoy, Kent, and one of the authors showed that such power series expansions associated to the quantum modular forms for partial theta functions satisfy many congruences. However, as also explained in [**246**], the results become much more elegant in the special cases when the partial theta functions have representations in the Habiro ring. Such connections are not yet fully understood. For example, it is unclear which partial theta functions are related via a "strange identity" like (21.8) to an element of the Habiro ring. All known examples seem to come in a natural sense from knot theory. For example, a beautiful family of examples was discovered by Hikami [**268**]. These are related to half-derivatives of infinite product modular forms on the partial theta function side, and to Kashaev invariants of torus knots on the Habiro ring side. For example, Zagier's strange identity (14.21) for Kontsevich's function connects the Dedekind eta-function (the simplest product modular form) with the trefoil knot (the simplest non-trivial knot).

APPENDIX A

Representations of Mock Theta Functions

In this appendix, we provide a list of all of Ramanujan's original mock theta functions, as well as some newer but still "classical" mock theta functions found in subsequent years. We define each mock theta function as a q-hypergeometric series and then express each in terms of the universal mock theta functions g_2 and g_3 (Definition 9.9), in terms of Zwegers' Appell-Lerch sum μ (Definition 8.1), in terms of indefinite theta functions (8.24), and list the first few terms in the q-expansions. We organize the mock theta functions in terms of their order. We use the original notation found in Ramanujan's letters and in the literature, noting that at times, the same notation is used to denote two different functions (e.g. the letter ψ is used to denote (different) mock theta functions of order 3 and of order 6, and the notation μ refers to both Zwegers' function and a mock theta function of order 6). Some external references for such representations include **[228, 265, 529]**.

A.1. Order 2 mock theta functions

$$
\begin{aligned}
A(q) :&= \sum_{n=0}^{\infty} \frac{q^{n+1}\left(-q^2;q^2\right)_n}{\left(q;q^2\right)_{n+1}} \\
&= q^{\frac{1}{2}} g_2\left(q^{\frac{1}{2}};q^2\right) - q^{\frac{1}{2}}\left(-q;q\right)_\infty \left(-q^2;q^2\right)_\infty^2 \left(q^4;q^4\right)_\infty \\
&= -i\mu\left(3\tau,2\tau;4\tau\right) \\
&= \frac{q^2\eta(4\tau)}{2\eta\left(2\tau\right)^2}\Theta^+_{\left(\begin{smallmatrix}1&1\\1&0\end{smallmatrix}\right),\left(\begin{smallmatrix}0\\1\end{smallmatrix}\right),\left(\begin{smallmatrix}-1\\1\end{smallmatrix}\right)}\left(3\tau,\tau+\frac{1}{2};4\tau\right) \\
&= q + 2q^2 + 3q^3 + 5q^4 + 8q^5 + 11q^6 + 16q^7 + 23q^8 + 31q^9 + \cdots, \\
B(q) :&= \sum_{n=0}^{\infty} \frac{q^{n}\left(-q;q^2\right)_n}{\left(q;q^2\right)_{n+1}} \\
&= g_2\left(q;q^2\right) \\
&= -iq^{-\frac{1}{2}}\mu\left(3\tau,3\tau;4\tau\right) \\
&= \frac{q^{\frac{17}{8}}\eta(2\tau)}{2\eta(\tau)\eta(4\tau)}\Theta^+_{\left(\begin{smallmatrix}1&1\\1&0\end{smallmatrix}\right),\left(\begin{smallmatrix}0\\1\end{smallmatrix}\right),\left(\begin{smallmatrix}-1\\1\end{smallmatrix}\right)}\left(3\tau,2\tau+\frac{1}{2};4\tau\right) \\
&= 1 + 2q + 4q^2 + 6q^3 + 9q^4 + 14q^5 + 20q^6 + 28q^7 + 40q^8 + 54q^9 + \cdots,
\end{aligned}
$$

$$
\begin{aligned}
\mu(q) &:= \sum_{n=0}^{\infty} \frac{(-1)^n q^{n^2} \left(q;q^2\right)_n}{\left(-q^2;q^2\right)_n^2} \\
&= -2q^{\frac{1}{4}} g_2\left(q^{\frac{1}{4}};q^{\frac{1}{2}}\right) + \frac{\left(q^{\frac{1}{2}};q^{\frac{1}{2}}\right)_\infty (q;q)_\infty^3 \left(q^2;q^2\right)_\infty}{\left(q^{\frac{1}{4}};q^{\frac{1}{4}}\right)_\infty^2 \left(q^4;q^4\right)_\infty^2} \\
&= -4\mu\left(\tau, \frac{1}{2}; 4\tau\right) - q^{\frac{1}{8}} \frac{\eta(2\tau)^8}{\eta(\tau)^3\eta(4\tau)^4} \\
&= \frac{q^{\frac{1}{2}}\eta(4\tau)}{\eta\left(8\tau\right)^2} \Theta^+_{\left(\begin{smallmatrix}1&1\\1&0\end{smallmatrix}\right),\left(\begin{smallmatrix}0\\1\end{smallmatrix}\right),\left(\begin{smallmatrix}-1\\1\end{smallmatrix}\right)}(\tau,\tau;4\tau) - q^{\frac{1}{8}} \frac{\eta(2\tau)^8}{\eta(\tau)^3\eta(4\tau)^4} \\
&= 1 - q + q^2 + 2q^3 - q^4 - 4q^5 + q^6 + 5q^7 - 2q^8 - 5q^9 + \cdots.
\end{aligned}
$$

A.2. Order 3 mock theta functions

$$
\begin{aligned}
f(q) &:= \sum_{n=0}^{\infty} \frac{q^{n^2}}{(-q;q)_n^2} \\
&= 2 - 2g_3(-1;q) \\
&= 4q^{-\frac{1}{8}}\mu\left(2\tau + \frac{1}{2}, \tau; 3\tau\right) + q^{\frac{1}{24}} \frac{\eta(3\tau)^4}{\eta(\tau)\eta(6\tau)^2} \\
&= -\frac{2q^{\frac{25}{24}}}{\eta\left(\tau\right)} \Theta^+_{\left(\begin{smallmatrix}1&1\\1&0\end{smallmatrix}\right),\left(\begin{smallmatrix}0\\1\end{smallmatrix}\right),\left(\begin{smallmatrix}-1\\1\end{smallmatrix}\right)}\left(2\tau + \frac{1}{2}, \frac{1}{2}\tau; 3\tau\right) + q^{\frac{1}{24}} \frac{\eta(3\tau)^4}{\eta(\tau)\eta(6\tau)^2} \\
&= 1 + q - 2q^2 + 3q^3 - 3q^4 + 3q^5 - 5q^6 + 7q^7 - 6q^8 + 6q^9 + \cdots, \\
\phi(q) &:= \sum_{n=0}^{\infty} \frac{q^{n^2}}{\left(-q^2;q^2\right)_n} \\
&= (1-i)\left(1 + ig_3(i;q)\right) \\
&= 2ie^{\frac{\pi i}{8}} q^{-\frac{1}{8}}\mu\left(2\tau, 3\tau; 3\tau + \frac{1}{2}\right) + 2q^{\frac{1}{24}} \frac{\eta(6\tau)\eta(12\tau)^2}{\eta(3\tau)\eta(4\tau)} \\
&= \frac{e^{-\frac{\pi i}{12}} q^{\frac{1}{24}}\eta(\tau)}{2\eta(2\tau)^2} \Theta^+_{\left(\begin{smallmatrix}4&0\\0&-3\end{smallmatrix}\right),\left(\begin{smallmatrix}1/4\\-1/6\end{smallmatrix}\right),\left(\begin{smallmatrix}1/8\\-1/6\end{smallmatrix}\right),\left(\begin{smallmatrix}3\\4\end{smallmatrix}\right),\left(\begin{smallmatrix}-3\\4\end{smallmatrix}\right)}(\tau) \\
&= 1 + q - q^3 + q^4 + q^5 - q^6 - q^7 + 2q^9 + \cdots, \\
\psi(q) &:= \sum_{n=1}^{\infty} \frac{q^{n^2}}{\left(q;q^2\right)_n} \\
&= qg_3\left(q;q^4\right) \\
&= -ie^{\frac{\pi i}{8}} q^{-\frac{1}{8}}\mu\left(2\tau, \tau; 3\tau + \frac{1}{2}\right) + q^{\frac{1}{24}} \frac{\eta(6\tau)\eta(12\tau)^2}{\eta(3\tau)\eta(4\tau)} \\
&= \frac{e^{-\frac{\pi i}{4}} q^{\frac{1}{24}}}{4\eta(\tau)} \Theta^+_{\left(\begin{smallmatrix}4&0\\0&-1\end{smallmatrix}\right),\left(\begin{smallmatrix}1/4\\1/2\end{smallmatrix}\right),\left(\begin{smallmatrix}1/8\\0\end{smallmatrix}\right),\left(\begin{smallmatrix}1\\4\end{smallmatrix}\right),\left(\begin{smallmatrix}-1\\4\end{smallmatrix}\right)}(\tau) \\
&= q + q^2 + q^3 + 2q^4 + 2q^5 + 2q^6 + 3q^7 + 3q^8 + 4q^9 + \cdots,
\end{aligned}
$$

$$\begin{aligned}
\chi(q) &:= \sum_{n=0}^{\infty} \frac{q^{n^2}(-q;q)_n}{(-q^3;q^3)_n} \\
&= (1+\zeta_3)\left(1-\zeta_3 g_3(-\zeta_3;q)\right) \\
&= q^{-\frac{1}{8}}\mu\left(2\tau+\frac{1}{2},\tau;3\tau\right) + q^{\frac{1}{24}}\frac{\eta(3\tau)^4}{\eta(\tau)\eta(6\tau)^2} \\
&= -\frac{q^{\frac{25}{24}}}{2\eta(\tau)}\Theta^{+}_{\left(\begin{smallmatrix}1&1\\1&0\end{smallmatrix}\right),\left(\begin{smallmatrix}0\\1\end{smallmatrix}\right),\left(\begin{smallmatrix}-1\\1\end{smallmatrix}\right)}\left(2\tau+\frac{1}{2},\frac{\tau}{2};3\tau\right) + q^{\frac{1}{24}}\frac{\eta(3\tau)^4}{\eta(\tau)\eta(6\tau)^2} \\
&= 1+q+q^2+q^6+q^7+\cdots,
\end{aligned}$$

$$\begin{aligned}
\omega(q) &:= \sum_{n=0}^{\infty} \frac{q^{2n(n+1)}}{(q;q^2)_{n+1}^2} \\
&= g_3\left(q;q^2\right) \\
&= -2iq^{-\frac{3}{4}}\mu\left(3\tau,2\tau;6\tau\right) + q^{-\frac{2}{3}}\frac{\eta(6\tau)^4}{\eta(2\tau)\eta(3\tau)^2} \\
&= \frac{q^{\frac{13}{12}}}{\eta(2\tau)}\Theta^{+}_{\left(\begin{smallmatrix}1&1\\1&0\end{smallmatrix}\right),\left(\begin{smallmatrix}0\\1\end{smallmatrix}\right),\left(\begin{smallmatrix}-1\\1\end{smallmatrix}\right)}\left(3\tau,2\tau+\frac{1}{2};6\tau\right) + q^{-\frac{2}{3}}\frac{\eta(6\tau)^4}{\eta(2\tau)\eta(3\tau)^2} \\
&= 1+2q+3q^2+4q^3+6q^4+8q^5+10q^6+14q^7+18q^8+22q^9+\cdots,
\end{aligned}$$

$$\begin{aligned}
\nu(q) &:= \sum_{n=0}^{\infty} \frac{q^{n(n+1)}}{(-q;q^2)_{n+1}} \\
&= g_3\left(i\sqrt{q};q\right) \\
&= 2iq^{-\frac{1}{2}}\mu\left(5\tau+\frac{1}{2},3\tau+\frac{1}{2};12\tau\right) + q^{-\frac{1}{3}}\frac{\eta(\tau)\eta(3\tau)\eta(12\tau)}{\eta(2\tau)\eta(6\tau)} \\
&= \frac{e^{-\frac{\pi i}{3}}q^{-\frac{1}{3}}\eta(\tau)}{4\eta(2\tau)^2}\Theta^{+}_{\left(\begin{smallmatrix}4&0\\0&-3\end{smallmatrix}\right),\left(\begin{smallmatrix}1/2\\-1/6\end{smallmatrix}\right),\left(\begin{smallmatrix}1/8\\-1/6\end{smallmatrix}\right),\left(\begin{smallmatrix}3\\4\end{smallmatrix}\right),\left(\begin{smallmatrix}-3\\4\end{smallmatrix}\right)}(\tau) \\
&= 1-q+2q^2-2q^3+2q^4-3q^5+4q^6-4q^7+5q^8-6q^9+\cdots,
\end{aligned}$$

$$\begin{aligned}
\rho(q) &:= \sum_{n=0}^{\infty} \frac{q^{n^2}}{(q;q^2)_n} \\
&= g_3\left(\zeta_3 q;q^2\right) \\
&= iq^{-\frac{3}{4}}\mu\left(2\tau+\frac{1}{2},\tau+\frac{1}{2};6\tau\right) \\
&= \frac{q^{\frac{1}{3}}\eta(\tau)\eta(4\tau)\eta(6\tau)}{2\eta(2\tau)^2\eta(3\tau)\eta(12\tau)}\Theta^{+}_{\left(\begin{smallmatrix}1&1\\1&0\end{smallmatrix}\right),\left(\begin{smallmatrix}0\\1\end{smallmatrix}\right),\left(\begin{smallmatrix}-1\\1\end{smallmatrix}\right)}\left(2\tau+\frac{1}{2},2\tau+\frac{1}{2};6\tau\right) \\
&= 1-q+q^3-q^5+q^6-q^7+q^9+\cdots.
\end{aligned}$$

A.3. Order 5 mock theta functions

$$
\begin{aligned}
f_0(q) :&= \sum_{n=0}^{\infty} \frac{q^{n^2}}{(-q;q)_n} \\
&= -2q^2 g_3\left(q^2;q^{10}\right) + \frac{\left(q^5;q^5\right)_\infty \left(q^5;q^{10}\right)_\infty}{\left(q;q^5\right)_\infty \left(q^4;q^5\right)_\infty} \\
&= 2iq^{-\frac{13}{4}} \mu\left(18\tau, 4\tau; 30\tau\right) + 2iq^{-\frac{1}{4}} \mu\left(8\tau, 4\tau; 30\tau\right) + iq^{\frac{5}{12}} \frac{\eta(5\tau)^2 \vartheta(2\tau;5\tau)}{\eta(\tau)\eta(10\tau)} \\
&= \frac{q^{\frac{1}{60}}}{2\eta(\tau)} \Theta^+_{\left(\begin{smallmatrix}5&0\\0&-2\end{smallmatrix}\right),\left(\begin{smallmatrix}1/10\\0\end{smallmatrix}\right),\left(\begin{smallmatrix}0\\1/4\end{smallmatrix}\right),\left(\begin{smallmatrix}2\\5\end{smallmatrix}\right),\left(\begin{smallmatrix}-2\\5\end{smallmatrix}\right)}(\tau) \\
&= 1 + q - q^2 + q^3 - q^6 + q^7 + q^9 + \cdots,
\end{aligned}
$$

$$
\begin{aligned}
f_1(q) :&= \sum_{n=0}^{\infty} \frac{q^{n(n+1)}}{(-q;q)_n} \\
&= -2q^3 g_3\left(q^4;q^{10}\right) + \frac{\left(q^5;q^5\right)_\infty \left(q^5;q^{10}\right)_\infty}{\left(q^2;q^5\right)_\infty \left(q^3;q^5\right)_\infty} \\
&= 2iq^{-\frac{5}{4}} \mu\left(16\tau, 8\tau; 30\tau\right) + 2iq^{-\frac{1}{4}} \mu\left(6\tau, 8\tau; 30\tau\right) + iq^{-\frac{1}{12}} \frac{\eta(5\tau)^2 \vartheta(\tau;5\tau)}{\eta(\tau)\eta(10\tau)} \\
&= \frac{q^{-\frac{11}{60}}}{2\eta(\tau)} \Theta^+_{\left(\begin{smallmatrix}5&0\\0&-2\end{smallmatrix}\right),\left(\begin{smallmatrix}3/10\\0\end{smallmatrix}\right),\left(\begin{smallmatrix}0\\1/4\end{smallmatrix}\right),\left(\begin{smallmatrix}2\\5\end{smallmatrix}\right),\left(\begin{smallmatrix}-2\\5\end{smallmatrix}\right)}(\tau) \\
&= 1 + q^2 - q^3 + q^4 - q^5 + 2q^6 - 2q^7 + q^8 - q^9 + \cdots,
\end{aligned}
$$

$$
\begin{aligned}
F_0(q) :&= \sum_{n=0}^{\infty} \frac{q^{2n^2}}{\left(q;q^2\right)_n} \\
&= 1 + q g_3\left(q;q^5\right) - q \frac{\left(-q^5;q^{20}\right)_\infty \left(-q^{15};q^{20}\right)_\infty \left(q^{20};q^{20}\right)_\infty}{\left(q^4;q^{10}\right)_\infty \left(q^6;q^{10}\right)_\infty} \\
&= -iq^{-\frac{1}{8}} \mu\left(3\tau, \tau; 15\tau\right) + iq^{-\frac{17}{8}} \mu\left(12\tau, 4\tau; 15\tau\right) + iq^{-\frac{19}{24}} \frac{\eta(10\tau)^3}{\eta(5\tau)\vartheta(4\tau;10\tau)} \\
&= 1 + \frac{q^{\frac{1}{120}}}{2\eta(2\tau)} \Theta^+_{\left(\begin{smallmatrix}5&0\\0&-2\end{smallmatrix}\right),\left(\begin{smallmatrix}1/5\\1/4\end{smallmatrix}\right),\left(\begin{smallmatrix}1/2\\1\end{smallmatrix}\right),\left(\begin{smallmatrix}2\\5\end{smallmatrix}\right)}(2\tau) \\
&= 1 + q^2 + q^3 + q^4 + q^5 + q^6 + q^7 + 2q^8 + 2q^9 + \cdots,
\end{aligned}
$$

$$
\begin{aligned}
F_1(q) :&= \sum_{n=0}^{\infty} \frac{q^{2n(n+1)}}{\left(q;q^2\right)_{n+1}} \\
&= q g_3\left(q^2;q^5\right) + \frac{\left(-q^5;q^{20}\right)_\infty \left(-q^{15};q^{20}\right)_\infty \left(q^{20};q^{20}\right)_\infty}{\left(q^2;q^{10}\right)_\infty \left(q^8;q^{10}\right)_\infty} \\
&= -iq^{-\frac{5}{8}} \mu\left(3\tau, 4\tau; 15\tau\right) - iq^{-\frac{9}{8}} \mu\left(8\tau, 4\tau; 15\tau\right) - iq^{-\frac{19}{24}} \frac{\eta(10\tau)^3}{\eta(5\tau)\vartheta(2\tau;10\tau)} \\
&= \frac{q^{-\frac{71}{120}}}{2\eta(2\tau)} \Theta^+_{\left(\begin{smallmatrix}5&0\\0&-2\end{smallmatrix}\right),\left(\begin{smallmatrix}2/5\\1/4\end{smallmatrix}\right),\left(\begin{smallmatrix}1/2\\2\end{smallmatrix}\right),\left(\begin{smallmatrix}2\\5\end{smallmatrix}\right),\left(\begin{smallmatrix}-2\\5\end{smallmatrix}\right)}(2\tau) \\
&= 1 + q + q^2 + q^3 + 2q^4 + 2q^5 + 2q^6 + 3q^7 + 3q^8 + 3q^9 + \cdots,
\end{aligned}
$$

$$\begin{aligned}
\phi_0(q) :&= \sum_{n=0}^{\infty} q^{n^2} \left(-q;q^2\right)_n \\
&= qg_3\left(-q;-q^5\right) + \frac{\left(-q^2;-q^5\right)_\infty \left(q^3;-q^5\right)_\infty \left(-q^5;-q^5\right)_\infty}{\left(q^2;q^{10}\right)_\infty \left(q^8;q^{10}\right)_\infty} \\
&= e^{\frac{\pi i}{8}} q^{-\frac{13}{8}} \mu\left(16\tau+\frac{1}{2}, 9\tau; 15\tau+\frac{1}{2}\right) - ie^{\frac{\pi i}{8}} q^{-\frac{1}{8}} \mu\left(11\tau, 9\tau; 15\tau+\frac{1}{2}\right) \\
&= \frac{e^{\frac{\pi i}{8}} q^{\frac{1}{120}} \eta\left(\tau+\frac{1}{2}\right)}{2\eta(2\tau)^2} \Theta^+_{\left(\begin{smallmatrix}5&0\\0&-3\end{smallmatrix}\right),\left(\begin{smallmatrix}1/5\\1/6\end{smallmatrix}\right),\left(\begin{smallmatrix}0\\1/6\end{smallmatrix}\right),\left(\begin{smallmatrix}3\\5\end{smallmatrix}\right),\left(\begin{smallmatrix}-3\\5\end{smallmatrix}\right)}(2\tau) \\
&= 1+q+q^2+q^4+q^5+q^7+q^8+q^9+\cdots,
\end{aligned}$$

$$\begin{aligned}
\phi_1(q) :&= \sum_{n=0}^{\infty} q^{(n+1)^2} \left(-q;q^2\right)_n \\
&= q^2 g_3\left(q^2;-q^5\right) + q\frac{\left(q;-q^5\right)_\infty \left(-q^4;-q^5\right)_\infty \left(-q^5;-q^5\right)_\infty}{\left(q^4;q^{10}\right)_\infty \left(q^6;q^{10}\right)_\infty} \\
&= -e^{\frac{\pi i}{8}} q^{\frac{3}{8}} \mu\left(2\tau+\frac{1}{2}, 3\tau; 15\tau+\frac{1}{2}\right) - ie^{\frac{\pi i}{8}} q^{-\frac{1}{8}} \mu\left(7\tau, 3\tau; 15\tau+\frac{1}{2}\right) \\
&= \frac{e^{\frac{\pi i}{8}} q^{\frac{49}{120}} \eta\left(\tau+\frac{1}{2}\right)}{2\eta(2\tau)^2} \Theta^+_{\left(\begin{smallmatrix}5&0\\0&-3\end{smallmatrix}\right),\left(\begin{smallmatrix}2/5\\1/6\end{smallmatrix}\right),\left(\begin{smallmatrix}0\\1/6\end{smallmatrix}\right),\left(\begin{smallmatrix}3\\5\end{smallmatrix}\right),\left(\begin{smallmatrix}-3\\5\end{smallmatrix}\right)}(2\tau) \\
&= q+q^4+q^5+q^9+\cdots,
\end{aligned}$$

$$\begin{aligned}
\psi_0(q) :&= \sum_{n=0}^{\infty} q^{\frac{1}{2}(n+1)(n+2)} (-q;q)_n \\
&= q^2 g_3\left(q^2;q^{10}\right) + q\frac{\left(q;q^{10}\right)_\infty \left(q^9;q^{10}\right)_\infty \left(q^{10};q^{10}\right)_\infty}{\left(q^2;q^5\right)_\infty \left(q^3;q^5\right)_\infty} \\
&= -iq^{-\frac{13}{4}} \mu(17\tau, 3\tau; 30\tau) - iq^{-\frac{1}{4}} \mu(7\tau, 3\tau; 30\tau) \\
&= \frac{e^{\frac{\pi i}{6}} q^{\frac{1}{60}} \eta(2\tau)}{2\eta(\tau)^2} \Theta^+_{\left(\begin{smallmatrix}5&0\\0&-3\end{smallmatrix}\right),\left(\begin{smallmatrix}1/10\\1/6\end{smallmatrix}\right),\left(\begin{smallmatrix}0\\1/6\end{smallmatrix}\right),\left(\begin{smallmatrix}3\\5\end{smallmatrix}\right),\left(\begin{smallmatrix}-3\\5\end{smallmatrix}\right)}(\tau) \\
&= q+q^3+q^4+q^6+q^7+q^8+q^9+\cdots,
\end{aligned}$$

$$\begin{aligned}
\psi_1(q) :&= \sum_{n=0}^{\infty} q^{\frac{1}{2}n(n+1)} (-q;q)_n \\
&= q^3 g_3\left(q^4;q^{10}\right) + \frac{\left(q^3;q^{10}\right)_\infty \left(q^7;q^{10}\right)_\infty \left(q^{10};q^{10}\right)_\infty}{\left(q;q^5\right)_\infty \left(q^4;q^5\right)_\infty} \\
&= -iq^{-\frac{5}{4}} \mu(\tau, 9\tau; 30\tau) - iq^{-\frac{1}{4}} \mu(11\tau, 9\tau; 30\tau) \\
&= \frac{e^{\frac{\pi i}{6}} q^{-\frac{11}{60}} \eta(2\tau)}{2\eta(\tau)^2} \Theta^+_{\left(\begin{smallmatrix}5&0\\0&-3\end{smallmatrix}\right),\left(\begin{smallmatrix}3/10\\1/6\end{smallmatrix}\right),\left(\begin{smallmatrix}0\\1/6\end{smallmatrix}\right),\left(\begin{smallmatrix}3\\5\end{smallmatrix}\right),\left(\begin{smallmatrix}-3\\5\end{smallmatrix}\right)}(\tau) \\
&= 1+q+q^2+q^3+q^4+q^5+2q^6+q^7+q^8+2q^9+\cdots,
\end{aligned}$$

$$
\begin{aligned}
\chi_0(q) :&= \sum_{n=0}^{\infty} \frac{q^n}{(q^{n+1};q)_n} \\
&= 2 + 3qg_3\left(q;q^5\right) - \frac{\left(q^2;q^5\right)_\infty \left(q^3;q^5\right)_\infty \left(q^5;q^5\right)_\infty}{\left(q;q^5\right)_\infty^2 \left(q^4;q^5\right)_\infty^2} \\
&= 2 - 3iq^{-\frac{13}{8}}\mu\left(\tau,-6\tau;15\tau\right) - 3iq^{-\frac{1}{8}}\mu\left(6\tau,4\tau;15\tau\right) - 2iq^{\frac{29}{24}}\frac{\vartheta(2\tau;5\tau)^3}{\eta(\tau)^2} \\
&= 2 + \frac{3iq^{-\frac{9}{8}}}{2\vartheta\left(6\tau;15\tau\right)}\Theta^+_{\left(\begin{smallmatrix}1&1\\1&0\end{smallmatrix}\right),\left(\begin{smallmatrix}0\\1\end{smallmatrix}\right),\left(\begin{smallmatrix}-1\\1\end{smallmatrix}\right)}\left(\tau,\frac{1}{2}\tau+\frac{1}{2};15\tau\right) \\
&\quad - \frac{3iq^{\frac{23}{8}}}{2\vartheta\left(4\tau;15\tau\right)}\Theta^+_{\left(\begin{smallmatrix}1&1\\1&0\end{smallmatrix}\right),\left(\begin{smallmatrix}0\\1\end{smallmatrix}\right),\left(\begin{smallmatrix}-1\\1\end{smallmatrix}\right)}\left(6\tau,\frac{11}{2}\tau+\frac{1}{2};15\tau\right) - 2iq^{\frac{29}{24}}\frac{\vartheta(2\tau;5\tau)^3}{\eta(\tau)^2} \\
&= 1 + q + q^2 + 2q^3 + q^4 + 3q^5 + 2q^6 + 3q^7 + 3q^8 + 5q^9 + \cdots, \\
\chi_1(q) :&= \sum_{n=0}^{\infty} \frac{q^n}{(q^{n+1};q)_{n+1}} \\
&= 3qg_3\left(q^2;q^5\right) + \frac{\left(q,q^5\right)_\infty \left(q^4;q^5\right)_\infty \left(q^5;q^5\right)_\infty}{\left(q^2;q^5\right)_\infty^2 \left(q^3;q^5\right)_\infty^2} \\
&= -3iq^{-\frac{9}{8}}\mu(7\tau,3\tau;15\tau) - 3iq^{-\frac{5}{8}}\mu(3\tau,2\tau;15\tau) + 2iq^{-\frac{7}{24}}\frac{\vartheta(\tau;5\tau)^3}{\eta(\tau)^2} \\
&= -\frac{3iq^{\frac{19}{8}}}{2\vartheta\left(3\tau;15\tau\right)}\Theta^+_{\left(\begin{smallmatrix}1&1\\1&0\end{smallmatrix}\right),\left(\begin{smallmatrix}0\\1\end{smallmatrix}\right),\left(\begin{smallmatrix}-1\\1\end{smallmatrix}\right)}\left(7\tau,\frac{7}{2}\tau+\frac{1}{2};15\tau\right) \\
&\quad - \frac{3iq^{\frac{7}{8}}}{2\vartheta\left(2\tau;15\tau\right)}\Theta^+_{\left(\begin{smallmatrix}1&1\\1&0\end{smallmatrix}\right),\left(\begin{smallmatrix}0\\1\end{smallmatrix}\right),\left(\begin{smallmatrix}-1\\1\end{smallmatrix}\right)}\left(3\tau,\frac{13}{2}\tau+\frac{1}{2};15\tau\right) + 2iq^{-\frac{7}{24}}\frac{\vartheta(\tau;5\tau)^3}{\eta(\tau)^2} \\
&= 1 + 2q + 2q^2 + 3q^3 + 3q^4 + 4q^5 + 4q^6 + 6q^7 + 5q^8 + 7q^9 + \cdots, \\
\Psi_0(q) :&= -1 + \sum_{n=0}^{\infty} \frac{q^{5n^2}}{\left(q;q^5\right)_{n+1}\left(q^4;q^5\right)_n} \\
&= qg_3\left(q;q^5\right) \\
&= -iq^{-\frac{1}{8}}\mu\left(4\tau,2\tau;15\tau\right) - iq^{-\frac{13}{8}}\mu\left(9\tau,2\tau;15\tau\right) \\
&= -\frac{iq^{\frac{15}{8}}}{2\vartheta\left(2\tau;15\tau\right)}\Theta^+_{\left(\begin{smallmatrix}1&1\\1&0\end{smallmatrix}\right),\left(\begin{smallmatrix}0\\1\end{smallmatrix}\right),\left(\begin{smallmatrix}-1\\1\end{smallmatrix}\right)}\left(4\tau,\frac{11}{2}\tau+\frac{1}{2};15\tau\right) \\
&\quad - \frac{iq^{\frac{23}{8}}}{2\vartheta\left(2\tau;15\tau\right)}\Theta^+_{\left(\begin{smallmatrix}1&1\\1&0\end{smallmatrix}\right),\left(\begin{smallmatrix}0\\1\end{smallmatrix}\right),\left(\begin{smallmatrix}-1\\1\end{smallmatrix}\right)}\left(9\tau,\frac{\tau}{2}+\frac{1}{2};15\tau\right) \\
&= q + q^2 + q^3 + q^4 + 2q^5 + 2q^6 + 2q^7 + 2q^8 + 3q^9 + \cdots, \\
\Psi_1(q) :&= -1 + \sum_{n=0}^{\infty} \frac{q^{5n^2}}{\left(q^2;q^5\right)_{n+1}\left(q^3;q^5\right)_n} \\
&= q^2g_3\left(q^2;q^5\right) \\
&= -iq^{\frac{3}{8}}\mu\left(3\tau,4\tau;15\tau\right) - iq^{-\frac{1}{8}}\mu\left(8\tau,4\tau;15\tau\right) \\
&= -\frac{iq^{\frac{15}{8}}}{2\vartheta\left(4\tau;15\tau\right)}\Theta^+_{\left(\begin{smallmatrix}1&1\\1&0\end{smallmatrix}\right),\left(\begin{smallmatrix}0\\1\end{smallmatrix}\right),\left(\begin{smallmatrix}-1\\1\end{smallmatrix}\right)}\left(3\tau,\frac{17}{2}\tau+\frac{1}{2};15\tau\right)
\end{aligned}
$$

$$- \frac{iq^{\frac{31}{8}}}{2\vartheta(4\tau;15\tau)} \Theta^+_{\left(\begin{smallmatrix}1&1\\1&0\end{smallmatrix}\right),\left(\begin{smallmatrix}0\\1\end{smallmatrix}\right),\left(\begin{smallmatrix}-1\\1\end{smallmatrix}\right)} \left(8\tau, \frac{7}{2}\tau + \frac{1}{2}; 15\tau\right)$$
$$= q^2 + q^4 + q^5 + q^6 + q^7 + 2q^8 + q^9 + \cdots .$$

A.4. Order 6 mock theta functions

$$\phi(q) := \sum_{n=0}^{\infty} \frac{(-1)^n q^{n^2} \left(q;q^2\right)_n}{(-q;q)_{2n}}$$
$$= -2q^{\frac{1}{4}} g_2\left(q^{\frac{1}{4}}; q^{\frac{3}{2}}\right) + \frac{\left(q^{\frac{1}{2}};q^{\frac{1}{2}}\right)_\infty^3 \left(q^{\frac{3}{4}};q^{\frac{3}{4}}\right)_\infty^2 \left(q^3;q^3\right)_\infty^3}{\left(q^{\frac{1}{4}};q^{\frac{1}{4}}\right)_\infty^2 \left(q^{\frac{3}{2}};q^{\frac{3}{2}}\right)_\infty^3 \left(q^2;q^2\right)_\infty \left(q^6;q^6\right)_\infty}$$
$$= 2iq^{-\frac{1}{8}} \mu\left(\tau + \frac{1}{2}, \frac{1}{2}; 3\tau\right)$$
$$= \frac{q^{\frac{3}{8}}\eta(3\tau)}{2\eta(6\tau)^2} \Theta^+_{\left(\begin{smallmatrix}1&1\\1&0\end{smallmatrix}\right),\left(\begin{smallmatrix}0\\1\end{smallmatrix}\right),\left(\begin{smallmatrix}-1\\1\end{smallmatrix}\right)} \left(\tau + \frac{1}{2}, \frac{1}{2}\tau + \frac{1}{2}; 3\tau\right)$$
$$= 1 - q + 2q^2 - q^3 + q^4 - 3q^5 + 3q^6 - 3q^7 + 4q^8 - 4q^9 + \cdots ,$$

$$\psi(q) := \sum_{n=0}^{\infty} \frac{(-1)^n q^{(n+1)^2} \left(q;q^2\right)_n}{(-q;q)_{2n+1}}$$
$$= -q^{\frac{3}{4}} g_2\left(q^{\frac{3}{4}}; q^{\frac{3}{2}}\right) + q^{\frac{3}{4}} \frac{\left(q^{\frac{1}{2}};q^{\frac{1}{2}}\right)_\infty^2 (q;q)_\infty \left(q^6;q^6\right)_\infty^2}{\left(q^{\frac{1}{4}};q^{\frac{1}{4}}\right)_\infty \left(q^{\frac{3}{4}};q^{\frac{3}{4}}\right)_\infty \left(q^2;q^2\right)_\infty^2}$$
$$= iq^{\frac{3}{8}} \mu\left(\tau + \frac{1}{2}, \tau + \frac{1}{2}; 3\tau\right)$$
$$= \frac{q^{\frac{25}{24}}\eta(\tau)\eta(6\tau)}{2\eta(2\tau)\eta(3\tau)^2} \Theta^+_{\left(\begin{smallmatrix}1&1\\1&0\end{smallmatrix}\right),\left(\begin{smallmatrix}0\\1\end{smallmatrix}\right),\left(\begin{smallmatrix}-1\\1\end{smallmatrix}\right)} \left(\tau + \frac{1}{2}, \frac{3}{2}\tau + \frac{1}{2}; 3\tau\right)$$
$$= q - q^2 + q^3 - 2q^4 + 3q^5 - 2q^6 + 2q^7 - 4q^8 + 5q^9 + \cdots ,$$

$$\rho(q) := \sum_{n=0}^{\infty} \frac{q^{\frac{n(n+1)}{2}} (-q;q)_n}{\left(q;q^2\right)_{n+1}}$$
$$= q^{\frac{1}{2}} g_2\left(q^{\frac{3}{2}}; q^3\right) - q^{\frac{1}{2}} \frac{(q;q)_\infty^2 \left(q^2;q^2\right)_\infty \left(q^{12};q^{12}\right)_\infty^2}{\left(q^{\frac{1}{2}};q^{\frac{1}{2}}\right)_\infty \left(q^{\frac{3}{2}};q^{\frac{3}{2}}\right)_\infty \left(q^4;q^4\right)_\infty^2}$$
$$+ \frac{\left(q^2;q^2\right)_\infty^6 \left(q^3;q^3\right)_\infty \left(q^{12};q^{12}\right)_\infty}{(q;q)_\infty^3 \left(q^4;q^4\right)_\infty^3 \left(q^6;q^6\right)_\infty}$$
$$= -iq^{-\frac{1}{4}} \mu(\tau, \tau; 6\tau)$$
$$= \frac{q^{\frac{1}{3}}\eta(2\tau)\eta(3\tau)}{2\eta(\tau)\eta(6\tau)^2} \Theta^+_{\left(\begin{smallmatrix}1&1\\1&0\end{smallmatrix}\right),\left(\begin{smallmatrix}0\\1\end{smallmatrix}\right),\left(\begin{smallmatrix}-1\\1\end{smallmatrix}\right)} \left(\tau, 3\tau + \frac{1}{2}; 6\tau\right)$$
$$= 1 + 2q + 3q^2 + 4q^3 + 6q^4 + 8q^5 + 11q^6 + 14q^7 + 18q^8 + 24q^9 + \cdots ,$$

$$\sigma(q) := \sum_{n=0}^{\infty} \frac{q^{\frac{(n+1)(n+2)}{2}} (-q;q)_n}{(q;q^2)_{n+1}}$$

$$= q^{\frac{1}{2}} g_2\left(q^{\frac{1}{2}};q^3\right) - \frac{1}{2} \frac{(q;q)_\infty^3 \left(q^{\frac{3}{2}};q^{\frac{3}{2}}\right)_\infty^2 (q^6;q^6)_\infty^3}{\left(q^{\frac{1}{2}};q^{\frac{1}{2}}\right)_\infty^2 (q^3;q^3)_\infty^3 (q^4;q^4)_\infty (q^{12};q^{12})_\infty}$$

$$+ \frac{1}{2} \frac{\left(q^2;q^2\right)_\infty^4 (q^6;q^6)_\infty^5}{(q;q)_\infty^2 (q^3;q^3)_\infty^2 (q^4;q^4)_\infty^2 (q^{12};q^{12})_\infty^2}$$

$$= -iq^{-\frac{1}{4}} \mu(3\tau,\tau;6\tau)$$

$$= \frac{q^{\frac{4}{3}} \eta(2\tau)\eta(3\tau)}{2\eta(\tau)\eta(6\tau)^2} \Theta^+_{\left(\begin{smallmatrix}1&1\\1&0\end{smallmatrix}\right),\left(\begin{smallmatrix}0\\1\end{smallmatrix}\right),\left(\begin{smallmatrix}-1\\1\end{smallmatrix}\right)}\left(3\tau,\tau+\frac{1}{2};6\tau\right)$$

$$= q + q^2 + 2q^3 + 3q^4 + 3q^5 + 5q^6 + 7q^7 + 8q^8 + 11q^9 + \cdots,$$

$$\lambda(q) := \sum_{n=0}^{\infty} \frac{(-1)^n q^n \left(q;q^2\right)_n}{(-q;q)_n}$$

$$= -2q^{\frac{1}{2}} g_2\left(q^{\frac{3}{2}};q^3\right) + 2q^{\frac{1}{2}} \frac{(q;q)_\infty^2 (q^2;q^2)_\infty \left(q^{12};q^{12}\right)_\infty^2}{\left(q^{\frac{1}{2}};q^{\frac{1}{2}}\right)_\infty \left(q^{\frac{3}{2}};q^{\frac{3}{2}}\right)_\infty (q^4;q^4)_\infty^2}$$

$$+ \frac{(q;q)_\infty^3 \left(q^6;q^6\right)_\infty^2}{(q^2;q^2)_\infty^3 (q^3;q^3)_\infty}$$

$$= 2iq^{-\frac{1}{4}} \mu\left(2\tau+\frac{1}{2}, 2\tau+\frac{1}{2};6\tau\right) + q^{-\frac{1}{4}} \frac{\eta(\tau)^3\eta(6\tau)^2}{\eta(2\tau)^3\eta(3\tau)}$$

$$= \frac{q^{\frac{13}{12}} \eta(2\tau)\eta(12\tau)}{\eta(4\tau)\eta(6\tau)^2} \Theta^+_{\left(\begin{smallmatrix}1&1\\1&0\end{smallmatrix}\right),\left(\begin{smallmatrix}0\\1\end{smallmatrix}\right),\left(\begin{smallmatrix}-1\\1\end{smallmatrix}\right)}\left(2\tau+\frac{1}{2}, 3\tau+\frac{1}{2};6\tau\right) + q^{-\frac{1}{4}} \frac{\eta(\tau)^3\eta(6\tau)^2}{\eta(2\tau)^3\eta(3\tau)}$$

$$= 1 - q + 3q^2 - 5q^3 + 6q^4 - 7q^5 + 11q^6 - 16q^7 + 18q^8 - 21q^9 + \cdots,$$

$$\mu(q) := \frac{1}{2} + \frac{1}{2} \sum_{n=0}^{\infty} \frac{(-1)^n q^{n+1} (1+q^n) \left(q;q^2\right)_n}{(-q;q)_{n+1}}$$

$$= -2q^{\frac{1}{2}} g_2\left(q^{\frac{1}{2}};q^3\right) + \frac{(q;q)_\infty^3 \left(q^{\frac{3}{2}};q^{\frac{3}{2}}\right)_\infty^2 (q^6;q^6)_\infty^3}{\left(q^{\frac{1}{2}};q^{\frac{1}{2}}\right)_\infty^2 (q^3;q^3)_\infty^3 (q^4;q^4)_\infty (q^{12};q^{12})_\infty}$$

$$- \frac{1}{2} \frac{(q;q)_\infty^2 \left(q^3;q^3\right)_\infty^2}{(q^2;q^2)_\infty^2 (q^6;q^6)_\infty}$$

$$= 2iq^{-\frac{1}{4}} \mu\left(2\tau+\frac{1}{2}, \frac{1}{2};6\tau\right) - \frac{1}{2} q^{\frac{1}{12}} \frac{\eta(\tau)^2\eta(3\tau)^2}{\eta(2\tau)^2\eta(6\tau)}$$

$$= \frac{q^{\frac{3}{4}} \eta(6\tau)}{2\eta(12\tau)^2} \Theta^+_{\left(\begin{smallmatrix}1&1\\1&0\end{smallmatrix}\right),\left(\begin{smallmatrix}0\\1\end{smallmatrix}\right),\left(\begin{smallmatrix}-1\\1\end{smallmatrix}\right)}\left(2\tau+\frac{1}{2}, \tau+\frac{1}{2};6\tau\right) - \frac{1}{2} q^{\frac{1}{12}} \frac{\eta(\tau)^2\eta(3\tau)^2}{\eta(2\tau)^2\eta(6\tau)}$$

$$= \frac{1}{2} + q - \frac{3}{2}q^2 + 2q^3 - 2q^4 + 3q^5 - \frac{11}{2}q^6 + 7q^7 - \frac{15}{2}q^8 + 11q^9 + \cdots,$$

$$\gamma(q) := \sum_{n=0}^{\infty} \frac{q^{n^2}(q;q)_n}{(q^3;q^3)_n}$$

$$= -3q^{\frac{1}{4}} g_2\left(q^{\frac{1}{4}}; q^{\frac{3}{2}}\right) + \frac{3}{2} \frac{\left(q^{\frac{1}{2}}; q^{\frac{1}{2}}\right)_\infty^3 \left(q^{\frac{3}{4}}; q^{\frac{3}{4}}\right)_\infty^2 (q^3;q^3)_\infty^3}{\left(q^{\frac{1}{4}}; q^{\frac{1}{4}}\right)_\infty^2 \left(q^{\frac{3}{2}}; q^{\frac{3}{2}}\right)_\infty^3 (q^2;q^2)_\infty (q^6;q^6)_\infty}$$

$$- \frac{1}{2} \frac{(q;q)_\infty^5 (q^6;q^6)_\infty}{(q^2;q^2)_\infty^3 (q^3;q^3)_\infty^2}$$

$$= 3iq^{-\frac{1}{8}} \mu\left(2\tau + \frac{1}{2}, \tau + \frac{1}{2}; 3\tau\right) + q^{\frac{1}{24}} \frac{\eta(\tau)^5 \eta(6\tau)}{\eta(2\tau)^3 \eta(3\tau)^2}$$

$$= \frac{3q^{\frac{25}{24}} \eta(\tau)\eta(6\tau)}{2\eta(2\tau)\eta(3\tau)^2} \Theta^+_{\left(\begin{smallmatrix}1&1\\1&0\end{smallmatrix}\right),\left(\begin{smallmatrix}0\\1\end{smallmatrix}\right),\left(\begin{smallmatrix}-1\\1\end{smallmatrix}\right)} \left(2\tau + \frac{1}{2}, \frac{\tau}{2} + \frac{1}{2}; 3\tau\right) + q^{\frac{1}{24}} \frac{\eta(\tau)^5 \eta(6\tau)}{\eta(2\tau)^3 \eta(3\tau)^2}$$

$$= 1 + q - q^2 + 2q^4 - 2q^5 - q^6 + 3q^7 - 2q^8 + \cdots,$$

$$\phi_-(q) := \sum_{n=1}^{\infty} \frac{q^n(-q;q)_{2n-1}}{(q;q^2)_n}$$

$$= iq^{\frac{1}{4}} g_2\left(iq^{\frac{1}{4}}; -q^{\frac{3}{2}}\right) - \frac{1}{2} \frac{\left(-q^{\frac{1}{2}}; -q^{\frac{1}{2}}\right)_\infty^3 \left(-iq^{\frac{3}{4}}; -iq^{\frac{3}{4}}\right)_\infty^2 (q^3;q^3)_\infty^3}{\left(iq^{\frac{1}{4}}; iq^{\frac{1}{4}}\right)_\infty^2 \left(-q^{\frac{3}{2}}; -q^{\frac{3}{2}}\right)_\infty^3 (q^2;q^2)_\infty (q^6;q^6)_\infty}$$

$$+ \frac{1}{2} \frac{(q;q)_\infty^4 (q^3;q^3)_\infty^5}{\left(-q^{\frac{1}{2}}; -q^{\frac{1}{2}}\right)_\infty^2 \left(-q^{\frac{3}{2}}; -q^{\frac{3}{2}}\right)_\infty^2 (q^2;q^2)_\infty^2 (q^6;q^6)_\infty^2}$$

$$+ q^{\frac{1}{2}} \frac{(q^2;q^2)_\infty^2 (q^6;q^6)_\infty^2}{(q;q)_\infty^2 (q^3;q^3)_\infty}$$

$$= -iq^{-\frac{1}{8}} \mu(2\tau, \tau; 3\tau) - q^{\frac{1}{24}} \frac{\eta(6\tau)^6}{\eta(2\tau)^2 \eta(3\tau)^3}$$

$$= \frac{q^{\frac{25}{24}}}{2\eta(\tau)} \Theta^+_{\left(\begin{smallmatrix}1&1\\1&0\end{smallmatrix}\right),\left(\begin{smallmatrix}0\\1\end{smallmatrix}\right),\left(\begin{smallmatrix}-1\\1\end{smallmatrix}\right)} \left(2\tau, \frac{\tau}{2} + \frac{1}{2}; 3\tau\right) - q^{\frac{1}{24}} \frac{\eta(6\tau)^6}{\eta(2\tau)^2 \eta(3\tau)^3}$$

$$= q + 3q^2 + 5q^3 + 8q^4 + 14q^5 + 22q^6 + 33q^7 + 51q^8 + 74q^9 + \cdots,$$

$$\psi_-(q) := \sum_{n=1}^{\infty} \frac{q^n(-q;q)_{2n-2}}{(q;q^2)_n}$$

$$= -\frac{i}{2} q^{\frac{3}{4}} g_2\left(-iq^{\frac{3}{4}}; -q^{\frac{3}{2}}\right) + \frac{i}{2} q^{\frac{3}{4}} \frac{\left(-q^{\frac{1}{2}}; -q^{\frac{1}{2}}\right)_\infty^2 (q;q)_\infty (q^6;q^6)_\infty^2}{\left(iq^{\frac{1}{4}}; iq^{\frac{1}{4}}\right)_\infty \left(-iq^{\frac{3}{4}}; -iq^{\frac{3}{4}}\right)_\infty (q^2;q^2)_\infty^2}$$

$$- \frac{1}{2} q^{\frac{1}{2}} \frac{(q;q)_\infty^6 \left(-q^{\frac{3}{2}}; -q^{\frac{3}{2}}\right)_\infty (q^6;q^6)_\infty}{\left(-q^{\frac{1}{2}}; -q^{\frac{1}{2}}\right)_\infty^3 (q^2;q^2)_\infty^3 (q^3;q^3)_\infty} + \frac{1}{2} q^{\frac{1}{2}} \frac{(q^2;q^2)_\infty^3 (q^3;q^3)_\infty^2}{(q;q)_\infty^3 (q^6;q^6)_\infty}$$

$$= -\frac{i}{2} q^{\frac{3}{8}} \mu(\tau, \tau; 3\tau) + \frac{1}{2} q^{\frac{3}{8}} \frac{\eta(6\tau)^3}{\eta(\tau)\eta(2\tau)}$$

$$= \frac{q^{\frac{25}{24}}}{4\eta(\tau)} \Theta^{+}_{\left(\begin{smallmatrix}1&1\\1&0\end{smallmatrix}\right),\left(\begin{smallmatrix}0\\1\end{smallmatrix}\right),\left(\begin{smallmatrix}-1\\1\end{smallmatrix}\right)}\left(\tau, \frac{3}{2}\tau + \frac{1}{2}; 3\tau\right) + \frac{1}{2} q^{\frac{3}{8}} \frac{\eta(6\tau)^3}{\eta(\tau)\eta(2\tau)}$$
$$= q + 2q^2 + 4q^3 + 7q^4 + 12q^5 + 19q^6 + 29q^7 + 44q^8 + 65q^9 + \cdots .$$

A.5. Order 7 mock theta functions

$$\begin{aligned}
\mathcal{F}_0(q) :&= \sum_{n=0}^{\infty} \frac{q^{n^2}}{(q^{n+1};q)_n} \\
&= 2 + 2q g_3\left(q;q^7\right) - \frac{\left(q^3;q^7\right)_\infty^2 \left(q^4;q^7\right)_\infty^2 \left(q^7;q^7\right)_\infty^2}{(q;q)_\infty} \\
&= 2iq^{-\frac{19}{8}} \mu(19\tau, 9\tau; 21\tau) - 2iq^{-\frac{3}{8}} \mu(5\tau, 9\tau; 21\tau) - q^{\frac{31}{24}} \frac{\vartheta(3\tau;7\tau)^2}{\eta(\tau)} \\
&= \frac{e^{-\frac{\pi i}{7}} q^{\frac{1}{168}}}{2\eta(\tau)} \Theta^{+}_{\left(\begin{smallmatrix}3&4\\4&3\end{smallmatrix}\right),\left(\begin{smallmatrix}1/14\\1/14\end{smallmatrix}\right),\left(\begin{smallmatrix}1/14\\1/14\end{smallmatrix}\right),\left(\begin{smallmatrix}-3\\4\end{smallmatrix}\right),\left(\begin{smallmatrix}-4\\3\end{smallmatrix}\right)}(\tau) \\
&= 1 + q + q^3 + q^4 + q^5 + 2q^7 + q^8 + 2q^9 + \cdots ,\\
\mathcal{F}_1(q) :&= \sum_{n=1}^{\infty} \frac{q^{n^2}}{(q^{n};q)_n} \\
&= 2q^2 g_3\left(q^2;q^7\right) + q \frac{\left(q;q^7\right)_\infty^2 \left(q^6;q^7\right)_\infty^2 \left(q^7;q^7\right)_\infty^2}{(q;q)_\infty} \\
&= -2iq^{-\frac{11}{8}} \mu(11\tau, 3\tau; 21\tau) - 2iq^{\frac{1}{8}} \mu(4\tau, 3\tau; 21\tau) + q^{\frac{7}{24}} \frac{\vartheta(\tau;7\tau)^2}{\eta(\tau)} \\
&= \frac{e^{-\frac{\pi i}{7}} q^{\frac{25}{168}}}{2\eta(\tau)} \Theta^{+}_{\left(\begin{smallmatrix}3&4\\4&3\end{smallmatrix}\right),\left(\begin{smallmatrix}5/14\\5/14\end{smallmatrix}\right),\left(\begin{smallmatrix}3/14\\3/14\end{smallmatrix}\right),\left(\begin{smallmatrix}-3\\4\end{smallmatrix}\right),\left(\begin{smallmatrix}-4\\3\end{smallmatrix}\right)}(\tau) \\
&= q + q^2 + q^3 + 2q^4 + q^5 + 2q^6 + 2q^7 + 2q^8 + 3q^9 + \cdots ,\\
\mathcal{F}_2(q) :&= \sum_{n=0}^{\infty} \frac{q^{n(n+1)}}{(q^{n+1};q)_{n+1}} \\
&= 2q^2 g_3\left(q^3;q^7\right) + \frac{\left(q^2;q^7\right)_\infty^2 \left(q^5;q^7\right)_\infty^2 \left(q^7;q^7\right)_\infty^2}{(q;q)_\infty} \\
&= -2iq^{-\frac{7}{8}} \mu(11\tau, 6\tau; 21\tau) - 2iq^{-\frac{3}{8}} \mu(4\tau, 6\tau; 21\tau) - q^{\frac{7}{24}} \frac{\vartheta(2\tau;7\tau)^2}{\eta(\tau)} \\
&= \frac{e^{-\frac{\pi i}{7}} q^{-\frac{47}{168}}}{2\eta(\tau)} \Theta^{+}_{\left(\begin{smallmatrix}3&4\\4&3\end{smallmatrix}\right),\left(\begin{smallmatrix}3/14\\3/14\end{smallmatrix}\right),\left(\begin{smallmatrix}5/14\\5/14\end{smallmatrix}\right),\left(\begin{smallmatrix}-3\\4\end{smallmatrix}\right),\left(\begin{smallmatrix}-4\\3\end{smallmatrix}\right)}(\tau) \\
&= 1 + q + 2q^2 + q^3 + 2q^4 + 2q^5 + 3q^6 + 2q^7 + 3q^8 + 3q^9 + \cdots .
\end{aligned}$$

A.6. Order 8 mock theta functions

$$
\begin{aligned}
S_0(q) :&= \sum_{n=0}^{\infty} \frac{q^{n^2}\left(-q;q^2\right)_n}{\left(-q^2;q^2\right)_n} \\
&= -2iq^{\frac{1}{2}} g_2\left(iq^{\frac{1}{2}}, q^4\right) + \frac{\left(-iq^{\frac{1}{2}};-q\right)_\infty^2 (-q;-q)_\infty \left(-q^3;q^8\right)_\infty \left(-q^5;q^8\right)_\infty}{\left(-q;q^4\right)_\infty \left(-q^3;q^4\right)_\infty \left(q^4;q^4\right)_\infty} \\
&= -2q^{-\frac{1}{2}} \mu\left(3\tau, \frac{1}{2}; 8\tau\right) + q^{-1} \frac{\eta(2\tau)^2 \eta(8\tau)^2 \vartheta\left(\tau+\frac{1}{2};8\tau\right)}{\eta(4\tau)^2 \vartheta(3\tau;8\tau)^2} \\
&= \frac{q\eta(8\tau)}{2\eta(16\tau)^2} \Theta^+_{\left(\begin{smallmatrix}1&1\\1&0\end{smallmatrix}\right),\left(\begin{smallmatrix}0\\1\end{smallmatrix}\right),\left(\begin{smallmatrix}-1\\1\end{smallmatrix}\right)}(3\tau,\tau;8\tau) + q^{-1} \frac{\eta(2\tau)^2 \eta(8\tau)^2 \vartheta\left(\tau+\frac{1}{2};8\tau\right)}{\eta(4\tau)^2 \vartheta(3\tau;8\tau)^2} \\
&= 1 + q + q^2 - q^3 + 2q^5 - q^7 + q^9 + \cdots,
\end{aligned}
$$

$$
\begin{aligned}
S_1(q) :&= \sum_{n=0}^{\infty} \frac{q^{n(n+2)}\left(-q;q^2\right)_n}{\left(-q^2;q^2\right)_n} \\
&= -2iq^{\frac{1}{2}} g_2\left(-iq^{\frac{3}{2}}; q^4\right) + \frac{\left(-iq^{\frac{1}{2}};-q\right)_\infty^2 (-q;-q)_\infty \left(-q;q^8\right)_\infty \left(-q^7;q^8\right)_\infty}{\left(-q;q^4\right)_\infty \left(-q^3;q^4\right)_\infty \left(q^4;q^4\right)_\infty} \\
&= 2q^{-\frac{1}{2}} \mu\left(\tau, \frac{1}{2}; 8\tau\right) + \frac{\eta(2\tau)^2 \eta(8\tau)^2 \vartheta\left(3\tau+\frac{1}{2};8\tau\right)}{\eta(4\tau)^2 \vartheta(\tau;8\tau)^2} \\
&= -\frac{\eta(8\tau)}{2\eta(16\tau)^2} \Theta^+_{\left(\begin{smallmatrix}1&1\\1&0\end{smallmatrix}\right),\left(\begin{smallmatrix}0\\1\end{smallmatrix}\right),\left(\begin{smallmatrix}-1\\1\end{smallmatrix}\right)}(\tau,3\tau;8\tau) + \frac{\eta(2\tau)^2 \eta(8\tau)^2 \vartheta\left(3\tau+\frac{1}{2};8\tau\right)}{\eta(4\tau)^2 \vartheta(\tau;8\tau)^2} \\
&= 1 + q^3 + q^4 - q^5 - q^6 + q^7 + 2q^8 + \cdots,
\end{aligned}
$$

$$
\begin{aligned}
T_0(q) :&= \sum_{n=0}^{\infty} \frac{q^{(n+1)(n+2)}\left(-q^2;q^2\right)_n}{\left(-q;q^2\right)_{n+1}} \\
&= iq^{\frac{1}{2}} g_2\left(iq^{\frac{1}{2}}; q^4\right) - \frac{\left(-iq^{\frac{1}{2}};-q\right)_\infty^2 (-q;-q)_\infty \left(-q^3;q^8\right)_\infty \left(-q^5;q^8\right)_\infty}{2\left(-q;q^4\right)_\infty \left(-q^3;q^4\right)_\infty \left(q^4;q^4\right)_\infty} \\
&\quad + \frac{1}{4} \frac{\left(q^{\frac{1}{2}};q^{\frac{1}{2}}\right)_\infty^3}{(q;q)_\infty \left(q^2;q^2\right)_\infty} + \frac{1}{4} \frac{(q;q)_\infty^8}{\left(q^{\frac{1}{2}};q^{\frac{1}{2}}\right)_\infty^3 \left(q^2;q^2\right)_\infty^4} \\
&= -q^{-\frac{1}{2}} \mu\left(5\tau+\frac{1}{2}, 2\tau; 8\tau\right) \\
&= \frac{q^{\frac{9}{4}}\eta(4\tau)}{2\eta(2\tau)\eta(8\tau)} \Theta^+_{\left(\begin{smallmatrix}1&1\\1&0\end{smallmatrix}\right),\left(\begin{smallmatrix}0\\1\end{smallmatrix}\right),\left(\begin{smallmatrix}-1\\1\end{smallmatrix}\right)}\left(5\tau+\frac{1}{2},\tau;8\tau\right) \\
&= q^2 - q^3 + q^4 - q^5 + 2q^6 - 2q^7 + 3q^8 - 4q^9 + \cdots,
\end{aligned}
$$

$$T_1(q) := \sum_{n=0}^{\infty} \frac{q^{n(n+1)}\left(-q^2;q^2\right)_n}{(-q;q^2)_{n+1}}$$

$$= iq^{\frac{1}{2}} g_2\left(-iq^{\frac{3}{2}};q^4\right) - \frac{\left(-iq^{\frac{1}{2}};-q\right)_\infty^2 (-q;-q)_\infty \left(-q;q^8\right)_\infty \left(-q^7;q^8\right)_\infty}{2\left(-q;q^4\right)_\infty \left(-q^3;q^4\right)_\infty \left(q^4;q^4\right)_\infty}$$

$$- \frac{1}{4} q^{-\frac{1}{2}} \frac{\left(q^{\frac{1}{2}};q^{\frac{1}{2}}\right)_\infty^3}{(q;q)_\infty \left(q^2;q^2\right)_\infty} + \frac{1}{4} q^{-\frac{1}{2}} \frac{(q;q)_\infty^8}{\left(q^{\frac{1}{2}};q^{\frac{1}{2}}\right)_\infty^3 \left(q^2;q^2\right)_\infty^4}$$

$$= q^{-\frac{1}{2}} \mu\left(7\tau + \frac{1}{2}, 6\tau; 8\tau\right)$$

$$= -\frac{q^{\frac{21}{4}}\eta(4\tau)}{2\eta(2\tau)\eta(8\tau)} \Theta^+_{\left(\begin{smallmatrix}1&1\\1&0\end{smallmatrix}\right),\left(\begin{smallmatrix}0\\1\end{smallmatrix}\right),\left(\begin{smallmatrix}-1\\1\end{smallmatrix}\right)} \left(7\tau + \frac{1}{2}, 3\tau; 8\tau\right)$$

$$= 1 - q + 2q^2 - 2q^3 + 3q^4 - 4q^5 + 5q^6 - 6q^7 + 8q^8 - 9q^9 + \cdots,$$

$$U_0(q) := \sum_{n=0}^{\infty} \frac{q^{n^2}\left(-q;q^2\right)_n}{(-q^4;q^4)_n}$$

$$= -q^{\frac{1}{4}} g_2\left(q^{\frac{1}{4}};q^{\frac{1}{2}}\right) + \frac{1}{2}(q;q)_\infty \left(-q;q^2\right)_\infty^4 + \frac{1}{2} \frac{\left(q^{\frac{1}{2}};q^{\frac{1}{2}}\right)_\infty (q;q)_\infty^3 \left(q^2;q^2\right)_\infty}{\left(q^{\frac{1}{4}};q^{\frac{1}{4}}\right)_\infty^2 \left(q^4;q^4\right)_\infty^2}$$

$$= -2\mu\left(\tau, \frac{1}{2}; 4\tau\right)$$

$$= \frac{q^{\frac{1}{2}}\eta(4\tau)}{2\eta(8\tau)^2} \Theta_{\left(\begin{smallmatrix}1&1\\1&0\end{smallmatrix}\right),\left(\begin{smallmatrix}0\\1\end{smallmatrix}\right),\left(\begin{smallmatrix}-1\\1\end{smallmatrix}\right)} (\tau, \tau; 4\tau)$$

$$= 1 + q + q^2 + q^4 - q^6 + q^7 + q^9 + \cdots,$$

$$U_1(q) := \sum_{n=0}^{\infty} \frac{q^{(n+1)^2}\left(-q;q^2\right)_n}{(-q^2;q^4)_{n+1}}$$

$$= \frac{1}{2} q^{\frac{1}{4}} g_2\left(q^{\frac{1}{4}};q^{\frac{1}{2}}\right) + \frac{1}{4}(q;q)_\infty \left(-q;q^2\right)_\infty^4 - \frac{1}{4} \frac{\left(q^{\frac{1}{2}};q^{\frac{1}{2}}\right)_\infty (q;q)_\infty^3 \left(q^2;q^2\right)_\infty}{\left(q^{\frac{1}{4}};q^{\frac{1}{4}}\right)_\infty^2 \left(q^4;q^4\right)_\infty^2}$$

$$= \mu\left(3\tau, 2\tau + \frac{1}{2}; 4\tau\right)$$

$$= -\frac{q^2\eta(2\tau)^2\eta(8\tau)^2}{2\eta(4\tau)^5} \Theta_{\left(\begin{smallmatrix}1&1\\1&0\end{smallmatrix}\right),\left(\begin{smallmatrix}0\\1\end{smallmatrix}\right),\left(\begin{smallmatrix}-1\\1\end{smallmatrix}\right)} (3\tau, \tau; 4\tau)$$

$$= q - q^3 + q^4 + 2q^5 - q^6 - 2q^7 + q^8 + 3q^9 + \cdots,$$

$$V_0(q) := -1 + 2\sum_{n=0}^{\infty} \frac{q^{n^2}\left(-q;q^2\right)_n}{(q;q^2)_n}$$

$$= 2qg_2\left(q^2;q^4\right) + \left(-q^2;q^4\right)_\infty^4 \left(q^8;q^8\right)_\infty$$

$$= -2i\mu(\tau, \tau; 8\tau) - \frac{\eta(2\tau)^3\eta(4\tau)}{\eta(\tau)^2\eta(8\tau)}$$

$$= -\frac{iq^{\frac{1}{2}}}{\vartheta(\tau;8\tau)}\Theta_{\left(\begin{smallmatrix}1&1\\1&0\end{smallmatrix}\right),\left(\begin{smallmatrix}0\\1\end{smallmatrix}\right),\left(\begin{smallmatrix}-1\\1\end{smallmatrix}\right)}\left(\tau,4\tau+\frac{1}{2};8\tau\right) - \frac{\eta(2\tau)^3\eta(4\tau)}{\eta(\tau)^2\eta(8\tau)}$$
$$= 1 + 2q + 4q^2 + 4q^3 + 6q^4 + 8q^5 + 8q^6 + 12q^7 + 16q^8 + 18q^9 + \cdots,$$
$$V_1(q) := \sum_{n=0}^{\infty} \frac{q^{(n+1)^2}\left(-q;q^2\right)_n}{\left(q;q^2\right)_{n+1}}$$
$$= qg_2\left(q;q^4\right)$$
$$= -i\mu\left(3\tau,\tau;8\tau\right)$$
$$= -\frac{iq^{\frac{3}{2}}}{2\vartheta(\tau;8\tau)}\Theta_{\left(\begin{smallmatrix}1&1\\1&0\end{smallmatrix}\right),\left(\begin{smallmatrix}0\\1\end{smallmatrix}\right),\left(\begin{smallmatrix}-1\\1\end{smallmatrix}\right)}\left(3\tau,2\tau+\frac{1}{2};8\tau\right)$$
$$= q + q^2 + q^3 + 2q^4 + 3q^5 + 3q^6 + 4q^7 + 5q^8 + 6q^9 + \cdots.$$

A.7. Order 10 mock theta functions

$$\phi(q) := \sum_{n=0}^{\infty} \frac{q^{\frac{n(n+1)}{2}}}{\left(q;q^2\right)_{n+1}}$$
$$= 2qg_2\left(q^2;q^5\right) + \frac{\left(-q^2;q^5\right)_\infty\left(-q^3;q^5\right)_\infty\left(q^{10};q^{10}\right)_\infty}{\left(q^2;q^{10}\right)_\infty\left(q^8;q^{10}\right)_\infty}$$
$$= -2iq^{-\frac{1}{4}}\mu\left(3\tau,2\tau;10\tau\right) + iq^{-1}\frac{\eta(2\tau)\eta(5\tau)\eta(10\tau)^2}{\vartheta(2\tau;5\tau)\,\vartheta(2\tau;10\tau)^2}$$
$$= -\frac{iq^{\frac{5}{4}}}{\vartheta(2\tau;10\tau)}\Theta^+_{\left(\begin{smallmatrix}1&1\\1&0\end{smallmatrix}\right),\left(\begin{smallmatrix}0\\1\end{smallmatrix}\right),\left(\begin{smallmatrix}-1\\1\end{smallmatrix}\right)}\left(3\tau,4\tau+\frac{1}{2};10\tau\right) + iq^{-1}\frac{\eta(2\tau)\eta(5\tau)\eta(10\tau)^2}{\vartheta(2\tau;5\tau)\,\vartheta(2\tau;10\tau)^2}$$
$$= 1 + 2q + 2q^2 + 3q^3 + 4q^4 + 4q^5 + 6q^6 + 7q^7 + 8q^8 + 10q^9 + \cdots,$$
$$\psi(q) := \sum_{n=0}^{\infty} \frac{q^{\frac{(n+1)(n+2)}{2}}}{\left(q;q^2\right)_{n+1}}$$
$$= 2qg_2\left(q;q^5\right) - q\frac{\left(-q;q^5\right)_\infty\left(-q^4;q^5\right)_\infty\left(q^{10};q^{10}\right)_\infty}{\left(q^4;q^{10}\right)_\infty\left(q^6;q^{10}\right)_\infty}$$
$$= -2iq^{-\frac{1}{4}}\mu\left(4\tau,\tau;10\tau\right) - iq^{-\frac{3}{2}}\frac{\eta(2\tau)\eta(5\tau)\eta(10\tau)^2}{\vartheta(\tau;5\tau)\,\vartheta(4\tau;10\tau)^2}$$
$$= -\frac{iq^{\frac{7}{4}}}{\vartheta(\tau;10\tau)}\Theta^+_{\left(\begin{smallmatrix}1&1\\1&0\end{smallmatrix}\right),\left(\begin{smallmatrix}0\\1\end{smallmatrix}\right),\left(\begin{smallmatrix}-1\\1\end{smallmatrix}\right)}\left(4\tau,2\tau+\frac{1}{2};10\tau\right) - iq^{-\frac{3}{2}}\frac{\eta(2\tau)\eta(5\tau)\eta(10\tau)^2}{\vartheta(\tau;5\tau)\,\vartheta(4\tau;10\tau)^2}$$
$$= q + q^2 + 2q^3 + 2q^4 + 2q^5 + 4q^6 + 4q^7 + 4q^8 + 6q^9 + \cdots,$$
$$X(q) := \sum_{n=0}^{\infty} \frac{(-1)^n q^{n^2}}{(-q;q)_{2n}}$$
$$= -2iq^{\frac{1}{2}}g_2\left(iq^{\frac{1}{2}};q^{10}\right) + 2iq^{\frac{5}{2}}g_2\left(iq^{\frac{9}{2}},q^{10}\right)$$

$$+ 2iq^{\frac{1}{2}} \frac{\left(q^2;q^2\right)_\infty^2}{(-q;-q)_\infty \left(q^4;q^{20}\right)_\infty \left(q^{10};q^{20}\right)_\infty \left(q^{16};q^{20}\right)_\infty}$$

$$+ \frac{\left(q^2;q^2\right)_\infty^2 \left(q;q^{10}\right)_\infty^2 \left(q^9;q^{10}\right)_\infty^2 \left(q^6;q^{20}\right)_\infty \left(q^{10};q^{20}\right)_\infty \left(q^{14};q^{20}\right)_\infty}{(-q;-q)_\infty \left(q^4;q^{20}\right)_\infty \left(q^{16};q^{20}\right)_\infty}$$

$$= 2q^{-\frac{3}{8}} \mu\left(\tau + \frac{1}{2}, -\tau; 5\tau\right) - q^{\frac{3}{8}} \frac{\eta(5\tau)^2 \vartheta(3\tau;10\tau)}{\eta(10\tau)\vartheta(\tau;5\tau)}$$

$$= \frac{iq^{\frac{1}{8}}}{\vartheta(-\tau;5\tau)} \Theta^+_{\left(\begin{smallmatrix}1&1\\1&0\end{smallmatrix}\right),\left(\begin{smallmatrix}0\\1\end{smallmatrix}\right),\left(\begin{smallmatrix}-1\\1\end{smallmatrix}\right)}\left(\tau + \frac{1}{2}, \frac{1}{2}\tau; 5\tau\right) - q^{\frac{3}{8}} \frac{\eta(5\tau)^2 \vartheta(3\tau;10\tau)}{\eta(10\tau)\vartheta(\tau;5\tau)}$$

$$= 1 - q + q^2 + q^4 - 2q^5 + q^6 - q^7 + q^8 - 2q^9 + \cdots,$$

$$\chi(q) := \sum_{n=0}^{\infty} \frac{(-1)^n q^{(n+1)^2}}{(-q;q)_{2n+1}}$$

$$= 2iq^{\frac{3}{2}} g_2\left(-iq^{\frac{3}{2}}; q^{10}\right) - 2iq^{\frac{5}{2}} g_2\left(-iq^{\frac{7}{2}}, q^{10}\right)$$

$$- 2iq^{\frac{3}{2}} \frac{\left(q^2;q^2\right)_\infty^2}{(-q;-q)_\infty \left(q^8;q^{20}\right)_\infty \left(q^{10};q^{20}\right)_\infty \left(q^{12};q^{20}\right)_\infty}$$

$$+ q \frac{\left(q^2;q^2\right)_\infty^2 \left(q^3;q^{10}\right)_\infty^2 \left(q^7;q^{10}\right)_\infty^2 \left(q^2;q^{20}\right)_\infty \left(q^{10};q^{20}\right)_\infty \left(q^{18};q^{20}\right)_\infty}{(-q;-q)_\infty \left(q^8;q^{20}\right)_\infty \left(q^{12};q^{20}\right)_\infty}$$

$$= 2q^{\frac{1}{8}} \mu\left(3\tau + \frac{1}{2}, 2\tau; 5\tau\right) + q^{-\frac{1}{8}} \frac{\eta(5\tau)^2 \vartheta(\tau;10\tau)}{\eta(10\tau)\vartheta(2\tau;5\tau)}$$

$$= \frac{iq^{\frac{13}{8}}}{\vartheta(2\tau;5\tau)} \Theta^+_{\left(\begin{smallmatrix}1&1\\1&0\end{smallmatrix}\right),\left(\begin{smallmatrix}0\\1\end{smallmatrix}\right),\left(\begin{smallmatrix}-1\\1\end{smallmatrix}\right)}\left(3\tau + \frac{1}{2}, \frac{3}{2}\tau; 5\tau\right) + q^{-\frac{1}{8}} \frac{\eta(5\tau)^2 \vartheta(\tau;10\tau)}{\eta(10\tau)\vartheta(2\tau;5\tau)}$$

$$= q - q^2 + q^3 - 2q^4 + 2q^5 - q^6 + 2q^7 - 3q^8 + 3q^9 + \cdots.$$

Bibliography

1. Milton Abramowitz and Irene A. Stegun, *Handbook of mathematical functions with formulas, graphs, and mathematical tables*, National Bureau of Standards Applied Mathematics Series, vol. 55, Superintendent of Documents, U.S. Government Printing Office, Washington, D.C., 1964.
2. Dražen Adamović and Antun Milas, *The doublet vertex operator superalgebras $\mathcal{A}(p)$ and $\mathcal{A}_{2,p}$*, Recent developments in algebraic and combinatorial aspects of representation theory, Contemp. Math., vol. 602, Amer. Math. Soc., Providence, RI, 2013, pp. 23–38.
3. M. K. Agrawal, J. H. Coates, D. C. Hunt, and A. J. van der Poorten, *Elliptic curves of conductor* 11, Math. Comp. **35** (1980), no. 151, 991–1002.
4. Scott Ahlgren, *Distribution of the partition function modulo composite integers M*, Math. Ann. **318** (2000), no. 4, 795–803.
5. Scott Ahlgren and Nickolas Andersen, *Kloosterman sums and Maass cusp forms of half integral weight for the modular group*, To appear in Int. Math. Res. Not. IMRN.
6. Scott Ahlgren and Byungchan Kim, *Mock modular grids and Hecke relations for mock modular forms*, Forum Math. **26** (2014), no. 4, 1261–1287.
7. ———, *Dissections of a "strange" function*, Int. J. Number Theory **11** (2015), no. 5, 1557–1562.
8. ———, *Mock theta functions and weakly holomorphic modular forms modulo 2 and 3*, Math. Proc. Cambridge Philos. Soc. **158** (2015), no. 1, 111–129.
9. Scott Ahlgren and Ken Ono, *Congruence properties for the partition function*, Proc. Natl. Acad. Sci. USA **98** (2001), no. 23, 12882–12884.
10. Sergei Alexandrov, Sibasish Banerjee, Jan Manschot, and Boris Pioline, *Indefinite theta series and generalized error functions*, Preprint.
11. Claudia Alfes, *CM values and Fourier coefficients of harmonic Maass forms*, Diss. TU Darmstadt. 2015.
12. ———, *Formulas for the coefficients of half-integral weight harmonic Maaß forms*, Math. Z. **277** (2014), no. 3-4, 769–795.
13. Claudia Alfes and Stephan Ehlen, *Twisted traces of CM values of weak Maass forms*, J. Number Theory **133** (2013), no. 6, 1827–1845.
14. Claudia Alfes, Michael Griffin, Ken Ono, and Larry Rolen, *Weierstrass mock modular forms and elliptic curves*, Res. Number Theory **1** (2015), Art. 24, 31.
15. Krishnaswami Alladi, *A new combinatorial study of the Rogers-Fine identity and a related partial theta series*, Int. J. Number Theory **5** (2009), no. 7, 1311–1320.
16. Nickolas Andersen, *Arithmetic of Maass forms of half-integral weight*, Ph.D. thesis. University of Illinios. 2016.
17. ———, *Classification of congruences for mock theta functions and weakly holomorphic modular forms*, Q. J. Math. **65** (2014), no. 3, 781–805.
18. ———, *Periods of the j-function along infinite geodesics and mock modular forms*, Bull. Lond. Math. Soc. **47** (2015), no. 3, 407–417.
19. ———, *Vector-valued modular forms and the mock theta conjectures*, Res. Number Theory **2** (2016), Art. 32, 14.
20. George E. Andrews, *On the theorems of Watson and Dragonette for Ramanujan's mock theta functions*, Amer. J. Math. **88** (1966), 454–490.
21. ———, *Partitions: yesterday and today*, New Zealand Mathematical Society, Wellington, 1979, With a foreword by J. C. Turner.
22. ———, *Ramanujan's "lost" notebook. I. Partial θ-functions*, Adv. in Math. **41** (1981), no. 2, 137–172.

23. ______, *The fifth and seventh order mock theta functions*, Trans. Amer. Math. Soc. **293** (1986), no. 1, 113–134.
24. ______, *Ramanujan's fifth order mock theta functions as constant terms*, Ramanujan revisited (Urbana-Champaign, Ill., 1987), Academic Press, Boston, MA, 1988, pp. 47–56.
25. ______, *The theory of partitions*, Cambridge Mathematical Library, Cambridge University Press, Cambridge, 1998, Reprint of the 1976 original.
26. ______, *Partitions: at the interface of q-series and modular forms*, Ramanujan J. **7** (2003), no. 1-3, 385–400, Rankin memorial issues.
27. ______, *Partitions with short sequences and mock theta functions*, Proc. Natl. Acad. Sci. USA **102** (2005), no. 13, 4666–4671.
28. ______, *The number of smallest parts in the partitions of n*, J. Reine Angew. Math. **624** (2008), 133–142.
29. ______, *Concave and convex compositions*, Ramanujan J. **31** (2013), no. 1-2, 67–82.
30. George E. Andrews and Bruce C. Berndt, *Ramanujan's lost notebook. Part II*, Springer, New York, 2009.
31. George E. Andrews, Song Heng Chan, and Byungchan Kim, *The odd moments of ranks and cranks*, J. Combin. Theory Ser. A **120** (2013), no. 1, 77–91.
32. George E. Andrews, Freeman J. Dyson, and Dean Hickerson, *Partitions and indefinite quadratic forms*, Invent. Math. **91** (1988), no. 3, 391–407.
33. George E. Andrews and F. G. Garvan, *Dyson's crank of a partition*, Bull. Amer. Math. Soc. (N.S.) **18** (1988), no. 2, 167–171.
34. ______, *Ramanujan's "lost" notebook. VI. The mock theta conjectures*, Adv. in Math. **73** (1989), no. 2, 242–255.
35. George E. Andrews, Frank G. Garvan, and Jie Liang, *Combinatorial interpretations of congruences for the spt-function*, Ramanujan J. **29** (2012), no. 1-3, 321–338.
36. ______, *Self-conjugate vector partitions and the parity of the spt-function*, Acta Arith. **158** (2013), no. 3, 199–218.
37. George E. Andrews and Richard Lewis, *The ranks and cranks of partitions moduli 2, 3, and 4*, J. Number Theory **85** (2000), no. 1, 74–84.
38. George E. Andrews, Robert C. Rhoades, and Sander P. Zwegers, *Modularity of the concave composition generating function*, Algebra Number Theory **7** (2013), no. 9, 2103–2139.
39. George E. Andrews and James A. Sellers, *Congruences for the Fishburn numbers*, J. Number Theory **161** (2016), 298–310.
40. M. P. Appell, *Sur les fonctions doublement périodiques de troisième espèce*, Ann. Sci. École Norm. Sup. (3) **1** (1884), 135–164.
41. Tetsuya Asai, Masanobu Kaneko, and Hirohito Ninomiya, *Zeros of certain modular functions and an application*, Comment. Math. Univ. St. Paul. **46** (1997), no. 1, 93–101.
42. A. O. L. Atkin, *Proof of a conjecture of Ramanujan*, Glasgow Math. J. **8** (1967), 14–32.
43. ______, *Multiplicative congruence properties and density problems for $p(n)$*, Proc. London Math. Soc. (3) **18** (1968), 563–576.
44. A. O. L. Atkin and F. G. Garvan, *Relations between the ranks and cranks of partitions*, Ramanujan J. **7** (2003), no. 1-3, 343–366, Rankin memorial issues.
45. A. O. L. Atkin and P. Swinnerton-Dyer, *Some properties of partitions*, Proc. London Math. Soc. (3) **4** (1954), 84–106.
46. F. C. Auluck, *On some new types of partitions associated with generalized Ferrers graphs*, Proc. Cambridge Philos. Soc. **47** (1951), 679–686.
47. Tom Barnet-Lamb, David Geraghty, Michael Harris, and Richard Taylor, *A family of Calabi-Yau varieties and potential automorphy II*, Publ. Res. Inst. Math. Sci. **47** (2011), no. 1, 29–98.
48. Wolf P. Barth, Klaus Hulek, Chris A. M. Peters, and Antonius Van de Ven, *Compact complex surfaces*, second ed., Ergebnisse der Mathematik und ihrer Grenzgebiete. 3. Folge. A Series of Modern Surveys in Mathematics [Results in Mathematics and Related Areas. 3rd Series. A Series of Modern Surveys in Mathematics], vol. 4, Springer-Verlag, Berlin, 2004.
49. Olivia Beckwith, *Asymptotic bounds for special values of shifted convolution Dirichlet series*, Proc. Amer. Math. Soc., Accepted for publication.
50. ______, *Indivisibility of class numbers of imaginary quadratic fields*, Res. Math. Sci., Accepted for publication.
51. Paloma Bengoechea, *Corps quadratiques et formes modulaires*, Ph.D. thesis. 2013.

52. Alexander Berkovich and Frank G. Garvan, *Some observations on Dyson's new symmetries of partitions*, J. Combin. Theory Ser. A **100** (2002), no. 1, 61–93.
53. Bruce C. Berndt, Paul R. Bialek, and Ae Ja Yee, *Formulas of Ramanujan for the power series coefficients of certain quotients of Eisenstein series*, Int. Math. Res. Not. (2002), no. 21, 1077–1109.
54. Bruce C. Berndt and Robert A. Rankin, *Ramanujan*, History of Mathematics, vol. 9, American Mathematical Society, Providence, RI; London Mathematical Society, London, 1995, Letters and commentary.
55. Bruce C. Berndt and Ae Ja Yee, *Combinatorial proofs of identities in Ramanujan's lost notebook associated with the Rogers-Fine identity and false theta functions*, Ann. Comb. **7** (2003), no. 4, 409–423.
56. Rolf Berndt and Ralf Schmidt, *Elements of the representation theory of the Jacobi group*, Progress in Mathematics, vol. 163, Birkhäuser Verlag, Basel, 1998.
57. Manjul Bhargava and Arul Shankar, *The average size of the* 5*-Selmer group of elliptic curves is* 6*, and the average rank of elliptic curves is less than* 1, Preprint.
58. ———, *Binary quartic forms having bounded invariants, and the boundedness of the average rank of elliptic curves*, Ann. of Math. (2) **181** (2015), no. 1, 191–242.
59. ———, *Ternary cubic forms having bounded invariants, and the existence of a positive proportion of elliptic curves having rank 0*, Ann. of Math. (2) **181** (2015), no. 2, 587–621.
60. Manjul Bhargava and Christopher Skinner, *A positive proportion of elliptic curves over* $\mathbb{Q}$ *have rank one*, J. Ramanujan Math. Soc. **29** (2014), no. 2, 221–242.
61. Manjul Bhargava, Christopher Skinner, and Wei Zhang, *A majority of elliptic curves over* $\mathbb{Q}$ *satisfy the Birch and Swinnerton-Dyer conjecture*, Preprint.
62. Paul Richard Bialek, *Ramanujan's formulas for the coefficients in the power series expansions of certain modular forms*, ProQuest LLC, Ann Arbor, MI, 1995, Thesis (Ph.D.)–University of Illinois at Urbana-Champaign.
63. B. J. Birch and H. P. F. Swinnerton-Dyer, *Notes on elliptic curves. I*, J. Reine Angew. Math. **212** (1963), 7–25.
64. ———, *Notes on elliptic curves. II*, J. Reine Angew. Math. **218** (1965), 79–108.
65. Valentin Blomer, *Shifted convolution sums and subconvexity bounds for automorphic L-functions*, Int. Math. Res. Not. (2004), no. 73, 3905–3926.
66. Valentin Blomer and Gergely Harcos, *Hybrid bounds for twisted L-functions*, J. Reine Angew. Math. **621** (2008), 53–79.
67. ———, *The spectral decomposition of shifted convolution sums*, Duke Math. J. **144** (2008), no. 2, 321–339.
68. ———, *Twisted L-functions over number fields and Hilbert's eleventh problem*, Geom. Funct. Anal. **20** (2010), no. 1, 1–52.
69. Valentin Blomer, Gergely Harcos, and Philippe Michel, *A Burgess-like subconvex bound for twisted L-functions*, Forum Math. **19** (2007), no. 1, 61–105, Appendix 2 by Z. Mao.
70. Andrew R. Booker and Andreas Strömbergsson, *Numerical computations with the trace formula and the Selberg eigenvalue conjecture*, J. Reine Angew. Math. **607** (2007), 113–161.
71. Richard E. Borcherds, *Monstrous moonshine and monstrous Lie superalgebras*, Invent. Math. **109** (1992), no. 2, 405–444.
72. ———, *Automorphic forms on* $\mathrm{O}_{s+2,2}(\mathbf{R})$ *and infinite products*, Invent. Math. **120** (1995), no. 1, 161–213.
73. ———, *Automorphic forms on* $\mathrm{O}_{s+2,2}(\mathbf{R})^+$ *and generalized Kac-Moody algebras*, Proceedings of the International Congress of Mathematicians, Vol. 1, 2 (Zürich, 1994), Birkhäuser, Basel, 1995, pp. 744–752.
74. ———, *Automorphic forms with singularities on Grassmannians*, Invent. Math. **132** (1998), no. 3, 491–562.
75. ———, *What is Moonshine?*, Proceedings of the International Congress of Mathematicians, Vol. I (Berlin, 1998), no. Extra Vol. I, 1998, pp. 607–615.
76. Lev A. Borisov and Anatoly Libgober, *Elliptic genera of toric varieties and applications to mirror symmetry*, Invent. Math. **140** (2000), no. 2, 453–485.
77. Hatice Boylan, Nils-Peter Skoruppa, and Haigang Zhou, *The arithmetic theory of skew-holomorphic Jacobi forms*, Preprint.
78. Christophe Breuil, Brian Conrad, Fred Diamond, and Richard Taylor, *On the modularity of elliptic curves over* $\mathbf{Q}$*: wild 3-adic exercises*, J. Amer. Math. Soc. **14** (2001), no. 4, 843–939.

79. Kathrin Bringmann, *On the explicit construction of higher deformations of partition statistics*, Duke Math. J. **144** (2008), no. 2, 195–233.
80. ______, *Asymptotics for rank partition functions*, Trans. Amer. Math. Soc. **361** (2009), no. 7, 3483–3500.
81. Kathrin Bringmann, Thomas Creutzig, and Larry Rolen, *Negative index Jacobi forms and quantum modular forms*, Res. Math. Sci. **1** (2014), Art. 11, 32.
82. Kathrin Bringmann, Nikolaos Diamantis, and Stephan Ehlen, *Regularized inner products and errors of modularity*, To appear in Int. Math. Res. Not. IMRN.
83. Kathrin Bringmann, Nikolaos Diamantis, and Martin Raum, *Mock period functions, sesquiharmonic Maass forms, and non-critical values of L-functions*, Adv. Math. **233** (2013), 115–134.
84. Kathrin Bringmann and Jehanne Dousse, *On Dyson's crank conjecture and the uniform asymptotic behavior of certain inverse theta functions*, Transaction of the AMS, **368** (2016), 3141–3155.
85. Kathrin Bringmann and Amanda Folsom, *Quantum Jacobi forms and finite evaluations of unimodal rank generating functions*, Archiv der Mathematik, 107 (2016), 367–378.
86. ______, *Almost harmonic Maass forms and Kac-Wakimoto characters*, J. Reine Angew. Math. **694** (2014), 179–202.
87. Kathrin Bringmann, Amanda Folsom, and Karl Mahlburg, *Quasimodular forms and $s\ell(m|m)^\wedge$ characters*, Ramanujan J. **36** (2015), no. 1-2, 103–116.
88. ______, *Corrigendum to Quasimodular forms and $s\ell(m|m)^\wedge$ characters*, Ramanujan J. **36** (2015), no. 1-2, 103–116, Preprint.
89. Kathrin Bringmann, Amanda Folsom, and Robert C. Rhoades, *Unimodal sequences and "strange" functions: a family of quantum modular forms*, Pacific J. Math. **274** (2015), no. 1, 1–25.
90. Kathrin Bringmann, Karl-Heinz Fricke, and Zachary A. Kent, *Special L-values and periods of weakly holomorphic modular forms*, Proc. Amer. Math. Soc. **142** (2014), no. 10, 3425–3439.
91. Kathrin Bringmann, Pavel Guerzhoy, and Ben Kane, *Mock modular forms as p-adic modular forms*, Trans. Amer. Math. Soc. **364** (2012), no. 5, 2393–2410.
92. ______, *Shintani lifts and fractional derivatives for harmonic weak Maass forms*, Adv. Math. **255** (2014), 641–671.
93. ______, *On cycle integrals of weakly holomorphic modular forms*, Math. Proc. Cambridge Philos. Soc. **158** (2015), no. 3, 439–449.
94. Kathrin Bringmann, Pavel Guerzhoy, Zachary Kent, and Ken Ono, *Eichler-Shimura theory for mock modular forms*, Math. Ann. **355** (2013), no. 3, 1085–1121.
95. Kathrin Bringmann, Paul Jenkins, and Ben Kane, *Differential operators on polar harmonic Maass forms and elliptic duality*, Preprint.
96. Kathrin Bringmann and Ben Kane, *Fourier coefficients of meromorphic modular forms and a question of Petersson*, Submitted for publication.
97. ______, *Modular local polynomials*, Math. Res. Lett. **23** (2016), no. 4, 973–987.
98. ______, *A problem of Petersson about weight 0 meromorphic modular forms*, Res. Math. Sci. **3** (2016), Paper No. 24, 31.
99. Kathrin Bringmann, Ben Kane, and Winfried Kohnen, *Locally harmonic Maass forms and the kernel of the Shintani lift*, Int. Math. Res. Not. IMRN (2015), no. 11, 3185–3224.
100. Kathrin Bringmann, Ben Kane, Steffen Löbrich, Ken Ono, and Larry Rolen, *On divisors of modular forms*, To appear in J. Phys. A.
101. Kathrin Bringmann, Ben Kane, and Anna-Maria von Pippich, *Cycle integrals of meromorphic modular forms and CM-values of automorphic forms*, Submitted for publication.
102. Kathrin Bringmann and Stephen Kudla, *A classification of harmonic Maass forms*, Math. Ann. (2017). https://doi.org/10.1007/s00208-017-1563-x
103. Kathrin Bringmann and Karl Mahlburg, *Asymptotic inequalities for positive crank and rank moments*, Trans. Amer. Math. Soc. **366** (2014), no. 2, 1073–1094.
104. Kathrin Bringmann and Jan Manschot, *From sheaves on $\mathbb{P}^2$ to a generalization of the Rademacher expansion*, Amer. J. Math. **135** (2013), no. 4, 1039–1065.
105. Kathrin Bringmann, Michael H. Mertens, and Ken Ono, *p-adic properties of modular shifted convolution Dirichlet series*, Proc. Amer. Math. Soc. **144** (2016), no. 4, 1439–1451.
106. Kathrin Bringmann and Antun Milas, *W-algebras, false theta functions and quantum modular forms, I*, Int. Math. Res. Not. IMRN (2015), no. 21, 11351–11387.

107. Kathrin Bringmann and Ken Ono, *The $f(q)$ mock theta function conjecture and partition ranks*, Invent. Math. **165** (2006), no. 2, 243–266.
108. ———, *Arithmetic properties of coefficients of half-integral weight Maass-Poincaré series*, Math. Ann. **337** (2007), no. 3, 591–612.
109. ———, *Some characters of Kac and Wakimoto and nonholomorphic modular functions*, Math. Ann. **345** (2009), no. 3, 547–558.
110. ———, *Dyson's ranks and Maass forms*, Ann. of Math. (2) **171** (2010), no. 1, 419–449.
111. ———, *Coefficients of harmonic Maass forms*, Partitions, q-series, and modular forms, Dev. Math., vol. 23, Springer, New York, 2012, pp. 23–38.
112. Kathrin Bringmann, Ken Ono, and Robert C. Rhoades, *Eulerian series as modular forms*, J. Amer. Math. Soc. **21** (2008), no. 4, 1085–1104.
113. Kathrin Bringmann, Martin Raum, and Olav K. Richter, *Harmonic Maass-Jacobi forms with singularities and a theta-like decomposition*, Trans. Amer. Math. Soc. **367** (2015), no. 9, 6647–6670.
114. Kathrin Bringmann and Olav K. Richter, *Zagier-type dualities and lifting maps for harmonic Maass-Jacobi forms*, Adv. Math. **225** (2010), no. 4, 2298–2315.
115. Kathrin Bringmann and Larry Rolen, *Radial limits of mock theta functions*, Res. Math. Sci. **2** (2015), Art. 17, 18.
116. ———, *Half-integral weight Eichler integrals and quantum modular forms*, J. Number Theory **161** (2016), 240–254.
117. Kathrin Bringmann, Larry Rolen, and Sander Zwegers, *On the Fourier coefficients of negative index meromorphic Jacobi forms*, Res. Math. Sci. **3** (2016), Paper No. 5, 9.
118. R. Bruggeman, *Quantum Maass forms*, The Conference on L-Functions, World Sci. Publ., Hackensack, NJ, 2007, pp. 1–15.
119. Jan H. Bruinier, *Borcherds products and Chern classes of Hirzebruch-Zagier divisors*, Invent. Math. **138** (1999), no. 1, 51–83.
120. ———, *Harmonic Maass forms and periods*, Math. Ann. **357** (2013), no. 4, 1363–1387.
121. Jan H. Bruinier and Jens Funke, *On two geometric theta lifts*, Duke Math. J. **125** (2004), no. 1, 45–90.
122. ———, *Traces of CM values of modular functions*, J. Reine Angew. Math. **594** (2006), 1–33.
123. Jan H. Bruinier, Jens Funke, and Özlem Imamoḡlu, *Regularized theta liftings and periods of modular functions*, J. Reine Angew. Math. **703** (2015), 43–93.
124. Jan H. Bruinier, Benjamin Howard, and Tonghai Yang, *Heights of Kudla-Rapoport divisors and derivatives of L-functions*, Invent. Math. **201** (2015), no. 1, 1–95.
125. Jan H. Bruinier, Winfried Kohnen, and Ken Ono, *The arithmetic of the values of modular functions and the divisors of modular forms*, Compos. Math. **140** (2004), no. 3, 552–566.
126. Jan H. Bruinier and Yingkun Li, *Heegner divisors in generalized Jacobians and traces of singular moduli*, Algebra Number Theory **10** (2016), no. 6, 1277–1300.
127. Jan H. Bruinier and Ken Ono, *Heegner divisors, L-functions and harmonic weak Maass forms*, Ann. of Math. (2) **172** (2010), no. 3, 2135–2181.
128. ———, *Algebraic formulas for the coefficients of half-integral weight harmonic weak Maass forms*, Adv. Math. **246** (2013), 198–219.
129. Jan H. Bruinier, Ken Ono, and Robert C. Rhoades, *Differential operators for harmonic weak Maass forms and the vanishing of Hecke eigenvalues*, Math. Ann. **342** (2008), no. 3, 673–693.
130. Jan H. Bruinier and Markus Schwagenscheidt, *Algebraic formulas for the coefficients of mock theta functions and Weyl vectors of Borcherds products*, J. Algebra, **478** (2017), 38–57.
131. Jan H. Bruinier and Fredrik Strömberg, *Computation of harmonic weak Maass forms*, Exp. Math. **21** (2012), no. 2, 117–131.
132. Jan H. Bruinier and Tonghai Yang, *Faltings heights of CM cycles and derivatives of L-functions*, Invent. Math. **177** (2009), no. 3, 631–681.
133. Jennifer Bryson, Ken Ono, Sarah Pitman, and Robert C. Rhoades, *Unimodal sequences and quantum and mock modular forms*, Proc. Natl. Acad. Sci. USA **109** (2012), no. 40, 16063–16067.
134. Daniel Bump, *Automorphic forms and representations*, Cambridge Studies in Advanced Mathematics, vol. 55, Cambridge University Press, Cambridge, 1997.
135. Daniel Bump, Solomon Friedberg, and Jeffrey Hoffstein, *Nonvanishing theorems for L-functions of modular forms and their derivatives*, Invent. Math. **102** (1990), no. 3, 543–618.

136. V. A. Bykovskiĭ, *A trace formula for the scalar product of Hecke series and its applications*, Zap. Nauchn. Sem. S.-Peterburg. Otdel. Mat. Inst. Steklov. (POMI) **226** (1996), no. Anal. Teor. Chisel i Teor. Funktsiĭ. 13, 14–36, 235–236.
137. Luca Candelori, *Harmonic weak Maass forms of integral weight: a geometric approach*, Math. Ann. **360** (2014), no. 1-2, 489–517.
138. Luca Candelori and Francesc Castella, *A geometric perspective on p-adic properties of mock modular forms*, Res. Math. Sci. **4** (2017), Paper No. 5, 15.
139. Song Heng Chan, Atul Dixit, and Frank G. Garvan, *Rank-crank-type PDEs and generalized Lambert series identities*, Ramanujan J. **31** (2013), no. 1-2, 163–189.
140. Song Heng Chan and Renrong Mao, *Inequalities for ranks of partitions and the first moment of ranks and cranks of partitions*, Adv. Math. **258** (2014), 414–437.
141. Miranda C. N. Cheng, *K3 surfaces, $\mathcal{N}=4$ dyons and the Mathieu group M_{24}*, Commun. Number Theory Phys. **4** (2010), no. 4, 623–657.
142. Miranda C. N. Cheng, John F. R. Duncan, and Jeffrey A. Harvey, *Umbral moonshine*, Commun. Number Theory Phys. **8** (2014), no. 2, 101–242.
143. ______, *Umbral moonshine and the Niemeier lattices*, Res. Math. Sci. **1** (2014), Art. 3, 81.
144. Miranda C. N. Cheng and Sarah Harrison, *Umbral moonshine and K3 surfaces*, Comm. Math. Phys. **339** (2015), no. 1, 221–261.
145. Dohoon Choi, Subong Lim, and Robert C. Rhoades, *Mock modular forms and quantum modular forms*, Proc. Amer. Math. Soc. **144** (2016), no. 6, 2337–2349.
146. Youn-Seo Choi, *The basic bilateral hypergeometric series and the mock theta functions*, Ramanujan J. **24** (2011), no. 3, 345–386.
147. YoungJu Choie, *Correspondence among Eisenstein series $E_{2,1}(\tau,z)$, $H_{3/2}(\tau)$ and $E_2(\tau)$*, Manuscripta Math. **93** (1997), no. 2, 177–187.
148. YoungJu Choie and Seokho Jin, *Periods of Jacobi forms and the Hecke operator*, J. Math. Anal. Appl. **408** (2013), no. 1, 345–354.
149. YoungJu Choie, Marvin Isadore Knopp, and Wladimir de Azevedo Pribitkin, *Niebur integrals, mock automorphic forms and harmonic Maass forms: a retrospective on the work of the Rademacher school*, Preprint (2010).
150. Henri Cohen, *Sums involving the values at negative integers of L-functions of quadratic characters*, Math. Ann. **217** (1975), no. 3, 271–285.
151. ______, *q-identities for Maass waveforms*, Invent. Math. **91** (1988), no. 3, 409–422.
152. Brian Conrad, Fred Diamond, and Richard Taylor, *Modularity of certain potentially Barsotti-Tate Galois representations*, J. Amer. Math. Soc. **12** (1999), no. 2, 521–567.
153. J. B. Conrey and H. Iwaniec, *The cubic moment of central values of automorphic L-functions*, Ann. of Math. (2) **151** (2000), no. 3, 1175–1216.
154. J. H. Conway and S. P. Norton, *Monstrous moonshine*, Bull. London Math. Soc. **11** (1979), no. 3, 308–339.
155. David A. Cox, *Primes of the form x^2+ny^2*, second ed., Pure and Applied Mathematics (Hoboken), John Wiley & Sons, Inc., Hoboken, NJ, 2013, Fermat, class field theory, and complex multiplication.
156. Thomas Creutzig and Antun Milas, *False theta functions and the Verlinde formula*, Adv. Math. **262** (2014), 520–545.
157. C. J. Cummins and T. Gannon, *Modular equations and the genus zero property of moonshine functions*, Invent. Math. **129** (1997), no. 3, 413–443.
158. Atish Dabholkar, Sameer Murthy, and Don Zagier, *Quantum black holes, wall crossing, and mock modular forms*, to appear in Cambridge Monographs in Mathematical Physics, Cambridge Univ. Press.
159. Pierre Deligne, *La conjecture de Weil. I*, Inst. Hautes Études Sci. Publ. Math. (1974), no. 43, 273–307.
160. Pierre Deligne and Jean-Pierre Serre, *Formes modulaires de poids* 1, Ann. Sci. École Norm. Sup. (4) **7** (1974), 507–530 (1975).
161. Michael Dewar and M. Ram Murty, *A derivation of the Hardy-Ramanujan formula from an arithmetic formula*, Proc. Amer. Math. Soc. **141** (2013), no. 6, 1903–1911.
162. Fred Diamond and Jerry Shurman, *A first course in modular forms*, Graduate Texts in Mathematics, vol. 228, Springer-Verlag, New York, 2005.
163. Jehanne Dousse and Michael H. Mertens, *Asymptotic formulae for partition ranks*, Acta Arith. **168** (2015), no. 1, 83–100.

164. Leila A. Dragonette, *Some asymptotic formulae for the mock theta series of Ramanujan*, Trans. Amer. Math. Soc. **72** (1952), 474–500.
165. W. Duke, *Hyperbolic distribution problems and half-integral weight Maass forms*, Invent. Math. **92** (1988), no. 1, 73–90.
166. ———, *Almost a century of answering the question: what is a mock theta function?*, Notices Amer. Math. Soc. **61** (2014), no. 11, 1314–1320.
167. W. Duke, J. B. Friedlander, and H. Iwaniec, *Bounds for automorphic L-functions*, Invent. Math. **112** (1993), no. 1, 1–8.
168. ———, *Bounds for automorphic L-functions. II*, Invent. Math. **115** (1994), no. 2, 219–239.
169. ———, *Bounds for automorphic L-functions. III*, Invent. Math. **143** (2001), no. 2, 221–248.
170. W. Duke, Ö. Imamoḡlu, and Á. Tóth, *Katok–Sarnak formula for higher weight*, In preparation.
171. ———, *Rational period functions and cycle integrals*, Abh. Math. Semin. Univ. Hambg. **80** (2010), no. 2, 255–264.
172. ———, *Cycle integrals of the j-function and mock modular forms*, Ann. of Math. (2) **173** (2011), no. 2, 947–981.
173. ———, *Real quadratic analogs of traces of singular moduli*, Int. Math. Res. Not. IMRN (2011), no. 13, 3082–3094.
174. ———, *Geometric invariants for real quadratic fields*, Ann. of Math. (2) **184** (2016), no. 3, 949–990.
175. W. Duke and P. Jenkins, *Integral traces of singular values of weak Maass forms*, Algebra Number Theory **2** (2008), no. 5, 573–593.
176. W. Duke and Y. Li, *Harmonic Maass forms of weight 1*, Duke Math. J. **164** (2015), no. 1, 39–113.
177. John F. R. Duncan, Michael J. Griffin, and Ken Ono, *Moonshine*, Res. Math. Sci. **2** (2015), Art. 11, 57.
178. ———, *Proof of the umbral moonshine conjecture*, Res. Math. Sci. **2** (2015), Art. 26, 47.
179. John F. R. Duncan, Michael H. Mertens, and Ken Ono, *O'Nan moonshine and arithmetic*, Preprint.
180. F. J. Dyson, *Some guesses in the theory of partitions*, Eureka (1944), no. 8, 10–15.
181. ———, *Mappings and symmetries of partitions*, J. Combin. Theory Ser. A **51** (1989), no. 2, 169–180.
182. Tohru Eguchi and Kazuhiro Hikami, *Superconformal algebras and mock theta functions*, J. Phys. A **42** (2009), no. 30, 304010, 23.
183. ———, *Note on twisted elliptic genus of $K3$ surface*, Phys. Lett. B **694** (2011), no. 4-5, 446–455.
184. Tohru Eguchi, Hirosi Ooguri, and Yuji Tachikawa, *Notes on the $K3$ surface and the Mathieu group M_{24}*, Exp. Math. **20** (2011), no. 1, 91–96.
185. Tohru Eguchi, Hirosi Ooguri, Anne Taormina, and Sung-Kil Yang, *Superconformal algebras and string compactification on manifolds with* $\mathrm{SU}(n)$ *holonomy*, Nuclear Phys. B **315** (1989), no. 1, 193–221.
186. Tohru Eguchi and Anne Taormina, *On the unitary representations of $N = 2$ and $N = 4$ superconformal algebras*, Phys. Lett. B **210** (1988), no. 1-2, 125–132.
187. Stephan Ehlen, *CM values of regularized theta lifts and harmonic Maass forms of weight one*, Duke Math. J., Accepted for publication.
188. ———, *Singular moduli of higher level and special cycles*, Res. Math. Sci. **2** (2015), Art. 16, 27.
189. Martin Eichler, *On the class of imaginary quadratic fields and the sums of divisors of natural numbers*, J. Indian Math. Soc. (N.S.) **19** (1955), 153–180 (1956).
190. ———, *Eine Verallgemeinerung der Abelschen Integrale*, Math. Z. **67** (1957), 267–298.
191. Martin Eichler and Don Zagier, *The theory of Jacobi forms*, Progress in Mathematics, vol. 55, Birkhäuser Boston, Inc., Boston, MA, 1985.
192. John D. Fay, *Fourier coefficients of the resolvent for a Fuchsian group*, J. Reine Angew. Math. **293/294** (1977), 143–203.
193. Nathan J. Fine, *Basic hypergeometric series and applications*, Mathematical Surveys and Monographs, vol. 27, American Mathematical Society, Providence, RI, 1988, With a foreword by George E. Andrews.

194. Amanda Folsom, *A short proof of the mock theta conjectures using Maass forms*, Proc. Amer. Math. Soc. **136** (2008), no. 12, 4143–4149.
195. ______, *What is ... a mock modular form?*, Notices Amer. Math. Soc. **57** (2010), no. 11, 1441–1443.
196. ______, *Kac-Wakimoto characters and universal mock theta functions*, Trans. Amer. Math. Soc. **363** (2011), no. 1, 439–455.
197. ______, *Arizona Winter School 2013, Exercises: Weak Maass forms, mock modular forms, and q-hypergeometric series*, (2013), `http://swc.math.arizona.edu/aws/2013/2013FolsomProblems.pdf`.
198. ______, *Perspectives on mock modular forms*, J. Number Theory **176** (2017), 500–540.
199. Amanda Folsom, Caleb Ki, Yen N. Truong Vu, and Bowen Yang, *Strange combinatorial quantum modular forms*, J. Number Theory 170 (2017), 315–346.
200. Amanda Folsom and Riad Masri, *Equidistribution of Heegner points and the partition function*, Math. Ann. **348** (2010), no. 2, 289–317.
201. Amanda Folsom and Ken Ono, *Duality involving the mock theta function $f(q)$*, J. Lond. Math. Soc. (2) **77** (2008), no. 2, 320–334.
202. ______, *The spt-function of Andrews*, Proc. Natl. Acad. Sci. USA **105** (2008), no. 51, 20152–20156.
203. Amanda Folsom, Ken Ono, and Robert C. Rhoades, *Mock theta functions and quantum modular forms*, Forum Math. Pi **1** (2013), e2, 27.
204. ______, *Ramanujan's radial limits*, Ramanujan 125, Contemp. Math., vol. 627, Amer. Math. Soc., Providence, RI, 2014, pp. 91–102.
205. E. Freitag, *Siegelsche Modulfunktionen*, Grundlehren der Mathematischen Wissenschaften [Fundamental Principles of Mathematical Sciences], vol. 254, Springer-Verlag, Berlin, 1983.
206. I. B. Frenkel, J. Lepowsky, and A. Meurman, *A natural representation of the Fischer-Griess Monster with the modular function J as character*, Proc. Nat. Acad. Sci. U.S.A. **81** (1984), no. 10, Phys. Sci., 3256–3260.
207. ______, *Vertex operator algebras and the Monster*, Pure and Applied Mathematics, vol. 134, Academic Press, Inc., Boston, MA, 1988.
208. Karl-Heinz Fricke, *Analytische und p-adische Aspekte von klassischen und Mock-Modulformen*, Universität Bonn. 2013.
209. Jens Funke, *Cohomological aspects of weakly holomorphic modular forms*, In preparation.
210. ______, *Heegner divisors and nonholomorphic modular forms*, Compositio Math. **133** (2002), no. 3, 289–321.
211. Matthias R. Gaberdiel, Stefan Hohenegger, and Roberto Volpato, *Mathieu Moonshine in the elliptic genus of $K3$*, J. High Energy Phys. (2010), no. 10, 062, 24.
212. ______, *Mathieu twining characters for $K3$*, J. High Energy Phys. (2010), no. 9, 058, 20.
213. Terry Gannon, *Monstrous moonshine: the first twenty-five years*, Bull. London Math. Soc. **38** (2006), no. 1, 1–33.
214. ______, *Much ado about Mathieu*, Adv. Math. **301** (2016), 322–358.
215. Sharon Anne Garthwaite, *The coefficients of the $\omega(q)$ mock theta function*, Int. J. Number Theory **4** (2008), no. 6, 1027–1042.
216. ______, *Vector-valued Maass-Poincaré series*, Proc. Amer. Math. Soc. **136** (2008), no. 2, 427–436.
217. F. G. Garvan, *Combinatorial interpretations of Ramanujan's partition congruences*, Ramanujan revisited (Urbana-Champaign, Ill., 1987), Academic Press, Boston, MA, 1988, pp. 29–45.
218. ______, *New combinatorial interpretations of Ramanujan's partition congruences mod* 5, 7 *and* 11, Trans. Amer. Math. Soc. **305** (1988), no. 1, 47–77.
219. ______, *The crank of partitions mod* 8, 9 *and* 10, Trans. Amer. Math. Soc. **322** (1990), no. 1, 79–94.
220. ______, *Higher order spt-functions*, Adv. Math. **228** (2011), no. 1, 241–265.
221. ______, *Congruences for Andrews' spt-function modulo powers of* 5, 7 *and* 13, Trans. Amer. Math. Soc. **364** (2012), no. 9, 4847–4873.
222. ______, *Congruences and relations for r-Fishburn numbers*, J. Combin. Theory Ser. A **134** (2015), 147–165.
223. Frank Garvan, Dongsu Kim, and Dennis Stanton, *Cranks and t-cores*, Invent. Math. **101** (1990), no. 1, 1–17.

224. George Gasper and Mizan Rahman, *Basic hypergeometric series*, Encyclopedia of Mathematics and its Applications, vol. 35, Cambridge University Press, Cambridge, 1990, With a foreword by Richard Askey.
225. D. Goldfeld and P. Sarnak, *Sums of Kloosterman sums*, Invent. Math. **71** (1983), no. 2, 243–250.
226. Dorian Goldfeld, *Conjectures on elliptic curves over quadratic fields*, Number theory, Carbondale 1979 (Proc. Southern Illinois Conf., Southern Illinois Univ., Carbondale, Ill., 1979), Lecture Notes in Math., vol. 751, Springer, Berlin, 1979, pp. 108–118.
227. ———, *Automorphic forms and L-functions for the group* $\mathrm{GL}(n, \mathbf{R})$, Cambridge Studies in Advanced Mathematics, vol. 99, Cambridge University Press, Cambridge, 2006, With an appendix by Kevin A. Broughan.
228. Basil Gordon and Richard J. McIntosh, *A survey of classical mock theta functions*, Partitions, q-series, and modular forms, Dev. Math., vol. 23, Springer, New York, 2012, pp. 95–144.
229. Lothar Göttsche, *The Betti numbers of the Hilbert scheme of points on a smooth projective surface*, Math. Ann. **286** (1990), no. 1-3, 193–207.
230. Lothar Göttsche and Don Zagier, *Jacobi forms and the structure of Donaldson invariants for 4-manifolds with* $b_+ = 1$, Selecta Math. (N.S.) **4** (1998), no. 1, 69–115.
231. Robert L. Griess, Jr., *The structure of the "monster" simple group*, (1976), 113–118.
232. ———, *A construction of* F_1 *as automorphisms of a* $196,883$*-dimensional algebra*, Proc. Nat. Acad. Sci. U.S.A. **78** (1981), no. 2, part 1, 686–691.
233. ———, *The friendly giant*, Invent. Math. **69** (1982), no. 1, 1–102.
234. Michael J. Griffin, *Private communication.*
235. Michael J. Griffin, *p-adic harmonic Maass functions*, Preprint.
236. Michael J. Griffin and Michael H. Mertens, *A proof of the Thompson moonshine conjecture*, Res. Math. Sci. **3** (2016), Paper No. 36, 32.
237. Michael J. Griffin, Ken Ono, and Larry Rolen, *Ramanujan's mock theta functions*, Proc. Natl. Acad. Sci. USA **110** (2013), no. 15, 5765–5768.
238. V. Gritsenko, *Elliptic genus of Calabi-Yau manifolds and Jacobi and Siegel modular forms*, Algebra i Analiz **11** (1999), no. 5, 100–125.
239. B. Gross, W. Kohnen, and D. Zagier, *Heegner points and derivatives of L-series. II*, Math. Ann. **278** (1987), no. 1-4, 497–562.
240. Benedict H. Gross, *The classes of singular moduli in the generalized Jacobian*, Geometry and arithmetic, EMS Ser. Congr. Rep., Eur. Math. Soc., Zürich, 2012, pp. 137–141.
241. Benedict H. Gross and Don B. Zagier, *On singular moduli*, J. Reine Angew. Math. **355** (1985), 191–220.
242. ———, *Heegner points and derivatives of L-series*, Invent. Math. **84** (1986), no. 2, 225–320.
243. P. Guerzhoy, *Hecke operators for weakly holomorphic modular forms and supersingular congruences*, Proc. Amer. Math. Soc. **136** (2008), no. 9, 3051–3059.
244. Pavel Guerzhoy, *On Zagier's adele*, Res. Math. Sci. **1** (2014), Art. 7, 19.
245. Pavel Guerzhoy, Zachary A. Kent, and Ken Ono, *p-adic coupling of mock modular forms and shadows*, Proc. Natl. Acad. Sci. USA **107** (2010), no. 14, 6169–6174.
246. Pavel Guerzhoy, Zachary A. Kent, and Larry Rolen, *Congruences for Taylor expansions of quantum modular forms*, Res. Math. Sci. **1** (2014), Art. 17, 17.
247. Klaus Haberland, *Perioden von Modulformen einer Variabler and Gruppencohomologie. I, II, III*, Math. Nachr. **112** (1983), 245–282, 283–295, 297–315.
248. Kazuo Habiro, *Cyclotomic completions of polynomial rings*, Publ. Res. Inst. Math. Sci. **40** (2004), no. 4, 1127–1146.
249. Gergely Harcos, *An additive problem in the Fourier coefficients of cusp forms*, Math. Ann. **326** (2003), no. 2, 347–365.
250. ———, *Twisted Hilbert modular L-functions and spectral theory*, Automorphic forms and L-functions, Adv. Lect. Math. (ALM), vol. 30, Int. Press, Somerville, MA, 2014, pp. 49–67.
251. Gergely Harcos and Philippe Michel, *The subconvexity problem for Rankin-Selberg L-functions and equidistribution of Heegner points. II*, Invent. Math. **163** (2006), no. 3, 581–655.
252. G. H. Hardy and S. Ramanujan, *Asymptotic formulæ for the distribution of integers of various types [Proc. London Math. Soc. (2)* ***16*** *(1917), 112–132]*, Collected papers of Srinivasa Ramanujan, AMS Chelsea Publ., Providence, RI, 2000, pp. 245–261.

253. ______, *On the coefficients in the expansions of certain modular functions [Proc. Roy. Soc. A* **95** *(1919), 144–155]*, Collected papers of Srinivasa Ramanujan, AMS Chelsea Publ., Providence, RI, 2000, pp. 310–321.

254. Jeffrey A. Harvey and Gregory Moore, *Algebras, BPS states, and strings*, Nuclear Phys. B **463** (1996), no. 2-3, 315–368.

255. Jeffrey A. Harvey and Brandon C. Rayhaun, *Traces of singular moduli and moonshine for the Thompson group*, Commun. Number Theory Phys. **10** (2016), no. 1, 23–62.

256. D. R. Heath-Brown, *The size of Selmer groups for the congruent number problem*, Invent. Math. **111** (1993), no. 1, 171–195.

257. ______, *The size of Selmer groups for the congruent number problem. II*, Invent. Math. **118** (1994), no. 2, 331–370, With an appendix by P. Monsky.

258. E. Hecke, *Eine neue Art von Zetafunktionen und ihre Beziehungen zur Verteilung der Primzahlen*, Math. Z. **6** (1920), no. 1-2, 11–51.

259. ______, *Analytische funktionen und algebraische zahlen*, Abh. Math. Sem. Univ. Hamburg **3** (1924), no. 1, 213–236.

260. Erich Hecke, *Lectures on Dirichlet series, modular functions and quadratic forms*, Vandenhoeck & Ruprecht, Göttingen, 1983, Edited by Bruno Schoeneberg, With the collaboration of Wilhelm Maak.

261. Dennis A. Hejhal, *The Selberg trace formula for* PSL(2, **R**). *Vol. 2*, Lecture Notes in Mathematics, vol. 1001, Springer-Verlag, Berlin, 1983.

262. Harald A. Helfgott, *The ternary Goldbach conjecture is true*, Preprint.

263. Dean Hickerson, *On the seventh order mock theta functions*, Invent. Math. **94** (1988), no. 3, 661–677.

264. ______, *A proof of the mock theta conjectures*, Invent. Math. **94** (1988), no. 3, 639–660.

265. Dean R. Hickerson and Eric T. Mortenson, *Hecke-type double sums, Appell-Lerch sums, and mock theta functions, I*, Proc. Lond. Math. Soc. (3) **109** (2014), no. 2, 382–422.

266. Kazuhiro Hikami, *Difference equation of the colored Jones polynomial for torus knot*, Internat. J. Math. **15** (2004), no. 9, 959–965.

267. ______, *Mock (false) theta functions as quantum invariants*, Regul. Chaotic Dyn. **10** (2005), no. 4, 509–530.

268. ______, *q-series and L-functions related to half-derivatives of the Andrews-Gordon identity*, Ramanujan J. **11** (2006), no. 2, 175–197.

269. Kazuhiro Hikami and Anatol N. Kirillov, *Torus knot and minimal model*, Phys. Lett. B **575** (2003), no. 3-4, 343–348.

270. Kazuhiro Hikami and Jeremy Lovejoy, *Torus knots and quantum modular forms*, Res. Math. Sci. **2** (2015), Art. 2, 15.

271. F. Hirzebruch and D. Zagier, *Intersection numbers of curves on Hilbert modular surfaces and modular forms of Nebentypus*, Invent. Math. **36** (1976), 57–113.

272. Jeff Hoffstein and Thomas A. Hulse, *Multiple Dirichlet series and shifted convolutions*, J. Number Theory **161** (2016), 457–533, With an appendix by Andre Reznikov.

273. Alexander E. Holroyd, Thomas M. Liggett, and Dan Romik, *Integrals, partitions, and cellular automata*, Trans. Amer. Math. Soc. **356** (2004), no. 8, 3349–3368.

274. Christopher Hooley, *On the number of divisors of a quadratic polynomial*, Acta Math. **110** (1963), 97–114.

275. Martin Hövel, *Automorphe Formen mit Singularitäten auf dem hyperbolischen Raum*, Ph.D. thesis. 2012.

276. Adolf Hurwitz, *Ueber Relationen zwischen Classenanzahlen binärer quadratischer Formen von negativer Determinante*, Math. Ann. **25** (1885), no. 2, 157–196.

277. Özlem Imamoḡlu and Cormac O'Sullivan, *Parabolic, hyperbolic and elliptic Poincaré series*, Acta Arith. **139** (2009), no. 3, 199–228.

278. Özlem Imamoḡlu, Martin Raum, and Olav K. Richter, *Holomorphic projections and Ramanujan's mock theta functions*, Proc. Natl. Acad. Sci. USA **111** (2014), no. 11, 3961–3967.

279. A. E. Ingham, *A Tauberian theorem for partitions*, Ann. of Math. (2) **42** (1941), 1075–1090.

280. H. Iwaniec and P. Sarnak, *Perspectives on the analytic theory of L-functions*, Geom. Funct. Anal. (2000), no. Special Volume, Part II, 705–741, GAFA 2000 (Tel Aviv, 1999).

281. Henryk Iwaniec, *Fourier coefficients of modular forms of half-integral weight*, Invent. Math. **87** (1987), no. 2, 385–401.

282. ———, *Topics in classical automorphic forms*, Graduate Studies in Mathematics, vol. 17, American Mathematical Society, Providence, RI, 1997.
283. ———, *Spectral methods of automorphic forms*, second ed., Graduate Studies in Mathematics, vol. 53, American Mathematical Society, Providence, RI; Revista Matemática Iberoamericana, Madrid, 2002.
284. Henryk Iwaniec and Emmanuel Kowalski, *Analytic number theory*, American Mathematical Society Colloquium Publications, vol. 53, American Mathematical Society, Providence, RI, 2004.
285. Kenkichi Iwasawa, *On p-adic L-functions*, Ann. of Math. (2) **89** (1969), 198–205.
286. Carl G. J. Jacobi, *Fundamenta nova theoriae functionum ellipticarum*, Königsberg. 1829.
287. Min-Joo Jang and Steffen Löbrich, *Radial limits of the universal mock theta function g_3*, Proc. Amer. Math. Soc. **145** (2017), no. 3, 925–935.
288. V. G. Kac, *Infinite-dimensional Lie algebras, and the Dedekind η-function*, Funkcional. Anal. i Priložen. **8** (1974), no. 1, 77–78.
289. Victor G. Kac and Minoru Wakimoto, *Integrable highest weight modules over affine superalgebras and number theory*, Lie theory and geometry, Progr. Math., vol. 123, Birkhäuser Boston, Boston, MA, 1994, pp. 415–456.
290. ———, *Integrable highest weight modules over affine superalgebras and Appell's function*, Comm. Math. Phys. **215** (2001), no. 3, 631–682.
291. ———, *Representations of affine superalgebras and mock theta functions*, Transform. Groups **19** (2014), no. 2, 383–455.
292. ———, *Representations of affine superalgebras and mock theta functions II*, Adv. Math. **300** (2016), 17–70.
293. ———, *Representations of affine superalgebras and mock theta functions. III*, Izv. Ross. Akad. Nauk Ser. Mat. **80** (2016), no. 4, 65–122.
294. Ben Kane and Matthias Waldherr, *Explicit congruences for mock modular forms*, J. Number Theory **166** (2016), 1–18.
295. Daniel M. Kane, *Resolution of a conjecture of Andrews and Lewis involving cranks of partitions*, Proc. Amer. Math. Soc. **132** (2004), no. 8, 2247–2256.
296. ———, *On the ranks of the 2-Selmer groups of twists of a given elliptic curve*, Algebra Number Theory **7** (2013), no. 5, 1253–1279.
297. Masanobu Kaneko, *Observations on the 'values' of the elliptic modular function $j(\tau)$ at real quadratics*, Kyushu J. Math. **63** (2009), no. 2, 353–364.
298. Masanobu Kaneko and Don Zagier, *A generalized Jacobi theta function and quasimodular forms*, The moduli space of curves (Texel Island, 1994), Progr. Math., vol. 129, Birkhäuser Boston, Boston, MA, 1995, pp. 165–172.
299. Soon-Yi Kang, *Mock Jacobi forms in basic hypergeometric series*, Compos. Math. **145** (2009), no. 3, 553–565.
300. Ernst Kani, *The space of binary theta series*, Ann. Sci. Math. Québec **36** (2012), no. 2, 501–534 (2013).
301. Svetlana Katok, *Closed geodesics, periods and arithmetic of modular forms*, Invent. Math. **80** (1985), no. 3, 469–480.
302. Svetlana Katok and Peter Sarnak, *Heegner points, cycles and Maass forms*, Israel J. Math. **84** (1993), no. 1-2, 193–227.
303. Nicholas M. Katz and Peter Sarnak, *Random matrices, Frobenius eigenvalues, and monodromy*, American Mathematical Society Colloquium Publications, vol. 45, American Mathematical Society, Providence, RI, 1999.
304. Toshiya Kawai, Yasuhiko Yamada, and Sung-Kil Yang, *Elliptic genera and $N = 2$ superconformal field theory*, Nuclear Phys. B **414** (1994), no. 1-2, 191–212.
305. Zachary A. Kent, *Periods of Eisenstein series and some applications*, Proc. Amer. Math. Soc. **139** (2011), no. 11, 3789–3794.
306. Byungchan Kim and Jeremy Lovejoy, *Ramanujan-type partial theta identities and rank differences for special unimodal sequences*, Ann. Comb. **19** (2015), no. 4, 705–733.
307. Chang Heon Kim, *Borcherds products associated with certain Thompson series*, Compos. Math. **140** (2004), no. 3, 541–551.
308. ———, *Traces of singular values and Borcherds products*, Bull. London Math. Soc. **38** (2006), no. 5, 730–740.

309. Henry H. Kim, *Functoriality for the exterior square of* GL_4 *and the symmetric fourth of* GL_2, J. Amer. Math. Soc. **16** (2003), no. 1, 139–183, With appendix 1 by Dinakar Ramakrishnan and appendix 2 by Kim and Peter Sarnak.
310. Susanna Kimport, *Quantum modular forms, mock modular forms, and partial theta functions*, ProQuest LLC, Ann Arbor, MI, 2015, Thesis (Ph.D.)–Yale University.
311. Anthony W. Knapp, *Elliptic curves*, Mathematical Notes, vol. 40, Princeton University Press, Princeton, NJ, 1992.
312. Marvin Isadore Knopp, *On abelian integrals of the second kind and modular functions*, Amer. J. Math. **84** (1962), 615–628.
313. ______, *On generalized abelian integrals of the second kind and modular forms of dimension zero*, Amer. J. Math. **86** (1964), 430–440.
314. ______, *Polynomial automorphic forms and nondiscontinuous groups*, Trans. Amer. Math. Soc. **123** (1966), 506–520.
315. ______, *Some new results on the Eichler cohomology of automorphic forms*, Bull. Amer. Math. Soc. **80** (1974), 607–632.
316. Neal Koblitz, *Introduction to elliptic curves and modular forms*, second ed., Graduate Texts in Mathematics, vol. 97, Springer-Verlag, New York, 1993.
317. Lisa Kodgis, *Zeros of the modular parametrization of rational elliptic curves*, M.A. thesis. University of Hawaii. 2011.
318. Max Koecher and Aloys Krieg, *Elliptische Funktionen und Modulformen*, revised ed., Springer-Verlag, Berlin, 2007.
319. Günter Köhler, *Eta products and theta series identities*, Springer Monographs in Mathematics, Springer, Heidelberg, 2011.
320. W. Kohnen and D. Zagier, *Values of L-series of modular forms at the center of the critical strip*, Invent. Math. **64** (1981), no. 2, 175–198.
321. ______, *Modular forms with rational periods*, Modular forms (Durham, 1983), Ellis Horwood Ser. Math. Appl.: Statist. Oper. Res., Horwood, Chichester, 1984, pp. 197–249.
322. Winfried Kohnen, *Modular forms of half-integral weight on* $\Gamma_0(4)$, Math. Ann. **248** (1980), no. 3, 249–266.
323. ______, *Newforms of half-integral weight*, J. Reine Angew. Math. **333** (1982), 32–72.
324. ______, *Fourier coefficients of modular forms of half-integral weight*, Math. Ann. **271** (1985), no. 2, 237–268.
325. ______, *On the average growth of Fourier coefficients of Siegel cusp forms of genus* 2, Pacific J. Math. **179** (1997), no. 1, 119–121.
326. ______, *On the proportion of quadratic character twists of L-functions attached to cusp forms not vanishing at the central point*, J. Reine Angew. Math. **508** (1999), 179–187.
327. Winfried Kohnen and Ken Ono, *Indivisibility of class numbers of imaginary quadratic fields and orders of Tate-Shafarevich groups of elliptic curves with complex multiplication*, Invent. Math. **135** (1999), no. 2, 387–398.
328. M. Koike, *On replication formula and Hecke operators*, Nagoya University, preprint.
329. V. A. Kolyvagin, *Finiteness of* $E(\mathbf{Q})$ *and the Tate-Shafarevich group of* $(E, \mathbf{Q})$ *for a subclass of Weil curves*, Izv. Akad. Nauk SSSR Ser. Mat. **52** (1988), no. 3, 522–540, 670–671.
330. E. Kowalski, P. Michel, and J. VanderKam, *Rankin-Selberg L-functions in the level aspect*, Duke Math. J. **114** (2002), no. 1, 123–191.
331. Daniel Kriz and Chao Li, *Goldfeld's conjecture and congruences between Heegner points*, Preprint.
332. L. Kronecker, *Ueber die Anzahl der verschiedenen Classen quadratischer Formen von negativer Determinante*, J. Reine Angew. Math. **57** (1860), 248–255.
333. Tomio Kubota and Heinrich-Wolfgang Leopoldt, *Eine p-adische Theorie der Zetawerte. I. Einführung der p-adischen Dirichletschen L-Funktionen*, J. Reine Angew. Math. **214/215** (1964), 328–339.
334. Stephen S. Kudla, *On the integrals of certain singular theta-functions*, J. Fac. Sci. Univ. Tokyo Sect. IA Math. **28** (1981), no. 3, 439–463 (1982).
335. ______, *Special cycles and derivatives of Eisenstein series*, Heegner points and Rankin L-series, Math. Sci. Res. Inst. Publ., vol. 49, Cambridge Univ. Press, Cambridge, 2004, pp. 243–270.
336. Stephen S. Kudla and John J. Millson, *The theta correspondence and harmonic forms. I*, Math. Ann. **274** (1986), no. 3, 353–378.

337. Stephen S. Kudla, Michael Rapoport, and Tonghai Yang, *On the derivative of an Eisenstein series of weight one*, Internat. Math. Res. Notices (1999), no. 7, 347–385.
338. ———, *Derivatives of Eisenstein series and Faltings heights*, Compos. Math. **140** (2004), no. 4, 887–951.
339. Stephen S. Kudla and Tonghai Yang, *Eisenstein series for SL(2)*, Sci. China Math. **53** (2010), no. 9, 2275–2316.
340. Nobushige Kurokawa, *Examples of eigenvalues of Hecke operators on Siegel cusp forms of degree two*, Invent. Math. **49** (1978), no. 2, 149–165.
341. T. Y. Lam, *Introduction to quadratic forms over fields*, Graduate Studies in Mathematics, vol. 67, American Mathematical Society, Providence, RI, 2005.
342. Serge Lang, *Elliptic functions*, second ed., Graduate Texts in Mathematics, vol. 112, Springer-Verlag, New York, 1987, With an appendix by J. Tate.
343. Eric Larson and Larry Rolen, *Integrality properties of the CM-values of certain weak Maass forms*, Forum Math., **27** no. 2 (2015), 961–972.
344. Ruth Lawrence and Don Zagier, *Modular forms and quantum invariants of 3-manifolds*, Asian J. Math. **3** (1999), no. 1, 93–107, Sir Michael Atiyah: a great mathematician of the twentieth century.
345. D. H. Lehmer, *On the series for the partition function*, Trans. Amer. Math. Soc. **43** (1938), no. 2, 271–295.
346. ———, *On the remainders and convergence of the series for the partition function*, Trans. Amer. Math. Soc. **46** (1939), 362–373.
347. Joseph Lehner, *The Eichler cohomology of a Kleinian group*, Math. Ann. **192** (1971), 125–143.
348. J. Lepowsky and S. Milne, *Lie algebraic approaches to classical partition identities*, Adv. in Math. **29** (1978), no. 1, 15–59.
349. ———, *Lie algebras and classical partition identities*, Proc. Nat. Acad. Sci. U.S.A. **75** (1978), no. 2, 578–579.
350. James Lepowsky and Robert Lee Wilson, *The Rogers-Ramanujan identities: Lie theoretic interpretation and proof*, Proc. Nat. Acad. Sci. U.S.A. **78** (1981), no. 2, part 1, 699–701.
351. M. Lerch, *Bemerkungen zur Theorie der elliptischen Funktionen*, Rozpravyceske Akademie Cisare Frantiska Josefa pro vedy slovesnost, a umeni (II Cl) Prag. (Bohmisch) I (1892), no. 24, Abstract in the Jahrbuch über die Fortschritte der Mathematik, 24 (1892), 442-445.
352. ———, *Poznamky k theorii funkci ellipickych*, Rozpravy Ceske Akademie Cisare Frantiska Josefa pro vedy, slovesnost, a umeni (1892), no. 24.
353. J. Lewis and D. Zagier, *Period functions for Maass wave forms. I*, Ann. of Math. (2) **153** (2001), no. 1, 191–258.
354. Richard Lewis, *On some relations between the rank and the crank*, J. Combin. Theory Ser. A **59** (1992), no. 1, 104–110.
355. Richard Lewis and Nicolas Santa-Gadea, *On the rank and the crank modulo* 4 *and* 8, Trans. Amer. Math. Soc. **341** (1994), no. 1, 449–465.
356. Yingkun Li, Hieu T. Ngo, and Robert C. Rhoades, *Renormalization and quantum modular forms, Part I: Maass wave forms*, Preprint.
357. R. Lipschitz, *Untersuchung der Eigenschaften einer Gattung von unendlichen Reihen*, J. Reine Angew. Math. **105** (1889), 127–156.
358. Jianya Liu, *A quick introduction to Maass forms*, Number theory, Ser. Number Theory Appl., vol. 6, World Sci. Publ., Hackensack, NJ, 2010, pp. 184–216.
359. Jianya Liu and Yangbo Ye, *Subconvexity for Rankin-Selberg L-functions of Maass forms*, Geom. Funct. Anal. **12** (2002), no. 6, 1296–1323.
360. Sheng-Chi Liu, Riad Masri, and Matthew P. Young, *Subconvexity and equidistribution of Heegner points in the level aspect*, Compos. Math. **149** (2013), no. 7, 1150–1174.
361. Jeremy Lovejoy, *Rank and conjugation for the Frobenius representation of an overpartition*, Ann. Comb. **9** (2005), no. 3, 321–334.
362. ———, *Rank and conjugation for a second Frobenius representation of an overpartition*, Ann. Comb. **12** (2008), no. 1, 101–113.
363. Jeremy Lovejoy and Robert Osburn, *M_2-rank differences for partitions without repeated odd parts*, J. Théor. Nombres Bordeaux **21** (2009), no. 2, 313–334.
364. ———, *The Bailey chain and mock theta functions*, Adv. Math. **238** (2013), 442–458.

365. ______, *Mixed mock modular q-series*, J. Indian Math. Soc. (N.S.) (2013), no. Special volume to commemorate the 125th birth anniversary of Srinivasa Ramanujan, 45–61.
366. Hans Maass, *Automorphe Funktionen und indefinite quadratische Formen*, S.-B. Heidelberger Akad. Wiss. Math.-Nat. Kl. **1949** (1949), no. 1, 42.
367. ______, *Über eine neue Art von nichtanalytischen automorphen Funktionen und die Bestimmung Dirichletscher Reihen durch Funktionalgleichungen*, Math. Ann. **121** (1949), 141–183.
368. ______, *Die Differentialgleichungen in der Theorie der elliptischen Modulfunktionen*, Math. Ann. **125** (1952), 235–263 (1953).
369. ______, *Differentialgleichungen und automorphe Funktionen*, Proceedings of the International Congress of Mathematicians, 1954, Amsterdam, vol. III, Erven P. Noordhoff N.V., Groningen; North-Holland Publishing Co., Amsterdam, 1956, pp. 34–39.
370. ______, *Über die räumliche Verteilung der Punkte in Gittern mit indefiniter Metrik*, Math. Ann. **138** (1959), 287–315.
371. I. G. Macdonald, *Affine root systems and Dedekind's η-function*, Invent. Math. **15** (1972), 91–143.
372. Péter Maga, *Shifted convolution sums and Burgess type subconvexity over number fields*, To appear in J. Reine Angew. Math.
373. Karl Mahlburg, *Partition congruences and the Andrews-Garvan-Dyson crank*, Proc. Natl. Acad. Sci. USA **102** (2005), no. 43, 15373–15376.
374. Yuri I. Manin, *Periods of parabolic forms and p-adic Hecke series*, Mat. Sb. (N.S.) **21** (1973), 371–393.
375. ______, *Cyclotomy and analytic geometry over $\mathbb{F}_1$*, Quanta of maths, Clay Math. Proc., vol. 11, Amer. Math. Soc., Providence, RI, 2010, pp. 385–408.
376. Jan Manschot, *BPS invariants of semi-stable sheaves on rational surfaces*, Lett. Math. Phys. **103** (2013), no. 8, 895–918.
377. Riad Masri, *The asymptotic distribution of traces of cycle integrals of the j-function*, Duke Math. J. **161** (2012), no. 10, 1971–2000.
378. David Masser, *Elliptic functions and transcendence*, Lecture Notes in Mathematics, Vol. 437, Springer-Verlag, Berlin-New York, 1975.
379. B. Mazur, *Modular curves and the Eisenstein ideal*, Inst. Hautes Études Sci. Publ. Math. (1977), no. 47, 33–186 (1978).
380. Richard J. McIntosh, *Second order mock theta functions*, Canad. Math. Bull. **50** (2007), no. 2, 284–290.
381. ______, *Some identities for Appell-Lerch sums and a universal mock theta function*, Ramanujan J. (2017), Accepted for publication.
382. Anton Mellit, *Higher Green's functions for modular forms*, Ph.D thesis. Rheinische Friedrich-Wilhelms-Universität Bonn, 2008.
383. Loïc Merel, *Bornes pour la torsion des courbes elliptiques sur les corps de nombres*, Invent. Math. **124** (1996), no. 1-3, 437–449.
384. Michael H. Mertens, *Mock modular forms and class number relations*, Res. Math. Sci. **1** (2014), Art. 6, 16.
385. ______, *Eichler-Selberg type identities for mixed mock modular forms*, Adv. Math. **301** (2016), 359–382.
386. Michael H. Mertens and Ken Ono, *Special values of shifted convolution Dirichlet series*, Mathematika **62** (2016), no. 1, 47–66.
387. Philippe Michel and Akshay Venkatesh, *The subconvexity problem for* GL_2, Publ. Math. Inst. Hautes Études Sci. (2010), no. 111, 171–271.
388. Toshitsune Miyake, *Modular forms*, english ed., Springer Monographs in Mathematics, Springer-Verlag, Berlin, 2006, Translated from the 1976 Japanese original by Yoshitaka Maeda.
389. Sergey Mozgovoy, *Invariants of moduli spaces of stable sheaves on ruled surfaces*, Preprint.
390. David Mumford, *Tata lectures on theta. I*, Progress in Mathematics, vol. 28, Birkhäuser Boston, Inc., Boston, MA, 1983, With the assistance of C. Musili, M. Nori, E. Previato and M. Stillman.
391. M. Ram Murty and V. Kumar Murty, *Mean values of derivatives of modular L-series*, Ann. of Math. (2) **133** (1991), no. 3, 447–475.
392. ______, *The mathematical legacy of Srinivasa Ramanujan*, Springer, New Delhi, 2013.

393. Werner Nahm, *Conformal field theory, dilogarithms, and three-dimensional manifolds*, Interface between physics and mathematics (Hangzhou, 1993), World Sci. Publ., River Edge, NJ, 1994, pp. 154–165.
394. ———, *Conformal field theory and the dilogarithm*, XIth International Congress of Mathematical Physics (Paris, 1994), Int. Press, Cambridge, MA, 1995, pp. 662–667.
395. ———, *Conformal field theory and torsion elements of the Bloch group*, Frontiers in number theory, physics, and geometry. II, Springer, Berlin, 2007, pp. 67–132.
396. Caner Nazaroglu, *r-tuple error functions and indefinite theta series of higher-depth*, Preprint.
397. Douglas Niebur, *A class of nonanalytic automorphic functions*, Nagoya Math. J. **52** (1973), 133–145.
398. ———, *Construction of automorphic forms and integrals*, Trans. Amer. Math. Soc. **191** (1974), 373–385.
399. Hans-Volker Niemeier, *Definite quadratische Formen der Dimension* 24 *und Diskriminante* 1, J. Number Theory **5** (1973), 142–178.
400. NIST Digital Library of Mathematical Functions, `http://dlmf.nist.gov/`, Release 1.0.6 of 2013-05-06.
401. Andrew P. Ogg, *Automorphismes de courbes modulaires*, (1975), 8.
402. René Olivetto, *A family of mock theta functions of weight* 1/2 *and* 3/2 *and their congruence properties*, Preprint.
403. ———, *On the Fourier coefficients of meromorphic Jacobi forms*, Int. J. Number Theory **10** (2014), no. 6, 1519–1540.
404. Ken Ono, *Distribution of the partition function modulo m*, Ann. of Math. (2) **151** (2000), no. 1, 293–307.
405. ———, *The web of modularity: arithmetic of the coefficients of modular forms and q-series*, CBMS Regional Conference Series in Mathematics, vol. 102, Published for the Conference Board of the Mathematical Sciences, Washington, DC; by the American Mathematical Society, Providence, RI, 2004.
406. ———, *A mock theta function for the Delta-function*, Combinatorial number theory, Walter de Gruyter, Berlin, 2009, pp. 141–155.
407. ———, *Unearthing the visions of a master: harmonic Maass forms and number theory*, Current developments in mathematics, 2008, Int. Press, Somerville, MA, 2009, pp. 347–454.
408. ———, *The last words of a genius*, Notices Amer. Math. Soc. **57** (2010), no. 11, 1410–1419.
409. ———, *Congruences for the Andrews spt function*, Proc. Natl. Acad. Sci. USA **108** (2011), no. 2, 473–476.
410. Ken Ono, Larry Rolen, and Sarah Trebat-Leder, *Classical and umbral moonshine: connections and p-adic properties*, J. Ramanujan Math. Soc. **30** (2015), no. 2, 135–159.
411. Ken Ono and Christopher Skinner, *Non-vanishing of quadratic twists of modular L-functions*, Invent. Math. **134** (1998), no. 3, 651–660.
412. Vicentiu Paşol and Alexandru A. Popa, *Modular forms and period polynomials*, Proc. Lond. Math. Soc. (3) **107** (2013), no. 4, 713–743.
413. Edward Allen Parberry, *On a partition function related to a theorem of Schur*, ProQuest LLC, Ann Arbor, MI, 1969, Thesis (Ph.D.)–The Pennsylvania State University.
414. Daniel Parry and Robert C. Rhoades, *On Dyson's crank distribution conjecture and its generalizations*, Proc. Amer. Math. Soc. **145** (2017), no. 1, 101–108.
415. Sarah Peluse, *On zeros of Eichler integrals*, Arch. Math. (Basel) **102** (2014), no. 1, 71–81.
416. A. Perelli and J. Pomykala, *Averages of twisted elliptic L-functions*, Acta Arith. **80** (1997), no. 2, 149–163.
417. Daniel Persson and Roberto Volpato, *Second-quantized Mathieu moonshine*, Commun. Number Theory Phys. **8** (2014), no. 3, 403–509.
418. Hans Petersson, *Theorie der automorphen Formen beliebiger reeller Dimension und ihre Darstellung durch eine neue Art Poincaréscher Reihen*, Math. Ann. **103** (1930), no. 1, 369–436.
419. ———, *Konstruktion der Modulformen und der zu gewissen Grenzkreisgruppen gehörigen automorphen Formen von positiver reeller Dimension und die vollständige Bestimmung ihrer Fourierkoeffizienten*, S.-B. Heidelberger Akad. Wiss. Math.-Nat. Kl. **1950** (1950), 417–494.
420. Boris Pioline, *Rankin-Selberg methods for closed string amplitudes*, String-Math 2013, Proc. Sympos. Pure Math., vol. 88, Amer. Math. Soc., Providence, RI, 2014, pp. 119–143.

421. Alexander Polishchuk, *Abelian varieties, theta functions and the Fourier transform*, Cambridge Tracts in Mathematics, vol. 153, Cambridge University Press, Cambridge, 2003.

422. I. I. Pyateskii-Shapiro, *Automorphic functions and the geometry of classical domains*, Translated from the Russian. Mathematics and Its Applications, Vol. 8, Gordon and Breach Science Publishers, New York-London-Paris, 1969.

423. Hans Rademacher, *On the Partition Function* $p(n)$, Proc. London Math. Soc. **S2-43**, no. 4, 241.

424. ———, *On the expansion of the partition function in a series*, Ann. of Math. (2) **44** (1943), 416–422.

425. ———, *On the Selberg formula for* $A_k(n)$, J. Indian Math. Soc. (N.S.) **21** (1957), 41–55 (1958).

426. Hans Rademacher and Herbert S. Zuckerman, *On the Fourier coefficients of certain modular forms of positive dimension*, Ann. of Math. (2) **39** (1938), no. 2, 433–462.

427. Cristian-Silviu Radu, *A proof of Subbarao's conjecture*, J. Reine Angew. Math. **672** (2012), 161–175.

428. S. Ramanujan, *Congruence properties of partitions*, Math. Z. **9** (1921), no. 1-2, 147–153.

429. ———, *The lost notebook and other unpublished papers*, Springer-Verlag, Berlin; Narosa Publishing House, New Delhi, 1988, With an introduction by George E. Andrews.

430. ———, *On certain arithmetical functions [Trans. Cambridge Philos. Soc.* **22** *(1916), no. 9, 159–184]*, Collected papers of Srinivasa Ramanujan, AMS Chelsea Publ., Providence, RI, 2000, pp. 136–162.

431. ———, *Some properties of* $p(n)$*, the number of partitions of* n *[Proc. Cambridge Philos. Soc.* **19** *(1919), 207–210]*, Collected papers of Srinivasa Ramanujan, AMS Chelsea Publ., Providence, RI, 2000, pp. 210–213.

432. R. A. Rankin, *Contributions to the theory of Ramanujan's function* $\tau(n)$ *and similar arithmetical functions. I. The zeros of the function* $\sum_{n=1}^{\infty} \tau(n)/n^s$ *on the line* $\Re(s) = 13/2$*. II. The order of the Fourier coefficients of integral modular forms*, Proc. Cambridge Philos. Soc. **35** (1939), 351–372.

433. ———, *The construction of automorphic forms from the derivatives of a given form*, J. Indian Math. Soc. (N.S.) **20** (1956), 103–116.

434. Dillon Reihill, *Private communication.*

435. Robert C. Rhoades, *On Ramanujan's definition of mock theta function*, Proc. Natl. Acad. Sci. USA **110** (2013), no. 19, 7592–7594.

436. ———, *Asymptotics for the number of strongly unimodal sequences*, Int. Math. Res. Not. IMRN (2014), no. 3, 700–719.

437. L. J. Rogers, *On two theorems of combinatory analysis and some allied identities*, Proceedings of the London Mathematical Society **s2-16** (1917), no. 1, 315–336.

438. Larry Rolen, *Notes from: "School on mock modular forms and related topics"*, Kyushu University. November 21-25, 2016. http://www.maths.tcd.ie/~lrolen/Notes.html.

439. ———, *Maass Forms and Quantum Modular Forms*, ProQuest LLC, Ann Arbor, MI, 2013, Thesis (Ph.D.)–Emory University.

440. ———, *A new construction of Eisenstein's completion of the Weierstrass zeta function*, Proc. Amer. Math. Soc. **144** (2016), no. 4, 1453–1456.

441. Larry Rolen and Robert P. Schneider, *A "strange" vector-valued quantum modular form*, Arch. Math. (Basel) **101** (2013), no. 1, 43–52.

442. Nathan C. Ryan, Nicolás Sirolli, Nils-Peter Skoruppa, and Gonzalo Tornaría, *Computing Jacobi forms*, LMS J. Comput. Math. **19** (2016), no. suppl. A, 205–219.

443. Nicolas Santa-Gadea, *On some relations for the rank moduli* 9 *and* 12, J. Number Theory **40** (1992), no. 2, 130–145.

444. Peter Sarnak, *Estimates for Rankin-Selberg L-functions and quantum unique ergodicity*, J. Funct. Anal. **184** (2001), no. 2, 419–453.

445. Bruno Schoeneberg, *Elliptic modular functions: an introduction*, Springer-Verlag, New York-Heidelberg, 1974, Translated from the German by J. R. Smart and E. A. Schwandt, Die Grundlehren der mathematischen Wissenschaften, Band 203.

446. A. J. Scholl, *Fourier coefficients of Eisenstein series on noncongruence subgroups*, Math. Proc. Cambridge Philos. Soc. **99** (1986), no. 1, 11–17.

447. Rainer Schulze-Pillot, *Weak Maaß forms and* $(\mathfrak{g}, K)$*-modules*, Ramanujan J. **26** (2011), no. 3, 437–445.

448. Atle Selberg, *Bemerkungen über eine Dirichletsche Reihe, die mit der Theorie der Modulformen nahe verbunden ist*, Arch. Math. Naturvid. **43** (1940), 47–50.
449. ———, *On the estimation of Fourier coefficients of modular forms*, Proc. Sympos. Pure Math., Vol. VIII, Amer. Math. Soc., Providence, R.I., 1965, pp. 1–15.
450. J.-P. Serre and H. M. Stark, *Modular forms of weight* 1/2, (1977), 27–67. Lecture Notes in Math., Vol. 627.
451. Jean-Pierre Serre, *A course in arithmetic*, Springer-Verlag, New York-Heidelberg, 1973, Translated from the French, Graduate Texts in Mathematics, No. 7.
452. ———, *Formes modulaires et fonctions zêta p-adiques*, (1973), 191–268. Lecture Notes in Math., Vol. 350.
453. ———, *Divisibilité de certaines fonctions arithmétiques*, Enseignement Math. (2) **22** (1976), no. 3-4, 227–260.
454. ———, *Algebraic groups and class fields*, Graduate Texts in Mathematics, vol. 117, Springer-Verlag, New York, 1988, Translated from the French.
455. Goro Shimura, *Introduction to the arithmetic theory of automorphic functions*, Publications of the Mathematical Society of Japan, No. 11. Iwanami Shoten, Publishers, Tokyo; Princeton University Press, Princeton, N.J., 1971, Kanô Memorial Lectures, No. 1.
456. ———, *On modular forms of half integral weight*, Ann. of Math. (2) **97** (1973), 440–481.
457. ———, *On the holomorphy of certain Dirichlet series*, Proc. London Math. Soc. (3) **31** (1975), no. 1, 79–98.
458. ———, *On a class of nearly holomorphic automorphic forms*, Ann. of Math. (2) **123** (1986), no. 2, 347–406.
459. Carl Ludwig Siegel, *Indefinite quadratische Formen und Funktionentheorie. I*, Math. Ann. **124** (1951), 17–54.
460. ———, *Die Funktionalgleichungen einiger Dirichletscher Reihen*, Math. Z. **63** (1956), 363–373.
461. ———, *Advanced analytic number theory*, second ed., Tata Institute of Fundamental Research Studies in Mathematics, vol. 9, Tata Institute of Fundamental Research, Bombay, 1980.
462. Joseph H. Silverman, *The arithmetic of elliptic curves*, Graduate Texts in Mathematics, vol. 106, Springer-Verlag, New York, 1986.
463. ———, *Advanced topics in the arithmetic of elliptic curves*, Graduate Texts in Mathematics, vol. 151, Springer-Verlag, New York, 1994.
464. Joseph H. Silverman and John Tate, *Rational points on elliptic curves*, Undergraduate Texts in Mathematics, Springer-Verlag, New York, 1992.
465. Nils-Peter Skoruppa, *Über den Zusammenhang zwischen Jacobiformen und Modulformen halbganzen Gewichts*, Bonner Mathematische Schriften [Bonn Mathematical Publications], vol. 159, Universität Bonn, Mathematisches Institut, Bonn, 1985, Dissertation, Rheinische Friedrich-Wilhelms-Universität, Bonn, 1984.
466. ———, *Developments in the theory of Jacobi forms*, Automorphic functions and their applications (Khabarovsk, 1988), Acad. Sci. USSR, Inst. Appl. Math., Khabarovsk, 1990, pp. 167–185.
467. Nils-Peter Skoruppa and Don Zagier, *Jacobi forms and a certain space of modular forms*, Invent. Math. **94** (1988), no. 1, 113–146.
468. Stephen D. Smith, *On the head characters of the Monster simple group*, Finite groups—coming of age (Montreal, Que., 1982), Contemp. Math., vol. 45, Amer. Math. Soc., Providence, RI, 1985, pp. 303–313.
469. Richard P. Stanley, *Unimodal sequences arising from Lie algebras*, Combinatorics, representation theory and statistical methods in groups, Lecture Notes in Pure and Appl. Math., vol. 57, Dekker, New York, 1980, pp. 127–136.
470. ———, *Log-concave and unimodal sequences in algebra, combinatorics, and geometry*, Graph theory and its applications: East and West (Jinan, 1986), Ann. New York Acad. Sci., vol. 576, New York Acad. Sci., New York, 1989, pp. 500–535.
471. ———, *Enumerative combinatorics. Vol. 1*, Cambridge Studies in Advanced Mathematics, vol. 49, Cambridge University Press, Cambridge, 1997, With a foreword by Gian-Carlo Rota, Corrected reprint of the 1986 original.

472. William Stein, *Modular forms, a computational approach*, Graduate Studies in Mathematics, vol. 79, American Mathematical Society, Providence, RI, 2007, With an appendix by Paul E. Gunnells.
473. Henning Stichtenoth, *Algebraic function fields and codes*, second ed., Graduate Texts in Mathematics, vol. 254, Springer-Verlag, Berlin, 2009.
474. Armin Straub, *Congruences for Fishburn numbers modulo prime powers*, Int. J. Number Theory **11** (2015), no. 5, 1679–1690.
475. Jacob Sturm, *Projections of C^∞ automorphic forms*, Bull. Amer. Math. Soc. (N.S.) **2** (1980), no. 3, 435–439.
476. ______, *On the congruence of modular forms*, Number theory (New York, 1984–1985), Lecture Notes in Math., vol. 1240, Springer, Berlin, 1987, pp. 275–280.
477. Richard Taylor and Andrew Wiles, *Ring-theoretic properties of certain Hecke algebras*, Ann. of Math. (2) **141** (1995), no. 3, 553–572.
478. Stephanie Treneer, *Congruences for the coefficients of weakly holomorphic modular forms*, Proc. London Math. Soc. (3) **93** (2006), no. 2, 304–324.
479. J. B. Tunnell, *A classical Diophantine problem and modular forms of weight* 3/2, Invent. Math. **72** (1983), no. 2, 323–334.
480. Cumrun Vafa and Edward Witten, *A strong coupling test of S-duality*, Nuclear Phys. B **431** (1994), no. 1-2, 3–77.
481. R. C. Vaughan, *The Hardy-Littlewood method*, second ed., Cambridge Tracts in Mathematics, vol. 125, Cambridge University Press, Cambridge, 1997.
482. Maryna Viazovska, *CM values of higher Green's functions*, Preprint.
483. ______, *Petersson inner products of weight one modular forms*, To appear in J. Reine Angew. Math.
484. Marie-France Vignéras, *Séries thêta des formes quadratiques indéfinies*, (1977), 3.
485. Ian Wagner, *Harmonic Maass form eigencurves*, Preprint.
486. Michel Waldschmidt, *Nombres transcendants et groupes algébriques*, Astérisque, vol. 69, Société Mathématique de France, Paris, 1979, With appendices by Daniel Bertrand and Jean-Pierre Serre, With an English summary.
487. J.-L. Waldspurger, *Sur les coefficients de Fourier des formes modulaires de poids demi-entier*, J. Math. Pures Appl. (9) **60** (1981), no. 4, 375–484.
488. Xueli Wang and Dingyi Pei, *Modular forms with integral and half-integral weights*, Science Press Beijing, Beijing; Springer, Heidelberg, 2012.
489. G. N. Watson, *The Final Problem: An Account of the Mock Theta Functions*, J. London Math. Soc. **S1-11**, no. 1, 55.
490. ______, *Ramanujans Vermutung über Zerfällungszahlen*, J. Reine Angew. Math. **179** (1938), 97–128.
491. André Weil, *Sur la formule de Siegel dans la théorie des groupes classiques*, Acta Math. **113** (1965), 1–87.
492. ______, *Elliptic functions according to Eisenstein and Kronecker*, Springer-Verlag, Berlin-New York, 1976, Ergebnisse der Mathematik und ihrer Grenzgebiete, Band 88.
493. Martin Westerholt-Raum, *Indefinite theta series on tetrahedral cones*, Preprint.
494. ______, *$\mathfrak{H}$-harmonic Maaß-Jacobi forms of degree 1*, Res. Math. Sci. **2** (2015), Art. 12, 34.
495. Gabor Wiese, *On modular symbols and the cohomology of Hecke triangle surfaces*, Int. J. Number Theory **5** (2009), no. 1, 89–108.
496. Andrew Wiles, *Modular elliptic curves and Fermat's last theorem*, Ann. of Math. (2) **141** (1995), no. 3, 443–551.
497. ______, *On class groups of imaginary quadratic fields*, J. Lond. Math. Soc. (2) **92** (2015), no. 2, 411–426.
498. E. M. Wright, *Asymptotic Partition Formulae: (II) Weighted Partitions*, Proc. London Math. Soc. **S2-36**, no. 1, 117.
499. ______, *Stacks*, Quart. J. Math. Oxford Ser. (2) **19** (1968), 313–320.
500. ______, *Stacks. II*, Quart. J. Math. Oxford Ser. (2) **22** (1971), 107–116.
501. Han Wu, *Burgess-like subconvex bounds for $GL_2 \times GL_1$*, Geom. Funct. Anal. **24** (2014), no. 3, 968–1036.
502. Tonghai Yang, *Faltings heights and the derivative of Zagier's Eisenstein series*, Heegner points and Rankin L-series, Math. Sci. Res. Inst. Publ., vol. 49, Cambridge Univ. Press, Cambridge, 2004, pp. 271–284.

503. ———, *The second term of an Eisenstein series*, Second International Congress of Chinese Mathematicians, New Stud. Adv. Math., vol. 4, Int. Press, Somerville, MA, 2004, pp. 233–248.
504. Gang Yu, *Rank 0 quadratic twists of a family of elliptic curves*, Compositio Math. **135** (2003), no. 3, 331–356.
505. Don Zagier, *Mock theta functions of order one*, Unpublished note.
506. ———, *Private communication.*
507. ———, *Modular forms associated to real quadratic fields*, Invent. Math. **30** (1975), no. 1, 1–46.
508. ———, *Nombres de classes et formes modulaires de poids* 3/2, C. R. Acad. Sci. Paris Sér. A-B **281** (1975), no. 21, Ai, A883–A886.
509. ———, *Modular forms whose Fourier coefficients involve zeta-functions of quadratic fields*, (1977), 105–169. Lecture Notes in Math., Vol. 627.
510. ———, *Zetafunktionen und quadratische Körper*, Springer-Verlag, Berlin-New York, 1981, Eine Einführung in die höhere Zahlentheorie. [An introduction to higher number theory], Hochschultext. [University Text].
511. ———, *L-series of elliptic curves, the Birch-Swinnerton-Dyer conjecture, and the class number problem of Gauss*, Notices Amer. Math. Soc. **31** (1984), no. 7, 739–743.
512. ———, *Modular forms of one variable*, Lecture Notes (1991), Universiteit Utrecht. http://people.mpim-bonn.mpg.de/zagier/files/tex/UtrechtLectures/UtBook.pdf.
513. ———, *Periods of modular forms and Jacobi theta functions*, Invent. Math. **104** (1991), no. 3, 449–465.
514. ———, *Introduction to modular forms*, From number theory to physics (Les Houches, 1989), Springer, Berlin, 1992, pp. 238–291.
515. ———, *Vassiliev invariants and a strange identity related to the Dedekind eta-function*, Topology **40** (2001), no. 5, 945–960.
516. ———, *Traces of singular moduli*, Motives, polylogarithms and Hodge theory, Part I (Irvine, CA, 1998), Int. Press Lect. Ser., vol. 3, Int. Press, Somerville, MA, 2002, pp. 211–244.
517. ———, *The Mellin transform and other useful analytic techniques*, Appendix to E. Zeidler, Quantum Field Theory I: Basics in Mathematics and Physics. A Bridge Between Mathematicians and Physicists, Springer-Verlag, Berlin-Heidelberg-New York, 2006.
518. ———, *The dilogarithm function*, Frontiers in number theory, physics, and geometry. II, Springer, Berlin, 2007, pp. 3–65.
519. ———, *Elliptic modular forms and their applications*, The 1-2-3 of modular forms, Universitext, Springer, Berlin, 2008, pp. 1–103.
520. ———, *Ramanujan's mock theta functions and their applications (after Zwegers and Ono-Bringmann)*, Astérisque (2009), no. 326, Exp. No. 986, vii–viii, 143–164 (2010), Séminaire Bourbaki. Vol. 2007/2008.
521. ———, *Quantum modular forms*, Quanta of maths, Clay Math. Proc., vol. 11, Amer. Math. Soc., Providence, RI, 2010, pp. 659–675.
522. Don Zagier and Sander Zwegers, *Private communication.*
523. Jose Miguel Zapata Rolon, *Asymptotics of crank generating functions and Ramanujan congruences*, Ramanujan J. **38** (2015), no. 1, 147–178.
524. Herbert S. Zuckerman, *On the coefficients of certain modular forms belonging to subgroups of the modular group*, Trans. Amer. Math. Soc. **45** (1939), no. 2, 298–321.
525. ———, *On the expansions of certain modular forms of positive dimension*, Amer. J. Math. **62** (1940), 127–152.
526. Wadim Zudilin, *On three theorems of Folsom, Ono and Rhoades*, Proc. Amer. Math. Soc. **143** (2015), no. 4, 1471–1476.
527. Sander Zwegers, *Multivariable Appell functions*, Preprint.
528. ———, *Mock θ-functions and real analytic modular forms*, q-series with applications to combinatorics, number theory, and physics (Urbana, IL, 2000), Contemp. Math., vol. 291, Amer. Math. Soc., Providence, RI, 2001, pp. 269–277.
529. ———, *Mock theta functions*, 2002, Ph.D. Thesis, Universiteit Utrecht.
530. ———, *Rank-crank type PDE's for higher level Appell functions*, Acta Arith. **144** (2010), no. 3, 263–273.

Index